Group ie same e- in outer ring

					7A	0
					1 **H** 1.008	2 **He** 4.003

	1B	2B	3A	4A	5A	6A	7A	0
			5 **B** 10.81	6 **C** 12.01	7 **N** 14.01	8 **O** 16.00	9 **F** 19.00	10 **Ne** 20.18
			13 **Al** 26.98	14 **Si** 28.09	15 **P** 30.97	16 **S** 32.06	17 **Cl** 35.45	18 **Ar** 39.95
28 **Ni** 58.70	29 **Cu** 63.55	30 **Zn** 65.38	31 **Ga** 69.72	32 **Ge** 72.59	33 **As** 74.92	34 **Se** 78.96	35 **Br** 79.90	36 **Kr** 83.80
46 **Pd** 106.4	47 **Ag** 107.9	48 **Cd** 112.4	49 **In** 114.8	50 **Sn** 118.7	51 **Sb** 121.8	52 **Te** 127.6	53 **I** 126.9	54 **Xe** 131.3
78 **Pt** 195.1	79 **Au** 197.0	80 **Hg** 200.6	81 **Tl** 204.4	82 **Pb** 207.2	83 **Bi** 209.0	84 **Po** (209)	85 **At** (210)	86 **Rn** (222)

63 **Eu** 152.0	64 **Gd** 157.3	65 **Tb** 158.9	66 **Dy** 162.5	67 **Ho** 164.9	68 **Er** 167.3	69 **Tm** 168.9	70 **Yb** 173.0	71 **Lu** 175.0
95 **Am** (243)	96 **Cm** (247)	97 **Bk** (247)	98 **Cf** (251)	99 **Es** (252)	100 **Fm** (257)	101 **Md** (258)	102 **No** (259)	103 **Lr** (260)

§ The International Union for Pure and Applied Chemistry has not adopted official names or symbols for these elements. All atomic weights have been rounded off to four significant figures.

introduction to chemical principles

third edition

Edward I. Peters

Department of Chemistry, West Valley College, Saratoga, California

SAUNDERS GOLDEN SUNBURST SERIES
SAUNDERS COLLEGE PUBLISHING

Philadelphia   New York   Chicago
San Francisco   Montreal   Toronto
London   Sydney   Tokyo   Mexico City
Rio de Janeiro   Madrid

Address orders to:
383 Madison Avenue
New York, NY 10017

Address editorial correspondence to:
West Washington Square
Philadelphia, PA 19105

This book was set in Electra by York Graphic Services.
The editors were John Vondeling, Patrice Smith, Michael Fisher, Merry Post, Louise Robinson.
The art & design director was Richard L. Moore.
The text design was done by Adrianne Dudden.
The cover design was done by Richard L. Moore.
The production manager was Tom O'Connor.
This book was printed by Fairfield Graphics.

INTRODUCTION TO CHEMICAL PRINCIPLES 3/e   ISBN 0-03-058432-9

234 144   9876543

CBS COLLEGE PUBLISHING
Saunders College Publishing
Holt, Rinehart and Winston
The Dryden Press

preface

Design for learning. Design for teaching. These have been the two primary focal points in writing the third edition of *Introduction to Chemical Principles*.

The goal of this text is the preparation of students for college-level general chemistry. After a one-semester or one-quarter course with this book, the successful student will be able to do the following: (a) read, write, and talk about chemistry, using a basic chemical vocabulary; (b) write routine chemical formulas and write names when formulas are given; (c) write balanced chemical equations for several kinds of reactions; (d) set up and solve elementary chemical problems, using dimensional analysis where applicable; (e) "think" chemistry in some of the simpler theoretical areas, to visualize what is happening on the atomic or molecular level.

One of the "design for learning" improvements in this edition is its readability. Many regard poor reading skills as second only to poor math skills among the common obstacles to success in chemistry. Reading level is usually estimated in terms of the number of syllables and sentences in a given length of text. But reading level is only a part of readability. Logical thought patterns and the development of topics in a way that students understand are more important. A conscious effort has been made to combine logical development with a minimum reading level, while continuing to respect students as young adults who think beyond Dick-and-Jane sentences containing mostly monosyllables. If we have succeeded, neither instructors nor students will notice the effort, but teachers will observe improved learning among students for whom English is a second language and others with below normal reading ability.

This edition continues the most successful learning aid of earlier editions, the semi-programmed format for most example problems. If these are used properly, the student thinks through and solves each example, rather than trying to understand the author's solution. Learning how to solve problems is the desirable and unavoidable result. Following the procedure is encouraged in this edition by a small periodic table, printed on the end sheet. It is intended to be torn from the book and used as a shield to cover intermediate answers in the example while the student solves the problem. The periodic table not only gives the student the information needed, but also constantly directs attention to the table as a source of all kinds of information. The reverse side of the card lists the steps by which examples should be solved for maximum learning, a ready reminder to take full advantage of the format.

Attention is given to the learning process throughout the text. Many new tables, summaries, and suggestions have been added to encourage the student to pause and assimilate the material just studied. The main ideas are learned immediately, rather than merely introduced for later learning.

A *Student Guide for Introduction to Chemical Principles*, written by Dr. Robert Kowerski of the College of San Mateo and myself, is available. Unlike many student guides, this one is organized in manageable assignments, rather than a whole chapter as the smallest study unit. (The teacher may find this handy for making assignments, too.) This guide does not rewrite the text, but helps the student learn from the text. When appropriate, it reminds the reader of earlier topics that are about to be used again. Specific page references are given. The guide also identifies main ideas in the new assignment, suggests how to avoid common pitfalls, and asks questions that are answered elsewhere with prescriptions for improvement. Coupled with the text, it has all the elements of a PSI study guide; in fact, it will be used for that purpose in our classes. Keller Plan jargon is avoided, however, so the book is equally useful in the conventional course format. If it helps the PSI student to learn independently, why should it not do the same for the student who attends formal lectures and takes typical midterm exams?

Other "design for learning" features in the third edition include the following. "Performance Goals," or objectives, appear in two locations. First, they appear at the beginning of each section where a concept or skill is introduced, the most logical position if they are to enhance learning. Second, they are used as chapter summaries to help in studying for a test. Open space is provided for student notes and calculations. Notes in the student's handwriting, rather than highlighting, improve learning. New terms are introduced in bold face type and listed at the end of each chapter for review. Appendix I is a comprehensive review of all the mathematics used in this book. A new section gives detailed instructions on the proper use of a hand calculator. The Glossary gives each student a dictionary of chemical and other scientific terms. It is not necessary to search the index or text pages to clarify the meaning of words that have been forgotten.

A book that is designed for teaching must include all the topics an instructor intends to teach. Not all teachers choose to teach the same topics. This was strikingly apparent when one reviewer of the manuscript for this book wrote that the idea of normality is obsolete and should be dropped, while another said the book could never be used on his campus unless the same topic was expanded and treated in greater detail. Both of these positions deserve respect, and both needs must be met. We have therefore included nearly all the topics that may logically be considered for an introductory chemistry course, leaving it to the instructor to select those that will be assigned and to omit the rest.

All textbooks present topics in a logical order, though it may not be the one an individual teacher may prefer. If the book is truly designed for teaching, it must be flexible enough to allow an instructor to modify the order of topics without breaking continuity. We have endeavored to anticipate different sequences. Sections and chapters may be rearranged in several ways, and the student will never encounter a sentence that cannot be understood because it depends on something that was skipped earlier in the book. For example, if quantitative laboratory experiments are used early in the semester, the quantum model of the atom (Chapter 5) may be postponed until formulas, equations, and stoichiometry (Chapters 6-9) have been studied.

Chapter 19, on energy, has been written in three parts. The part on specific heat may be introduced as early as Chapter 3 as a measurable property of matter; the part of heat of reaction fits with Chapter 9 on stoichiometry; the entire chapter may be considered as a unit any time after Chapter 9; or the chapter may be omitted entirely. Instructors of nonlaboratory courses sometimes choose to omit significant figures. To the extent that it is necessary in the text, this topic is presented at the beginning of Chapter 7. In Part H of Appendix I the subject receives full textbook treatment and may logically be given as a regular assignment. Order-of-topics options, optional chapters and sections, and other instructor choices that are built into the book are discussed in detail in the instructor's manual.

As in previous editions, questions and problems at the end of each chapter are printed in two matched columns. Questions in the left column are answered in the text, including complete setups for problems. Questions in the right column are not answered, so they may be used for assignments. The instructor's manual includes one section that has only answers for all right column questions and problems, and another section that shows all problem setups. The instructor may choose how much information about right column problems will be given to students, if any.

A set of three matched tests for each chapter is available to any instructor who adopts this text. Though written specifically for our PSI courses, they are equally useful as a test bank for midterm examinations. A copy, including complete answer keys, is available from your local Saunders College Publishing field representative.

One name may appear as the author of a textbook, but rarely is the book the work of one individual. The remarks of associates and students contribute much to the revision of each edition. There are always a few colleagues who read earlier editions and manuscripts and criticize them in detail. This exchange of ideas is one of the most stimulating byproducts of writing a book, and one that never fails to improve an author's writing and teaching. This valuable service was performed by Jay and Ellen Bardole of Vincennes University Junior College, Vincennes, Indiana; Dr. Keith Berry, University of Puget Sound, Tacoma, Washington; and Dr. O. H. Bezirjian, College of Marin, Kentfield, California. Dr. Robert Kowerski, my coauthor on the student guide, offered numerous helpful suggestions as we worked together on that project. The most comprehensive critique came from Dr. Peter Berlow of Dawson College, Montreal, Quebec. His penetrating analysis was particularly helpful. Ideas from all these friends and colleagues pervade these pages, and I thank them all for their contributions.

My appreciation goes also to my proofreading assistants: my daughter-in-law, Zandra Peters; and a former student, Melinda Torres, who taught me, even at this late date, some fine points about the English language that she gleaned from her English classes at West Valley College. As always, my biggest proofreading support came from my wife, Geb. But that was only the visible manifestation of her part in this book. Her greater contribution was her patience and encouragement throughout the eighteen months or so that have been devoted to this project. This isn't why I love her, but it surely would be an added reason if there weren't more than enough already.

Edward I. Peters

contents

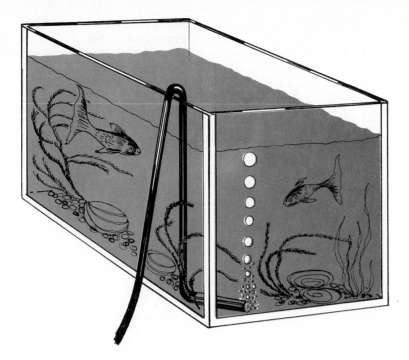

about bubbles in fish tanks–a prologue

Have you ever noticed that bubbles rising from the bottom of a fish tank become larger as they approach the surface? At least I think they do. I must confess that I've never consciously observed and measured bubbles in a fish tank, but I still have full confidence that they grow larger as they rise. To explain why I have this confidence, let's consider a story—a bit of fiction that tells us something about chemistry and, for that matter, science in general.

Once there was a man who personally *observed* that bubbles increase in size as they rise from the bottom of a fish tank. Being a Curious Man, he *wondered why*. His *curiosity* drove him to find out what caused this to happen. So he *thought* about it. Eventually he developed a *hypothesis*. He guessed that the volume of the bubble was smaller at the bottom because there was more

Figure P.1
The ''bubble'' experiment. The size of the bubble (colored) depends on the depth of the water in which it is submerged.

water on top of the bubble to "push" it into a smaller size. As the bubble rises there is less pressure, or "push," over its surface, so it gets bigger.

Having figured out a possible explanation for his observation, the Curious Man wanted to check it. Being a practical man, as well as curious, he built a bubble. His bubble had the form of a sealed cylinder (Fig. P.1). It had a snug-fitting piston that was both air- and watertight, but was free to move up and down so the pressure of the air inside would equal the pressure on the outside. The volume of the bubble was therefore governed by the pressure on top of the piston.

Gleefully our Curious Man took his bubble to the nearest lake, where he *experimented* with it. He lowered it into the water to different depths, measuring the volume at each depth. His efforts were rewarded. He found that, like real bubbles, his artificial bubble became larger as it approached the surface. His hypothesis was correct: the volume of a bubble does depend upon the pressure on the outside. For that matter, he could say the pressure exerted by the gas inside the bubble was related to the volume, inasmuch as the inside and outside pressures were equal.

Excitedly, our Curious Man *communicated* his findings to anyone who would listen. Not many did; not many men are curious about bubbles. But one was. He was also *skeptical*. He didn't believe the reported results. Therefore he *repeated the experiments*. Curious Man No. 2 found that the experiments of Curious Man No. 1 were correct and gave *reproducible results*. Being an *intellectually honest* person, Curious Man No. 2 freely admitted his error in doubting Curious Man No. 1. The next time they gathered with their Curious

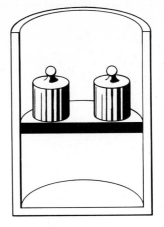

Figure P.2
The relationship between pressure and volume of a confined gas. At low pressure the volume is larger.

Friends—curious in the same sense we have been using the term so far!—Curious Man No. 2 reported to all that he had checked the results himself.

One of the Curious Friends, Curious Woman, was not entirely satisfied. She thought there should be more data. She suggested the *hypothesis* that if you *measured* the pressure of the gas and its volume, you would find a quantitative relationship between them—a relationship between the amount of pressure and the amount of volume. She designed an experiment to test her hypothesis. Her experiment didn't require her to go down to the lake and get all wet, incidentally. She reasoned that you could measure the pressure more easily simply by putting weights on top of the piston (Fig. P.2).

One morning Curious Woman conducted her experiment and *drew a graph* of her results (Fig. P.3). She found that if you multiply gas pressure by gas volume, you always get the same answer. Wisely she *checked her results* before telling them to her Curious Friends. She checked them many times, in fact, until she was quite sure of her findings.

At the next gathering of the Curious Friends, Curious Woman reported that the product of pressure times volume of a gas is a constant. Other Curious Persons picked up the idea and tried it in their laboratories. They got the same results. Improvements were made. They found that temperature and amount of gas were variables that also had to be controlled. Experimenting with them as well, new relationships were found. Finally one Curious Person put them all together and proposed an "explanation" for these *experimental facts*. His *theory* pictured a gas as made up of many tiny particles moving wildly about, causing all the experimental results recorded by the other Curious Persons.

It's been over a hundred years since Curious Woman first proposed that pressure × volume is constant (at fixed quantity and temperature, of course). Recently the Society of Curious Persons has honored Curious Woman's relationship by elevating it to the status of a *law* of Science. It is now called Curious Woman's Law.

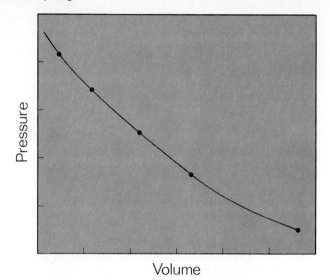

Figure P.3
Graph of pressure vs. volume of a gas.

Now you know why I am so sure that bubbles become larger as they rise in a fish tank, even though I have not personally made this observation. It's the law. But even here, I must be cautious. Scientific laws are rarely found to be in error, but it has happened. To be absolutely certain, I'm going to conduct my own experiment. Where can I find a fish tank?

*   *   *   *   *

In this little story we have tried to provide a small glimpse of the character of chemistry. The observing, hypothesizing, experimenting, testing and retesting, theorizing and, finally, reaching conclusions have been going on for centuries. And they continue today more actively than ever before. Collectively they are often called the *scientific method*.

There is really no rigid order to the scientific method. Looking back over the history of science, though, the features listed above always appear. They are the outcome of the day-to-day thinking of the scientifically curious person as he continually asks himself, "What do I already know that can be applied here? What is the next logical step I can take?" Out of such questions come new hypotheses, new experiments, new theories. Ultimately new laws are discovered.

Our story also lists in italicized words some of the qualities and actions of the scientist. He surely is an *observer*, and he is *curious* about and *thinks* about what he sees. He *develops a hypothesis* as a tentative explanation of his observations. He *conducts experiments* to test the hypothesis. If he finds something new, he *communicates* with his fellow scientists, usually through scientific journals. He combines the qualities of *skepticism* with *intellectual honesty*, both of which leave him free to receive and evaluate new information that reaches him through many sources.

Our story furnishes one more important insight into the nature of chemistry. Notice that the first two curious men concerned themselves with *What* was taking place and *How*. Answers to these questions are considered to be the *qualitative* part of chemistry. It was not until the Curious Woman entered the picture that measurements appeared. She recognized that *What* and *How* furnish only some of the answers, but they cannot be used to make reliable predictions about the extent of chemical activity. She added the vital question, *How much*? We see, then, that the study of chemistry is both qualitative and *quantitative*. In this course we will consider both of these essential areas.

While the characters and events in this story are obviously fictitious, the law around which it was built is very real. It is the product of one person, not three, and was first proposed in the 17th century by Robert Boyle (1627–1691). Boyle was one of the first scientists to devote himself to orderly and careful experimental investigation. You will study Boyle's Law in Chapter 13 of this text.

Figure P.4
Bubbles of gas from high pressure tanks expand as the pressure decreases on the way to the surface.

1

learning chemistry

The next few pages may be among the most important pages you will read in this book. Nothing in them will be on a test question. They will teach you no chemistry. You will receive no grade for having read them. But they will have a lot to do with the chemistry you learn, your performance on tests, and the grades you will receive. Therefore, you are urged to read these pages carefully and consider what they suggest about the way you learn chemistry.

You and *learn* are the two most important words in the above paragraph. Your teacher may teach, but only you can learn. The responsibility belongs to you alone. This is why the first chapter is about learning chemistry. If you learn *how to learn* chemistry, the rest is much easier.

No matter what you may have heard, it is not hard to learn chemistry at the beginning level. Chemistry does require you to develop some new thought processes. It requires you to think in ways you have not thought before. You will have to figure out how to apply ideas you have read about. But if you are a person of normal intelligence, and if—this is important—you do what every student should do in every course, you will not find chemistry too difficult. Time-consuming, perhaps, but not hard. If you decide to succeed in chemistry, and do what must be done, you will succeed.

What must be done? Three things. First, you must spend the time required. Learning chemistry takes time, often more time than most other subjects. Furthermore the time must be spent regularly, doing each assignment each day. You cannot learn chemistry all at once, just before a test. It must be learned a bit at a time, because the ideas you learn in today's lecture depend upon what you learned doing yesterday's assignment. If you haven't finished yesterday's assignment, today's lecture will make no sense. You must keep up; falling behind is the biggest problem of all when it comes to learning chemistry.

The second thing you must do to succeed in chemistry is to concentrate. You will have to think, and think hard, while learning. Then you must practice what you have learned. Neither of these is difficult if you concentrate while studying. This means studying without distractions—without sounds, sights, people, or thoughts that take your attention away from chemistry. Every minute your mind wanders while you are studying must be added to your total study time, making that time one minute longer. If you have a limited amount of time for study, that minute is lost forever. What you learn has been reduced, and your grade will suffer from that lost learning.

The third requirement for success is that you follow the study procedures described in the rest of this chapter. They have been tried and tested. They work. You may later develop procedures that are better for you, but it is unlikely that you will begin with better procedures. It is best to start with something good and then to improve it as you go along.

These three requirements for success in chemistry are related. If you concentrate while studying and use good study procedures, you will learn more in less time. This means more time for fun and games, and perhaps a part-time job. It also means a better grade in chemistry and more satisfaction with your academic work.

STUDY TOOLS

This textbook is one of several tools you will use while learning chemistry. It must be supported by other tools. You do not just read when you study chemistry. You do things. You answer questions, not only in your mind but also by writing down the answers. This means you need a pencil. Some paper helps too, although there is enough open space in this book for you to make notes, to answer questions, and to solve problems. There are many calculation problems in chemistry. To solve them in any reasonable amount of time you will need a calculator. On page 572 you will find suggestions about the kind of calculator you should buy, if you do not already have one.

You simply cannot honestly claim to be studying chemistry without these tools. If you try, you will spend more time studying while you learn less.

PERFORMANCE GOALS

PG 1 A   Read the performance goal or goals at the beginning of each section; study the section; reread the performance goals; satisfy yourself that the goals have been reached; if so, go on; if not, go back—restudy the text until the goal has been met.

As you approach most sections in this text, you will encounter one or more "performance goals," as you did here. In many cases the performance goal will contain language that is unfamiliar, but about to be introduced in the section. It gives you some idea of what to expect and what to look for as you study the section. This should help you to recognize major points. You can then direct your attention to them. This is the key to efficient study.

Notice that the performance goal above is written as an action. It describes *doing* something—performing some act. Each performance goal in this book should be thought of as the close of a sentence that begins, "After studying this section, you will be able to. . ." The goal describes something you should learn to do in your study. You should find out whether or not it has been learned immediately after completing the section. Return to the performance goals, asking yourself the question, "Am I now able to do what is expected?" Demand of yourself that the honest answer to this question be "Yes!" If it is not, go over the material again until you can do it.

Performance goals are designated in the text by PG, followed by a number and a letter. The number is the chapter number, and the letter identifies the performance goal within the chapter.

The performance goals in the book are those the author has in mind for each section. They may or may not correspond to those of your instructor. If not, you should be alert to changes he or she may introduce, and adjust your study plan to what is required.

TEXT AND EXAMPLES

Most paragraphs in this book are written like any other textbook. Nearly all of the example problems, however, are written in a self-teaching style, a series of questions and answers that guide you to an understanding of a concept. It will succeed only if you "play the game by the rules." The rules call for you to answer each question *yourself* before looking at the answer you know is a bit further down the page. This way you force yourself to *reason* through the question and to gain an understanding of the chemical concept it illustrates.

Example problems follow most of the principles covered in this text. Suggestions are made for applying the principle to the example, ending with a question that leads you to the first step in the solution of the problem. Space is provided for you to calculate and write your answer to the question. At that point short broken lines appear on each side of the page:

‒‒‒‒‒   ‒‒‒‒‒

Right below the line is the answer to the question, including the mathematical setup if one is required. An explanation of the answer also may appear. More suggestions lead to a second question, space for your answer, and another set of broken lines. The pattern repeats until the example is completed.

To learn by these examples you must answer each question without looking at the answer. To do this you will have to cover the answer. Elsewhere in this book you will find shields that can easily be torn out of the book. One side of each shield is a periodic table from which you will obtain information needed for solving problems. The other side lists the steps you should follow when working on examples. These steps are:

1. When you come to an example, use the shield to cover the part of the page below the first set of dashed lines in the example.
2. Read the example question. Write in the open space above the shield, or on separate paper, any answers or calculations needed.
3. Move the shield down to the next broken lines.
4. Compare your answer with the one you can now read in the book. Be sure you understand the example up to that point before going on.
5. Repeat the procedure until you finish the example.

CHAPTER QUESTIONS AND PROBLEMS

At the end of most chapters you will find a two-column set of questions and/or problems. They are matched side-by-side; they involve similar reasoning and, in the case of problems, similar calculations. Some questions are easy, requiring only the basic information presented in the chapter. Others are more demanding and require that you analyze a situation, apply chemical principles, and then explain or predict some event or calculate some result. The more difficult of these, or those clearly extending beyond the stated performance goals, are identified by an asterisk (*). Answers appear in the back of the book for all questions and problems in the left-hand column. These include calculation setups for most problems. Questions in the right-hand column are without answers; they may be used for assignments by your instructor.

If you have difficulty working any problem from either column, you will have reached a critical point. You will be tempted to return to the examples in the chapter, find one that matches your problem, and then solve the assigned problem step-by-step as in the example. Resist that temptation! If you get stuck on a problem, it means you did not understand the corresponding example in the text. Study the example again by itself until you understand it thoroughly. Then return to the assigned problem with a fresh start and work it to the end without looking at the text example again.

You should approach your problem work in chemistry with a clear sense of purpose. This purpose is not simply to get the right answer. It is, rather, to understand (1) the principle that underlies each problem and (2) the method by which it is solved. This is why you should not copy examples from the book. You should, rather, look at the problem as a whole so you can *reason* through the steps unaided. More than anything else you do, demanding such understanding of each problem you attempt will assure you success in solving chemistry problems.

GLOSSARY

As with any area of study, chemistry has its own special language. One of the main purposes of a first course in chemistry is to make you familiar with this language so that you will understand and use it properly.

A large part of any language is its vocabulary. At the end of each chapter you will find a list of the scientific terms introduced or used in the chapter. There is a number showing the page on which each term first appears. The presence of a word on this list does not mean its definition must be memorized. It means you should understand the term so it makes sense to you if you read or hear it, and so you can use it properly in your own speaking and writing. Any definitions or concepts you must be able to memorize or explain are identified in the performance goals.

You will no doubt run across technical words in this book whose meanings are not quite clear. A word may have been introduced earlier, and you need to refresh yourself on it, or it may be a word out of the general vocabulary that has a scientific meaning you have had no reason to know before. The Glossary is provided so you can look it up easily; it begins on page 600. The Glossary includes nearly all words in the chapter summaries, and many others. *You should use the Glossary regularly.*

A CHOICE

In this chapter you have read about the procedures by which you will learn chemistry. You now have a choice to make. For the study of chemistry you may choose to schedule the time it requires on a regular basis, to establish good study habits, and to use the text as it should be used. If you make this choice, you will learn chemistry. You won't find it particularly difficult. If, on the other hand, you choose to study chemistry only now and then, mainly just before a test, if you do not solve problems right after they are assigned, if you "study" in the middle of distractions, if you take short cuts, or if you get behind, you will make chemistry harder than it really is.

If you ever begin to feel that chemistry is a difficult subject, remember these lines. Read this chapter again, and see if your study habits are at fault. Perhaps improving those habits can solve your problem.

At all stages of our lives we make choices. We then live with the consequences of those choices. Choose wisely—and enjoy chemistry.

2

matter and energy

In a broad sense, chemistry is the study of matter and the energy associated with chemical change. It is appropriate, then, that we begin the study of chemistry by an examination of these well-known but not widely understood parts of the physical universe.

2.1 PHYSICAL AND CHEMICAL PROPERTIES AND CHANGES

PG **2 A** Distinguish between physical and chemical properties.
2 B Distinguish between physical and chemical changes.

Matter has mass. Matter also occupies space. These two characteristics combine to define matter. In order to examine matter, to understand what it does, or appears to do, we must be able to describe it clearly. Normally a sample of matter is described by listing its **physical properties.** Certain physical properties can be detected directly by our senses. They tell us how a material *looks* (the blackness of charcoal compared with the yellow of sulfur), *feels* (the hardness of glass compared with the softness of putty), *smells* (the odor of sour milk or the scent of a rose), or *tastes* (salt vs. sugar). Other physical properties can be measured in the laboratory. Among them are the temperatures at which materials boil or melt, called the *boiling* and *melting points*, or the *density*, or relative "heaviness," of the material.

Changes that alter the physical form of matter *without changing its chemical identity* are called **physical changes.** The melting of ice is a physical change. The substance is water both before and after the change. Dissolving sugar in water is another example of a physical change. The form of the sugar changes, but it is still sugar. The dissolved sugar may be recovered by simply evaporating the water, another physical change.

The chemist is interested in more than the physical properties of matter. He or she wants to know what sort of chemical reactions it can have. Paper burns. Iron rusts. Milk becomes sour. Eggs become rotten. Each of these is a **chemical change.** A chemical change can be recognized when one or more substances are chemically destroyed, and one or more new substances are formed. The **chemical properties** of a substance are simply a list of the chemical changes possible for that substance.

Chemical and physical properties lead to several ways in which matter may be classified, some of which will be considered now.

2.2 STATES OF MATTER: GASES, LIQUIDS, SOLIDS

PG 2 C Identify and explain the differences between gases, liquids, and solids in terms of (1) visible properties and (2) particle movement.

The air we breathe, the water we drink, and the food we eat are examples of the gaseous, liquid, and solid **states of matter.** Water is the only substance we normally meet in all three states, as suggested in Figure 2.1. The differences among gases, liquids, and solids can be explained in terms of what is called the **kinetic molecular theory of matter.** According to this theory, all matter consists of extremely tiny bits or particles that are in constant motion. (*Kinetic* refers to motion.) In solids this movement is limited by strong attractions between the individual particles. These particles are thought to have a vibrating or shaking motion in which they remain in fixed positions relative to each other. This is why a solid has a definite shape and volume (Fig. 2.1).

Attractions between particles are weaker in liquids. They are strong enough to hold the particles together in a definite volume, but not strong enough to hold them in fixed positions. The particles are free to move past each other, to tumble and slip about almost at will, just so they all remain together. Because the liquid particles remain together the volume they occupy is definite, but their freedom to move allows them to adjust to the shape of the bottom of the vessel that holds them.

The attractions between gas particles are very weak—they are close to

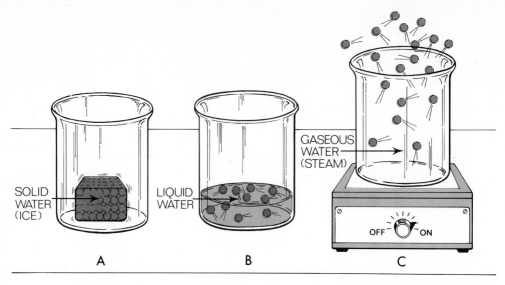

Figure 2.1
The three states of matter, illustrated by water. A, Solid water, ice, has a definite volume and shape. Particles vibrate in place, holding fixed positions relative to each other. B, Liquid water has a definite volume, but takes the shape of the bottom of its container to the depth necessary to hold that volume. Molecules are free to move relative to each other within that volume. C, Gaseous water, steam, has neither definite volume nor shape, but expands to fill the container if closed, and to escape from it if open. Particles move in random fashion, completely independent of each other.

zero. The particles separate from each other, move independently, and fill the vessel that holds them. This is why a gas has no definite shape or volume. Particle movement is described as random, without order, limited only by the walls of the container.

The kinetic molecular theory suggests that the "amount" of particle movement, or the speed at which the particles move, depends upon temperature. At higher temperatures the movement is faster, more vigorous. The particles have more energy. At lower temperatures the movement is less energetic. This agrees with the picture of solids, liquids, and gases. In a piece of ice, for example, the limited movement at low temperatures permits the attractions between particles to hold them in the structure of a solid. As temperature is increased, more vigorous movement overcomes those attractive forces, destroys the rigid structure of a solid, and the ice becomes liquid water. If the temperature is raised still more, the movement becomes so vigorous that the attractions are completely overcome and the particles fly apart in the gaseous state.

IN SUMMARY:

a. *A solid has definite shape and volume. Particle movement consists of vibration in place within a rigid structure.*

b. *A liquid has definite volume, but takes the shape of its container up to that volume. Particles are free to move among themselves so long as they remain together.*

c. *A gas fills its container, taking both its shape and volume. Movement of particles is completely random.*

2.3 PURE SUBSTANCES AND MIXTURES

PG 2 D   Distinguish between a pure substance and a mixture.

A **pure substance** is a single chemical, one kind of matter. It may be an *element* or a *compound* (Section 2.4). A pure substance has its own set of physical and chemical properties, not exactly the same as any other pure substance. These properties may be used to identify the substance.

A **mixture** is a sample of matter that contains two or more pure substances, either elements or compounds. The properties of a mixture are determined by the substances in it, and they vary as the percentages of the different parts change.

Pure water and salt water offer a good example of the differences between the properties of pure substances and mixtures. At normal pressure, pure water boils at 100°C. Salt water boils at a higher temperature; how much higher depends upon the amount of salt in the mixture. The salt and water can be separated by a process called **distillation** (Fig. 2.2). When the mixture is heated,

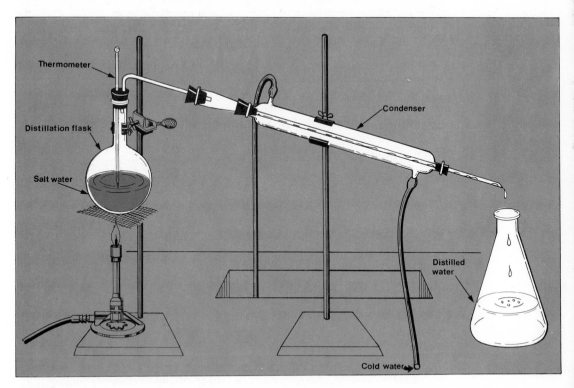

Figure 2.2
Laboratory distillation apparatus. Only the water in the solution is boiled—changed into steam. The steam passes into the condenser where it cools and changes back into pure liquid water. The distilled water is collected in the flask.

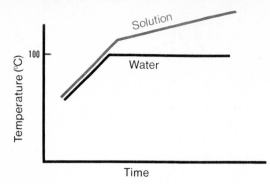

Figure 2.3
Comparison of boiling temperatures of a pure substance and an impure substance (solution). As water boils off, the solution that remains becomes more concentrated. This change in composition causes the boiling temperature to increase.

the water boils off, is cooled and condensed (changed back to the liquid state), and collected as pure water. The boiling temperature of the remaining salt water gradually increases because removal of water from the mixture increases the percentage of salt (Fig. 2.3). If boiling is continued long enough the salt and water will be separated completely. This is another property of mixtures: they may be separated into pure components by physical means alone.

2.4 ELEMENTS AND COMPOUNDS

PG 2 E Distinguish between elements and compounds.

If a pure substance cannot be broken down chemically—*decomposed* is the scientific term—into other pure substances, it is said to be an **element.** The fact that nobody has ever been able to decompose sulfur into two or more other substances shows that sulfur is an element.

Elements can combine chemically to form other pure substances called **compounds.** Water, for example, is a compound made up of the elements hydrogen and oxygen. Unlike elements, compounds can be separated into other pure substances, either elements or compounds. Water may be decomposed into its elements by passing an electric current through it (Fig. 2.4).

Nature provides us with 88 elements. Nuclear research as of the time of this writing has yielded 19 more, bringing the total to 107. Most of our environment is made up of compounds containing a relatively small number of these elements (Table 2.1). It is natural that elements such as oxygen, silicon, and aluminum, which are so abundant in nature, have become "familiar" elements. However, some elements that make up only a very small percentage of the crust of the earth are readily recognized and have become vital to man. For example, such well-known elements as chromium, copper, nickel, tin, silver, and gold are absent from Table 2.1. Other elements that occur in lower percentages but are essential to modern technology include cobalt, vanadium, cadmium, and molybdenum.

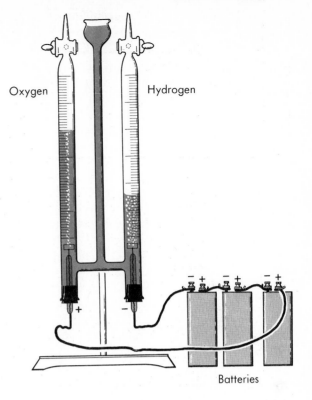

Oxygen · Hydrogen

Batteries

Figure 2.4
Decomposition of water into hydrogen and oxygen by means of an electric current.

Only a few elements occur uncombined in nature. Most of these are gases (nitrogen, oxygen, argon) that make up over 99% of the earth's atmosphere. Gold, silver, sulfur, and copper are among the few solid elements that exist in nature in elemental form. To obtain pure samples of nearly all other solid elements, it is necessary to obtain them from naturally occurring compounds or ores.

Most elements are solids at normal temperatures and pressures. At 25°C and one atmosphere, only two are liquids (mercury and bromine), while 11 elements, including hydrogen, fluorine, chlorine, oxygen, and nitrogen, are gases. The elements may also be divided into metals (iron, copper, magnesium, aluminum) and nonmetals (carbon, iodine). A few (boron, silicon) are often

Table 2.1
Composition of Earth's Crust*

Element	Per Cent by Weight	Element	Per Cent by Weight
Oxygen	49.2	Sodium	2.6
Silicon	25.7	Potassium	2.4
Aluminum	7.5	Magnesium	1.9
Iron	4.7	Hydrogen	0.9
Calcium	3.4	All others	1.7

*The earth's "crust" includes the atmosphere and surface waters.

referred to as "metalloids" because they have properties between those of metals and nonmetals.

The properties of compounds are different from the properties of the elements from which they are formed. To illustrate, powdered iron is black in appearance and powdered sulfur is yellow. A mixture of these two elements, which appears gray, is readily separated into the elements. The iron in the mixture may be removed with a magnet. Sulfur can be taken from the mixture by shaking it in a liquid called carbon disulfide. If the iron-sulfur mixture is heated strongly, a chemical reaction occurs to produce a compound, iron sulfide. The compound is not attracted by a magnet, nor does it dissolve in carbon disulfide. In other words, iron sulfide has none of the properties of iron or sulfur.

Sodium chloride is another example. Neither sodium, a metal that reacts with both water and oxygen, nor chlorine, a poisonous gas with a suffocating odor, is pleasant to work with. Yet the compound formed by these two elements, sodium chloride (table salt), is used to flavor food.

An important fact about compounds is summarized in the **Law of Definite Composition.** This law states that the percentage by weight of the elements in a compound is always the same, regardless of the source or method of preparation of the compound. Water, for example, whether it comes from a pond, river, or lake, from Europe, America, or Asia, or is the product of a chemical reaction, always contains 11.1% hydrogen and 88.9% oxygen.

2.5 HOMOGENEOUS AND HETEROGENEOUS MATTER

PG 2 F Distinguish between homogeneous and heterogeneous samples of matter.

The prefixes *homo-* for *same* and *hetero-* for *different* begin many words in the English language. A **homogeneous** sample of matter has the same appearance, composition or make-up, and physical and chemical properties throughout. **Heterogeneous** materials are made up of visibly different parts, or **phases.** Composition varies from one place in the sample to another. Each phase has its own unique properties and appearance. Therefore, a sample of matter generally may be classified as homogeneous or heterogeneous by its appearance alone.

Oil and water are both homogeneous liquids. When mixed in the same container, they quickly separate into two distinct liquid phases, forming a heterogeneous mixture (Fig. 2.5). Sugar and water, on the other hand, mix completely with each other to form a single liquid phase, a homogeneous mixture called a **solution.** Even a pure substance can be heterogeneous if it is present in two visibly distinct states. An ice cube floating in water is an example.

*   *   *   *   *

The differences between pure substances and mixtures, elements and compounds, and homogeneous and heterogeneous matter are summarized in Figure 2.6.

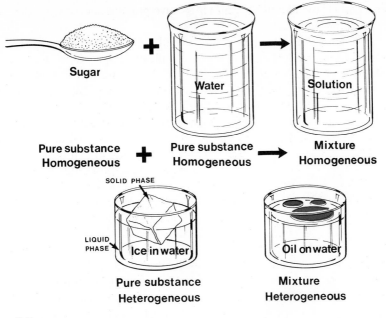

Figure 2.5
Examples of homogeneous and heterogeneous samples of matter.

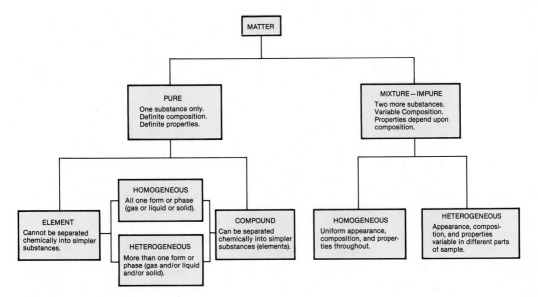

Figure 2.6
Summary of classification systems for matter.

2.6 THE LAW OF CONSERVATION OF MASS

PG 2 G　State the meaning of, or draw conclusions based upon, the Law of Conservation of Mass.

Early chemists who studied *burning,* one of the most familiar of chemical changes, concluded that since the ash remaining was so much lighter than the object burned, something was 'lost' in the reaction. Their reasoning was faulty because they did not realize that gases, which they could not see, took part in the reaction, both as reactants, or things used up, and as products, or things made. When a piece of wood burns, gaseous oxygen is a reactant that is lost in the reaction. The products, in addition to ash, include two gases, carbon dioxide and water vapor. If we take careful account of all the reactants and products, we find that

$$\begin{array}{c} \text{total weight of reactants} \\ \text{(wood + oxygen)} \end{array} = \begin{array}{c} \text{total weight of products} \\ \text{(ash + carbon dioxide + water vapor)} \end{array}$$

This type of weight balance applies to all ordinary chemical reactions. It was once called the law of conservation of matter, but now is referred to more accurately as the **Law of Conservation of Mass.** In other words, **in a chemical change, mass is conserved; it is neither created nor destroyed.** The word *mass* refers to quantity of matter, and is closely associated with the more familiar term *weight.* The difference between them will be pointed out in Chapter 3.

2.7 ELECTRICAL CHARACTER OF MATTER

PG 2 H　Match electrostatic forces of attraction and repulsion with combinations of positive and negative charges.

If you release an object held above the floor, it falls to the floor. This is caused by gravity, an invisible attractive force between the object and the earth. Although the force is invisible, its effect is very evident. There are two other invisible forces, both capable not only of attraction but also of repulsion. They are magnetic and **electrostatic forces.**

One of the most common electrostatic events occurs when you scrape your feet across a carpet on a dry day and then turn on a light switch, or perhaps "shock" another person by touching him. In the foot-scraping operation you become "charged with electricity." In the laboratory, we can charge a rubber rod by rubbing it with fur. If a pith ball* hung on a string is touched with the rod, the pith ball gains the same charge as the rod (Fig. 2.7). Two pith balls charged in the same manner push each other apart, indicating a *repulsion force between objects carrying the same kind of electrical charge.* Similarly, if two pith balls are touched with glass rods rubbed by silk, they repel each other, again illustrating repulsion between like charges. If one of the balls touched by the rubber rod is now brought near a ball touched by the glass rod, the balls are attracted toward each other. Obviously, the balls do not have like charges, which repel each other; instead they must have opposite charges. *Objects which are oppositely charged are attracted toward each other.*

These and numerous other examples lead to the conclusion that there are two, and only two, kinds of **electrical charges.** One is called **positive,** the other **negative**—adjectives assigned long before the nature of the charges was known. The attractive and repulsive forces between charged objects are called electrostatic forces. The electrostatic force between two like charges (+ and +, or − and −) is one of repulsion; that between a positive and a negative charge is one of attraction. These relationships are illustrated in Figure 2.7.

Electrostatic forces show that matter has electrical properties. We now know that all matter consists of tiny particles called atoms, and that these atoms contain positively charged particles called protons and negatively charged particles called electrons. In rubbing operations of the type we have described, electrons (never protons) are moved from one object to another. The object that ends up with more electrons than protons has a net negative charge. The posi-

*Pith is a soft, spongy, substance made from plant fibers. Pith balls used in this experiment are typically $\frac{1}{4}$ inch in diameter.

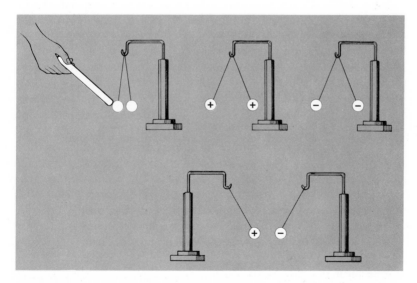

Figure 2.7
Electrostatic attraction and repulsion. If each of two pith balls is given a positive charge, or if each is given a negative charge (top views), they repel each other. If one ball is given a positive charge and the other a negative charge, they attract each other (bottom view).

tively charged object has fewer electrons than protons. If the number of protons is equal to the number of electrons, the object is said to be electrically **neutral.**

2.8 ENERGY IN CHEMICAL CHANGE

PG 2 I   Distinguish between exothermic and endothermic changes.
2 J   Distinguish between potential energy and kinetic energy.

Changes in matter don't "just happen." Something causes each change. Frequently change occurs because matter absorbs or releases energy. For example, if you supply heat energy to a pan of water, it boils. Figure 2.4 showed that water may be chemically decomposed into its elements by means of electrical energy. **Energy** is absorbed—taken in— in each of these examples. Chemical or physical changes in which energy is *absorbed* are called **endothermic changes.**

By contrast, chemical or physical changes in which energy is *released* to the surroundings are called **exothermic changes.** If you strike a match and put your finger in the flame you learn very quickly that the chemical change in burning wood is releasing heat energy. It also releases light energy. Chemical changes in flashlight and automobile batteries release electrical energy. The physical change when steam condenses to water releases exactly the same energy that was required to boil the water in the first place. All these are examples of exothermic changes.

Most of the energy changes we will be concerned with will be in the form of heat. We will leave other forms of energy for more advanced courses. To understand the source of chemical energy, however, we will consider briefly a physicist's definition of energy and its closely associated concept, **work.**

Work, according to the physicist, is the application of a force over a distance. If you raise a book from the floor to a desk, you do work (Fig. 2.8). You exert the force required to lift the book against the gravitational attraction of the earth. You exert this force over the distance from the floor to the desk. **Energy may be defined as the ability to do work.** You had the energy required to do the work of raising the book. In fact, you transferred this energy to the book, which has more energy on the desk than it had on the floor. The book now has **potential energy** with respect to the floor (Fig. 2.8B). **Potential energy is energy possessed by an object because of its position.** That an object having potential energy is capable of doing work may be shown by pushing the book off the desk and allowing it to fall toward the box still on the floor. In the act of falling (Fig. 2.8C) the potential energy is converted into **kinetic energy, the energy possessed by an object because of its motion.** When the book hits the box (Fig. 2.8D), it does work on it by crushing it.

Potential and kinetic energy are two forms of mechanical energy. While it is not necessary for you to know about these in detail in an introductory chemistry course, you should recognize that *potential energy is related to position* and

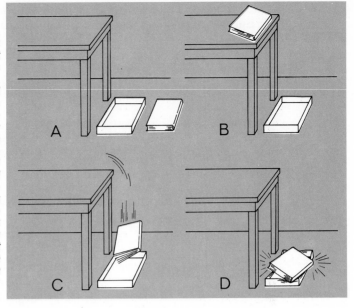

Figure 2.8
Potential and kinetic energy. A, Book lies on floor next to box from which it has been removed. It has no potential energy relative to floor, and no kinetic energy. B, Book on table has potential energy equal to the work done to raise it from the floor. It has no kinetic energy. C, Falling book now has kinetic energy because of its motion, plus some remaining potential energy because it is still above the floor. The combined energies equal the work done raising the book to the table. D, Book now has neither kinetic nor potential energy, but has done work crushing the box and creating small amount of frictional heat when striking floor, all equal to the work done raising the book from the floor to the table.

kinetic energy is energy of motion. What is called chemical energy comes mainly from the positions of positively and negatively charged objects. The energy associated with chemical changes comes from changes in these positions, just as there is a change in potential energy of a book as it is moved between the floor and a desk.

2.9 THE LAW OF CONSERVATION OF ENERGY

> **PG 2 K** State the meaning of, or draw conclusions based upon, the Law of Conservation of Energy.

The various processes that take place when we drive an automobile offer a good illustration of energy conversions, changing energy from one form to another. Starting uses the chemical energy of the storage battery to produce electrical energy. As we drive along, chemical energy of the fuel is being converted into both heat energy and "mechanical" energy of the engine parts. This mechanical energy is transferred to the car as kinetic energy, or energy of motion. Some of this energy is converted to heat through friction between the tires and the road. The extent of this conversion increases dramatically if we slam on the brakes at an intersection. Potential energy, the energy associated with position, that the car acquires when we climb a hill is converted back into kinetic energy if we coast back down.

Careful investigation of energy conversions such as these show that the energy "lost" in one form is always exactly equal to the energy "gained" in another form. This leads to another conservation law, the **Law of Conservation of Energy,** which states that, **in any ordinary change, energy is neither created nor destroyed.**

2.10 THE MODIFIED CONSERVATION LAW

Early in this century, Albert Einstein recognized a "sameness" between mass and energy. He suggested that it should be possible to convert one of these into the other. Such conversions are indeed going on in the world around us. In a few cases, enough mass is converted to energy to become dramatically apparent. This happens, for example, when a hydrogen bomb explodes or a nuclear reactor is used to produce electrical energy. Einstein's famous equation relates the amounts of mass and energy:

$$\Delta E = \Delta mc^2,$$

where ΔE is the energy change, Δm is the mass change, and c is the speed of light.

The amount of energy that could be produced from the conversion of mass is enormous. If it were possible to convert all of a given mass of coal to energy, that energy would be about 2.5 billion times as great as the energy derived from burning that same amount of coal! This is why nuclear energy is such an attractive alternative to the traditional sources of energy, an attraction clouded by serious questions of safety.

The suggestion that matter can be converted into energy, or energy into matter, might lead one to believe that the laws of conservation of mass and energy are not true. In ordinary (non-nuclear) chemical reactions the matter-energy conversion is so small it cannot be measured. For all practical purposes, the laws are as good today as when they were first proposed. For *all* chemical changes, both nuclear and non-nuclear, the laws may be combined into a single conservation law which states that the total amount of mass and energy in the universe is a constant. Expressed as an equation this becomes

$$(\text{mass} + \text{energy})_{\text{reactants}} = (\text{mass} + \text{energy})_{\text{products}}$$

CHAPTER 2 IN REVIEW

2.1   PHYSICAL AND CHEMICAL PROPERTIES AND CHANGES

2 A Distinguish between physical and chemical properties. (12)
2 B Distinguish between physical and chemical changes. (12)

2.2   STATES OF MATTER: GASES, LIQUIDS, SOLIDS

2 C Identity and explain the differences between gases, liquids, and solids in terms of (1) visible properties and (2) particle movement. (13)

2.3   PURE SUBSTANCES AND MIXTURES

    2 D Distinguish between a pure substance and a mixture. (15)

2.4   ELEMENTS AND COMPOUNDS

    2 E Distinguish between elements and compounds. (16)

2.5   HOMOGENEOUS AND HETEROGENEOUS MATTER

    2 F Distinguish between homogeneous and heterogeneous samples of matter. (18)

2.6   THE LAW OF CONSERVATION OF MASS

    2 G State the meaning of, or draw conclusions based upon, the Law of Conservation of Mass. (20)

2.7   ELECTRICAL CHARACTER OF MATTER

    2 H Match electrostatic forces of attraction and repulsion with combinations of positive and negative charges. (20)

2.8   ENERGY IN CHEMICAL CHANGE

    2 I Distinguish between exothermic and endothermic changes. (22)
    2 J Distinguish between potential energy and kinetic energy. (22)

2.9   THE LAW OF CONSERVATION OF ENERGY

    2 K State the meaning of, or draw conclusions based upon, the Law of Conservation of Energy. (23)

2.10   THE MODIFIED CONSERVATION LAW (24)

TERMS AND CONCEPTS

Matter (12)
Physical property (12)
Physical change (13)
Chemical property (13)
Chemical change (13)
States of matter (13)
Kinetic Molecular Theory (13)
Pure substance (15)
Mixture (15)
Element (16)
Compound (16)
Law of Definite Composition (18)
Homogeneous (18)

Heterogeneous (18)
Phase (18)
Solution (18)
Law of Conservation of Mass (20)
Electrostatic forces (20)
Electrical charge (21)
Endothermic change (22)
Exothermic change (22)
Energy (22)
Potential energy (22)
Kinetic energy (22)
Law of Conservation of Energy (24)

Most of these terms and many others appear in the Glossary.

QUESTIONS AND PROBLEMS

Section 2.1

2.1) Is each of the following a physical property or a chemical property? (a) hardness of a diamond; (b) burning ability of coal; (c) discoloration, or tarnishing, of silver; (d) "springiness" of spring steel; (e) electrical conductivity of copper.

2.2) Is each of the following changes physical or chemical? (a) spoiling of food; (b) changing of "dry ice" to gas; (c) stretching of a rubber band; (d) explsion of dynamite; (e) shattering of glass.

2.3) Explain how you would separate a mixture of sand and salt. Identify and classify as chemical or physical the property on which your separation is based.

2.19) Is each of the following a physical property or a chemical property? (a) darkening of sugar when heated; (b) yellow color of sulfur; (c) nitrogen does not react readily; (d) "slipperiness" of soap; (e) wood floating on water.

2.20) Identify each change as physical or chemical: (a) decay of dead plants; (b) production of iron from ore; (c) melting snow; (d) burning of oily rags (spontaneous combustion); (e) grinding of wheat.

2.21) Suggest two ways to separate iron shavings from sawdust. What property would you be using in each method? Is that property physical or chemical?

Section 2.2

2.4) List the principal differences between a gas, a liquid, and a solid.

2.5) Why is it easier to carry a plate of solid food around a corner than a shallow bowl of soup? Answer in terms of the differences between solids and liquids.

2.6) To which one or more of the three states of matter may the word "fluid" be properly applied? Explain.

2.22) Explain why a container that holds a solid or liquid may be open, but a container that holds a gas must be closed.

2.23) Hydrogen has been used as a fuel in space. Explain why it is more convenient to carry it as a liquid at very low temperatures than as a gas at room temperature.

2.24) We speak logically of "pouring" a liquid. Is it possible to pour a solid? or a gas? Explain why in each case.

Section 2.3

2.7) Identify the difference between a pure substance and a mixture.

2.8) A liquid is heated until it begins to boil at temperature X. Five minutes later it is still boiling, but the temperature has increased to Y. Is the liquid a pure substance or a mixture? Explain.

2.9) The density of a liquid is measured at its boiling point. It is boiled in an open container for 15 minutes, and the density of the remaining liquid is determined again. The densities are identical. Was the beginning liquid a mixture or a pure substance? Explain.

2.25) How do the properties of pure substances differ from the properties of mixtures?

2.26) The temperature at which a liquid first begins to freeze is determined. Frozen solid is removed, and the freezing temperature of the remaining liquid is measured again. The second temperature is lower than the first. Was the beginning liquid a mixture or a pure substance? Explain.

2.27) A colored liquid evaporates in an open container. Over a period of time the color becomes darker. Was the original liquid a pure substance or a mixture? Explain.

Section 2.4

2.10) What is the essential difference between an element and a compound?

2.11) Identify each of the following pure substances as an element or a compound: tin; nitrogen; carbon dioxide; iron; baking soda.

2.28) A pure red powder darkens on heating, releases a gas, and leaves behind a silvery liquid. Is the powder a compound or an element? How do you know?

2.29) Identify the elements and the compounds among the following familiar pure substances: sugar; helium (used to fill balloons); alcohol; lead; steam.

Section 2.5

2.12) What is the meaning of the term *homogeneous*? List several homogeneous substances not mentioned in the text.

2.13) State whether each of the following samples of matter is homogeneous or heterogeneous: (a) sand on a beach; (b) filtered air; (c) fog; (d) salt water; (e) copper wire.

2.30) Define the term *heterogeneous*. List several examples of heterogeneous substances not mentioned in the text.

2.31) State whether each of the following substances is heterogeneous or homogeneous: (a) boiling water; (b) window glass; (c) ground beef; (d) pepper; (e) flawless diamond.

Section 2.6

2.14) State the Law of Conservation of Mass. Illustrate its meaning in reference to the chemical change that occurs when carbon combines with oxygen to form carbon monoxide.

2.32) Explain in terms of the Law of Conservation of Mass why, when paper combines with oxygen in burning, the remaining ash weighs less than the original paper, but when iron combines with oxygen in rusting, it gains in weight.

Section 2.7

2.15) Explain what is meant by an "electrostatic force," and under what conditions it may exist. Identify an observable difference between electrostatic force and gravitational force.

2.33) Certain chemical particles called ions are electrically charged. Would the following pairs attract or repel each other? (a) positively charged sodium ion and negatively charged chloride ion; (b) negatively charged chloride ion and negatively charged bromide ion. State the rule on which your answers are based.

Section 2.8

2.16) State the difference between exothermic changes and endothermic changes. When water boils, is the change exothermic or endothermic? Give two examples of each, that are not in the textbook.

2.17) Determine whether the form of mechanical energy in each of the situations described is *mostly* potential or kinetic: (a) a bullet leaving the muzzle of a gun; (b) a set mouse trap; (c) a pendulum at the bottom of its swing; (d) water near the bottom of a waterfall; (e) the north poles of two magnets held near each other; (f) water in a lake formed by a dam; (g) compressed air.

2.34) Classify the following changes as exothermic or endothermic: (a) water freezes; (b) an explosion; (c) burning of gasoline in an automobile engine; (d) decomposing water by electrolysis (Fig. 2.5).

2.35) Determine whether the form of mechanical energy in each of the situations described is *mostly* potential or kinetic: (a) a windmill; (b) the wind that drives the windmill; (c) a wound wristwatch; (d) a speedboat pulling a water skier; (e) a baseball bat just before hitting a ball; (f) positively and negatively charged objects held close to each other; (g) a roller coaster at the top of the first hill.

Section 2.9

2.18) When wood burns, the production of heat energy is obvious. Explain the source of this energy if, according to the energy conservation law, energy may neither be created nor destroyed in an ordinary chemical change.

2.36) In order to decompose water by electrolysis a continuous input of electrical energy is required. In view of this, compare the amount of "chemical energy" held by a given amount of water with the chemical energy held by the hydrogen and oxygen produced. Which holds more? Explain.

3

measurements in chemistry

3.1 INTRODUCTION TO MEASUREMENT

Chemistry is both qualitative and quantitative. In its qualitative role it explains *how* and *why* chemical changes occur. Quantitatively it measures *how much* of a substance is used or produced.

To consider how much of a chemical is involved in a reaction a chemist must make measurements. Making measurements is nothing new to you; you measure your height and weight, or the distance to school or work. If you shop in the United States, you buy milk by the quart or gallon, butter by the pound, and fabric by the yard. If you live in Canada—or just about any other place in the world, for that matter—the metric units of liters, grams, and meters are more familiar to you.*

Most scientific measurements are expressed in so-called **SI units**. SI is an abbreviation for the French name for the International System of Units. (A summary of SI units appears in Appendix II.) Metric units for measurements of mass and length are part of the SI system. In this text, metric units will be used almost exclusively. The only exceptions will be in the present chapter, which will include some of the important conversions between English and metric

---

*If you live outside the United States you probably spell the volume unit *litre* and the length unit *metre*. These spellings correspond with the French pronunciations of the words, and it was the French who first used them. In this book we will use the American spellings, which match the English pronunciations.

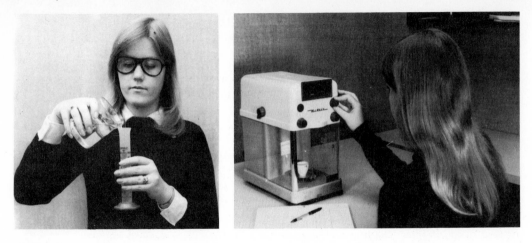

Figure 3.1
A student measures volume in graduated cylinder. A student measures mass on analytical balance. (Photographs by Gregory A. Peters.)

units. This is done to give you some "feeling" for the metric system in case you are not familiar with it. Several of the more important relationships between English and metric units are included in the conversion factor table in Appendix II, Table AP-5.

Two of the ways in which laboratory measurements are made are suggested in Figure 3.1. Liquid volume is measured in a number of ways, depending upon the precision required; a simple graduated cylinder is often satisfactory. Mass is measured on a balance; again the accuracy requirement determines which instrument is used. Length is ordinarily measured with a meter stick or "ruler." Most of the common one-foot rulers sold in bookstores of American colleges have a centimeter scale on one edge. A fourth fundamental measurement made in the laboratory is temperature. Thermometers are graduated in Celsius (formerly centigrade) degrees, rather than in the Fahrenheit degrees familiar to Americans.

At the time this book goes to press, the United States and no more than four minor and underdeveloped nations are the only countries in the world still clinging to the obsolete English system of units. In the United States the long overdue change has begun, but it is meeting strong resistance. It appears on food labels, where liters and grams are printed next to pints and pounds. The liter is gradually replacing the gallon as the unit in which gasoline is sold. Public signs and TV weather reports often give temperature in both Celsius and Fahrenheit degrees. Progress is slow, but metric measurements will no doubt become more apparent in the United States in the coming years.

3.2 DIMENSIONAL ANALYSIS

PG 3 A Identify the "given quantity" in the statement of a problem. Beginning with the given quantity, set up and solve the problem by the dimensional analysis method.

The way to solve chemistry problems is one of the most important things for you to learn in your beginning course. We therefore introduce the problem-solving method used in this book in much detail. The method is called **dimensional analysis.*** The thought process behind dimensional analysis will be shown by two simple problems, problems you can easily solve "in your head." Think carefully about these examples. If you master the method with simple problems, you will find it easy to use later when solving more difficult problems.

Dimensional analysis is based on the fact that if you multiply any quantity by 1, the product is the same as the original quantity; that is, $a \times 1 = a$. It also uses the fact that any fraction in which the numerator is equal to the denominator is equal to 1: $\dfrac{b}{b} = 1$. Or, if $c = d$, then $\dfrac{c}{d} = 1 = \dfrac{d}{c}$. Combining these facts,

$$\frac{em}{f} \times \frac{1}{m} = \frac{e}{f} \times \frac{m}{m} = \frac{e}{f} \times 1 = \frac{e}{f} \tag{3.1}$$

It is common to say that any factor that appears in both a numerator and denominator may be "canceled":

$$\frac{em}{f} \times \frac{1}{m} = \frac{e\cancel{m}}{f} \times \frac{1}{\cancel{m}} = \frac{e}{f} \tag{3.2}$$

First Problem: How many inches are in 3 feet?

We will use this problem to define some terms used in dimensional analysis. The first term is *given quantity.* Nearly all problems in this book begin with an amount of something, a quantity that can be counted or measured, such as 4 hours, 5 hot dogs, or 9 ounces. A given quantity is this kind of "thing." A given quantity always includes the *unit,* the thing that is counted or measured—the hours, hot dogs, or ounces. A given quantity always has only one kind of unit, not a combination, such as miles per hour, cents per hot dog, or ounces per pound. The given quantity in our sample problem is 3 feet.

Much of your success in solving chemistry problems will depend upon your ability to identify the given quantity. It tells you where to begin to solve the problem. Only in this chapter will we specifically mention the given quantity, but it will be used in the calculation setup of nearly all problems. In order

---

*Other names for dimensional analysis, or other problem-solving methods that include dimensional analysis, are the conversion factor, factor-label, unit cancellation, factor-unit and unary rate methods.

to remind you of its importance, the given quantity will be enclosed in a lightly colored box, ⬜ , in all calculation setups in this book.

In mentally solving the problem "How many inches are in 3 feet?" you no doubt thought, "There are 12 inches in 1 foot. Therefore in 3 feet there must be $3 \times 12 = 36$ inches." The "12 inches in 1 foot," or 12 inches = 1 foot, is called a *conversion relationship*, or *conversion factor*. Conversion factors show how two different measurements are related to each other, how many of one are equal to some number of the other. Any time two measured quantities are directly proportional to each other (see Appendix I, Part D, page 581), you can write a conversion relationship between them.

Some conversion factors are established by definition. 12 inches = 1 foot is an example. The relationship is permanent, never changing. The units of measurement are for the same thing—length, in this case. Other conversion relationships are temporary. They may have one value in one problem, and another value in another problem. Speed, measured in miles per hour, is an example. If we drive a car at an average of 50 miles per hour we can calculate the distance traveled in two, three, or any number of hours. As long as that speed is held, the number of miles driven is proportional to the number of hours, and 1 hour of driving is said to be "equivalent" to 50 miles. In this book these temporary equivalent relationships will be written with the symbol ≃, saving the equal sign, =, for true equalities. Thus the conversion relationship between miles and hours is 50 miles ≃ 1 hour *at 50 miles per hour.*

If you know the conversion relationship between two units, you can convert any quantity of one measurement to an equivalent amount of a second measurement. You follow a *unit path* between the units. In our sample problem we converted 3 feet to inches. The unit path is feet → inches. In calculating the miles traveled in 3 hours at 50 miles per hour, the unit path is hours → miles. The unit path always begins with the units of the given quantity and ends with the units in which the answer is to be expressed.

A conversion factor may be written as a fraction, such as 12 inches/1 foot or 1 foot/12 inches. The slash line, /, is read "per": 12 inches/1 foot is "12 inches per 1 foot," or simply 12 inches per foot, and 1 foot/12 inches is "1 foot per 12 inches." Each of these fractions is equal to 1 because the numerator is equal to the denominator. Similarly, at 50 miles per hour, the fractions 50 miles/1 hour and 1 hour/50 miles are both equal to 1 because the numerators and denominators are "equivalent." Every conversion relationship yields two fractions equal to 1.

Now we will apply these ideas to solving the problem, "How many inches are in 3 feet?" The dimensional analysis setup begins with the given quantity, which is multiplied by 1 in the form of 12 in./1 ft:

$$3 \, \cancel{ft} \times \frac{12 \text{ in.}}{1 \, \cancel{ft}} = 36 \text{ in.}$$

Notice that the units "ft" cancel, just the way the factors "m" cancel in Equation 3.2. This is the technique of dimensional analysis: you cancel units in exactly the same way you cancel variable factors in algebra.

Second Problem: How many feet are in 48 inches?

You have probably calculated the answer mentally by the time you read this sentence. You didn't think about a unit path, but in terms of dimensional analysis you used inches → feet from the same conversion relationship, 12 inches = 1 foot. In effect, you reasoned, "If there are 12 inches in 1 foot, how many 12-inch 'packages' are in 48 inches?" In other words, how many times does 12 go into 48? The answer, of course, is 4, so 48 inches = 4 feet.

In the second problem the given quantity, 48 inches, must be *divided* by the conversion factor 12 inches/1 foot. Do you remember how to divide by a fraction? You invert the fraction, and then multiply. In more mathematical terms, to divide by a fraction you multiply by its inverse, or reciprocal. Thus

$$48 \text{ in.} \div \frac{12 \text{ in.}}{1 \text{ ft}} = 48 \text{ in.} \times \frac{1 \text{ ft}}{12 \text{ in.}} = 4 \text{ ft}$$

Notice that 1 foot/12 inches is the second conversion fraction equal to 1 that comes from 12 inches = 1 foot. The units again cancel so the only unit left, feet, is the unit in which the answer is to be expressed.

One of the big advantages of dimensional analysis is the way it tells you if you have multiplied when you should have divided, or divided when you should have multiplied. Suppose in the second problem you had multiplied by mistake. Your setup would be

$$48 \text{ in.} \times \frac{12 \text{ in.}}{1 \text{ ft}} = 576 \text{ in.}^2/\text{ft}$$

Because of your familiarity with the numbers, you would have known 576 was wrong. But even if the numbers were not familiar to you, as they might not be later in this book, the "nonsense units" in.^{2}/ft would have alerted you to something being wrong. If the units of an answer are not correct, you can be almost positive that the numerical result is wrong. You should check your setup of the problem and correct the error.

We will now use the 50 miles per hour relationship for your first example to be solved by the question-and-answer method described on page 9. You may wish to review the procedure at this time. It is also summarized on the opaque shields found elsewhere in this book. We suggest you tear out one of these shields and use it to cover parts of the example as you work through it. The shield may also be used as a bookmark and as a reference periodic table when you work examples in later chapters.

EXAMPLE 3.1. How long will it take to drive 250 miles at an average speed of 50 miles per hour?

The first step in solving this problem is to identify the given quantity. Remember that a given quantity is always a measurement or count of a single "thing." It is not a relationship between two measurable quantities. Write the given quantity here.

$$250 \text{ miles } \times \underline{\hspace{3cm}} =$$

The number of miles, a measurable distance, is the given quantity. The speed, 50 miles per hour, is a relationship between two measurable quantities, distance and time. From that relationship, 50 miles ≃ 1 hour within this problem.

Before solving the problem by dimensional analysis, think through to the answer. Decide in your mind whether you must multiply or divide by 50 to find the number of hours for the trip. Express your decision as a fraction equal to 1, to be used as a multiplier for 250 miles. The line for that fraction is already printed above. Write the numerator and denominator in their proper places, or repeat the setup and complete the calculation on a separate sheet of paper. Include units. Be sure the units cancel correctly. Then calculate your answer—please don't reach for a calculator for such a simple problem—and write it after the equal sign.

- - - - - -   - - - - - -

$$250 \text{ ~~miles~~ } \times \frac{1 \text{ hour}}{50 \text{ ~~miles~~}} = 5 \text{ hours}$$

If you had multiplied by 50 miles/1 hour the units would not have canceled and your answer would have been 12 500 miles^2/hour. Both the number and the nonsense units would have told you that the answer was incorrect. (You may wish to check page 572 to see why there is no comma in 12 500.)

We didn't mention it in Example 3.1, but notice that your unit path was miles → hours. The unit path idea is obvious in the one-step conversions in the problems so far. Its value becomes apparent in problems that must be solved by a series of steps.

The dimensional analysis method of problem solving may be summarized as follows:

1. *Begin with the given quantity.*

2. *Establish the unit path from the given quantity to the wanted quantity. You must be given or otherwise know each conversion factor in the unit path.*

3. *Write the setup for the problem, multiplying or dividing in a logical sequence through each step of the unit path. Be sure to write all units.*

4. *Cancel units to be sure that the setup gives an answer expressed in the correct unit(s).*

5. *Multiply and/or divide to get the numerical value of the answer.*

EXAMPLE 3.2 How many times does the hour hand of a clock go around in two weeks?

Write the given quantity first.

- - - - - -   - - - - - -

2 weeks

Two weeks is the only quantity in the problem; it must be the given quantity.

The unit in which the answer is to be stated is revolutions of the hour hand. The unit path begins with weeks and ends with revolutions. Unless you happen to know how many times an hour hand goes around in one week, you cannot go from the beginning of the unit path to the end in one step. Additional information is required—information drawn from your knowledge about clocks, hours, weeks, and days. Can you suggest a unit path from weeks to revolutions that has a series of steps for which you know the conversion relationship between each pair of units in the path? Try it. Also write the required conversion relationship beneath each arrow.

weeks →   → revolutions

------   ------

weeks   →   days   →   hours   →   revolutions
  1 week = 7 days   1 day = 24 hours   12 hours = 1 revolution

You convert from weeks to days with the first relationship, then days may be changed to hours, and finally hours are changed to revolutions.

Now you are ready to take the first step. Below we have begun the setup. Decide whether you must multiply or divide by 7 days per week to change weeks to days. Write the necessary numerator and denominator in their proper places. Include units, and show the cancellation that is possible. Do not solve to a numerical answer.

$$2 \text{ weeks } \times \frac{\qquad\qquad}{\qquad\qquad}$$

------   ------

$$2 \text{ \cancel{weeks} } \times \frac{7 \text{ days}}{1 \text{ \cancel{week}}} \times \frac{\qquad\qquad}{\qquad\qquad}$$

Normally we don't work out intermediate answers in a dimensional analysis setup. If you did, however, and if your thought process was reasonable, the intermediate answer would be meaningful. In the setup this far, the "weeks" cancel and the calculation yields an intermediate answer of 14 days. An incorrect setup, $2 \text{ weeks } \times \frac{1 \text{ week}}{7 \text{ days}}$, would result in a nonsense unit, weeks²/day, which would have signaled an error at this point.

The setup through the first multiplication indicates the number of days. Now *extend* the setup by multiplying a conversion ratio that changes *days* into *hours*. Again, do not solve the setup.

------   ------

$$2 \text{ \cancel{weeks} } \times \frac{7 \text{ \cancel{days}}}{1 \text{ \cancel{week}}} \times \frac{24 \text{ hours}}{1 \text{ \cancel{day}}} \times \frac{\qquad\qquad}{\qquad\qquad} =$$

Again notice that calculation of the setup this far would yield a meaningful answer in hours. This would not be true if the incorrect ratio $\frac{1 \text{ day}}{24 \text{ hours}}$ had been used.

The final step converts hours to revolutions by the third equivalence relationship listed in the unit path. Complete the setup, and this time solve for the numerical answer as well.

------   ------

$$2 \text{ weeks} \times \frac{7 \text{ days}}{1 \text{ week}} \times \frac{24 \text{ hours}}{1 \text{ day}} \times \frac{1 \text{ revolution}}{12 \text{ hours}} = 28 \text{ revolutions}$$

Separating the arithmetic from the setup, we have

$$\frac{2 \times 7 \times 24}{12} = 28$$

which may be done mentally by first seeing that 24/12 is 2. This is another advantage you should seek in a dimensional analysis setup: setting up the entire problem frequently leads to simplification in calculations, and therefore fewer errors.

A word of caution—used without thinking, dimensional analysis can become a very mechanical process. Do not permit this. You will gain the most benefit from dimensional analysis if you think your way through each problem. Do not simply juggle units until they "come out right." Chart the unit path from the given quantity to the desired quantity, and then thoughtfully consider each conversion. This will make each step of the problem meaningful, and you will *understand* the overall solution. Understanding, rather than an answer, should be your goal for every problem.

Dimensional analysis setups will be used in unit conversions throughout this book. Additional practice problems may be found at the end of the chapter.

3.3 LENGTH AND VOLUME UNITS IN THE METRIC SYSTEM

PG 3 B   Given an English-metric conversion table, convert a length measurement expressed in (a) English units, (b) meters, (c) kilometers, (d) centimeters, or (e) millimeters to each of the other units.

3 C   Given an English-metric conversion table, convert a volume measurement expressed in (a) English units, (b) liters, (c) milliliters, or (d) cubic centimeters to each of the other units.

The standard unit of length is the **meter**. The meter was first defined as 1/10 000 000 of the distance from the North Pole to the equator. It is now identified more precisely as 1 650 763.73 times the wavelength of a certain line in the emission spectrum (see Chapter 5, p. 71) of an elemental gas, krypton. The advantage of the modern definition over the original may not be obvious, but precision measurements of modern technology require such a definition. The meter is 39.37 inches long—about three inches longer than a yard (Fig. 3.2).

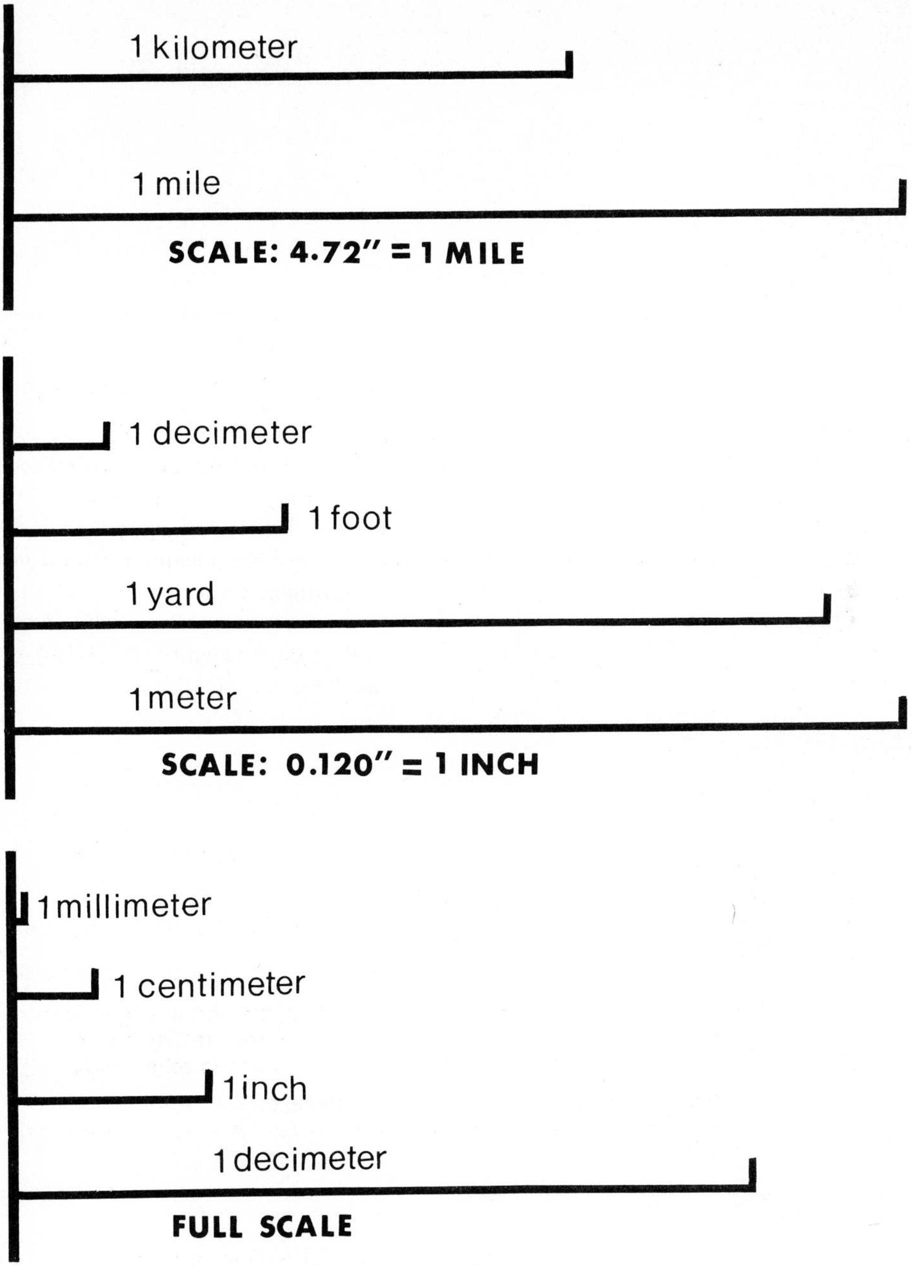

1 kilometer

1 mile

SCALE: 4.72″ = 1 MILE

1 decimeter

1 foot

1 yard

1 meter

SCALE: 0.120″ = 1 INCH

1 millimeter

1 centimeter

1 inch

1 decimeter

FULL SCALE

Figure 3.2
Comparison between English and metric units of length.

Table 3.1
*Metric Prefixes**

Large Units			Small Units		
Metric Prefix	*Metric Symbol*	*Multiple*	*Metric Prefix*	*Metric Symbol*	*Submultiple*
mega-	M	$1\ 000\ 000 = 10^6$	deci-	d	0.1† $\quad = 10^{-1}$
kilo-	**k**	$\mathbf{1,000} = 10^3$	**centi-**	**c**	**0.01** $\quad = 10^{-2}$
hecto-	h	$100 = 10^2$	**milli-**	**m**	**0.001** $\quad = 10^{-3}$
deka-	da	$10 = 10^1$	micro-	μ	$0.000\ 001 \quad = 10^{-6}$
(Unit: gram, meter, liter)		$1 = 10^0$	nano-	n	$0.000\ 000\ 001 = 10^{-9}$

*The most important metric prefixes are printed in boldface type.
†In scientific writing it is customary to place a zero before the decimal point of a number smaller than 1.

Because the English system uses both larger and smaller units than the yard for different purposes, so the meter must be modified for larger and smaller units. The metric system accomplishes this with multiples and submultiples of 10, each of which is identified by a prefix. The prefix for $\frac{1}{10}$ is *deci-*. Therefore $\frac{1}{10}$ of a meter, or 0.1 meter, is a decimeter. (In scientific writing it is customary to place a zero before the decimal point of a number smaller than 1.) The prefix for $\frac{1}{100}$ (or 0.01) metric unit is *centi-*; 0.01 meter is a centimeter. Similarly, using *milli-* for $\frac{1}{1000}$ (or 0.001) unit, 0.001 meter is a millimeter. The most important larger unit is 1000 times larger than the basic unit. Its prefix is *kilo-*; 1000 meters is a kilometer.

Centi-, *milli-*, and *kilo-* are the most important metric prefixes. They should become part of your vocabulary. Table 3.1 lists some of the other metric prefixes; a more complete list is on p. 596 in the Appendix. You may recognize common uses of some of these. *Megaton* (1 million tons) is used to describe the size of hydrogen bombs, and *kilobucks* and *megabucks* have become slang expressions for one thousand and one million dollars.

The metric symbols in Table 3.1 are used as abbreviations for metric units. The letter m identifies the meter. The symbol for kilometer is formed by placing a k for *kilo-* in front of the m for meter: 1 km = 1 kilometer. Similarly, 1 cm is 1 centimeter, and 1 mm is 1 millimeter. The relationships among these distances are summarized in the following three equations:*

$$1\ km = 1000\ m = 10^3\ m \quad or \quad 1\ m = 0.001\ km = 10^{-3}\ km \quad (3.3)$$

$$1\ cm = 0.01\ m = 10^{-2}\ m \quad or \quad 1\ m = 100\ cm = 10^2\ cm \quad (3.4)$$

$$1\ mm = 0.001\ m = 10^{-3}\ m \quad or \quad 1\ m = 1000\ mm = 10^3\ mm \quad (3.5)$$

The outstanding advantage of the metric system over the English system is the ease with which values may be converted from one unit to another. Conversions are all made by multiplying or dividing by multiples of 10—accomplished simply by moving the decimal point a certain number of places to the right or to the left. Until you become familiar with these conversions, you may

*See Appendix I, Part F, for information on exponential numbers.

be uncertain which direction the decimal point should move. A dimensional analysis setup of the problem helps remove this uncertainty.

EXAMPLE 3.3 How many meters are there in 28.6 cm?

First step: Identify the given quantity and unit path,

28.6 cm; cm → m

The given quantity is now to be multiplied by a conversion factor that will change centimeters to meters Equation 3.4 shows that there are 100 cm in one meter. From this, two factors are possible. Select the correct one and complete the problem.

$$28.6 \text{ cm} \times \frac{1 \text{ m}}{100 \text{ cm}} = 0.286 \text{ m}$$

If you had mistakenly multiplied by 100 instead of dividing, your units would have been meaningless (cm²/m), which would have alerted you to the error.

EXAMPLE 3.4 How many millimeters are in 3.04 cm?

Although the relationship between millimeters and centimeters has not been stated in an equation, you should be able to see it by comparing Equations 3.4 and 3.5. Supply the numbers to complete the following relationship:

_____ mm = _____ cm

1000 mm = 100 cm

From Equations 3.4 and 3.5, 100 cm and 1000 mm are both equal to 1 meter, and therefore equal to each other. The relationship reduces to 10 mm = 1 cm, an equally acceptable answer.

Now complete the problem: 3.04 cm × _____ =

$$3.04 \text{ cm} \times \frac{1000 \text{ mm}}{100 \text{ cm}} = 30.4 \text{ mm} \quad or \quad 3.04 \text{ cm} \times \frac{10 \text{ mm}}{1 \text{ cm}} = 30.4 \text{ mm}$$

To convert between the English and metric systems, one of several conversion factors is required. (Many of these appear in Appendix II, Table AP-5, page 597.) Perhaps the most common length conversion is the relationship between centimeters and inches:

$$2.54 \text{ cm} = 1 \text{ in.} \qquad (3.6)$$

EXAMPLE 3.5   How many meters are in 7.60 feet? (Use Equation 3.6 to get from English to metric units.)

Because you have been limited to Equation 3.6 for the conversion from English to metric units, your unit path must have several steps. Write that unit path.

ft → in. → cm → m

Now start with the given quantity, complete the entire dimensional analysis setup, and calculate the answer.

$$7.60 \, \cancel{ft} \times \frac{12 \, \cancel{in.}}{1 \, \cancel{ft}} \times \frac{2.54 \, \cancel{cm}}{1 \, \cancel{in.}} \times \frac{1 \text{ m}}{100 \, \cancel{cm}} = 2.32 \text{ m}$$

VOLUME

The most common volume unit in the chemistry laboratory is the **cubic centimeter,** or cm³. It is the volume of a cube one centimeter on an edge. (Compare with the English cubic inch, which represents the volume of a cube one inch on an edge.) Figure 3.3 illustrates metric volume measurements, including those used for capacity.

To convert a volume from one set of cubic units to another you must apply a length conversion three times. This is shown in the following example.

EXAMPLE 3.6   How many cubic inches are in a box 9.00 cm long by 6.00 cm wide by 4.00 cm high?

Solution: To find the volume of a rectangular box you multiply the length by the width by the height. The volume of the box in cubic centimeters is

$$9.00 \text{ cm} \times 6.00 \text{ cm} \times 4.00 \text{ cm} = 216 \text{ cm}^3$$

Figure 3.3
Metric volume units in the laboratory. The large block is a cube measuring 1 decimeter (slightly less than 4 inches) on each edge—a cubic decimeter. One decimeter is 10 centimeters, so the volume of the block is also 1000 cubic centimeters (10 cm × 10 cm × 10 cm). This is equal to the capacity unit 1 liter, shown by the beaker. Separated from the block is a piece 1 cm × 1 cm × 10 cm, from which is further separated a cube 1 centimeter on each edge—a cubic centimeters, 1/1000 of a cubic decimeter. One milliliter, 1/1000 of a liter, is equal to 1 cubic centimeter. (Photograph by Judith E. Serface.)

To find the volume in cubic inches, each centimeter measurement could be converted to inches:

$$9.00 \text{ cm} \times \frac{1 \text{ in}}{2.54 \text{ cm}} \times 6.00 \text{ cm} \times \frac{1 \text{ in}}{2.54 \text{ cm}} \times 4.00 \text{ cm} \times \frac{1 \text{ in}}{2.54 \text{ cm}} = 13.2 \text{ in}^3$$

$$\text{length} \qquad \times \qquad \text{width} \qquad \times \qquad \text{height} \qquad = \text{volume}$$

This expression may be simplified as

$$9.00 \text{ cm} \times 6.00 \text{ cm} \times 4.00 \text{ cm} \times \left(\frac{1 \text{ in}}{2.54 \text{ cm}}\right)^3 = 13.2 \text{ in}^3$$

or,

$$9.00 \text{ cm} \times 6.00 \text{ cm} \times 4.00 \text{ cm} \times \frac{1^3 \text{ in}^3}{2.54^3 \text{ cm}^3} = 13.2 \text{ in}^3$$

Notice that both the numbers and the units in both numerator and denominator are raised to a power when a fraction is to be raised to that power. (The algebra is explained in greater detail on page 580 in the Appendix, if you wish to review it.)

EXAMPLE 3.7   Calculate the number of cubic meters of earth in 15.0 cubic yards of fill.

Appendix II, Table AP-5, page 597, gives you the necessary length conversion. Identify the given quantity, set up the problem, and calculate the answer.

_ _ _ _ _ _   _ _ _ _ _ _

$$15.0 \text{ yd}^3 \times \frac{1^3 \text{ m}^3}{1.09^3 \text{ yd}^3} = 11.6 \text{ m}^3$$

This setup represents the unit path yd^3 → m^3, using the conversion factor 1 m = 1.09 yd three times, or 1^3 m^3 = 1.09^3 yd^3.

The English system uses special volume units for liquids, and sometimes gases. These units, which are sometimes called "capacity" units, are not derived from length units. The common metric capacity unit is the **liter,** which is exactly 1 cubic decimeter (1 dm³) or 1000 cubic centimeters (1000 cm³). This volume is very close to one U.S. quart*—1.06 quarts, to be more exact. The smaller unit commonly encountered in the laboratory is the **milliliter.** The prefix *milli-* again means 1/1000th part, so 1 milliliter = 0.001 liter, or 1000 milliliters = 1 liter. By definition there are also 1000 cubic centimeters in a liter. *This makes 1 cubic centimeter exactly equal to 1 milliliter.* The symbols for these terms are L for liter and mL for milliliter. (The older symbols, ℓ and ml, are still in common use.)

EXAMPLE 3.8   Calculate the number of liters in 0.500 cubic foot.

The necessary conversion factors are given in Appendix II, Table AP-5, page 597. Be careful about the final step in the setup. . . .

$$0.500 \ \cancel{ft^3} \ \times \ \frac{30.5^3 \ \cancel{cm^3}}{1^3 \ \cancel{ft^3}} \ \times \ \frac{1 \ L}{1000 \ \cancel{cm^3}} = 14.2 \ L$$

Notice that the 1000 and the 1 in the final step are *not* cubed. The liter is already a volume unit, not the cube of a length unit.

3.4 MASS AND WEIGHT

PG 3 D Distinguish between mass and weight.

3 E Given an English-metric conversion table, convert a mass measurement, expressed in (a) English units, (b) grams, (c) kilograms, (d) centigrams, or (e) milligrams, to each of the other units.

Consider a tool carried by astronauts on a trip to the moon. Suppose that tool, on earth, weighs six ounces. If it were on the surface of the moon, it would weigh about one ounce. But halfway between the earth and moon it would be essentially weightless: Released in "midair" it would remain there, floating, until moved by one of the astronauts to some other location. Yet in all three locations it would be the same tool, having a constant quantity of matter.

---

*In countries using the British or "imperial" measure, the pint, quart and gallon are larger than their corresponding U.S. measures by the factor of 6/5; i.e., 6 U.S. units = 5 imperial units. Thus, in Canada, 1 liter = 0.88 imperial quart.

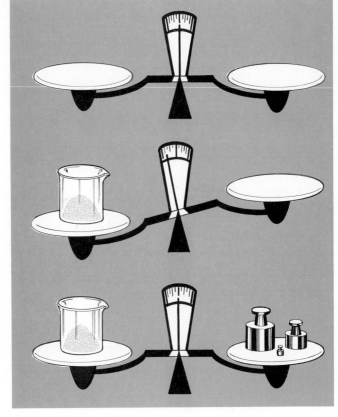

Figure 3.4
Use of a two-pan balance. (1) Balance is adjusted so pointer indicates zero on scale (top). (2) Object to be weighed is placed on left pan (middle). (3) Weights are placed on right pan until pointer again indicates zero (bottom). Mass of object on left pan is the same as mass of weights on right pan.

Mass is a measure of quantity of matter. Weight is a measure of the force of gravitational attraction. Weight is proportional to mass, and the ratio between them depends upon where in the universe you happen to be. Fortunately this proportionality is essentially constant over the surface of the earth. When you "weigh" something—measure the force of gravity on that object—you can express this weight in terms of mass. By common usage, *weighing* an object is the same as measuring its mass.

The instrument used to measure mass in the laboratory is the *balance*. Most balances are constructed so that masses on opposite sides of the balance point are compared. In use (Fig. 3.4), an unknown mass is placed on the left pan of the balance, and then "balanced" by placing known weights on the other pan. When the pointer comes to rest at the center of the scale, the masses on the two pans of the balance are equal. Most laboratory balances are more complex and more accurate than the two-pan decigram balance, but all operate on essentially the same principle.

The basic unit of mass in the metric system is the **kilogram**. A kilogram weighs about 2.2 pounds and, therefore, is too large a unit for small scale work in chemistry. The more common laboratory unit of mass is the gram, which is $\frac{1}{1000}$ kilogram. The balances commonly used in college chemistry laboratories are capable of measuring centigrams ($\frac{1}{100}$ g) or milligrams ($\frac{1}{1000}$ g).

Conversion between the metric and English units may be made with either of the following factors:

$$454 \text{ grams} = 1 \text{ pound} = 16 \text{ ounces}^* \qquad (3.7)$$

$$1 \text{ kilogram} = 2.20 \text{ pounds} \qquad (3.8)$$

EXAMPLE 3.9 Find the mass in kilograms of an object that weighs 13.4 ounces.

Solve the problem completely, using either metric-English conversion factor.

By Equation 3.7: $13.4 \text{ oz} \times \dfrac{1 \text{ lb}}{16 \text{ oz}} \times \dfrac{454 \text{ g}}{1 \text{ lb}} \times \dfrac{1 \text{ kg}}{1000 \text{ g}} = 0.380 \text{ kg}$

By Equation 3.8: $13.4 \text{ oz} \times \dfrac{1 \text{ lb}}{16 \text{ oz}} \times \dfrac{1 \text{ kg}}{2.20 \text{ lb}} = 0.380 \text{ kg}$

The first setup might be shortened slightly by moving directly from ounces to grams using the extension of Equation 3.7:

$$13.4 \text{ oz} \times \dfrac{454 \text{ g}}{16 \text{ oz}} \times \dfrac{1 \text{ kg}}{1000 \text{ g}} = 0.380 \text{ kg}$$

EXAMPLE 3.10 How many milligrams are in 34.9 grams?

This is a simple metric conversion using metric prefixes for mass measurements, just as you used them for length measurements earlier.

$$34.9 \text{ g} \times \dfrac{1000 \text{ mg}}{1 \text{ g}} = 34\,900 \text{ mg}$$

3.5 DENSITY AND SPECIFIC GRAVITY

PG 3 F Distinguish between density and specific gravity, including units of each.

3 G Given two of the following, calculate the third: mass of a sample; volume occupied by the sample; density, or specific gravity.

---

*Ounces referred to in this text are avoirdupois ounces, unless otherwise specified.

The concept of *density* is an example of the combination of fundamental measurements to express a physical property. The formal definition of **density is mass per unit volume.** Expressed mathematically, this becomes

$$\text{density} = \frac{\text{mass}}{\text{volume}} \tag{3.9}$$

We can think of density as a measure of the "heaviness" of a substance, in the sense that a block of iron is heavier than a block of aluminum of the same size. Table 3.2 lists the densities of some common materials.

The definition of density establishes the units in which it is expressed. Mass is measured in grams; volume is measured in cubic centimeters. Therefore, according to Equation 3.9, the units of density are grams/cubic centimeter. There are, of course, other units in which density can be expressed, but they all must reflect the definition in terms of mass/volume. Examples include grams/liter and pounds/cubic foot.

In order to find the density of a substance, it is necessary to know both the mass and volume of the same quantity of the substance. Dividing the mass by the volume yields the density.

EXAMPLE 3.11   A 12.0-cm³ piece of magnesium weighs 20.9 grams. Find the density of magnesium.

$$\text{density} = \frac{\text{mass}}{\text{volume}} = \frac{20.9 \text{ grams}}{12.0 \text{ cm}^3} = 1.74 \text{ g/cm}^3$$

Specific gravity tells us how dense a substance is compared to some standard, usually water. More precisely, **specific gravity is the ratio of the density of a substance to the density of water at 4°C.** Expressed as an equation,

$$\text{specific gravity} = \frac{\text{density of substance (g/cm}^3)}{\text{density of water at 4°C}} \tag{3.10}$$

Table 3.2
*Approximate Densities of Some Common Substances**

Helium (used in balloons)	0.000 17	Aluminum	2.7
Air	0.0012	Iron	7.8
Pine lumber	0.5	Copper	9.0
Maple lumber	0.6	Silver	10.5
Oak lumber	0.8	Lead	11.4
Water	1.0	Mercury	13.6
Glass	2.5	Gold	19.3

*Densities are given in grams per cubic centimeter at sea level and 20°C.

In Equation 3.10 the same units for density appear in both numerator and denominator. Therefore all units cancel, and specific gravity is a unitless or dimensionless term, a pure number.

In metric units the density of water at 4°C is 1.000 gram/cm³. Dividing the density of a substance by 1.000 gram/cm³ causes the specific gravity of a substance to be numerically equal to its density in grams per cubic centimeter. Accordingly, if the specific gravity of a substance is 6.3, we may conclude that its density is 6.3 grams/cm³.

Density furnishes a useful equivalence relationship for making conversions between mass and volume. Since the density of copper, for example, is 8.9 g/cm³, this means that 8.9 grams of copper represent the same quantity as 1.0 cm³ of copper. Therefore we may say 8.9 g copper ≃ 1.0 cm³ copper. Use of density as a connecting link between mass and volume is shown in the following example.

EXAMPLE 3.12   The specific gravity of a certain kind of wood is 0.860. Find the mass of a block of that wood measuring 16.0 cm × 5.00 cm × 3.00 cm.

Solution: In this problem the volume of the wood is not given, but, as in Example 3.6, it may be found simply by multiplying the length-width-height dimensions. This product is the given quantity. Because of the *numerical* equivalence between specific gravity and density, the specific gravity of 0.860 may be interpreted as a density of 0.860 g/cm³. This may be used as a conversion relationship in a dimensional analysis setup:

$$16.0 \text{ cm} \times 5.00 \text{ cm} \times 3.00 \text{ cm} \frac{0.860 \text{ g}}{1 \text{ cm}^3} = 206 \text{ g}$$

Many students like to solve density problems algebraically by substituting into the defining equation. For instance, in Example 3.13,

$$\text{density} = \frac{\text{mass}}{\text{volume}}$$

$$\frac{0.860 \text{ g}}{1 \text{ cm}^3} = \frac{\text{mass in grams}}{16.0 \text{ cm} \times 5.00 \text{ cm} \times 3.00 \text{ cm}}$$

$$\text{mass in grams} = 16.0 \text{ cm} \times 5.00 \text{ cm} \times 3.00 \text{ cm} \times \frac{0.860 \text{ g}}{1 \text{ cm}^3} = 206 \text{ g}$$

The resulting arithmetic is identical, and the algebraic approach is entirely correct. Part of the reason for introducing density at this point, however, is to give you practice in the use of dimensional analysis. We will continue to use it in the coming examples, and we recommend that you do too.

EXAMPLE 3.13   The density of a certain copper-plating solution is 1.18 g/cm³. How many milliliters will be occupied by 150 grams of the solution?

Recall that a cubic centimeter and a milliliter are identical volumes. Starting with the given quantity and interpreting density as an equivalence between mass and volume, the setup is straightforward. Complete the problem.

$$150 \, g \times \frac{1 \, \text{mL}}{1.18 \, g} = 127 \, \text{mL}$$

3.6 TEMPERATURE MEASUREMENT

> **PG 3 H**   Given a temperature in °F or °C, convert from one scale to the other.

The familiar temperature scale in the United States is the Fahrenheit scale. The temperature scale normally used for scientific purposes is the Celsius scale. The Celsius scale is based on two "fixed" points, the normal freezing and boiling points of water. The freezing point is assigned a value of 0°C, and the boiling point 100°C. The difference between the fixed points is divided into 100 identical degrees. The corresponding temperatures on the Fahrenheit scale are 32°F for the freezing point of water and 212°F for the boiling point (Fig. 3.5).

The Celsius and Fahrenheit temperature scales are related by the equation

$$°F - 32 = (1.8)(°C) \tag{3.11}$$

To solve a temperature conversion problem you substitute the given temperature into Equation 3.11 and solve for the unknown.

EXAMPLE 3.14   What is the Celsius temperature on a comfortable 72°F day? Solve to the nearest °C.

$$72 - 32 = (1.8)(°C)$$

$$°C = \frac{72 - 32}{1.8} = 22$$

EXAMPLE 3.15   It's a cold day, about $-25°C$. Calculate the corresponding Fahrenheit temperature to the nearest degree.

$$°F - 32 = (1.8)(°C)$$

$$°F = -45 + 32 = -13$$

SI units include a third temperature scale known as the Kelvin or absolute scale. Kelvin degrees are identical to Celsius degrees, but zero on the Kelvin scale is 273.16° below zero on the Celsius scale. The two scales are therefore related by the equation

$$K = °C + 273.16 \qquad (3.12)$$

Notice the degree sign, °, is not used with the Kelvin scale. Kelvin temperature is based on an absolute zero, theoretically the lowest temperature possible. We will not be concerned with Kelvin temperatures until we study gases.

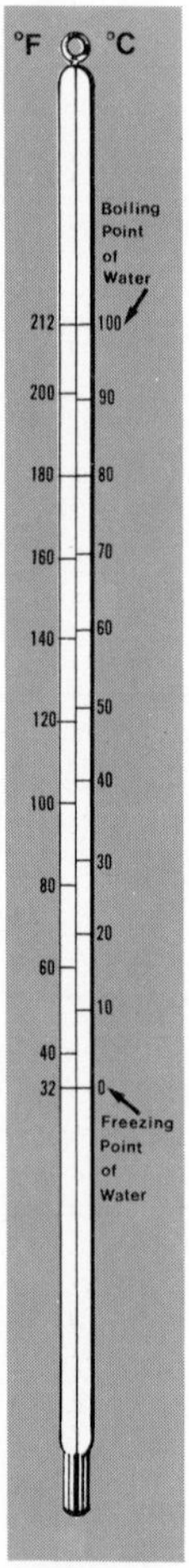

CHAPTER 3 IN REVIEW

3.1 INTRODUCTION TO MEASUREMENT (29)

3.2 DIMENSIONAL ANALYSIS

3 A Identify the "given quantity" in the statement of a problem. Beginning with the given quantity, set up and solve the problem by the dimensional analysis method. (31)

3.3 LENGTH AND VOLUME UNITS IN THE METRIC SYSTEM

3 B Given an English-metric conversion table, convert a length measurement expressed in (a) English units, (b) meters, (c) kilometers, (d) centimeters, or (e) millimeters to each of the other units. (36)

3 C Given an English-metric conversion table, convert a volume measurement expressed in (a) English units, (b) liters, (c) milliliters, or (d) cubic centimeters to each of the other units. (36)

Figure 3.5
Comparison between Fahrenheit and Celsius temperature scales.

3.4 MASS AND WEIGHT

wt = gravitational pull on mass
mass = measure of quantity

3 D Distinguish between mass and weight. (42)

3 E Given an English-metric conversion table, convert a mass measurement expressed in (a) English units, (b) grams, (c) kilograms, (d) centigrams or (e) milligrams to each of the other units. (42)

$- g/cm^3$

3.5 DENSITY AND SPECIFIC GRAVITY

$Density = \frac{mass}{unit\ volume}\ g/ml$

$SG = \frac{Density\ x}{Density\ H_2O}$

3 F Distinguish between density and specific gravity, including units of each. (44)

3 G Given two of the following, calculate the third: mass of a sample; volume occupied by the sample; density, or specific gravity. (44)

3.6 TEMPERATURE MEASUREMENT

3 H Given a temperature in °F or °C, convert from one scale to the other. (47)

TERMS AND CONCEPTS

SI units (29)	Mass (43)
Dimensional analysis (31)	Weight (43)
"Given quantity" (31)	Kilogram (43)
"Unit path" (32)	Density (45)
Meter (36)	Specific gravity (45)
Metric prefixes: centi-, milli-, kilo- (38)	Fahrenheit (47)
Liter (42)	Celsius (47)

Most of these terms and many others appear in the Glossary.

QUESTIONS AND PROBLEMS

For each problem that follows, write the calculation setup and solve for the numerical answer, even if either seems obvious. Remember that, at this point, the interpretation and setup of the problem are more important than the answer. Use dimensional analysis wherever possible.

Section 3.2

3.1) Express 48.5 gallons in quarts. (4 qt = 1 gal)

3.2) How many ounces are in 0.35 ton? (1 ton = 2000 lb; 1 lb = 16 oz)

3.3) A bicycle manufacturer plans to produce 1500 bicycles per month over a 5-month period. How many wheels must the production department schedule for the entire run?

3.4) How many minutes does it take a car traveling 35 miles per hour to cover a distance of 2500 feet?

3.34) If an American dollar = 1.14 Canadian dollars, express $1.50 Canadian in American dollars.

3.35) How many rulers are in 4.5 gross rulers? (1 gross = 12 dozen)

3.36) A laboratory hot plate has three identical heating elements. How long will an inventory of 608 elements last at a production rate of 40 hot plates per week?

3.37) A large reception is planned. Five cakes are to be baked as part of the refreshments. If the recipe for each cake calls for three eggs, and eggs are selling at 94¢ per dozen, calculate the cost of eggs in dollars.

3.5) Certain bolts are zinc-plated in batches averaging 75 pounds. If the plating cycle is 12 minutes for each batch, and the cost of operating the plating system is $28.00 per hour, what is the cost of plating 2.00 pounds of bolts, expressed in cents?

3.6) How many cubic feet of earth are in 3.6 cubic yards?

3.7) What is the area of a football field in square feet? (The field is 100 yards long × 53.3 yards wide.)

Section 3.3

3.8) A kitchen table is 28.0″ high. Express this measurement in centimeters.

3.9) A highway sign indicates that you are 148 kilometers from your destination. How many miles must you still travel?

3.10) A football kicker averages 39.8 yards per punt. Express this distance in meters.

3.11) What are the centimeter dimensions of a two-by-four (2″ × 4″) piece of lumber?

3.12) Fractional measurements that make up a large portion of the engineering design data in the United States will require a major adjustment when converted to metric units. How many millimeters, for example, are equal to $\frac{1}{8}$ inch?

3.13) A machine part is shown to be $3\frac{5}{8}$″ long on an engineering drawing. What is its millimeter measurement? Careful. . . .

3.14) How many millimeters are in 0.786 meter?

3.15) The wavelength of light is sometimes expressed in angstroms (Å). How many centimeters are in the wavelength of a 5.9×10^3-angstrom line in the yellow portion of the light spectrum? $(1 \text{ m} = 10^{10} \text{ Å})$

3.16) In planting a new lawn, the American home-owner today might order 2.50 cubic yards of top-soil. What equal amount in cubic meters would his son order for the same purpose in 2005?

3.17) How many cubic centimeters are in 14.2 m³?

3.18) All over the world, except in the United States, gasoline is sold by the liter. How many liters are in 5.00 gallons?

3.38) A certain type of copper rivet is sold at $2.15 per pound, and the average number of rivets in 1 pound is 34. What is the maximum number of rivets that can be purchased for $50.00?

3.39) A tank measures 84″ × 30″ × 24″. Calculate its volume in cubic feet.

3.40) How many square feet are in a field 14.5 rods long × 9.4 rods wide? (1 rod = 16.5 feet)

3.41) What will be the dimensions in meters of an 11.0′ × 13.0′ bedroom (11.0 feet wide, 13.0 feet long)?

3.42) A man's size 15 shirt has a neckband 15.0 inches in circumference. What size will he wear by metric measurement (centimeters)?

3.43) The speed limit on a highway is 55 miles per hour. To what number of kilometers per hour will the speedometer of a 1988 car point when moving at the same speed?

3.44) What are the dimensions of a sheet of 4′ × 8′ plywood, to the nearest centimeter?

3.45) Successful fashion models tend to be tall girls. How would you classify a girl who is 178 centimeters: short, tall, or in between? Express her height in inches.

3.46) It is easier to convert English measurement to metric than metric to English. To see the difference, write 195 mm to the nearest $\frac{1}{16}$ inch.

3.47) Find the number of kilometers in 4.29×10^5 centimeters.

3.48) An atom of copper has a diameter of 1.28×10^{-8} centimeter. Express this distance in angstroms. $(1 \text{ m} = 10^{10} \text{ Å})$

3.49) The area of the new lawn in Problem 3.16 is 247 square meters. How many square feet is this?

3.50) A glass holds 46 000 mm³ of water. What is this volume in cubic centimeters?

3.51) If gasoline today costs $1.55 per gallon, calculate its price in cents per liter.

Section 3.4

3.19) The terms *weight* and *mass* are commonly used interchangeably. We also say that an astronaut is "weightless" when traveling in space. Would it be proper to say instead that he or she is "massless?" Explain why or why not.

3.20) Calculate the number of grams in 4.80 ounces.

3.21) Convert 3.52 kilograms to grams.

3.22) How many grams are in 0.782 milligram?

3.23) The mass of a diamond is 0.218 g. How many carats is this? (1 carat = 200 mg)

3.24) Today the price of rice is 93¢ per pound. Assuming no inflation between now and 1990, what will you then pay in dollars per kilogram?

3.52) When you "weigh" yourself on a typical bathroom scale, are you actually measuring mass or force? Explain your answer.

3.53) Calculate the mass of a 2.8×10^3-pound automobile.

3.54) How many centigrams are equal to 46.9 grams?

3.55) The mass of an empty cup is 1560 milligrams. How many kilograms is this?

3.56) If the model of Problem 3.45 has a mass of 55.7 kg, what is her weight in pounds?

3.57) The former luxury ocean liner Queen Mary, now a tourist attraction in Long Beach, California, weighs 8.2×10^4 tons. Express her mass in megagrams. (1 megagram = 10^6 grams)

Section 3.5

3.25) The specific gravity of a certain substance is 0.89. What is its density? Compare the density of the substance with the density of water.

3.26) Find the mass of 50.0 mL of carbon tetrachloride if its specific gravity is 1.60.

3.27) Among natural minerals, gold is one of the most dense at 19.3 g/cm³. Find the volume occupied by 68.3 grams of gold.

3.28) Mercury is a dense material by any standard, particularly as a liquid. 150.0 mL of mercury weighs 2.04×10^3 grams. Calculate the specific gravity of mercury.

3.29) To fully appreciate the density of mercury, calculate the weight in pounds of one quart of mercury.

3.58) The density of bismuth is 608 pounds per cubic foot in English units. If the density of water is 62.4 pounds per cubic foot, calculate the specific gravity of bismuth.

3.59) What is the volume of 365 grams of ethyl alcohol if its specific gravity is 0.791?

3.60) Balsa wood, the soft, light wood used in making model airplanes, has a density of 0.125 g/cm³. Find the mass of 450 cm³ of balsa wood.

3.61) A typical ice cube from the refrigerator measures 4.0 cm $\times$ 3.5 cm $\times$ 3.0 cm, and weighs 38.5 grams. Calculate the density of ice.

3.62) Milk is among the heavier items carried home from the grocery store. Find the mass of $\frac{1}{2}$ gallon of milk (specific gravity = 1.03).

Section 3.6

3.30) Table salt melts at 805°C. What is this in Fahrenheit degrees?

3.31) Most people find it "uncomfortable" to touch a piece of metal if its temperature is above 120°F. What is the corresponding Celsius temperature?

3.32) Convert the following temperatures to degrees Celsius: −320°F; 205°F; 2460°F.

3.33) Find the Fahrenheit value of the following temperatures: −244°C; 46°C; 1520°C.

3.63) What Celsius temperature is the same as 0°F?

3.64) In the fractional distillation of petroleum, the components are separated by their boiling temperatures. One component of gasoline boils at 116°C. Express this in °F.

3.65) Express the following temperatures in Celsius degrees: 2190°F; −139°F; 326°F.

3.66) Change each of the following to degrees Fahrenheit: 1840°C; −102°C; 312°C.

4

atomic theory

4.1 INTRODUCTION TO THE ATOM

Atoms. Are they real? Is there such a thing as a particle that cannot be divided into smaller particles? The Greeks talked about such things as early as 400 B.C. It is from their word *atomos*, meaning indivisible or uncuttable, that the word *atom* comes. But atomic theory was not accepted by all Greeks. A rather influential gentleman named Aristotle didn't believe in atoms. Consequently, the idea lay quietly in the minds of philosophers for centuries. And that's exactly where "science" was during those unprogressive centuries—in the minds of philosophers.

It was not until the 17th century that *observation* and *experimentation* were added to pure thinking. Then things started to move, slowly at first, but quickening to a pace that, in recent times, has been difficult to keep up with. Today we find that the nucleus of the atom is at the same time our greatest threat for destruction and our greatest hope for meeting the world's growing demand for energy. We must hope that future research will reveal how to obtain the benefits of the atom in a way that is completely safe, and that mankind will never again release the power of the atom for the purpose of destruction as it was released in 1945.

In this chapter and the next we will outline the major concepts of the atom that have been developed over the past 170 years. We will concentrate more on the conclusions that have been reached than how they were reached. In Chapter 5 we will consider the structure of the atom and how it relates to the formation of chemical compounds.

4.2 DALTON'S ATOMIC THEORY

PG 4 A   List the main features of Dalton's atomic theory.

Early in the 19th century John Dalton, an English chemist and school-teacher, did a lot of thinking about two laws discussed in Chapter 2. The Law of Definite Composition (p. 18) states that the percentage by weight of each element in a compound is always the same. The Law of Conservation of Mass (p. 20) tells us that mass is conserved in a chemical change. In an attempt to explain these laws Dalton suggested his **atomic theory** in 1808. The main features of that theory are:

1. Each element is made up of tiny, individual particles called atoms.
2. Atoms are indivisible; they can neither be created nor destroyed.
3. All atoms of each element are identical in every respect.
4. Atoms of each element are different from atoms of any other element.
5. Atoms of one element may combine with atoms of another element, usually in the ratio of small, whole numbers, to form chemical compounds.

The agreement between Dalton's theory and the Law of Conservation of Mass is readily apparent. If a chemical change is simply a rearrangement of a fixed number of atoms, the total mass of those atoms must be the same before and after the rearrangement. The Law of Definite Composition also agrees with his theory. If there is a definite atom ratio by which two elements combine, and if all atoms of each element have the same mass, then the mass ratio must be constant.

As with many newly proposed explanations for natural events, Dalton's suggestions were not readily received. From them, however, came a prediction that *must* be true if atomic theory is correct. Now known as the **Law of Multiple Proportions,** it states that when two elements combine to form more than one compound, the different weights of one element that combine with a fixed weight of the other are in a simple ratio of whole numbers (Fig. 4.1). When experiments confirmed the prediction, the atomic theory on which it was based became generally accepted.

4.3 SYMBOLS OF THE ELEMENTS

As communication between chemists increased in the early 19th century, there was a need for an international shorthand to identify elements and compounds. John Dalton filled that need by developing simple symbols to represent the elements. Some of these reflected the character of the element. Carbon, for example, was a solid black circle, suggesting charcoal. Compounds were repre-

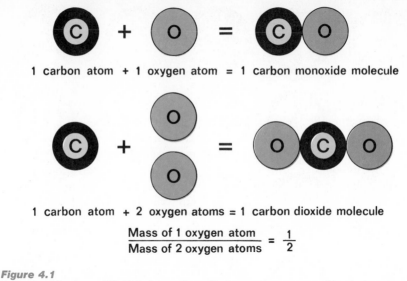

1 carbon atom + 1 oxygen atom = 1 carbon monoxide molecule

1 carbon atom + 2 oxygen atoms = 1 carbon dioxide molecule

$$\frac{\text{Mass of 1 oxygen atom}}{\text{Mass of 2 oxygen atoms}} = \frac{1}{2}$$

Figure 4.1
Example of the Law of Multiple Proportions. Carbon and oxygen combine to form two compounds, carbon monoxide and carbon dioxide. A unit particle of a chemical compound is called a molecule. A carbon monoxide molecule consists of one carbon atom and one oxygen atom; a carbon dioxide molecule has one carbon atom and two oxygen atoms. Considering both molecules, for a fixed number of carbon atoms—one in each molecule— the ratio of oxygen atoms is 1 to 2, or 1/2. If all oxygen atoms have the same mass, which will be represented by M, the mass ratio is also 1/2.

$$\frac{M \text{ grams (one atom)}}{2 M \text{ grams (two atoms)}} = \frac{1}{2}$$

The same ratio results from any number, N, of carbon atoms. N carbon atoms in carbon monoxide will combine with N oxygen atoms having a total mass of $1 \times N \times M$; N carbon atoms (an equal mass of carbon) in carbon dioxide will combine with $2 \times N \times M$ oxygen atoms having a total mass equal to $2 \times N \times M$. In the two compounds the ratio of oxygen masses

for any fixed number (mass) of carbon atoms is $\frac{1 \times N \times M}{2 \times N \times M} = \frac{1}{2}$.

sented by drawing the symbols touching each other. Some of Dalton's symbols are shown in Figure 4.2.

Dalton's symbols were reasonably satisfactory when the number of known elements was small. As more elements were discovered, however, more was needed than little sketches. The drawings were replaced by the first letters of the names of the elements. The symbol for hydrogen is H, for oxygen, O, and for nitrogen, N. When more than one element begins with the same letter, two letters are used for the additional elements. C is the symbol for carbon, Ca represents calcium, Cr is chromium, and Cl is chlorine.

The association of an element's symbol with one or two letters of its name makes it easy to learn most symbols, but not all. Some symbols come from Latin names. The symbol for iron, for example, comes from the Latin *ferrum*—Fe. The symbol for gold is Au, from *aurum*; for sodium, Na, from *natrium*; and for silver, Ag, from *argentum*.

Figure 4.2
Dalton's symbols for "elements" (1808). (*From Holton, G., et al:* Harvard Project Physics Text. *New York, Holt, Rinehart and Winston, 1975.*)

An alphabetical list of the elements and their symbols, as well as other information, appears inside the back cover of this book. A far more valuable reference is the **periodic table** inside the front cover. We will describe the periodic table in Section 4.8.

4.4 SUBATOMIC PARTICLES

PG 4 B Identify the features of Dalton's atomic theory that are no longer considered valid, and explain why.

4 C Identify the three major subatomic particles by charge and approximate atomic mass, expressed in atomic mass units.

Despite the general acceptance of the atomic theory, it was soon to be challenged in some of its details. As early as the 1830's the results of laboratory experiments suggested that the atom is made up of even smaller particles, called **subatomic particles.** Today we know that atoms of all elements are made up of

Table 4.1
Subatomic Particles

Subatomic Particle	Symbol	Funda-mental Charge	Mass		Location	Discovered
			Grams	amu (C^{12} = 12.000 00)		
Electron	e⁻	−1	9.107×10^{-28}	0.000 549 ≈ 0	Outside nucleus	1897 Thomson
Proton	p or p⁺	+1	1.672×10^{-24}	1.007 28 ≈ 1	Inside nucleus	1919 Rutherford
Neutron	n or n⁰	0	1.675×10^{-24}	1.008 67 ≈ 1	Inside nucleus	1932 Chadwick

different combinations of many kinds of subatomic particles. Only three are of interest in an introductory chemistry course. These are the electron, proton, and neutron.

The first hint that there were particles inside the atom appeared in the experiments of Michael Faraday. For want of a better name he called them "charged corpuscles." Many years later these "atoms of electricity" were given their present name, **electrons.** It was not until 1897 that J. J. Thompson isolated the electron, which has a negative charge. Its charge is apparently the smallest unit of electric charge possible, and it has been assigned a value of −1.

The **proton** was isolated in 1919 by Ernest Rutherford. It is about 1837 times as massive as the electron and carries a positive charge, +1, equal in size to the negative charge of the electron. The **neutron** was discovered by James Chadwick in 1932. As its name suggests, it is electrically neutral. Its mass is slightly greater than the mass of a proton. Masses of atoms and parts of atoms are often expressed in **atomic mass units (amu)** (defined in Section 4.7). Actual masses, symbols by which the subatomic particles are represented, and other information are summarized in Table 4.1.

4.5 THE NUCLEAR ATOM

PG 4 D Describe the nuclear model of the atom, based on the Rutherford scattering experiment.

In 1911 Ernest Rutherford and his students performed a series of experiments that are described in Figures 4.3 and 4.4. The results of these experiments led to the following conclusions:

1. Every atom contains an extremely small, extremely dense nucleus.
2. The nucleus accounts for all of the positive charge and nearly all of the mass of the atom.

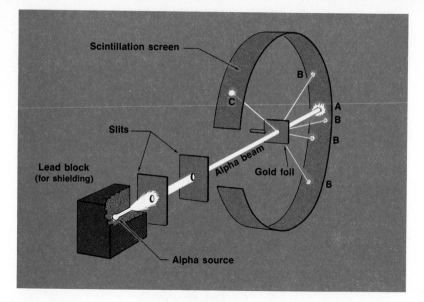

Figure 4.3
Rutherford scattering experiment. Using a natural radioactive source, a narrow beam of alpha particles (helium atoms stripped of their electrons) was directed at a very thin gold foil. Most of the particles passed right through the foil, striking a fluorescent screen at A, causing it to glow. Some of the particles were deflected, striking the screen at points such as those labeled B. The larger deflections were surprises, but the 0.001% of the total that were reflected at acute angles, C, were totally unexpected. Similar results were observed using foils of other metals.

3. The nucleus is surrounded by a much larger volume of nearly empty space, which makes up the rest of the atom.

4. The space outside the nucleus is very thinly populated by electrons, the total charge of which exactly balances the positive charge of the nucleus.

This description of the atom is called the **nuclear model of the atom.**

Rutherford could hardly believe the results of his experiments. He expected a slight deflection of the alpha particles, but to have some of them bounce *back* off the gold foil was a complete surprise. Describing the experiment, Rutherford is quoted as saying, "It was almost as incredible as if you fired a 15-inch shell at a piece of tissue paper and it came back and hit you." The conclusions suggested by the experiment about the size of the nucleus as compared to the size of the atom were no less amazing. If the nucleus were the size of a pea, the distance between it and its nearest neighbor would be about 0.6 mile, or nearly a kilometer. Furthermore, if it were possible to eliminate all the space around the nucleus and to fill a sphere the size of a period on this page with nuclei, then that sphere would weigh more than a million tons!

When protons and neutrons were later discovered, it was concluded that these relatively massive particles make up the nucleus of the atom. But electrons were already known in 1911, and it was natural to wonder what they did in the vast open space they occupied. The most widely held opinion was that they

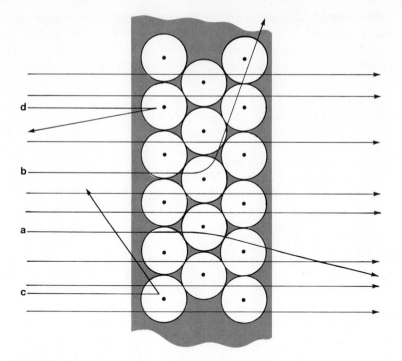

Figure 4.4
Interpretation of the Rutherford scattering technique. The atom is pictured as consisting mostly of open space. At the center is a tiny and extremely dense nucleus that contains all of the positive charge of the atom and nearly all of the mass. The electrons are thinly distributed throughout the open space. Most of the positively charged alpha particles pass through the open space undeflected, not coming near any gold nuclei. Those that would pass close to a nucleus (a and b) are repelled by electrostatic force and thereby deflected. The few particles that are on a "collision course" with gold nuclei are repelled backward at acute angles (c and d). Calculations based on the results of the experiment indicated that the diameter of the open-space atom is from 10 000 to 100 000 times greater than the diameter of the nucleus.

traveled in circular orbits around the nucleus, much as planets move in orbits around the sun. The atom would then have the character of a miniature solar system. This is called the **planetary model of the atom.** In Chapter 5 we will examine this model more closely—and find out why it is wrong.

4.6 ISOTOPES

PG 4 E Explain what isotopes of an element are and how they differ from each other.

4 F For an isotope of any element whose chemical symbol is known, given one of the following, state the other two: (a) nuclear symbol; (b) number of protons and neutrons in the nucleus; (c) atomic number and mass number.

After more than a hundred years, another feature of Dalton's atomic theory was shown to be incorrect. All atoms of an element are not identical; some have more mass than others. We now know that every atom of an element has the same number of protons. This number is called the **atomic number** and is represented by the symbol Z. The number of electrons in a neutral atom is the same as the number of protons, but electron masses are so small that their total contribution to atomic mass may usually be disregarded. We therefore conclude that differences in mass between atoms of an element must be caused by different numbers of neutrons in their nuclei. **Atoms of the same element that have different masses are called isotopes.**

An isotope of an element is identified by what we will call its **nuclear symbol.** The nuclear symbol of an isotope is written

$$\tfrac{A}{Z}Sy \quad or \quad ^{A}Sy$$

where Sy is the symbol of the element and A is the **mass number, the total number of protons and neutrons in the nucleus.** As the symbol alone is sufficient to identify the element, the atomic number, Z, may be omitted, but it is useful in writing nuclear equations (Chapter 21). Mass number may be calculated from the equation

$$\begin{aligned} \text{mass number} &= \text{number of protons} + \text{number of neutrons} \\ A &= Z \qquad\qquad + \text{number of neutrons} \end{aligned} \qquad (4.1)$$

The name of an isotope is the elemental name followed by the mass number. $^{16}_{8}O$, for example, is called "oxygen sixteen," and is written oxygen-16.

Two natural isotopes of carbon are $^{12}_{6}C$ and $^{13}_{6}C$, or carbon-12 and carbon-13. From the name and symbol of each isotope and from Equation 4.1 it is possible to find the number of neutrons in each nucleus. In carbon-12, if you subtract the atomic number (protons) from the mass number (protons + neutrons), you get the number of neutrons: $12 - 6 = 6$. In carbon-13 there are 7 neutrons: $13 - 6 = 7$.

If you know the number of protons and neutrons in an atom you can readily determine its mass number and nuclear symbol. A nucleus having 12 protons and 14 neutrons has an atomic number of 12, the same as the number of protons. The mass number, according to Equation 4.1, is $12 + 14 = 26$. You can find the symbol of the element in the periodic table inside the front cover of this book. The number at the top of each box in the table is the atomic number. The elemental symbol corresponding to $Z = 12$ is Mg, for magnesium. The nuclear symbol is $^{26}_{12}Mg$ or ^{26}Mg.

EXAMPLE 4.1   Write the name and number of protons, neutrons, and electrons in a calcium isotope if its nuclear symbol is $^{42}_{20}Ca$.

$_{20}^{42}$Ca: Calcium-42; 20 protons; 22 neutrons; 20 electrons

The name of the isotope is the elemental name followed by the mass number (super-script). The number of protons is the atomic number (subscript). The number of neutrons is, from Equation 4.1, the difference between the mass number and the atomic number. The number of electrons is equal to the number of protons.

EXAMPLE 4.2 Write the nuclear symbol for the sulfur isotope that has 18 neutrons. $Z = 16$ for sulfur, and its elemental symbol is S.

$_{16}^{34}$S or ^{34}S

The mass number is the sum of the number of protons (16) and neutrons (18).

EXAMPLE 4.3 Write the symbol and name of the lead isotope having 82 protons and 122 neutrons in its nucleus.

The mass number is found from Equation 4.1, as in Example 4.2. Find the elemental symbol from the periodic table inside the front cover. Remember that the number at the top of each box is the atomic number of the element.

$_{82}^{204}$Pb or ^{204}Pb; lead-204

The atomic number is the same as the number of protons, 82. In the periodic table the box for $Z = 82$ has the symbol Pb. This is from the Latin *plumbum*, meaning lead.

EXAMPLE 4.4 Write the number of protons, neutrons, and electrons in an atom of zinc-70.

Use the table inside the back cover of the book for the atomic number of zinc.

30 protons, 40 neutrons, and 30 electrons

The table inside the back cover gives 30 as the atomic number for zinc. This is the number of protons. As in Example 4.1, the number of neutrons is the mass number minus the atomic number, and the number of electrons is equal to the number of protons.

4.7 ATOMIC WEIGHT

> **PG 4 G** Define the atomic mass unit, amu.
>
> **4 H** Given the relative abundances of the natural isotopes of an element, calculate the atomic weight.

The mass of an atom is very small—much too small to be measured on a balance. Nevertheless, early chemists did find ways to isolate samples of elements that contained the same number of atoms. These samples were weighed and compared. It can be shown that the ratio of the masses of equal numbers of atoms of different elements is the same as the ratio of the masses of individual atoms.* This led to a scale of relative atomic weights. In its modern version, this scale compares the mass of atoms of different elements to the mass of an atom of carbon-12, to which is assigned the value of exactly 12 atomic mass units. Consequently, **one atomic mass unit is defined as exactly 1/12 of the mass of one atom of carbon-12.**

As a practical matter, almost every sample of any element consists of a mixture of the natural isotopes of that element.† Fortunately the percentage, or fraction, of each isotope is generally the same, regardless of the source of the sample. Table 4.2 lists the natural isotopes of some of the more common elements and their relative abundance in nature. From these it is possible to

Table 4.2
Percent Abundance of Some Natural Isotopes

Symbol	Mass (amu)	Per cent	Symbol	Mass (amu)	Per cent
$^{1}_{1}H$	1.007 825	99.985	$^{19}_{9}F$	18.998 40	100
$^{2}_{1}H$	2.014 0	0.015	$^{32}_{16}S$	31.972 07	95.0
$^{3}_{2}He$	3.016 03	0.000 13	$^{33}_{16}S$	32.971 46	0.76
$^{4}_{2}He$	4.002 60	100	$^{34}_{16}S$	33.967 86	4.22
$^{12}_{6}C$	12.000 00	98.89	$^{36}_{16}S$	35.967 09	0.014
$^{13}_{6}C$	13.003 35	1.11	$^{35}_{17}Cl$	34.968 85	75.53
$^{14}_{7}N$	14.003 07	99.63	$^{37}_{17}Cl$	36.965 90	24.47
$^{15}_{7}N$	15.000 11	0.37	$^{39}_{19}K$	38.963 71	93.1
$^{16}_{8}O$	15.994 91	99.759	$^{40}_{19}K$	39.974	0.001 18
$^{17}_{8}O$	16.994 74	0.037	$^{41}_{19}K$	40.961 84	6.88
$^{18}_{8}O$	17.994 77	0.204			

---

*If a is the mass of one atom of A, b is the mass of one atom of B, and N is any number of atoms, then $N \times a = $ mass of N atoms of A and $N \times b =$ mass of N atoms of B.

$$\frac{N \times a}{N \times b} = \frac{N}{N} \times \frac{a}{b} = \frac{a}{b}$$

† Isotopes of some elements do not occur in nature, but have been prepared in laboratories.

calculate the **atomic weight** of an element, which is defined as the **average mass of the atoms of an element compared to an atom of carbon-12 at exactly 12 atomic mass units.**

The following example shows how the atomic weight of an element is calculated.

EXAMPLE 4.5   The natural distribution of isotopes of magnesium is 78.70% $^{24}_{12}$Mg at a mass of 23.985 04 amu, 10.13% $^{25}_{12}$Mg at 24.985 84 amu, and 11.17% $^{26}_{12}$Mg at 25.982 59 amu. Calculate the atomic weight of magnesium.

Solution: The "average" magnesium atom consists of 78.70% of an atom that has a mass of 23.985 04 amu. It therefore contributes $0.7870 \times 23.985\ 04 = 18.88$ amu to the mass of the "average" atom. A similar calculation* for the other isotopes is added:

$$
\begin{array}{rl}
0.7870 \times 23.985\ 04 \text{ amu} = & 18.88 \text{ amu} \\
0.1013 \times 24.985\ 84 \text{ amu} = & 2.531 \text{ amu} \\
0.1117 \times 25.982\ 59 \text{ amu} = & \underline{2.902 \text{ amu}} \\
1.0000 \quad \text{average atom} \ = & 24.31 \text{ amu}
\end{array}
$$

The presently accepted value of the atomic weight of magnesium is 24.305. This matches the calculated value to four significant figures.

Now you try an atomic weight calculation.

EXAMPLE 4.6   Calculate the atomic weight of potassium (symbol K), using data from Table 4.2.

Proceed as in Example 4.5.

— — — — — —   — — — — — —

$$
\begin{array}{rl}
0.9310 \quad \times 38.963\ 71 \text{ amu} = 36.28 & \text{amu} \\
0.000\ 011\ 8 \times 39.974 \quad \text{amu} = 0.000\ 472 & \text{amu} \\
0.0688 \quad \times 40.961\ 84 \text{ amu} = \underline{2.82} & \text{amu} \\
39.10 & \text{amu}
\end{array}
$$

The accepted value of the atomic weight of potassium is 39.098 amu.

---

*Calculations are rounded off according to the rules of significant figures (p. 587).

A final comment: *Atomic weight* is really an incorrect name for the subject we have been considering. It should be *atomic mass*, consistent with the difference between *weight* and *mass* made on page 43. Some textbooks do refer to atomic mass, but atomic weight is used so much more widely that we will continue to use it in this text.

4.8 THE PERIODIC TABLE

PG 4 I Distinguish between groups and periods in a periodic table and identify them by number.

4 J Given the atomic number of an element, use a periodic table to find the symbol and atomic weight of that element, and identify the period and group in which it is found.

As knowledge of atoms and elements increased during the 19th century, chemists searched for an order among the elements. In 1869 it was found independently by two men at the same time. Dmitri Mendeleev and Lothar Meyer observed that when elements are arranged according to atomic weight, certain properties repeat at regular intervals. Mendeleev reached this conclusion by studying chemical properties, while Meyer examined physical properties.

Mendeleev and Meyer arranged the elements in tables so that the elements with similar properties were in the same column or row. These were the first **periodic tables** of the elements. The arrangements were not perfect; it was noticed that in order for all elements to fall into the proper groups it was necessary to switch a few, interrupting the orderly increase in atomic weight. This was partly because of errors in atomic weights as they were known in 1869. More important, nearly 50 years later it was found that the correct ordering property is atomic number (Z) rather than atomic weight.

As Dalton used atomic theory to predict the Law of Multiple Proportions correctly, so Mendeleev used his periodic table to predict the existence and properties of elements unknown in 1869. Noting certain "blanks" in the table, he reasoned that the blank spaces were there simply because nobody had yet discovered the elements that belonged there. By averaging the properties of the elements above and below or on each side of the unknown elements, he forecast the properties the elements would have when discovered. One element about which he made these predictions is germanium. The predictions and the presently accepted values of the properties of germanium are summarized in Table 4.3.

Figure 4.5 is a modern periodic table; another copy is inside the front cover of this book. A partial periodic table is printed on the opaque shield provided for working examples. You will find yourself referring to periodic tables throughout your study of chemistry.

Each box in the periodic table contains three items of information about the element it represents. The atomic number is on top, the chemical symbol is

1A	2A	3B	4B	5B	6B	7B	8B	8B	8B	1B	2B	3A	4A	5A	6A	7A	0
1 H 1.008																1 H 1.008	2 He 4.003
3 Li 6.941	4 Be 9.012											5 B 10.81	6 C 12.01	7 N 14.01	8 O 16.00	9 F 19.00	10 Ne 20.18
11 Na 22.99	12 Mg 24.31											13 Al 26.98	14 Si 28.09	15 P 30.97	16 S 32.06	17 Cl 35.45	18 Ar 39.95
19 K 39.10	20 Ca 40.08	21 Sc 44.96	22 Ti 47.90	23 V 50.94	24 Cr 52.00	25 Mn 54.94	26 Fe 55.85	27 Co 58.93	28 Ni 58.70	29 Cu 63.55	30 Zn 65.38	31 Ga 69.72	32 Ge 72.59	33 As 74.92	34 Se 78.96	35 Br 79.90	36 Kr 83.80
37 Rb 85.47	38 Sr 87.62	39 Y 88.91	40 Zr 91.22	41 Nb 92.91	42 Mo 95.94	43 Tc (98)	44 Ru 101.1	45 Rh 102.9	46 Pd 106.4	47 Ag 107.9	48 Cd 112.4	49 In 114.8	50 Sn 118.7	51 Sb 121.8	52 Te 127.6	53 I 126.9	54 Xe 131.3
55 Cs 132.9	56 Ba 137.3	57 *La 138.9	72 Hf 178.5	73 Ta 180.9	74 W 183.9	75 Re 186.2	76 Os 190.2	77 Ir 192.2	78 Pt 195.1	79 Au 197.0	80 Hg 200.6	81 Tl 204.4	82 Pb 207.2	83 Bi 209.0	84 Po (209)	85 At (210)	86 Rn (222)
87 Fr (223)	88 Ra 226.0	89 †Ac 227.0	104 § (261)	105 § (262)	106 § (263)												

*LANTHANIDE SERIES

58 Ce 140.1	59 Pr 140.9	60 Nd 144.2	61 Pm (145)	62 Sm 150.4	63 Eu 152.0	64 Gd 157.3	65 Tb 158.9	66 Dy 162.5	67 Ho 164.9	68 Er 167.3	69 Tm 168.9	70 Yb 173.0	71 Lu 175.0

†ACTINIDE SERIES

90 Th 232.0	91 Pa 231.0	92 U 238.0	93 Np 237.0	94 Pu (244)	95 Am (243)	96 Cm (247)	97 Bk (247)	98 Cf (251)	99 Es (252)	100 Fm (257)	101 Md (258)	102 No (259)	103 Lr (260)

§The International Union for Pure and Applied Chemistry has not adopted official names or symbols for these elements. All atomic weights have been rounded off to four significant figures.

Figure 4.5
Periodic table of the elements. Hydrogen (z = 1) appears twice in the table because it has properties similar to those of other elements in both columns.

Table 4.3
The Predicted and Observed Properties of Germanium

Property	Predicted by Mendeleev	Observed
Atomic weight	72	72.60
Density of metal	5.5 g/cm³	5.36 g/cm³
Color of metal	Dark gray	Gray
Formula of oxide	GeO_2	GeO_2
Density of oxide	4.7 g/cm³	4.703 g/cm³
Formula of chloride	$GeCl_4$	$GeCl_4$
Density of chloride	1.9 g/cm³	1.887 g/cm³
Boiling point of chloride	Below 100°C	86°C
Formula of ethyl compound	$Ge(C_2H_5)_4$	$Ge(C_2H_5)_4$
Boiling point of ethyl compound	160°C	160°C
Density of ethyl compound	0.96 g/cm³	Slightly less than 1.0 g/cm³

in the middle, and the atomic weight is on the bottom (Fig. 4.6). The boxes are arranged according to atomic number in horizontal rows called **periods.** Periods are identified by number from top to bottom, but the numbers are not usually printed. Periods vary in length. The first period has two elements, the second and third have eight elements each, and the fourth and fifth have eighteen. Period 6 has 32 elements, including atomic numbers 58 through 71, which are printed separately at the bottom to keep the table from becoming too wide. Period 7 also has 32 elements, but elements beyond Z = 106 have not been discovered. Elements with similar properties are arranged in vertical columns called **groups, or chemical families**. Groups are numbered across the top of the table. If you locate the element whose atomic number is 30, you will find it in the fourth period in Group 2B. Its symbol is Zn (zinc) and its atomic weight is

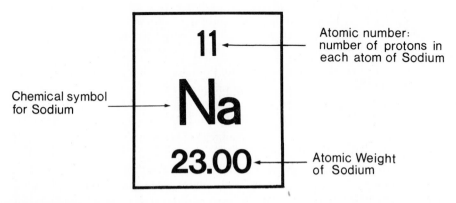

Figure 4.6
Sample box from the periodic table, showing the atomic number, symbol, and atomic weight of the element sodium.

65.38 amu. Elements in the A groups of the periodic table are called **representative elements,** and those in the B groups are **transition elements.**

EXAMPLE 4.7   List the atomic number, chemical symbol, and atomic weight of the fourth period element in Group 7A.

_____

_____

Z = 35; symbol, Br; atomic weight, 79.90 amu

In Group 7A, the second column from the right side of the table, you find Z = 1 in period 1, Z = 9 in period 2, Z = 17 in period 3, and Z = 35 in period 4. The element is bromine.

EXAMPLE 4.8   List the group, period, symbol and atomic weight for tin (Z = 50).

_____

_____

Group 4A; period, 5; symbol, Sn; atomic weight, 118.7 amu.

The first element in Group 4A, Z = 6, is in the second period. Counting down from there, Z = 50 is in Period 5.

The student who begins the study of chemistry in the 1980's may find contradictions between periodic tables in textbooks and on the walls of classrooms and lecture halls. In a few cases the letters A and B are used differently. In other cases, some recent textbooks do not identify the B groups at all. Group numbers continue to be used for the representative elements, but without the letter A. To minimize confusion for the beginning student, we have decided to continue using A and B groups for identification. This will make the periodic table in the book the same as the wall charts students will see in an overwhelming number of classrooms and lecture halls. If your teacher prefers one of the newer forms of the periodic table, you should, of course, follow his or her instructions.

You may wonder why the periodic table has such an unusual shape. Remember that the arrangement is based primarily on the periodic appearance of physical and chemical properties. Placing elements with similar properties in the same column dictates the shape. The _reason_ the properties produce the shape they do has to do with the arrangement of electrons in the atom. You will study this in Chapter 5.

Mendeleev and Meyer knew nothing about the protons and electrons—they hadn't even been discovered in 1869—which cause the periodic table to be what it is. The design of the table, based on observations alone, stands as one of the great achievements in the history of science. From this point in your study of chemistry you will reverse the process. You will use the table to predict properties of elements, to predict the formulas of chemical compounds, the shapes and properties of their molecules, and as a constant reference for atomic weights in solving problems. Make a friend of the periodic table. In chemistry, it may well be the best friend you can have.

CHAPTER 4 IN REVIEW

4.1 INTRODUCTION TO THE ATOM (52)

4.2 DALTON'S ATOMIC THEORY

    4 A List the main features of Dalton's atomic theory. (53)

4.3 SYMBOLS OF THE ELEMENTS (53)

4.4 SUBATOMIC PARTICLES

    4 B Identify the features of Dalton's atomic theory that are no longer considered valid, and explain why. (55)
    4 C Identify the three major subatomic particles by charge and approximate atomic mass, expressed in atomic mass units. (55)

4.5 THE NUCLEAR ATOM

    4 D Describe the nuclear model of the atom, based on the Rutherford scattering experiment. (56)

4.6 ISOTOPES

    4 E Explain what isotopes of an element are and how they differ from each other. (58)
    4 F For an isotope of any element whose chemical symbol is known, given one of the following, state the other two: (a) nuclear symbol; (b) number of protons and neutrons in the nucleus; (c) atomic number and mass number. (58)

4.7 ATOMIC WEIGHT

    4 G Define the atomic mass unit, amu. (61)
    4 H Given the masses and relative abundances of the natural isotopes of an element, calculate the atomic weight. (61)

4.8 THE PERIODIC TABLE

    4 I Distinguish between groups and periods in a periodic table and identify them by number. (63)
    4 J Given the atomic number of an element, use a periodic table to find the symbol and atomic weight of that element, and identify the period and group in which it is found. (63)

TERMS AND CONCEPTS

Dalton's atomic theory (53)
Law of Multiple Proportions (53)
Periodic table (53, 63)
Subatomic particles (55)
Electron (56)
Proton (56)
Neutron (56)
Atomic mass unit, amu (56, 61)
Nuclear model of the atom (57)
Planetary model of the atom (58)

Atomic number (59)
Isotope (59)
Nuclear symbol (59)
Mass number (59)
Atomic weight (61)
Period (in periodic table) (65)
Group (in periodic table) (65)
Chemical family (65)
Representative element (66)
Transition element (66)

Most of these terms and many others appear in the Glossary.

QUESTIONS AND PROBLEMS

An asterisk (*) identifies a question that is relatively difficult or that extends beyond the performance goals in the chapter.

Section 4.2

4.1) List the major points in Dalton's atomic theory.

4.2)* Sodium oxide and sodium peroxide are two compounds made up of the elements sodium and oxygen. 62 grams of sodium oxide consist of 46 grams of sodium and 16 grams of oxygen; 78 grams of sodium peroxide are made up of 46 grams of sodium and 32 grams of oxygen. Show how these figures confirm the Law of Multiple Proportions.

Section 4.4

4.3) Identify and explain that part of Dalton's atomic theory which was first proved to be incorrect.

Section 4.5

4.4) How can we account for the fact that most of the alpha particles in the Rutherford scattering experiment passed directly through a solid sheet of gold?

4.5) What major conclusions were drawn from the Rutherford scattering experiment?

4.6) The Rutherford experiment was performed and its conclusions reached before protons and neutrons were discovered. Why, when they were found, was it believed that they were in the nucleus of the atom?

4.22) Show how the atomic theory "explains" the Laws of Definite Composition and Conservation of Mass.

4.23)* Two compounds of mercury and chlorine are mercury(I) chloride and mercury(II) chloride. The amount of mercury(I) chloride that contains 71 grams of chlorine has 402 grams of mercury; the amount of mercury(II) chloride having 71 grams of chlorine has 201 grams of mercury. Show how the Law of Multiple Proportions is illustrated by these figures.

4.24) Compare the three major parts of an atom in charge and mass.

4.25) How can we account for the fact that, in the Rutherford scattering experiment, some of the alpha particles were deflected from their paths through the gold foil, and some even bounced back at various angles?

4.26) What name is given to the central part of an atom?

4.27) Describe the activity of electrons according to the planetary model of the atom that appeared after the Rutherford scattering experiment.

Section 4.6

4.7) In a neutral atom, compare the number of protons with the number of electrons. Do the same for protons and neutrons; electrons and neutrons.

4.8) A natural isotope of lithium has the symbol ^6_3Li. State the following about this isotope: atomic number; mass number; number of neutrons.

4.9) How many electrons, protons, and neutrons are in a neutral atom of $^{70}_{33}As$? What are the atomic number and mass number of this isotope?

4.28) A neutral atom of oxygen has 8 protons, 8 electrons, and 8 neutrons. State the number of protons, electrons, and neutrons that might exist in an isotope of the atom described.

4.29) The most common isotope of lithium (see Problem 4.8) has four neutrons in its nucleus. State the mass number and write the nuclear symbol of the isotope.

4.30) List in order the number of electrons, protons, and neutrons, and state the atomic number and mass number of an atom of $^{197}_{79}Au$.

You may refer to the periodic table inside the front cover and to the Table of Elements inside the back cover for any information you require in answering the next two questions in each column.

4.10) The atomic number of an element is 24, and the nucleus of one of its atoms contains 28 neutrons. State the name and mass number and write the symbol of the isotope.

4.11) The mass number of a certain isotope of iron is 57. For an atom of this isotope, list in order the number of protons, neutrons, and electrons, and state the nuclear symbol and the name.

4.31) 75 is the mass number of the only natural isotope of the element having Z = 33. State the number of neutrons in the nucleus, write the nuclear symbol of the isotope, and state its name.

4.32) One of the most hazardous species in nuclear fallout is strontium-90. State the nuclear symbol for this isotope and list the number of neutrons, protons, and electrons in the atom.

Section 4.7

4.12) What is an atomic mass unit?

4.13)* The average mass of boron atoms is 10.81 amu. How would you explain what this means to a friend who had never taken chemistry?

4.14) Calculate to two decimals the atomic weight of chlorine from the data in Table 4.2.

4.33) What advantage does the atomic mass unit have over grams when speaking of the mass of an atom or subatomic particle?

4.34)* The mass of an "average atom" of a certain element is 3.34 times as great as the mass of an atom of carbon-12. Using either the periodic table or the table of atoms inside the back cover, identify the element.

4.35) Using data from Table 4.2, find the atomic weight of sulfur to two decimal places.

In the next three problems in each column, you are given the atomic masses (amu) and percentage abundances of the natural isotopes of different elements. (1) Calculate the atomic weight of each element from these data. (2)* Using the tables inside the front and back covers, identify the element.

	ATOMIC MASS (amu)	PERCENTAGE ABUNDANCE		ATOMIC MASS (amu)	PERCENTAGE ABUNDANCE
4.15)	106.9041 108.9047	51.82 48.18	**4.36)**	62.9298 64.9278	69.09 30.91
4.16)	120.9038 122.9041	57.25 42.75	**4.37)**	184.9530 186.9560	37.07 62.93
4.17)	135.907 137.9057 139.9053 141.9090	0.193 0.250 88.48 11.07	**4.38)**	57.9353 59.9332 60.9310 61.9283 63.9280	67.88 26.23 1.19 3.66 1.08

Section 4.8

4.18) Write the symbols of the elements in Group 2A of the periodic table. Write the atomic numbers of the elements in Period 3.

4.19) Locate on the periodic table each pair of elements whose atomic numbers are given below. Does each pair belong to the same period or the same chemical family? (a) 12 and 16; (b) 7 and 33; (c) 2 and 10; (d) 42 and 51.

4.20) Referring only to a periodic table, list the atomic weights of the elements having atomic numbers 24, 50, and 77.

4.21)* What are the atomic weights of potassium and sulfur?

4.39) How many elements are in Period 5 of the periodic table? Write the atomic numbers of the elements in Group 1B.

4.40) Locate on a periodic table each element whose atomic number is given and identify first the number of the period it is in, and then the number of the group: (a) 20; (b) 14; (c) 43.

4.41) Using only a periodic table for reference, list the atomic weights of the elements whose atomic numbers are 29, 82, and 55.

4.42)* Write the atomic weights of helium and aluminum.

a modern view of the atom

5.1 THE BOHR MODEL OF THE HYDROGEN ATOM

In 1913, Niels Bohr, a Danish physicist, proposed a model of the hydrogen atom that opened the way to the understanding of atomic structure. His starting point was the planetary model of the atom. According to this model, an atom has an extremely dense nucleus that is responsible for all of the positive charge in the atom, and nearly all of its mass. Negatively charged electrons of very small mass travel in orbits around the nucleus. The orbits are huge compared to the nucleus, which means that most of the atom is empty space.

Bohr also drew on other scientific observations made late in the 19th century and the first decade of the 20th. Among them were:

1. When white light is passed through a prism, it produces a **continuous spectrum.** This is a rainbow-like spreading out of light in an orderly arrangement of colors. When light from a particular element passes through a prism, it gives a **line spectrum** consisting of separate, or *discrete*, lines of color (Figs. 5.1 and 5.2).

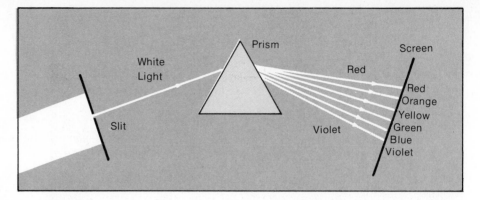

Figure 5.1
Spectrum of white light. White light is a combination of light waves having all the colors of the rainbow. When the light passes through a prism, the colors are separated into a spectrum, which may be projected onto a screen.

2. Light can be described by wave properties such as speed or velocity (c), wavelength (λ), and frequency (ν)* (Figs. 5.3 and 5.4).

3. Light has properties that suggest it is made up of individual bundles, or **quanta,** of energy. The energy of each quantum is calculated by the equation $E = h\nu$, where E is the energy, h is a fixed number known as *Planck's constant,* and ν is the frequency.

4. Physics describes mathematical relationships among radii, speed, energy, and forces when one object moves in an orbit around another, as the planets move around the sun. Physics also furnishes information about electrostatic attraction between plus and minus charges.

From these beginnings, Bohr made a bold assumption. He assumed that the energy possessed by the electron in a hydrogen atom and the radius of its orbit are **quantized. Quantized energy levels mean that the electron in the atom may have, at any instant, any one of several possible energies, but at no time may it have an energy between them.** An example of something that is quantized is shown in Figure 5.5.

Bohr calculated the values of his quantized energy levels using an equation that contains an integer, n—1, 2, 3 . . . and so forth. The results for the integers 1 to 4 are shown in Figure 5.6. The lowest energy level, when n = 1, is called the **ground state.** Higher energy levels, when n = 2 or more, are called **excited states.** The electron is normally found in the ground state. If the atom absorbs energy, the electron can be "excited," or raised to one of the higher energy levels. It cannot stay at that level, but falls back to the ground state. In doing this it releases light energy, hν, that is equal to the energy difference between the two levels. If this energy is in the visible part of the spectrum, it appears as one of the lines in the spectrum of hydrogen.

*λ is the Greek letter *lambda,* and ν is the Greek letter *nu.* Do not confuse ν with v.

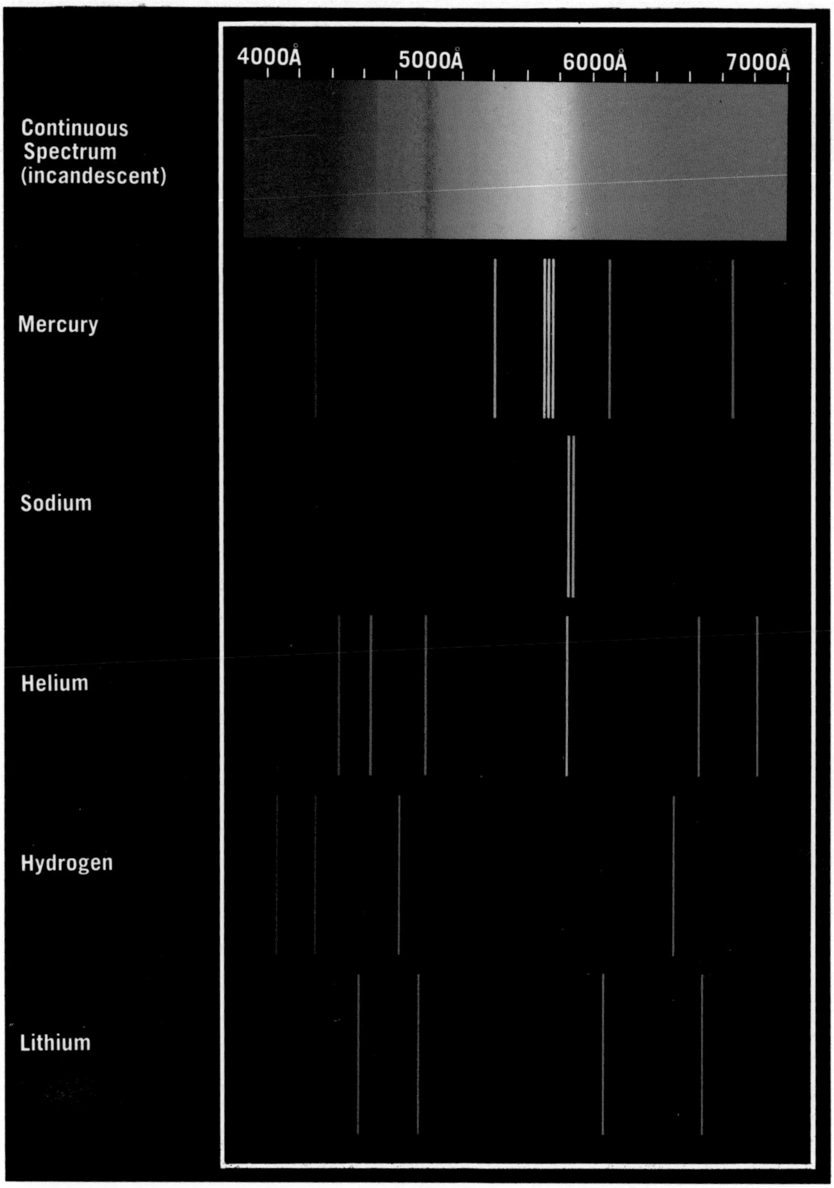

Figure 5.2
Spectrum chart. Continuous "rainbow" spectrum at top is from "white" light of an incandescent lamp. Beneath it are the bright line spectra that characterize certain elements.

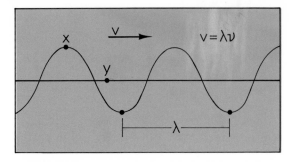

Figure 5.3
Wave properties. Waves may be represented mathematically by the curve shown. Water waves approximate the mathematical representation physically, while other wave phenomena have quite different appearances. All waves have measurements such as wavelength, λ (the Greek letter lambda), the distance between corresponding points on consecutive waves; velocity, v, or, in the case of light, c, the linear speed of a point x on a wave; and frequency, ν (the Greek letter nu), the number of wave cycles that pass a point y, each second. Velocity, wavelength, and frequency are related by the equation v × λν or, in the case of light, c = λν. Reflection, refraction, and diffraction are other wave properties.

Using different values of n in his equation, Bohr was able to calculate the energies of all known lines in the spectrum of the hydrogen atom. He also predicted additional lines and their energies. When the lines were found, his predictions were proved to be correct. In fact, all of Bohr's calculations correspond with measured values to within one part per thousand. It certainly seemed that Bohr had found the answer to the structure of the atom.

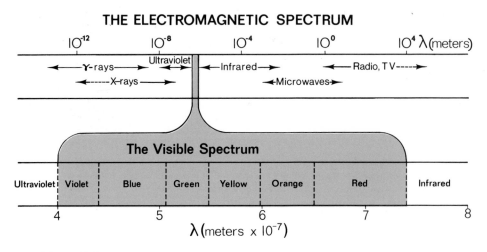

Figure 5.4
The electromagnetic spectrum. The electromagnetic spectrum covers a span from the very high energy, short wave gamma rays of 10^{-10} meters or less to the low energy long wave radio waves at 10^4 meters or more. The visible spectrum is but a small portion of the electromagnetic spectrum, covering wavelengths of 4×10^{-4} to 7×10^{-4} meters, approximately.

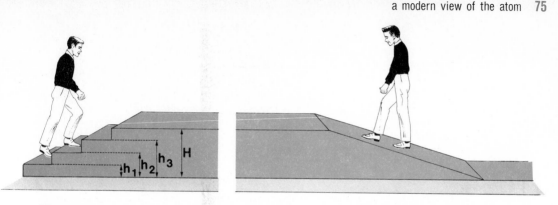

Figure 5.5
The quantization, or quantum, concept. Man on ramp can stop at any level above ground; his elevation above ground is not quantized. Man on stairs can stop only on a step; his elevation is quantized at h_1, h_2, h_3, or H.

There were problems, however. First, hydrogen is the *only* neutral atom that fits the Bohr model. The model fails for any atom with more than one electron. Second, it is a fact that a charged body moving in a circle radiates energy. This means the electron itself should lose energy and promptly—in about 0.000 000 000 01 second!—crash into the nucleus. To do otherwise is to violate the Law of Conservation of Energy (p. 23). So, for a period of 13 years, science did what it does so often when faced with contradictions. It accepted a theory that was known to be only partly correct. It worked on the faulty parts, modified them, improved them, and ultimately replaced them with more accurate concepts.

Niels Bohr made two huge contributions to the development of modern atomic theory. First, he suggested a reasonable explanation for atomic spectra in terms of electron energies. Second, he introduced the idea of quantized electron energy levels in the atom. These levels appear in modern theory as **principal energy levels**; they are identified by the **principal quantum number**, n.

5.2 THE QUANTUM MECHANICAL MODEL OF THE ATOM

In 1924, Louis de Broglie, a French physicist, suggested that matter in motion has properties normally associated with waves, and that these properties are important in subatomic particles. The period 1925 to 1928 saw the development of the **quantum mechanical model** of the atom, the heart of which is the wave equation proposed by Erwin Schrödinger. This model has been tested for half a century. It explains more satisfactorily than any other theory all experimental observations to date, and no exceptions to the theory have appeared. It is the theory that is generally accepted today.

Unfortunately, the quantum mechanical model does not give a very good idea of the appearance of an atom. One physicist has noted, ". . . that the

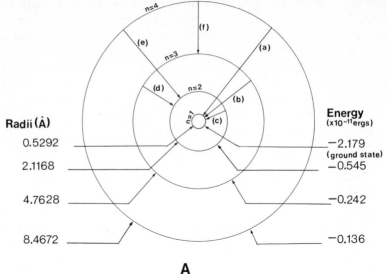

Radii (Å)

0.5292

2.1168

4.7628

8.4672

Energy
(×10⁻¹¹ergs)

−2.179
(ground state)

−0.545

−0.242

−0.136

A

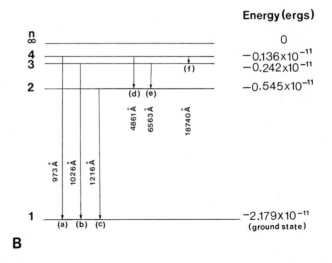

B

Figure 5.6

The hydrogen atom based on the Bohr model. According to this model the electron was "allowed" to circle to the nucleus only in certain radii, the first four of which are shown in A. At these radii the electron possesses energies indicated in A and plotted in B. An electron in the ground state—n = 1 level—can absorb the exact amount of energy to elevate it to any other level; e.g., n = 2, 3 or 4. An electron at an excited state is unstable and drops back to the n = 1 level by any combination of energy jumps possible, radiating energy with each jump. Energy jumps a to c are in the ultraviolet portion of the spectrum, d and e are in the visible range, and f is in the infrared region. (Å is the symbol for angstrom, a length unit; 1 nm = 10 Å. The erg is an energy unit; 1 J = 1 × 10⁷ ergs.)

question, 'What does an atom look like?' has no meaning, much less an answer."* The quantum concept is abstract and mathematical, rather than physical. You might keep this in mind as you read the following sections.

As its name suggests, the quantum mechanical model of the atom keeps
the quantized energy levels introduced by Bohr. It also states that more than
just the principal quantum number is needed to describe the electron energy. It
requires three quantum numbers that specify (1) the principal energy level, (2)
the sublevel, and (3) the orbital. These are summarized in Table 5.1, which you
may follow as they are examined separately.

PRINCIPAL ENERGY LEVELS

PG 5 D Identify the principal energy levels within an atom and state the
energy trend among them.

Following the Bohr model, principal energy levels are identified by the
principal quantum number, n. The first principal energy level is n = 1, the
second is n = 2, and so on. (An older system uses the letters, K, L, M. . . to
designate principal energy levels.) Mathematically, there is no end to the number of principal energy levels, but the seventh level is the highest occupied by
ground state electrons among the elements now known.

The energy possessed by an electron depends upon the principal energy
level in which it is found. An electron in the n = 2 level has more energy than
an n = 1 electron; an n = 3 electron is at higher energy than an electron in the
n = 2 level. This trend holds generally, but the principal energy levels for elements other than hydrogen represent a *range* of energies. Beginning with n = 3
and n = 4 the ranges overlap. We therefore find that a high energy n = 3
electron is at a higher level than a low energy n = 4 electron.

SUBLEVELS

PG 5 E For each principal energy level, state the number of sublevels,
identify them by letter, and state the energy trend among them.

According to the quantum mechanical model of the atom, principal energy levels consist of one or more **sublevels** (Table 5.1). These are identified by
the letters s, p, d, and f. (The letters come from terms formerly used in spectros-
copy.) *The total number of sublevels within a given principal energy level is
equal to n, the principal quantum number.* For n = 1 there is one sublevel,

*H. E. White: Modern College Physics. D. Van Nostrand, 1962.

Table 5.1
*Principal Energy Levels, Sublevels, and Orbitals in the Quantum Mechanical Model of the Atom**

Principal Energy Levels	Number of Sublevels	Identification of Sublevels						
n = 1	1	1s						
	Orbital per sublevel	1						
n = 2	2	2s	2p					
	Orbitals per sublevel	1	3					
n = 3	3	3s	3p	3d				
	Orbitals per sublevel	1	3	5				
n = 4	4	4s	4p	4d	4f			
	Orbitals per sublevel	1	3	5	7			
n = 5	5	5s	5p	5d	5f	5g		
	Orbitals per sublevel	1	3	5	7	9		
n = 6	6	6s	6p	6d	6f	6g	6h	
	Orbitals per sublevel	1	3	5	7	9	11	
n = 7	7	7s	7p	7d	7f	7g	7h	7i
	Orbitals per sublevel	1	3	5	7	9	11	13

*Principal energy levels are identified by numbers, 1, 2, 3, . . . 7. Each principal energy level has a number of sublevels equal to the identifying number of the principal energy level. Only four sublevels, s, p, d, and f, are required for the elements now known, but if more are discovered the electrons will be assigned to the g, h, and i sublevels. Sublevels are divided into orbitals. Each s sublevel has 1 orbital, each p sublevel has 3 orbitals, a d sublevel has 5 orbitals, and an f sublevel has 7 orbitals. The sublevels shown in color are not required to accommodate ground state electrons of elements now known.

designated 1s. At n = 2 there are two sublevels, 2s and 2p. When n = 3 there are three sublevels, 3s, 3p, and 3d; n = 4 has four sublevels, 4s, 4p, 4d, and 4f. Quantum theory describes sublevels beyond f when n = 5 or more, but these are not needed by elements known today. The energy of an electron is described by the sublevel in which it is found. An electron in the 2s sublevel is called a "2s electron," and an electron in the 3p sublevel is a "3p electron."

Just as electron energies increase through the principal energy levels, so do they increase through the sublevels. Within a given principal energy level, an electron in the s sublevel is at lower energy than one in the p sublevel, a p electron is lower in energy than a d electron, and a d is lower than an f. This is what makes it possible for principal energy levels to overlap. The highest n = 3 electron is a 3d electron, while the lowest n = 4 electron is a 4s electron. A 3d electron is at higher energy than a 4s electron. But, among s electrons, energy always increases from 1s to 2s to 3s, etc. Similarly, among p electrons, 2p < 3p < 4p, etc., in terms of electron energies. The same trends hold for d and f electrons.

ELECTRON ORBITALS

PG 5 F Sketch the shapes of s and p orbitals.

5 G State the number of orbitals in each sublevel.

Modern atomic theory states that we cannot know both the position of an electron in an atom at a given instant and the path it travels. Using mathematics, however, we can predict the *probability* of finding an electron in a particular region of space. The electron is pictured as forming a negatively charged "cloud" concentrated mostly in a certain location near the nucleus. The shape of the charged cloud depends upon the sublevel occupied by the electron. Three-dimensional representations of these **orbitals**, as they are called, are shown in Figure 5.7. The drawings show regions in which there is a "high probability" of finding an electron. Notice the uncertainty of the orbital description, stated in terms of probability, as compared with the Bohr description of the

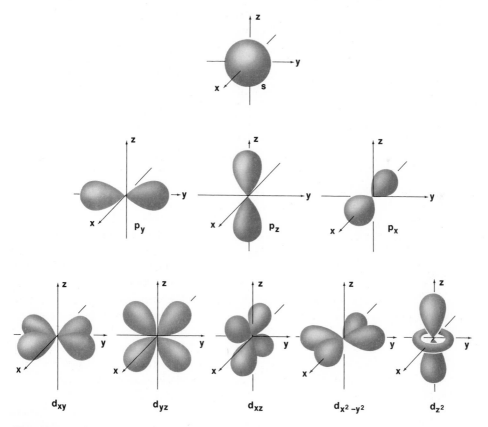

Figure 5.7
Shapes of electron orbitals according to the quantum mechanical model of the atom.

orbit that states exactly where the electron is, where it was, and where it is going.

Table 5.1 summarizes the sublevels and orbitals in the first seven principal energy levels. Each principal energy level has one s orbital. Figure 5.7 shows all s orbitals to be *spherical* in shape. Both the size of an s orbital and the electron energy increase as the principal quantum number increases from 1s to 2s to 3s... up to 7s, the largest s orbital required by any known element.

Table 5.1 and Figure 5.7 indicate three p orbitals when n is 2 or more. Each p orbital consists of a pair of lobes oriented at right angles to the other two pairs at their common center, the nucleus of the atom. These are described as lying along the x, y, and z axes. As with the s orbitals, both size and energy increase with increasing principal quantum number, n.

The d orbitals shown in Figure 5.7 are more complex, and we will not be concerned with their shapes. It is important to note, however, that the first d orbitals appear at n = 3, and there are five of them (Table 5.1). Furthermore, the energies of the d orbitals increase in the order 3d, 4d, 5d. . . .

Figure 5.7 does not include f orbitals. Their shapes are quite complex and difficult to illustrate on a page. Table 5.1 shows that there are seven f orbitals for each principal quantum number, beginning at n = 4. As with the s, p, and d orbitals, the sizes and energies of the f orbitals increase with increasing values of n.

THE PAULI EXCLUSION PRINCIPLE

PG 5 H   State the restrictions on the electron population of an orbital.

In the beginning of this section we said that the quantum mechanical model of the atom provides three quantum numbers by which the energy of an electron is specified. The principal quantum number, n, was identified. We did not name the other two quantum numbers, but we described them in the sublevels and orbitals. Furthermore, what is known as the **Pauli Exclusion Principle** requires a fourth quantum number, not coming directly from quantum mechanics, but added to make the theory match experimental observations. Its effect is to restrict the population of any orbital to two electrons. At any time, an orbital may be (a) unoccupied, (b) occupied by one electron, or (c) occupied by two electrons. No other occupancy condition is possible.

*   *   *   *   *   *   *   *   *   *

In summary, we note the following:

1. Generally speaking, energy increases with increasing principal quantum number: n = 1 < n = 2 < n = 3. . . . Specifically, for any given sublevel, s, p, d or f, energy increases with increasing principal quantum number: 1s < 2s < 3s . . . ; 2p < 3p < 4p . . . ; etc.

2. For any given value of n, energy increases through the sublevels in the order s
   p d f: 2s < 2p; 3s < 3p < 3d; 4s < 4p < 4d < 4f; etc.
3. For any value of n there are n sublevels:

n		*Sublevels*
1	s	(one sublevel)
2	s, p	(two sublevels)
3	s, p, d	(three sublevels)
4	s, p, d, f	(four sublevels)

4. The number of orbitals for each sublevel is

Sublevel	*Orbitals*
s	1
p	3
d	5
f	7

5. An orbital may be occupied by 0, 1 or 2 electrons, but never more than 2.

5.3 ELECTRON CONFIGURATION

DEVELOPMENT OF ELECTRON CONFIGURATIONS

Figure 5.8 is an electron energy level diagram that illustrates the above summary. Relative energies of sublevels are plotted vertically, and principal quantum numbers are plotted horizontally. Each box represents one orbital, which may contain 0, 1, or 2 electrons. Each s sublevel has one orbital, each p sublevel has three orbitals, each d has five, and each f has seven. The different backgrounds separate the sublevels by periods in the periodic table.

Many chemical properties of an element depend upon its **electron configuration,** the ground state distribution of electrons among the orbitals of a gaseous atom. Two rules guide the assignment of electrons to orbitals:

1. At ground state, the electrons will fill the *lowest* energy orbitals available, as shown in Figure 5.8.
2. No orbital can have more than two electrons.

1s An atom of hydrogen (H, Z = 1) has only one electron, and it occupies the lowest energy orbital in any atom. Figure 5.8 shows this is the 1s orbital. We indicate the total number of electrons in any sublevel by a superscript number. Therefore the electron configuration of hydrogen is written $1s^1$, showing one electron in the 1s orbital. Helium (He, Z = 2) has two electrons in the neutral atom, and both fit into the 1s orbital. The electron configuration of helium is therefore $1s^2$.

These and other electron configurations to be developed are shown in the first four periods of the periodic table in Figure 5.9.

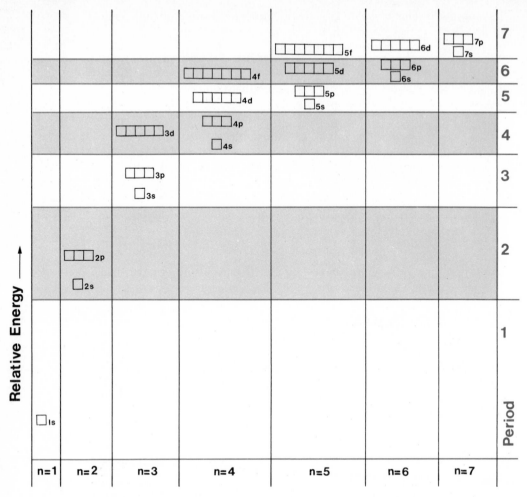

Figure 5.8
Electron energy level diagram. Each column contains one box for each orbital in the n = 1, n = 2, n = 3 . . . n = 7 principal energy levels. The boxes are grouped by sublevels s, p, d and f, as far as required within each principal energy level. All sublevels are positioned vertically on the page according to a general energy level scale shown at the left. Any orbital that is higher on the page than a second orbital is therefore higher in energy than the second orbital. In filling orbitals from the lowest energy level, each orbital will generally hold 2 electrons before any orbital higher in energy accepts an electron. The colored lines and the scale at the right correlate energies of the sublevels with the periods of the periodic table.

2s Lithium (Li, Z = 3) has three electrons. The first two fill the 1s orbital as before. The third electron goes to the next orbital up the energy scale with a vacancy, which Figure 5.8 shows to be the 2s orbital. The electron configuration for lithium is therefore $1s^2 2s^1$. In a similar manner beryllium (Be, Z = 4) divides its four electrons between the two lowest orbitals, filling them both: $1s^2 2s^2$. These two configurations are shown in Figure 5.9 also.

1	2	3	4	5	6	7	8	9	10	11	12	13	14	15	16	17	18
1 H $1s^1$																	2 He $1s^2$
3 Li $1s^2$ $2s^1$	4 Be $1s^2$ $2s^2$											5 B $1s^2\,2p^1$	6 C $1s^2\,2p^2$	7 N $1s^2\,2p^3$	8 O $1s^2\,2p^4$	9 F $1s^2\,2p^5$	10 Ne $1s^2\,2p^6$
11 Na [Ne] $3s^1$	12 Mg [Ne] $3s^2$											13 Al [Ne] $3s^2 3p^1$	14 Si [Ne] $3s^2 3p^2$	15 P [Ne] $3s^2 3p^3$	16 S [Ne] $3s^2 3p^4$	17 Cl [Ne] $3s^2 3p^5$	18 Ar [Ne] $3s^2 3p^6$
19 K [Ar] $4s^1$	20 Ca [Ar] $4s^2$	21 Sc [Ar]$4s^2$ $3d^1$	22 Ti [Ar]$4s^2$ $3d^2$	23 V [Ar]$4s^2$ $3d^3$	24 Cr [Ar]$4s^1$ $3d^5$	25 Mn [Ar]$4s^2$ $3d^5$	26 Fe [Ar]$4s^2$ $3d^6$	27 Co [Ar]$4s^2$ $3d^7$	28 Ni [Ar]$4s^2$ $3d^8$	29 Cu [Ar]$4s^1$ $3d^{10}$	30 Zn [Ar]$4s^2$ $3d^{10}$	31 Ga [Ar]$4s^2$ $3d^{10}4p^1$	32 Ge [Ar]$4s^2$ $3d^{10}4p^2$	33 As [Ar]$4s^2$ $3d^{10}4p^3$	34 Se [Ar]$4s^2$ $3d^{10}4p^4$	35 Br [Ar]$4s^2$ $3d^{10}4p^5$	36 Kr [Ar]$4s^2$ $3d^{10}4p^6$

Figure 5.9
Ground state electron configurations of neutral gaseous atoms.

2p The first four electrons of boron (B, Z = 5) fill 1s and 2s orbitals. The fifth electron goes to the next highest level, the 2p, according to Figure 5.8. The configuration for boron is $1s^2 2s^2 2p^1$. Similarly carbon (C, Z = 6) has a configuration $1s^2 2s^2 2p^2$. Though we will not emphasize the point, the two 2p electrons occupy different p orbitals. Electrons half-fill each orbital in a sublevel before completely filling any of them. The next four elements increase the number of electrons in the three 2p orbitals until they are filled with six electrons for neon (Ne, Z = 10): $1s^2 2s^2 2p^6$. All these configurations appear in Figure 5.9.

3s and 3p The first ten electrons of sodium (Na, Z = 11) distribute themselves among the 1s, 2s and 2p orbitals, as did those of neon. The eleventh is in the 3s orbital, giving the total configuration $1s^2 2s^2 2p^6 3s^1$. Because all electron configurations for elements having atomic numbers greater than 10 begin with the configuration for neon, $1s^2 2s^2 2p^6$, this part of those configurations is frequently shortened to the **neon core**, [Ne]. It is shown that way in Figure 5.9.

The development of the configuration for sodium at the 3s level repeats that of lithium at the 2s level. In fact, all the elements of the third period repeat the elements of the second period just above them, giving the configurations shown in Figure 5.9.

4s Potassium (K, Z = 19) repeats at the 4s level the development of sodium at the 3s level. Its complete configuration is $1s^2 2s^2 2p^6 3s^2 3p^6 4s^1$. All configurations for atomic numbers greater than 18 distribute their first 18 electrons in the configuration of argon (Ar, Z = 18), $1s^2 2s^2 2p^6 3s^2 3p^6$, which may be shortened to the **argon core**, [Ar]. Accordingly the configuration for potassium may be written [Ar]$4s^1$, and calcium (Ca, Z = 20) is [Ar]$4s^2$.

3d Figure 5.8 predicts that five 3d orbitals are next available for electron occupancy. At two electrons per orbital they should accommodate the next ten elements. The first three of these, scandium, Sc, titanium, Ti, and vanadium, V, atomic numbers 21 to 23, fill in order as predicted. The configuration for vanadium is $1s^2 2s^2 2p^6 3s^2 3p^6 4s^2 3d^3$, or $[Ar]4s^2 3d^3$.* Chromium (Cr, Z = 24) is the first element to break the orderly sequence in which lowest energy orbitals are filled. Its configuration is $[Ar]4s^1 3d^5$, rather than the expected $[Ar]4s^2 3d^4$. This is generally attributed to an extra stability found when a sublevel is half-filled or completely filled. Manganese (Mn, Z = 25) puts us back on the track, only to be derailed again at copper (Cu, Z = 29): $[Ar]4s^1 3d^{10}$. Zinc (Zn, Z = 30) has the expected configuration of $[Ar]4s^2 3d^{10}$. Examine the sequence for atomic numbers 21 to 30 in Figure 5.9 and note the two exceptions.

4p By now the pattern should be clear; you should expect atomic numbers 31 to 36 to fill in sequence the next orbitals available, which are the 4p orbitals. They do, as shown in Figure 5.9

Our consideration of electron configurations ends with atomic number 36, krypton. Were we to continue, we would find the higher s and p orbitals fill in perfect accord with the procedures used thus far. The 4d, 4f, 5d, and 5f orbitals have several variations like those with chromium and copper, so configurations involving those orbitals must be looked up. But you should be able to reproduce easily the configurations for the first 36 elements—not from memory, but from their correlation with the periodic table.

ELECTRON CONFIGURATIONS AND THE PERIODIC TABLE

PG 5 I   Using n for the highest occupied energy level, write the ground state ns and np electron configurations for an atom of any representative element.

Figure 5.9 shows that specific sublevels are being filled in different regions of the periodic table. This is indicated by color in Figure 5.10. In Groups 1A and 2A the s sublevels are the highest occupied energy levels. The p orbitals are filled in order across Groups 3A to 0. The d electrons appear in the B groups and Group 8. Finally, the f electrons show up in the lanthanide and actinide series beneath the table.

Notice that, when you read the periodic table from left to right across the periods in Figure 5.10, you get the order of increasing sublevel energy. The first period gives only the 1s sublevel. Period 2 takes you through 2s and 2p. Similarly, the third period covers the 3s and 3p sublevels. Period 4 starts with 4s, follows with 3d, and ends up with 4p, and so forth. Ignoring the minor variations in Periods 6 and 7, the complete list is

1s 2s 2p 3s 3p 4s 3d 4p 5s 4d 5p 6s 4f 5d 6p 7s 5f 6d 7p

---

*Some chemists prefer to write this configuration $1s^2 2s^2 2p^6 3s^2 3p^6 3d^3 4s^2$, or $[Ar]3d^3 4s^2$, putting the 3d before the 4s. This is an equally acceptable alternative. There is perhaps some advantage at this point in listing the sublevels in the order in which they fill, so we will continue to use this order in this text.

1A 2A   **0**

1s   **3A 4A 5A 6A 7A**   ls

2s   2p

3s   **3B**   **4B 5B 6B 7B**   — **8** —   **1B 2B**   3p

4s   3d   3d   4p

5s   4d   4d   5p

6s   5d   4f   5d   6p

7s   6d   5f   6d   7p

Lanthanide Series   4f

Actinide Series   5f

Figure 5.10
Arrangement of periodic table according to atomic sublevels. Highest energy sublevels oc-
cupied at ground state are s sublevels in Groups 1A and 2A; p sublevels in Groups 3A–O,
except for helium; d sublevels in Groups 1B–7B and 8; and f sublevels in the lanthanide and
actinide series.

Now compare this list with the order of increasing sublevel energy from Figure
5.8. They are exactly the same! *The periodic table therefore replaces the elec-*
tron energy diagram as a guide to the order of increasing sublevel energy.

If you are ever required to list the sublevels in order of increasing energy without
reference to a periodic table, the diagram produced by the sublevels in Table 5.1
is helpful and is easily recalled. Beginning at the upper left, the diagonal lines
pass through the sublevels in the sequence required.

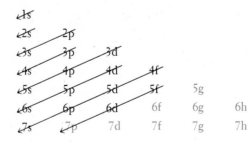

The sublevels shown in color are not needed for the elements known today, but
the mathematics of the quantum mechanical model predict the order of increasing energy indefinitely.

Look again at Figure 5.8. Observe that, with the exception of chromium and copper, $Z = 24$ and $Z = 29$, the number of electrons in each s, p, and d sublevel increases in numerical order from left to right. This means that every element in Group 1A has an electron configuration ending in ns^1, where n is the principal quantum number of the highest occupied energy level of the atom in its ground state. It also happens that n is the period number. For atomic number 3 this is $2s^1$ (n = 2), for atomic number 11, $3s^1$ (n = 3), and so on. Similarly, the general electron configuration for the highest occupied energy level of Group 2A elements is ns^2. For Group 3A it is $ns^2\, np^1$. It increases in order through the groups to $ns^2\, np^6$ for Group 0.

EXAMPLE 5.1   Referring only to a periodic table (not Fig. 5.8), (a) write the electron configuration for the highest occupied energy level for any Group 6A element, (b) identify the group whose electron configuration for the highest occupied energy level is $ns^2\, np^2$.

------   ------

(a) $ns^2\, np^4$; (b) Group 4A

Both groups have a filled s orbital: ns^2. Beginning with Group 3A there is one p electron, Group 4A has two p electrons (answer to part b), Group 5A has three p electrons, and Group 6A has four p electrons (answer to part a).

5.4 WRITING ELECTRON CONFIGURATIONS

PG 5 J   Referring only to a periodic table, write the ground state electron configuration of a gaseous atom of any element up to atomic number 36.

If you can list the sublevels in order of increasing energy, and if you can count, you can write electron configurations. To illustrate, use only the periodic table (not Fig. 5.8 or 5.10) while considering these examples:

EXAMPLE 5.2   Write the complete electron configuration for chlorine, Cl, $Z = 17$. (Do not use the neon core.)

Solution: It is helpful to go step-by-step in writing electron configurations:

1. *Locate the element (chlorine) in the periodic table.*
2. *List the sublevels in order of increasing energy until you reach the sublevel that contains the element. Leave space for superscripts.* Do this by reading across the periods in

the table from left to right, as in Figure 5.10. Atomic numbers 1 and 2 constitute the 1s sublevel. In the second period, atomic numbers 3 and 4 make up the 2s sublevel, and numbers 5 to 10 constitute the 2p sublevel. In row 3 we find atomic numbers 11 and 12 for 3s, and 13 to 18 for 3p. Chlorine is atomic number 17. Therefore the list is 1s 2s 2p 3s 3p.

3. *Fill in as superscripts the maximum electron populations of all sublevels except the last one* (or last two, if the element happens to be chromium or copper). You probably know the maximum electron populations of the s, p, and d sublevels by now (2, 6, and 10), but if you have forgotten, you can find them simply by counting across the periodic table. For chlorine this step yields $1s^22s^22p^63s^23p$.

4. *Determine the population of the final sublevel and insert that number as the last superscript.* Start at the beginning of that "sublevel group" in the table and count until reaching the element in question. In this instance chlorine is in the number 13 to 18 sublevel group, so atomic number 13 would be counted as "one," 14 as "two," 15 as "three," 16 as "four," and 17 as "five." There are five electrons in the 3p sublevel, so the complete configuration is $1s^22s^22p^63s^23p^5$.

5. *Add the superscripts and check the total against the atomic number.* They should be equal, since the number of protons is the same as the number of electrons in a neutral atom. $2 + 2 + 6 + 2 + 5 = 17$, the atomic number of chlorine.

In using a neon or argon core you would locate your element in the table and write the core symbol for that Group 0 element that lies *before* it in the atomic number sequence. For chlorine, atomic number 17, you would use neon, atomic number 10. Then list the sublevels as in the sequence above. For chlorine: $[Ne]3s^23p^5$.

If your element happens to be chromium (z = 24), you must remember that the 4s orbital has 1 electron and the 3d orbital has 5 electrons. For copper (Z = 29) there are one 4s and ten 4p electrons.

Now you try one:

EXAMPLE 5.3 Write the complete (no Group 0 core) electron configuration for the element potassium (K, Z = 19).

First locate potassium in the periodic table. Then list the sublevels in sequence until reaching the highest occupied sublevel for potassium.

_ _ _ _ _ _   _ _ _ _ _ _

1s   2s   2p   3s   3p   4s

Now fill in the maximum electron populations of all sublevels except the last.

_ _ _ _ _ _   _ _ _ _ _ _

$1s^22s^22p^63s^23p^64s$

Finally, starting with the beginning of the 4s sublevel in the periodic table, count off until reaching potassium. Fill in the last superscript.

------   ------

$1s^22s^22p^63s^23p^64s^1$

EXAMPLE 5.4 Write the complete electron configuration for oxygen, O, Z = 8.

You may be able to complete this example without intermediate steps.

------   ------

$1s^22s^22p^4$

Listing sublevels until reaching oxygen gives 1s 2s 2p. The 1s and 2s orbitals are filled with two electrons each, and counting from boron (Z = 5) to oxygen gives 4 electrons in the 2p sublevel.

EXAMPLE 5.5 Develop the electron configuration for cobalt, Co, Z = 27.

This is the first example in which you encounter d electrons. The procedure is the same; go all the way in one step, using a Group 0 core this time.

------   ------

$[Ar]4s^23d^7$

The noble gas prior to cobalt in the periodic table is argon. Reading from left to right through the sublevels of the fourth period until reaching cobalt we have 4s and 3d. The 4s orbital is filled with two electrons. Counting through the 3d sublevel we have seven, beginning with scandium (Z = 21) and ending with cobalt.

5.5 TRENDS IN THE PERIODIC TABLE

When we introduced the periodic table in Section 4.8, it was noted that Mendeleev and Meyer had been trying to "organize" some of the recurring chemical and physical properties of the elements. They suggested no reason for these regularities. We now recognize that modern atomic theory has provided the explanation that neither scientist could foresee, since they knew nothing of electrons, protons, nuclei, quantized energy levels, nor wave equations. It is truly remarkable that they were able to suggest a table that would, nearly 60 years later, correspond so perfectly with all of these things.

In this section some of the properties that recur according to the order of the periodic table will be identified. Chemical families, elements that have similar chemical properties, will be recognized. Other important regions in the table will be noted, and some trends across periods and down groups of the table will be pointed out.

IONIZATION ENERGY

A neutral atom of sodium, Na, $Z = 11$, has eleven protons in its nucleus and eleven electrons outside the nucleus. (We'll not be concerned with the twelve neutrons also present in most sodium atoms.) Mentally separate one of the electrons from the rest. Figure 5.11 will aid you in this exercise. The middle sketch shows the separated electron still as part of the atom. The electron has a charge of -1. The rest of the atom, consisting of eleven protons and ten electrons (11 plus charges and 10 minus charges) has a net charge of $+1$. There is an attractive force between the electron and the rest of the atom (p. 21). Work must be done against this attractive force (p. 22) to separate the charges and remove the electron completely from the atom. The "thing" that is left, with its net charge caused by an unequal number of protons and electrons, is called an **ion**. More generally, an ion may be defined as **an atom or group of atoms that has a positive or negative charge because it has more or fewer electrons than protons.** The ion that remains after removing an electron from a sodium atom is called a sodium ion. Its symbol is Na.$^{+}$.

The energy required to remove one electron from a neutral gaseous atom of an element is the ionization energy of that element. Ionization energy is one of the more striking examples of a periodic property, particularly when graphed (Fig. 5.12). Notice the similarity of the shape of the graph between atomic numbers 3 and 10 (Period 2 in the periodic table) and atomic numbers 11 and 18 (Period 3). Notice also that the three peaks are elements in Group 0 at the right end of the table, and the three low points are from Group 1A. Observe the trends: without exception, if you draw lines connecting elements in the same group in the periodic table the lines slant down to the right. This indicates

Figure 5.11
The formation of a sodium ion from a sodium atom.

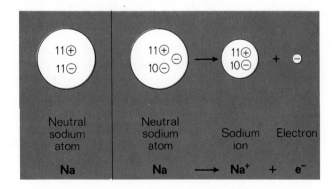

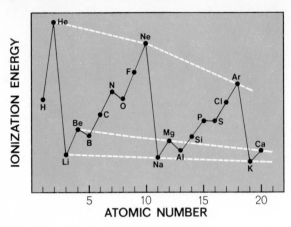

Figure 5.12
Ionization energy, plotted as a function of atomic number, to show periodic properties of elements. Dotted white lines show that ionization energies of elements in the same family decrease as atomic number increases.

lower ionization energies with increasing atomic number within the group. If the graph is extended to include higher atomic numbers, the same general shapes and trends are found through the s and p sublevel portions of all periods.

CHEMICAL FAMILIES

PG 5 K Explain, from the standpoint of electron configuration, why certain groups of elements make up chemical families.

5 L Identify in the periodic table the following chemical families: alkali metals; alkaline earths; halogens; noble gases.

We will see in Chapter 10 that the formation of chemical bonds involves the "high energy" electrons found in the outer energy levels of the bonded atoms. While d electrons do participate in bond formation, our consideration will be limited to s and p electrons. These are called **valence electrons.** We find that atoms having similar s and p electron configurations appear among the elements in the vertical groups of the periodic table. These groups are therefore called **chemical families.**

Alkali Metals With the exception of hydrogen, the elements in Group 1A, are known as **alkali metals** (Fig. 5.13). (Francium, Fr, Z = 87, is a radioactive element about which we know little, but which we presume behaves as an alkali metal.) The ns^1 electron in the highest occupied energy level is easily lost, yielding an ion having a +1 charge. This accounts for most of the chemical properties of the family. As you move down the column in the table, the highest energy s electron is found farther from the nucleus. Therefore it is lost more easily. This accounts for the lower ionization energy as atomic number increases, indicated above and in Figure 5.12. As a direct result of this, the reactiv-

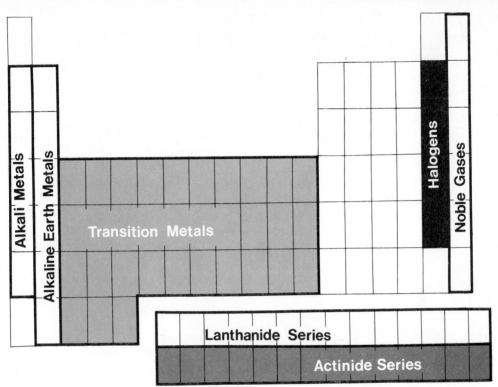

Figure 5.13
Chemical families and regions in the periodic table.

ity of the element—that is, its tendency to react with other elements and form compounds—increases as you go down the column.

Group 1A elements are metals, although they don't normally look like metals. This is because exposure to air leads to the formation of an oxide coating that hides the bright metallic luster that can be seen in a freshly cut sample. These elements possess other common metallic properties too, such as being good conductors of heat and electricity and being able to form wires and thin foils.

Table 5.2 lists some of the physical properties of the alkali metals, as well as those of other chemical families we will consider. Notice the steady increase in density as the atomic number increases. Melting and boiling points decrease, with one notable exception. (Exceptions are always challenging and frequently instructive when one seeks an explanation for them.)

Alkaline Earths Group 2A elements are called **alkaline earths**, or **alkaline earth metals.** Each member of Group 2A has an ns^2 configuration at its highest energy level: $3s^2$ for magnesium ($Z = 12$), $4s^2$ for calcium ($Z = 20$), and $6s^2$ for barium ($Z = 56$), to list the best known elements of the family. Although they

Table 5.2
Selected Physical Properties of Some Families of Elements

Family and Elements	Atomic Number	Density (g/cm³)	Melting Point (°C)	Boiling Point (°C)
Alkali Metals— Group 1A				
Lithium	3	0.53	180	1326
Sodium	11	0.97	98	889
Potassium	19	0.86	63	757
Rubidium	37	1.53	39	679
Cesium	55	1.87	29	690
Alkaline Earths— Group 2A				
Beryllium	4	1.86	1283	2970
Magnesium	12	1.74	650	1120
Calcium	20	1.54	850	1490
Strontium	38	2.60	770	1384
Barium	56	3.5	704	1140
Radium	88	5	700	1500
Halogens— Group 7A				
Fluorine	9	1.11 (liq.)	−218	−188
Chlorine	17	1.56 (liq.)	−101	−34
Bromine	35	3.12 (liq.)	−7	59
Iodine	53	4.93 (sol.)	114	184
Noble Gases— Group 0		*(g/L, STP)**		
Helium	2	0.18	−270[†]	−269
Neon	10	0.90	−249	−246
Argon	18	1.78	−189	−186
Krypton	36	3.74	−157	−153
Xenon	54	5.86	−112	−108
Radon	86	9.9	−71	−62

*STP refers to standard temperature and pressure, 0°C and 1.00 atmosphere.
[†]Measured at ten atmospheres of pressure.

are not as easily lost as the ns^1 electron of an alkali metal, the ns^2 electrons are given up readily and form ions with a +2 charge.

Trends like those noted with the alkali metals are seen also with the alkaline earth metals. Reactivity is again greater for the elements of higher atomic number. Physical property trends may also be noted (Table 5.2).

Halogens Four elements of Group 7A, fluorine ($Z = 9$), chlorine ($Z = 17$), bromine ($Z = 35$), and iodine ($Z = 53$) make up the family known as the **halogens,** *or salt-formers.* (Astatine, a radioactive element about which we know little, probably would fit into the group if we had enough information to classify it.) The highest energy electron configuration is ns^2np^5: $2s^22p^5$ for fluorine, $3s^23p^5$ for chlorine, etc. These elements form ions by *gaining* one electron, thereby acquiring a -1 charge.

Reactivity trends opposite to those of the alkali and alkaline earth metals may be observed. In general the reactivity of nonmetals decreases with increasing atomic number. Fluorine is thus the most reactive of the group, and iodine the least. Physical property trends are shown in Table 5.2.

Noble gases Group 0 in the periodic table makes up the **noble gas family,** so called because its elements generally do not form chemical compounds. In chemistry, the word *noble* means unreactive. Until the 1960's it was believed that these elements formed no compounds at all; then a few compounds of krypton $(Z = 36)$ and xenon $(Z = 54)$ were prepared. Nevertheless, as a group, their inactivity is their outstanding chemical property. Family trends are evident in the physical properties listed in Table 5.2.

The chemical inactivity of the noble gases is the result of an ns^2np^6 electron configuration at the highest occupied energy level. (Helium, with only two electrons, is obviously an exception. Its electron configuration is $1s^2$.) There appears to be an "extra" stability associated with filled s and p orbitals at the highest energy level. As we will see, this is the basis for much of our understanding of chemical bonding.

THE TRANSITION METALS

PG 5 M   Identify the transition elements and the representative elements in the periodic table.

Elements in the B groups and Group 8 of the periodic table are called **transition metals,** or **transition elements.** These are the elements whose 3d, 4d, and 5d sublevels are "filling" as you work through the electron configuration scheme. Included among the transition elements are some of our better known metals, such as iron, gold, copper, silver, chromium, zinc, nickel, and mercury. Although the elements within a given transition series differ from each other chemically, they are, as a group, distinguished from the 1A and 2A metals by being less reactive.

ATOMIC SIZE

PG 5 N   Predict how and explain why atomic size varies with position in the periodic table.

Figure 5.14 shows the sizes of Group A atoms. Moving across the table from left to right the atoms become smaller. This trend is perhaps most obvious

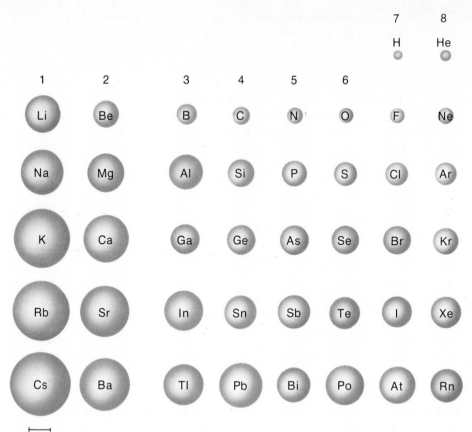

0.2 nm

Figure 5.14
Atomic radii of the A group elements. Atomic radii increase as one goes down a group and in general decrease in going across a row in the periodic table. Hydrogen has the smallest atom, cesium the largest.

in the second period, where the element at the far left, lithium, has a radius more than twice that of fluorine at the far right. As we go down the table, within a given group, atoms ordinarily increase in size. We see, for example, that cesium, the last element listed in Group 1A, has a radius nearly twice that of lithium.

These observations about atomic size are believed to be the result of three influences:

1. *Number of occupied principal energy levels.* As the number of occupied principal energy levels increases, the atom becomes larger. Electrons in higher energy levels are generally farther from the nucleus. This influence alone accounts for the increase in size as you move down a column in the

periodic table. For example, compare sodium (Z = 11) and lithium (Z = 3) atoms. Sodium atoms are larger. The highest electron of a sodium atom is a 3s electron, but for lithium it is a 2s electron.

2. *Nuclear charge.* Atoms generally become smaller from left to right across each row of the table. Within a period, the highest energy electrons are all in the same principal energy level. As the number of protons in the nucleus increases across the row, the positive nuclear charge increases. This causes the nucleus to exert a stronger attraction for the outermost electrons, thus pulling them closer to the nucleus. Compare sodium (Z = 11) and magnesium (Z = 12) atoms. Both have 3s electrons in their highest energy levels. The 12 protons in the nucleus of a magnesium atom attracts those 3s electrons more strongly than the 11 protons in the nucleus of a sodium atom. The magnesium electrons are therefore closer to the nucleus, making the magnesium atom smaller.

3. *Shielding effect.* The attraction of the positively charged nucleus for the negatively charged outermost electrons is partially canceled, or "shielded," by the repulsion of electrons in lower energy levels. This is one effect that makes nuclear charge less important than the number of occupied energy levels in determining atomic size. Compare sodium (Z = 11) and lithium (Z = 3) atoms again. Even though a sodium atom has nearly four times as much nuclear charge, it is larger than a lithium atom.

In summary, atomic size generally increases from right to left across any row of the periodic table, and from top to bottom in any column. The smallest atoms are toward the upper right corner of the table, and the largest are toward the bottom left corner.

EXAMPLE 5.6   Referring only to a periodic table, list atomic numbers 15, 16, and 33 in order of increasing atomic size.

The summary in the paragraph just before this example should tell you how to identify the smallest and the largest of the three atoms. . . .

‒ ‒ ‒ ‒ ‒ ‒   ‒ ‒ ‒ ‒ ‒ ‒

16 ‒ 15 ‒ 33

The smallest atom is toward the upper right (Z = 16) and the largest is toward the bottom left (Z = 33). Specifically, Z = 16 (sulfur) atoms are smaller than Z = 15 (phosphorus) atoms because sulfur atoms have a higher nuclear charge to attract the highest energy 3s and 3p electrons in both atoms. The phosphorus atom is smaller than the Z = 33 (arsenic) atom because the 3s and 3p electrons of phosphorus are at a lower energy than the 4s and 4p electrons of arsenic.

METALS AND NONMETALS

PG 50   Identify metals and nonmetals in the periodic table.

Both physically and chemically the alkali metals, alkaline earths, and transition elements are metals. Chemically, an element is known as a metal if it can lose one or more electrons and become a positively charged ion. The larger the atom, the more easily the outermost electron is removed. Therefore, the metallic character of elements in a group increases as you go down a column in the periodic table.

It has also been noted that the size of an atom becomes smaller as the nuclear charge increases across a period in the table. The larger number of protons also holds the outermost electrons more strongly, making it more difficult for them to be lost. This makes the metallic character of elements *decrease* as you go from left to right across the period.

The distinction between metals, elements that lose electrons in chemical reactions, and nonmetals, those that do not, is not a sharp one. It can be drawn roughly as a step-like line beginning between atomic numbers 4 and 5 in Period 2, and ending between 84 and 85 in Period 6 (Fig. 5.15). Most elements next to

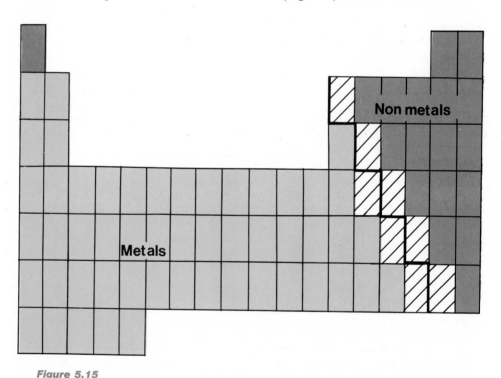

Figure 5.15
Metals and nonmetals. Shaded elements are metalloids, having properties intermediate between those of metals and nonmetals.

the line on either side appear in some compounds as metals and in others as nonmetals. They are often loosely classified as **metalloids.**

CHAPTER 5 IN REVIEW

5.1 THE BOHR MODEL OF THE HYDROGEN ATOM

5 A Describe the Bohr model of the hydrogen atom. (71)
5 B Explain the meaning of quantized energy levels in an atom and show how these levels are related to the discrete lines in the spectrum of that atom. (71)
5 C Distinguish between ground state and excited state. (71)

5.2 THE QUANTUM MECHANICAL MODEL OF THE ATOM.

5 D Identify the principal energy levels within an atom and state the energy trend among them. (77)
5 E For each principal energy level, state the number of sublevels, identify them by letter, and state the energy trend among them. (77)
5 F Sketch the shapes of s and p orbitals. (79)
5 G State the number of orbitals in each sublevel. (79)
5 H State the restrictions on the electron population of an orbital. (80)

5.3 ELECTRON CONFIGURATION

5 I Using n for the highest occupied energy level, write the ground state ns and np electron configurations for an atom of any representative element. (84)

5.4 WRITING ELECTRON CONFIGURATIONS

5 J Referring only to a periodic table, write the ground state electron configuration of a gaseous atom of any element up to atomic number 36. (86)

5.5 TRENDS IN THE PERIODIC TABLE

5 K Explain, from the standpoint of electron configuration, why certain groups of elements make up chemical families. (90)
5 L Identify in the periodic table the following chemical families: alkali metals; alkaline earths; halogens; noble gases. (90)
5 M Identify the transition elements and the representative elements in the periodic table. (93)
5 N Predict how and explain why atomic size varies with position in the periodic table. (93)
5 O Identify metals and nonmetals in the periodic table. (96)

TERMS AND CONCEPTS

Spectrum (continuous, line) (71)
Discrete (71)
Quantum, quanta (72)
Quantized energy level (72)
Ground state (72)
Excited state (72)
Principal energy level (75)
Principal quantum number (75)
Quantum mechanical model of the atom (75)
Sublevel (78)
s, p, d, f sublevels (78)
Orbital (79)
Pauli exclusion principle (80)
Electron configuration (84)

Neon core (87)
Argon core (87)
Ion (89)
Ionization energy (89)
Valence electrons (90)
Chemical family (90)
Alkali metals (family) (90)
Alkaline earths, alkaline earth metals (family) (91)
Halogens (family) (92)
Noble gases (family) (93)
Transition metal, transition element (96)
Metal, nonmetal (in terms of electron behavior) (96)
Metalloid (97)

Most of these terms and many others appear in the Glossary.

QUESTIONS AND PROBLEMS

Section 5.1

5.1) Distinguish between a continuous spectrum and a discrete line spectrum.

5.2)* What is meant by the statement that something behaves "like a wave"?

5.3) Distinguish between something that is quantized and something that is not. Give an example of each that is not in this book.

5.4) "Electron energies are quantized within the atom." What is the meaning of this statement?

5.5) What kind of light would be emitted by atoms if electron energy were not quantized? Explain.

5.6) Compare an atom in the ground state with an atom in an excited state.

5.7) Draw a sketch of an atom according to the Bohr model. Describe the atom with reference to the sketch.

5.8) Identify the shortcomings of the Bohr theory of the atom.

5.40) The visible spectrum is a small part of the whole electromagnetic spectrum. Name several other parts of the whole spectrum that are included in our everyday vocabulary.

5.41)* Identify measurable wave properties that are used in describing light.

5.42) State which of the following are quantized (a) water from a faucet; (b) heights of students in a class; (c) number of students in a class; (d) nails as purchased in a hardware store: (e) milk from a cow; (f) milk from a store; (g) temperature; (h) speed of a car.

5.43) What experimental evidence leads us to believe that electron energy levels are quantized? Explain.

5.44) What must be done to an atom, or what must happen to an atom, before it can emit light? Explain.

5.45) Which atom is most apt to emit light, one in the ground state or one in an excited state? Why?

5.46) Using a sketch of the Bohr model of an atom, explain the source of the observed lines in the spectrum of hydrogen.

5.47) Identify the major advances that came from the Bohr model of the atom.

Section 5.2

5.9) What is the meaning of the "principal energy levels" of an atom?

5.10) What are sublevels and how are they identified?

5.11) Distinguish between electron *orbits* and electron *orbitals*. Draw sketches of the s and p orbitals.

5.12) At n = 2 there are two p sublevels; at n = 3 there are three p sublevels. Comment on these statements.

5.13)* Although we may draw the 4s orbital with the shape of a ball, there is some probability of finding the electron outside the ball we draw. Comment on this statement.

5.14) What is the significance of the Pauli exclusion principle?

5.15) An electron in an orbital of a p sublevel follows a path similar to a figure 8. Comment on this statement.

5.16) What is the maximum number of d orbitals when n = 6?

5.17)* Is the quantum mechanical model of the atom consistent with the Bohr model?

5.48) Compare the relative energies of the principal energy levels within the same atom.

5.49) How many sublevels are present in each principal energy level?

5.50) How many orbitals are in the s sublevel? the p sublevel? the d sublevel? the f sublevel?

5.51) Each p sublevel contains six orbitals. Comment on this statement.

5.52) The principal energy level with n = 6 contains six sublevels, although not all six are occupied for any element now known. Comment on this statement.

5.53) At the first principal energy level, n = 1, an orbital may hold 0 electrons or 1 electron. At n = 2, an orbital may hold 0 electrons, 1 electron, or 2 electrons. At n = 3, an orbital may *not* hold 0 electrons, 1 electron, 2 electrons, or 3 electrons. Comment on these statements.

5.54) What is your opinion of the common picture showing one or more electrons whirling around the nucleus of an atom?

5.55) What is the largest number of electrons that can occupy the 4p orbitals? the 3d orbitals? the 5f orbitals?

5.56)* Which of the following statements is true? The quantum mechanical model of the atom includes (a) all of the Bohr model; (b) part of the Bohr model; (c) no part of the Bohr model. Justify your answer.

Section 5.3

5.18) What is the meaning of the symbol $3p^4$?

5.19) The electron configuration of an element is $1s^2 2s^2 2p^5$. What element is this? What period is it in, and what group is it in?

5.20) What is meant by [Ne] in [Ne] $3s^3 3p^1$?

5.21) Which group of the periodic table has $ns^2 np^2$ as its highest occupied energy level electron configuration?

5.57) What do you conclude about the symbol $2d^5$? (Careful. . .)

5.58) $1s^2 2s^2 2p^6 3s^2 3p^4$ is the electron configuration of an element. What element is it? In which period and group of the periodic table is it located?

5.59) What is the argon core? What is its symbol, and what do you use it for?

5.60) Using n for the principal quantum number, write the electron configuration of the highest occupied energy level for any Group 6A atom at ground state.

Section 5.4

For the next two questions in each column, write complete electron configurations for a neutral gaseous atom of the elements shown. Do not use a noble gas core.

5.22) Nitrogen (Z = 7) and titanium (Z = 22).

5.23) Potassium (Z = 19) and manganese (Z = 25).

5.24) If it can be done, rewrite the electron configurations in Questions 5.22 and 5.23 with a neon or argon core.

5.25) Use a noble gas core to write the electron configuration of germanium (Z = 32).

5.61) Magnesium (Z = 12) and nickel (Z = 28).

5.62) Vanadium (Z = 23) and selenium (Z = 34).

5.63) If it can be done, rewrite the electron configurations in Questions 5.61 and 5.62 with a neon or argon core.

5.64) Use a noble gas core to write the electron configuration of aluminum (Z = 13).

Section 5.5

5.26)* Explain the meaning of *ionization energy*.

5.27)* Compare the ionization energies of aluminum (Z = 13) and chlorine (Z = 17). Why are these values as they are?

5.28) What does it mean when two elements are in the same chemical family?

5.29) What is the highest energy electron configuration associated with the alkali metals?

5.30) Identify the chemical family of each of the following elements: iodine (Z = 53); rubidium (Z = 37).

5.31) Explain how the electron configurations of magnesium and calcium are responsible for the many chemical similarities between these elements.

5.32) Where in the periodic table do you find the representative elements? Are they metals or nonmetals?

5.33) Identify an atom in Group 5A that is (a) larger than an atom of arsenic (Z = 33) and (b) smaller than an atom of arsenic.

5.34) Why does atomic size increase as you go down a column in the periodic table?

5.35) Even though an atom of germanium (Z = 32) has more than twice the nuclear charge of an atom of silicon (Z = 14), the germanium atom is larger. Explain.

5.36) What property of atoms of an element determines that the element is a metal rather than a nonmetal?

5.65)* Why is the definition of atomic number based on the number of protons in an atom rather than the number of electrons?

5.66)* Suggest reasons why the ionization energy of magnesium (Z = 12) is greater than the ionization energies of sodium (Z = 11), aluminum (Z = 13), and calcium (Z = 20).

5.67) What is it about strontium (Z = 38) and barium (Z = 56) that makes them members of the same chemical family?

5.68) To what family does the electron configuration ns^2np^6 belong?

5.69) Identify the chemical family of each of the following elements: krypton (Z = 36); beryllium (Z = 4).

5.70) Account for the chemical similarities between chlorine and iodine in terms of their electron configurations.

5.71) Where in the periodic table do you find the transition elements? Are they metals or nonmetals?

5.72) Identify an atom in Period 4 that is (a) larger than an atom of arsenic (Z = 33) and (b) smaller than an atom of arsenic.

5.73) Why does atomic size decrease as you go left to right across a row in the periodic table?

5.74) Describe the shielding effect. Explain its influence on atomic size.

5.75) Do elements become more metallic or less metallic as you (a) go down a group in the periodic table, (b) move left to right across a period in the periodic table? Support your answers with examples.

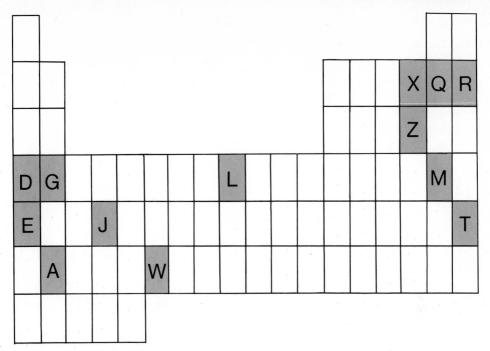

Figure 5.16
Chart for Questions 5.37 to 5.39 and 5.76 to 5.78.

Use the imaginary periodic table, Figure 5.16, to answer the remaining questions. Answer with letters (D, E, X, Z, etc.) from that table.

5.37) Give the letters that are in the positions of (a) halogens and (b) alkali metals.

5.38) Give the letters that correspond with transition elements.

5.39) List the elements D, E, and G in order of decreasing atomic size (largest atom first).

5.76) Give the letters that are in the positions of (a) alkaline earth metals and (b) noble gases.

5.77) List the letters that correspond with non-metals.

5.78) List the elements Q, X, and Z in order of increasing atomic size (smallest atom first).

6

introduction to chemical formulas

In this chapter you will be introduced to a small part of chemical nomenclature, the system of names and formulas of chemical substances. This introduction is intentionally limited. The idea is that a small sample of the system can be learned and mastered more readily than a large one. Once you learn the system and practice it with a limited number of compounds, you will be able to expand your skill easily when the topic is presented again in Chapter 12.

The important thing now is that you learn the *system*. Some names, formulas, and rules must be memorized. The names and formulas are examples of the rules, and the rules make up the system. If you just memorize some names and formulas, they are all you will know and they will soon be forgotten. But, if you learn the system, you will be able to apply it to names and formulas of substances you have never heard of before. Furthermore you will remember how to do this throughout your study of chemistry. Keep this in mind as you study this chapter. We cannot overemphasize this advice: *learn the system*.

6.1 SYMBOLS OF THE ELEMENTS

PG 6 A Given the name (or symbol) of an element in Figure 6.1, write the symbol (or name).

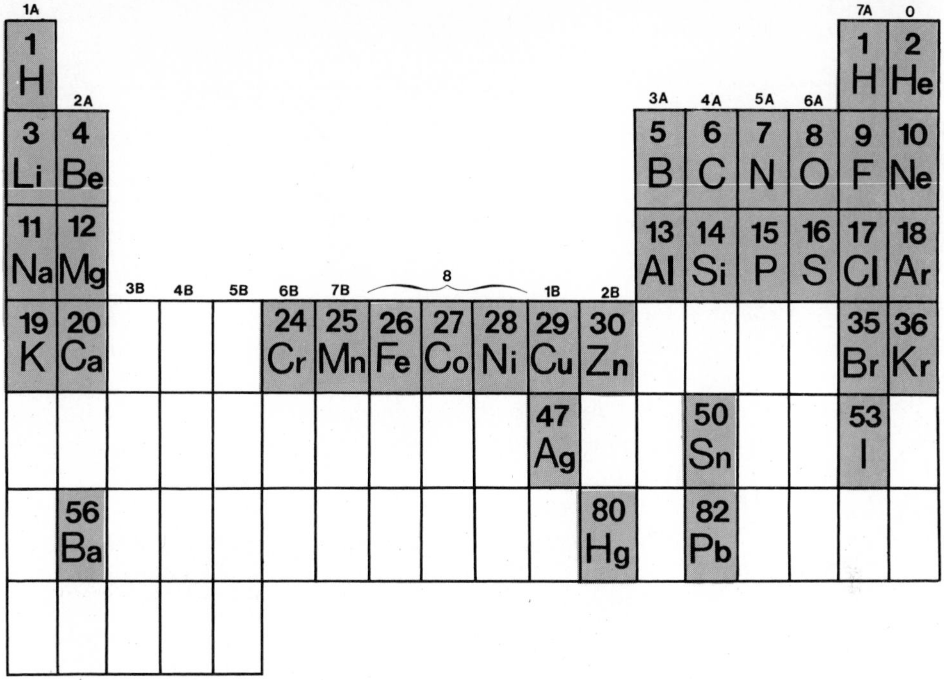

Figure 6.1
Partial periodic table showing the symbols and locations of the more common elements. The symbols above and the list that follows identify the elements you should be able to recognize or write, referring only to a complete periodic table. Associating the names and symbols with the table makes learning them much easier. The elemental names are:

aluminum	bromine	chromium	iodine	magnesium	nitrogen	silver
argon	calcium	copper	iron	manganese	oxygen	sodium
barium	carbon	fluorine	krypton	mercury	phosphorus	sulfur
beryllium	chlorine	helium	lead	neon	potassium	tin
boron	cobalt	hydrogen	lithium	nickel	silicon	zinc

Symbols of elements are used to write chemical formulas. You will work with about one third of the elements regularly, and you should know their symbols. Figure 6.1 highlights these important elements in a periodic table. A periodic table will nearly always be available for reference, so memorizing symbols is much easier if the location of each element in the table is learned at the same time. Familiarity with the table is also helpful when it comes to writing formulas and in finding atomic weights for solving problems.

Be careful when writing elemental symbols that have two letters, such as Cl for chlorine and Zn for zinc. The first letter is always capitalized, but *the second letter is always written in lower case*, or as a "small" letter. There is a world of difference between Co, the metal, cobalt, and CO, the deadly gas, carbon monoxide!

6.2 THE MEANING OF A CHEMICAL FORMULA

> **PG 6 B** Given a chemical formula, state the number of atoms of each element in the formula unit.
>
>   **6 C** Given the number of atoms of each element in a formula unit of a pure substance, write the chemical formula.

A **chemical formula** uses elemental symbols to identify the elements in a pure substance, either element or compound. A subscript number following the symbol of an element tells how many atoms of that element are in a "formula unit" of the substance. *Formula unit* is a general term by which all substances may be considered as independent particles, although in fact they may be parts of a group of particles.

The formula of water is H_2O. H is the elemental symbol for hydrogen, and O is the symbol for oxygen. The subscript 2 after H indicates that the formula unit of water contains two hydrogen atoms. The absence of a subscript after a symbol, as with oxygen, indicates that there is only one atom of that element in the formula unit. Thus, the formula unit of water contains two hydrogen atoms and one oxygen atom.

EXAMPLE 6.1   State the number of atoms of each element in the formula unit of the following common substances:

CO_2, a gas that you exhale:————————————————————————————

CH_4, the fuel for homes heated with natural gas:—————————————————

$NaHCO_3$, baking soda:————————————————————————————

$C_{12}H_{22}O_{11}$, table sugar:————————————————————————————

———   ———

CO_2: one carbon atom, two oxygen atoms
CH_4: one carbon atom, four hydrogen atoms
$NaHCO_3$: one sodium atom, one hydrogen atom, one carbon atom, three oxygen atoms
$C_{12}H_{22}O_{11}$: twelve carbon atoms, twenty-two hydrogen atoms, eleven oxygen atoms

EXAMPLE 6.2   Write the formula of a compound if its formula unit contains two sodium atoms, two sulfur atoms and three oxygen atoms.

———————   ——————

$Na_2S_2O_3$

6.3 THE FORMULAS OF ELEMENTS

PG 6 D   Given the name (or formula) of an element listed in Figure 6.1, write its formula (or name).

The formula unit of most elements is the single atom, even though the atoms may be bonded together in a crystal. The chemical formula is the symbol of the element. Examples are iron, Fe, aluminum, Al, and copper, Cu.

In nature, the stable form of some elements is molecular. **A molecule is the smallest particle of a pure substance, either element or compound.** Molecules behave physically and chemically as individual units, even though they generally contain more than one atom of one or more elements. The formula unit of a molecular substance is the molecule itself. Noble gases, the elements of Group 0 of the periodic table that form almost no compounds, exist as **monatomic molecules,** molecules consisting of only one atom (Fig. 6.2). (The prefixes *mono-* and *mon-* mean one.) The formula unit of a noble gas is therefore the individual atom, and its chemical formula is the symbol of the element. Examples: helium, He, the gas used in balloons; neon, Ne, used in red neon signs.

Seven other elements form **diatomic molecules,** in which two atoms are chemically bonded to each other as in Figure 6.2. (The prefix *di-* means two.) The symbol of the element has a subscript 2 to indicate two atoms in the formula unit. *These elements and their formulas must be memorized.* They are the most abundant element in air—nitrogen, N_2; the two elements in water—hydrogen, H_2, and oxygen, O_2; and the four elements called **halogens** in Group 7A of the periodic table—fluorine, F_2; chlorine, Cl_2; bromine, Br_2; and iodine, I_2.

CAUTION: *Forgetting to write the formulas of these seven elements as diatomic molecules is probably the most common mistake made by beginning students of chemistry. Be alert to it; avoid it. Also note that the*

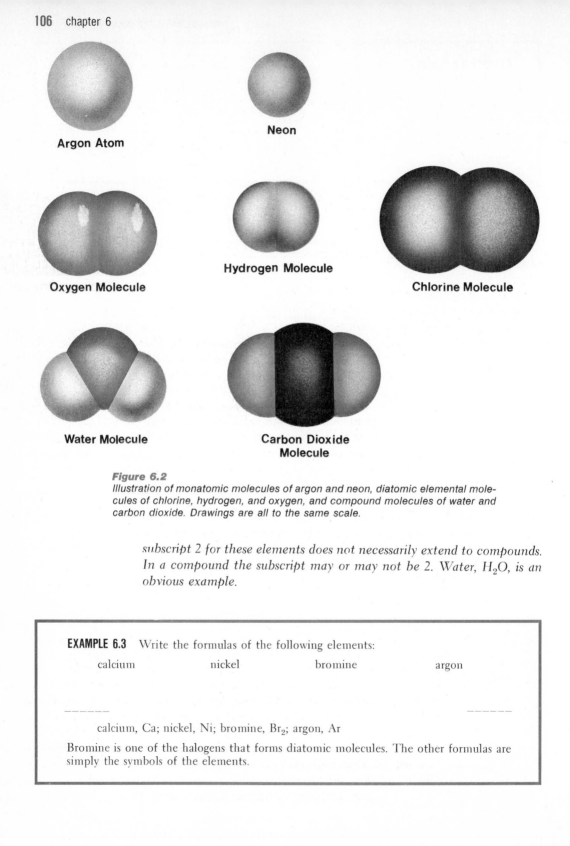

Figure 6.2
*Illustration of monatomic molecules of argon and neon, diatomic elemental mole-
cules of chlorine, hydrogen, and oxygen, and compound molecules of water and
carbon dioxide. Drawings are all to the same scale.*

*subscript 2 for these elements does not necessarily extend to compounds.
In a compound the subscript may or may not be 2. Water, H_2O, is an
obvious example.*

EXAMPLE 6.3   Write the formulas of the following elements:

    calcium   nickel   bromine   argon

――――――   ――――――

    calcium, Ca; nickel, Ni; bromine, Br_2; argon, Ar

Bromine is one of the halogens that forms diatomic molecules. The other formulas are
simply the symbols of the elements.

6.4 COMMON BINARY MOLECULAR COMPOUNDS

> **PG 6 E**   Given the name (or formula) of water, ammonia, carbon monoxide, or carbon dioxide, write its formula (or name).

There are two major types of compounds, molecular and ionic. Names and formulas of ionic compounds will be considered in Section 6.10. Molecular compounds consist of stable individual molecules of two or more elements. The formula unit is the molecule itself. A molecule of a **binary compound** has just two elements. (The prefix *bi-* means two.)

In Chapter 12 you will find the general nomenclature system for binary molecular compounds. Here we will include only four familiar substances, two of which are known by their common names and two by their chemical names. They are water, H_2O; ammonia, NH_3, a common household cleaner; carbon monoxide, CO, the deadly gas in automobile exhaust; and carbon dioxide, CO_2, a product of burning and a component of the atmosphere. The names and formulas of these compounds should be memorized. Notice that the prefixes before "oxide" in carbon monoxide and carbon dioxide tell us the number of oxygen atoms in the molecules. Water and carbon dioxide molecules are shown in Figure 6.2.

6.5 MONATOMIC IONS

> **PG 6 F**   Define or use the word *ion*. Distinguish among a neutral atom, a monatomic ion, and a polyatomic ion; between a cation and an anion.
>
> **6 G**   Given a periodic table and the name and atomic number of a Group A element, or a source from which that name and number may be found, write the name and formula of the monatomic ion formed by that element.

Ionic compounds are made up of positively and negatively charged particles called **ions.** The simplest kind of ion is a **monatomic ion,** which comes from a single atom. There are also **polyatomic ions,** ions made up of a group of atoms with a net electrical charge. (The prefix *poly-* means an indefinite number greater than one.) If the charge on an ion is positive, the ion is called a **cation** (pronounced cat′-ion, not ca-shun). A negatively charged ion is an **anion** (an′-ion). The names and formulas of ionic compounds include the names and formulas of their ions. You must therefore learn the system by which ions are identified. This is the purpose of this section and the four that follow.

In Chapter 8 you will learn how to write **chemical equations.** We introduce them here, however, because they will help you to see how monatomic ions are formed from atoms. A chemical equation uses chemical formulas to identify the substance in a chemical change (p. 13). Substances that enter the change are written first in the equation, followed by an arrow, →, and then the formulas of the substances formed are written. If there are two or more substances on either side of the arrow their formulas are separated by plus signs, +.

The formation of a sodium ion from a sodium atom may be represented by the equation

$$Na \rightarrow Na^+ + e^-$$
$$\text{sodium atom} \rightarrow \text{sodium ion} + \text{electron} \tag{6.1}$$

A sodium atom is electrically neutral; it has the same number of electrons (− charges) and protons (+ charges). If one electron—one negative charge—is removed from the atom, the ion that remains has one more proton than the number of electrons. The ion therefore has a +1 charge.

Equation 6.1 and the word equation beneath it illustrate two important rules about the names and formulas of ions:

The formula of a monatomic ion is the symbol of the element followed by the ionic charge written as a superscript.

The name of a monatomic cation is the name of the element, followed by the word ion.

Notice particularly the difference between the formula of an element and the formula of a monatomic ion of that element. The formula of the element has no charge; atoms are electrically neutral. The formula of the ion has a charge, written as a superscript. Ions are *always* charged. *It is not the formula of an ion if the charge is left off.* Atoms and ions must have different formulas because they are very different things with very different properties. You consume Na^+ daily as part of the salt in your food. If you ever swallowed any Na, you might not survive!

The formation of Na^+ ions from Na atoms is a typical property of the alkali metals, the elements below hydrogen in Group 1A of the periodic table. They all form monatomic ions with a +1 charge by losing one electron per atom. Atoms of Group 2A elements form ions by losing two electrons per atom. The ions therefore have a +2 charge. Magnesium is typical:

$$Mg \rightarrow Mg^{2+} + 2\,e^- \tag{6.2}$$

The coefficient 2 in front of e^- indicates that two electrons are released from the magnesium atom. Notice that in writing the formula of the magnesium ion, the +2 charge is written as a superscript 2+. A Group 3A element may form ions by losing three electrons per atom, producing ions with a +3 charge. Aluminum is the best example:

$$Al \rightarrow Al^{3+} + 3\,e^- \tag{6.3}$$

Carbon, silicon, and germanium, the top three elements in Group 4A, do not form monatomic ions. Compounds of these elements are molecular rather

than ionic. Tin (Sn) and lead (Pb) appear in both ionic and molecular compounds, but their monatomic ions do not fit into the pattern we are considering
here. These ions will be included in Chapter 12.

In Groups 5A, 6A, and 7A we generally find nonmetals that form monatomic ions by *gaining* electrons rather than by losing them. The added electrons
give the ions a negative charge. There are only a few compounds involving
monatomic anions of nitrogen and phosphorus (Group 5A), and we will not be
concerned with them. Elements in Group 6A produce anions with -2 charges
by gaining two electrons per atom. For sulfur this may be shown by the equation

$$S + 2\,e^- \rightarrow S^{2-} \tag{6.4}$$

The S^{2-} ion is called the sulfide ion, which illustrates the rule of naming monatomic anions:

> *The name of a monatomic anion is the name of the element changed to
> end in -ide.*

Elements in Group 7A form anions with a -1 charge. For an *atom* of
chlorine this is represented by the equation

$$Cl + e^- \rightarrow Cl^- \tag{6.5}$$

and the ion is called the chloride ion. (While this is a correct equation that
shows the chloride ion, atomic chlorine is not usually found in abundance. The
stable form of chlorine is in diatomic molecules, from which the formation of
chloride ions is shown by $Cl_2 + 2\,e^- \rightarrow 2\,Cl^-$.) Fluoride, bromide, and iodide
ions, F^-, Br^-, and I^-, respectively, are also formed by gaining one electron per
atom. Even hydrogen, unlike a halogen in other respects, forms a hydride ion,
H^-, in a small number of compounds.

Figure 6.3 highlights the monatomic ions you should know. From Figure
6.1 you already know the symbols of the elements and their location in the
periodic table. If you now relate the ionic charges to the groups in the table, you
will find that your mastery of Group A monatomic ions is complete.

EXAMPLE 6.4   Referring only to a periodic table (*not* Fig. 6.3), write the chemical formulas
of lithium atoms, lithium ions, oxygen atoms, and oxide ions.

———   ———

Lithium atoms, Li; lithium ions, Li^+; oxygen atoms, O; oxide ions, O^{2-}

All ions from Group 1A have a $+1$ charge and therefore carry a $+$ superscript. All Group
6A ions have a -2 charge, and therefore a $2-$ superscript. The question asks specifically
for the formula of oxygen *atoms*, O, rather than oxygen or oxygen molecules, O_2.

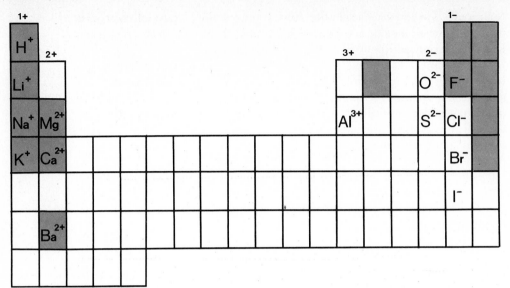

Figure 6.3
Partial periodic table showing the symbols and locations of some of the more important monatomic ions. This is the minimum list of monatomic ions you should be able to use in writing formulas of ionic compounds, referring only to a complete periodic table. Associating the charges on the ions with their location in the periodic table removes the need for memorizing ionic charges.

EXAMPLE 6.5   Referring only to a periodic table, write the name and formula of the monatomic ion formed by each of the following elements and classify the ion as a cation or an anion:

barium   sulfur

bromine   potassium

------   ------

barium: barium ion, Ba^{2+}, cation;   sulfur: sulfur ion, S^2, anion

bromine: bromide ion, Br^-, anion;   potassium: potassium ion, K^+, cation

---

6.6 THE HYDROGEN ION

The term "hydrogen ion" refers to H^+, as would be expected by the position of hydrogen in Group 1A of the periodic table. This species does not exist as such in pure compounds, but it is said to be present in **acids.** An acid may be considered as a solution containing the hydrogen ion, formed when certain hydrogen-

bearing compounds react with water. Hydrogen chloride, a gaseous compound having the formula HCl, is typical. When bubbled through water its reaction may be represented by

$$HCl + H_2O \rightarrow H_3O^+ + Cl^- \tag{6.6}$$

H_3O^+ is the **hydronium ion,** which may be thought of as a hydrogen ion connected to a water molecule: $H \cdot H_2O^+$. Removing a water molecule from both sides of Equation 6.6 leaves

$$HCl \rightarrow H^+ + Cl^- \tag{6.7}$$

In this form hydrogen appears to *ionize*, producing the hydrogen ion. When referring to acids it is understood that the hydrogen ion present is "hydrated," or attached to one or more water molecules, even though it may be written simply as H^+.*

The water solution of hydrogen chloride is hydrochloric acid. Both the compound and the acid have the formula HCl. The way the formula is used generally shows if it refers to the gas or the water solution. If more precise identification is needed, HCl(g) is used for the gaseous compound, and HCl(aq) is its **aqueous** solution, hydrochloric acid. (*Aqueous* is from the Latin *aqua*, meaning water.)

6.7 ACIDS AND THE IONS DERIVED FROM THEM

PG 6 H   Given the name (or formula) of an acid or ion in Table 6.1, write its formula (or name).

Table 6.1 gives the names of five acids, including hydrochloric acid, their total ionization equations, similar to Equation 6.7, and the names of the anions resulting from their total ionization. *Total ionization* is the removal of all the hydrogens that can separate from an acid molecule. A large part of the nomenclature system in this book is based on this table. *The names and formulas of the acids in the table should be memorized.*

Table 6.1 contains important features that illustrate the nomenclature system. Notice the following:

1. Each acid contains at least one hydrogen, written first in the formula, that can separate as H^+ in ionization. If a chemical formula begins with H, it is usually an acid.

2. The first acid, HCl, contains only two elements—hydrogen and a nonmetal. Acids having only two elements are called **binary acids**. *The name of a binary*

*Most chemists and chemistry textbooks use the hydrogen ion as H^+. A substantial number of chemists object to this simplification, however, and prefer to discuss only the hydronium ion. If you take chemistry courses from several teachers you are apt to meet champions of both causes.

Table 6.1
Acids and Anions

Acid	Ionization Equation	Ion Name
Hydrochloric acid	$HCl \rightarrow H^+ + Cl^-$	Chloride
Nitric acid	$HNO_3 \rightarrow H^+ + NO_3^-$	Nitrate
Sulfuric acid	$H_2SO_4 \rightarrow 2\,H^+ + SO_4^{2-}$	Sulfate
Carbonic acid*	$H_2CO_3 \rightarrow 2\,H^+ + CO_3^{2-}$	Carbonate
Phosphoric acid*	$H_3PO_4 \rightarrow 3\,H^+ + PO_4^{3-}$	Phosphate

*The carbonic and phosphoric acid ionizations occur only slightly in water solutions. They are used here to illustrate the derivation of the formulas and names of the carbonate and phosphate ions, both of which are quite abundant from sources other than their parent acids.

acid begins with hydro-, *followed by the name of the nonmetal, changed so it ends with* -ic. For HCl, chlorine → hydrochlor*ic*.

3. The anion from the total ionization of a binary acid is a monatomic anion, named by the rule for all monatomic anions. Example: chlorine → chlor*ide*.

4. Each of the other four acids have three elements—hydrogen, a nonmetal, and oxygen. Acids that contain oxygen are called **oxyacids**. *These particular oxyacids—the ones to be memorized—are named by changing the name of the nonmetal so it ends in* -ic. Examples: nitrogen → nitr*ic*; sulfur → sulfur*ic*.

5. The anion of an oxyacid is made up of the central nonmetal and oxygen. Because it contains oxygen it is called an **oxyanion**; because it has more than one atom it is a **polyatomic ion**. *The name of an anion derived from an oxyacid whose name ends in* -ic *is the name of the central element changed to end with* -ate. You might say the -*ic* of the acid name is replaced by -*ate*. Examples: nitr*ic* → nitr*ate*; sulfur*ic* → sulf*ate*.

6. The formula of the anion resulting from the total ionization of an acid is the formula of the acid without the hydrogen(s). The amount of negative charge, -1, -2, or -3, is equal to the number of hydrogen ions removed from the neutral molecule. Examples: *one* H^+ from $HNO_3 \rightarrow NO_3^-$, with a minus *one* charge; *two* H^+ from $H_2SO_4 \rightarrow SO_4^{2-}$, with a minus *two* charge.

Before leaving Table 6.1, we should point out that its only purpose is to show the relationship between the names and formulas of acids and ions. It is not intended to describe the chemical properties of the substances listed, although all of the ionizations shown do take place, some to a large extent, and some only slightly. The acids are not the usual sources of the anions, all of which are present in abundance in the crust of the earth.

6.8 OTHER POLYATOMIC IONS

The **ammonium ion, NH_4^+**, is the only polyatomic cation we will consider in this text. It forms when a hydrogen ion is added to an ammonia molecule:

$$H^+ \quad + \quad NH_3 \quad \rightarrow \quad NH_4^+$$

$$\text{hydrogen ion} + \text{ammonia} \rightarrow \text{ammonium ion} \qquad (6.8)$$

Be sure you know and use the small but important difference between the names and formulas of the compound ammonia, NH_3, and the ammonium ion, NH_4^+.

The **hydroxide ion, OH^-**, is responsible for the properties of some members of a large class of compounds known as **bases**. While available from many sources, it can be thought of as coming from the ionization of water into a hydrogen and hydroxide ion. Writing the formula of water as HOH, this takes the form of a typical acid ionization:

$$HOH \rightarrow H^+ + OH^- \qquad (6.9)$$

This ionization does, in fact, occur in water, but to a very, very small extent—about one molecule in ten million.

6.9 SUMMARY OF NAMES AND FORMULAS OF IONS

1. The name of a monatomic cation is the name of the element, followed by the word *ion*.

2. The name of a monatomic anion is the name of the element, changed to end with *-ide*, followed by the word *ion*.

3. The charge on a monatomic ion from an A group of the periodic table can be determined from the group it is in, as follows: 1A, +1; 2A, +2; 3A, +3; 5A, −3; 6A, −2; 7A, −1.

4. The hydrogen ion, H^+, comes from the ionization of an acid, a molecular compound having a formula beginning with H.

5. The formula for an anion coming from the total ionization of an acid is the formula of the acid minus the hydrogen(s), with a superscript charge that is the negative of the number of hydrogen(s) lost by the original acid.

6. The name of an oxyanion that comes from an oxyacid whose name ends in *-ic* is the name of the central element in the acid changed to end in *-ate*. In effect, the *-ic* of the acid name is replaced by *-ate*.

7. Ammonium ion is NH_4^+. Hydroxide ion is OH^+.

Once these nomenclature rules are firmly fixed in your mind, you are ready to write the names and formulas of ionic compounds.

6.10 NAMES AND FORMULAS OF IONIC COMPOUNDS

PG 6 J   Given a periodic table and the name (or formula) of any ionic compound whose ions are from the following, write its formula (or name):

Monatomic cation from Group 1A, 2A, or 3A of the periodic table, or the ammonium ion;

Monatomic anion from Group 6A or 7A of the periodic table, or the nitrate, sulfate, carbonate, phosphate, or hydroxide ion.

An **ionic compound** consists of a large group of positive and negative ions, usually arranged in a precise geometric pattern called a crystal. Sodium chloride, ordinary table salt, is a typical ionic compound. Its crystal form is that of a cube (Fig. 6.4). An ionic compound has no independent molecules. Instead, its formula unit is the simplest ratio in which the ions appear in the crystal. In sodium chloride the number of sodium ions is equal to the number of chloride ions, so the formula is NaCl.

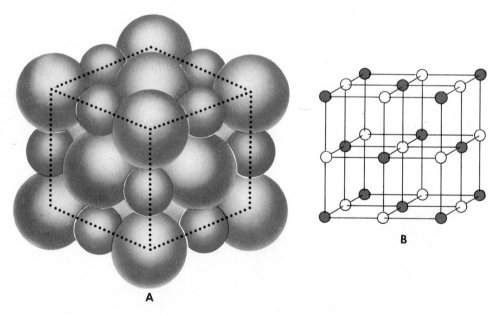

A   **B**

Figure 6.4
Two representations of the crystal structure of sodium chloride. A is a "space-filling" model showing the arrangement of the sodium (smaller) ions and chloride (larger) ions. B is a ball-and-stick model showing the cubic geometry of the crystal. The open circles represent sodium ions, and the colored circles represent chloride ions.

NAMES OF IONIC COMPOUNDS

The rule for naming ionic compounds is:

The name of an ionic compound is the name of the cation, followed by the name of the anion.

All you have to do is to recognize the ions that make up the compound. . . .

(handwritten in margin: NO_3^{1-} SO_4^{2-} CO_3^{2-} PO_4^{3-})

EXAMPLE 6.6 For each of the following ionic compounds, identify the cation and the anion, by name and by symbol, and write the name of the compound:

FORMULA	CATION	ANION	NAME
KCl	Potassium	cloride	
AlPO$_4$	Al^{3+}	PO$_4^{3-}$	Aluminum Phosphate
BaSO$_4$	Ba^{2+}	SO$_4^{2-}$	Barium Sulphate
NaF	Na^{1+}	F^{1-}	Sodium Flouride

FORMULA	CATION	ANION	NAME
KCl	potassium ion, K$^+$	chloride ion, Cl$^-$	potassium chloride
AlPO$_4$	aluminum ion, Al^{3+}	phosphate ion, PO$_4^{3-}$	aluminum phosphate
BaSO$_4$	barium ion, Ba^{2+}	sulfate ion, SO$_4^{2-}$	barium sulfate
NaF	sodium ion, Na$^+$	fluoride ion, F$^-$	sodium fluoride

FORMULAS OF IONIC COMPOUNDS

The rule for writing formulas of ionic compounds is:

The formula of an ionic compound is the formula of the cation followed by the formula of the anion, each taken as many times as is necessary to yield a total charge of zero. (Note: Individual ionic charges are not normally written in the formula of a compound.) Whenever a polyatomic ion appears more than once in a chemical formula, that ion is enclosed in parentheses.

Chemical compounds are electrically neutral. For ionic compounds this means that the formula unit must have an equal number of positive and negative charges. A net zero charge is achieved by taking the cations and anions in such numbers that the positive and negative charges are balanced. The cations generally have charges of +1, +2, and +3, and the anions −1, −2, and −3. This leads to four possibilities:

1. *Cation and anion charges are equal in magnitude.* The ions then combine on a one-to-one basis:

    One $+1$ in Na^+ is balanced by one -1 in Br^-
    $$Na^+ + Br^- \rightarrow NaI$$
    One $+2$ in Mg^{2+} is balanced by one -2 in CO_3^{2-}
    $$Mg^{2+} + CO_3^{2-} \rightarrow MgCO_3$$
    One $+3$ in Al^{3+} is balanced by one -3 in PO_4^{3-}
    $$Al^{3+} + PO_4^{3-} \rightarrow AlPO_4$$

2. *One ion charge has a magnitude of 1, and the other 2.* The ions combine on a one-to-two basis:

    One $+2$ in Mg^{2+} is balanced by two -1 in $2\ Br^-$
    $$Mg^{2+} + 2\ Br^- \rightarrow MgBr_2$$
    Two $+1$ in $2\ Na^+$ are balanced by one -2 in CO_3^{2-}
    $$2\ Na^+ + CO_3^{2-} \rightarrow Na_2CO_3$$

3. *One ion charge has a magnitude of 1, and the other 3.* The ions combine on a one-to-three basis:

    One $+3$ in Al^{3+} is balanced by three -1 in $3\ Br^-$
    $$Al^{3+} + 3\ Br^- \rightarrow AlBr_3$$
    Three $+1$ in Na^+ are balanced by one -3 in PO_4^{3-}
    $$3\ Na^+ + PO_4^{3-} \rightarrow Na_3PO_4$$

4. *One ion charge has a magnitude of 2, and the other 3.* The ions combine on a two-to-three basis:

    Two $+3$ in $2\ Al^{3+}$ are balanced by three -2 in $3\ CO_3^{2-}$
    $$2\ Al^{3+} + 3\ CO_3^{2-} \rightarrow Al_2(CO_3)_3$$
    Three $+2$ in $3\ Mg^{2+}$ are balanced by two $-\bar{3}$ in $2\ PO_4^{3-}$
    $$3\ Mg^{2+} + 2\ PO_4^{3-} \rightarrow Mg_3(PO_4)_2$$

It is neither necessary nor desirable to write equations when writing formulas, but at the beginning they may help you visualize the process. Use them temporarily if you wish, but discard them as soon as possible.

In the following example you may try all of the above combinations with hypothetical compounds. Try to complete the example without looking at the combinations above.

EXAMPLE 6.7   Assume that A^+, B^{2+}, and C^{3+} are three cations and that X^-, Y^{2-}, and Z^{3-} are three anions. Write the formulas of the compounds that may be formed by these ions in all possible combinations, placing the formulas in the appropriate boxes in the table. The compound that corresponds to B^{2+} and X^- is entered in its proper place as an example.

	X^-	Y^{2-}	Z^{3-}
A^+	AX	AY	A₃Z
B^{2+}	BX_2	BY	B₃Z₂
C^{3+}	CX₃	C₂Y₃	CZ

	X⁻	Y²⁻	Z³⁻
A⁺	AX	A_2Y	A_3Z
B²⁺	BX_2	BY	B_3Z_2
C³⁺	CX_3	C_2Y_3	CZ

Now we will apply this procedure to writing formulas of real compounds.

EXAMPLE 6.8   Write the formulas of potassium chloride and potassium hydroxide.

----

potassium chloride, KCl; potassium hydroxide, KOH

Both compounds are +1 and −1 combinations, corresponding to AX in Example 6.7.

EXAMPLE 6.9   Write the formulas of calcium chloride and calcium hydroxide. This question has a small catch to it—something you haven't encountered before in this text. Nevertheless, try to write the formulas without further instructions.

----

calcium chloride, $CaCl_2$; calcium hydroxide, $Ca(OH)_2$. Comments follow.

Did you by any chance write $CaOH_2$ for the second compound in Example 6.9? That formula indicates an atomic ratio of one atom of calcium, one atom of oxygen, and two atoms of hydrogen. The compound is actually made up of one calcium ion for every two *hydroxide ions*: Ca^{2+} and OH⁻ and OH⁻. There are two oxygen atoms in the formula unit. *To show more than one polyatomic ion in a formula, enclose that ion in parentheses and indicate the number of these ions with a subscript after the parentheses.* The subscript applies to everything in the parentheses immediately before it. Note that parentheses are used only to enclose polyatomic ions, never monatomic ions. The formula for calcium chloride is $CaCl_2$, not $Ca(Cl)_2$.

EXAMPLE 6.10   Write the formulas of potassium nitrate and calcium nitrate.

Think carefully about the formulas you have written already, and what has just been said about parentheses.

----

potassium nitrate, KNO_3; calcium nitrate, $Ca(NO_3)_2$

Potassium nitrate is a $+1$ and -1 combination, just like potassium hydroxide in Example 6.8. Calcium nitrate is a $+2$ and -1 combination, just like calcium hydroxide in Example 6.9. The subscript 3 at the end of NO_3^- does not make the handling of nitrate different from how you treat hydroxide. The entire symbol is enclosed in parentheses.

EXAMPLE 6.11   Write the formulas of ammonium nitrate and ammonium carbonate. Again think carefully as you apply the rules you have learned.

_____   _ _ _ _ _ _

ammonium nitrate, NH_4NO_3; ammonium carbonate, $(NH_4)_2CO_3$

Like all ionic compounds, the formula for ammonium nitrate has the formula of the cation followed by the formula of the anion. This is not changed just because both ions happen to contain the same element, nitrogen. In $(NH_4)_2CO_3$ it is a polyatomic cation that is taken twice and must be enclosed in parentheses.

EXAMPLE 6.12   Write the formula of barium phosphate.

_____   _ _ _ _ _ _

barium phosphate, $Ba_3(PO_4)_2$

This $+2$ and -3 combination corresponds to B_3Z_2 in Example 6.7.

EXAMPLE 6.13   Write the formula for each compound listed below:

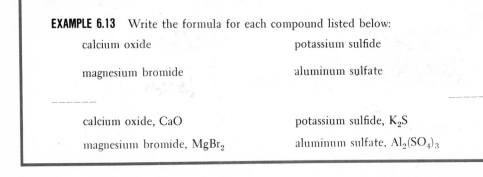

calcium oxide   potassium sulfide

magnesium bromide   aluminum sulfate

_____   _ _ _ _ _ _

calcium oxide, CaO   potassium sulfide, K_2S

magnesium bromide, $MgBr_2$ aluminum sulfate, $Al_2(SO_4)_3$

Practice is the only way to be sure you will master the skill of writing chemical formulas. On pages 125 and 127 there are two formula writing exercises covering all ionic compounds included in this chapter. We strongly recommend that you complete both of these exercises.

6.11 HYDRATES

PG 6 K Given the formula of a hydrate, state the number of water molecules associated with each formula unit of the anhydrous compound.

6 L Given the name (or formula) of a hydrate, write its formula (or name). (This performance goal is limited to hydrates of ionic compounds discussed in this chapter.)

Some compounds, when crystallized from water solutions, form solids that include water molecules as part of the crystal structure. Such water is referred to as **water of crystallization** or **water of hydration.** The compound is said to be **hydrated,** and is called a **hydrate.** Hydration water can usually be driven from a compound by heating, leaving the **anhydrous** compound.

We will use copper sulfate to illustrate the formulas of hydrates. Anhydrous copper sulfate, a nearly white powder, has the formula $CuSO_4$. The hydrate forms large dark blue crystals with a distinct shape. Its formula is $CuSO_4 \cdot 5\ H_2O$. The number of water molecules that crystallize with each formula unit of the anhydrous compound is shown after the anhydrous formula, separated by a dot. This number is 5 for the hydrate of copper sulfate. The equation for the dehydration of this compound is

$$CuSO_4 \cdot 5\ H_2O \rightarrow CuSO_4 + 5\ H_2O \qquad (6.10)$$

Prefixes are used to indicate the number of water molecules in a formula unit of a hydrate (Table 6.2). By this system, $CuSO_4 \cdot 5\ H_2O$ is called copper sulfate pentahydrate, since *penta-* is the prefix for 5. Prefixes are sometimes replaced by numbers, as copper sulfate 5-water, or copper sulfate 5-hydrate.

The number of water molecules associated with one formula unit of an anhydrous compound can vary depending upon temperature and pressure.

Table 6.2
Numerical Prefixes Used in Chemical Names

Number	Prefix	Number	Prefix
1	mono-	6	hexa-
2	di-	7	hepta-
3	tri-	8	octa-
4	tetra-	9	nona-
5	penta-	10	deca-

Some compounds form only one stable hydrate; other compounds form several hydrates. Sodium carbonate, for example, crystallizes from water solution at room temperature as a decahydrate, $Na_2CO_3 \cdot 10\,H_2O$. At higher temperatures the heptahydrate, $Na_2CO_3 \cdot 7\,H_2O$, and monohydrate, $Na_2CO_3 \cdot H_2O$, are stable.

EXAMPLE 6.14   (a) How many molecules of water are associated with each formula unit of anhydrous barium nitrate in $Ba(NO_3)_2 \cdot H_2O$? Name the hydrate.

   (b) Write the formula of the hexahydrate of nickel chloride if the formula of the anhydrous compound is $NiCl_2$.

- - - - - -   - - - - - -

   (a) One; barium nitrate monohydrate.
   (b) $NiCl_2 \cdot 6\,H_2O$; *hexa-* is the prefix for six.

6.12   SUMMARY

Figure 6.5 summarizes the pure substances whose names and formulas you have learned in this chapter. The circled numbers that follow refer to Figure 6.5. The formula unit of most elements is the individual atom, so the formula of the element is its symbol ①. Gaseous elements exist as individual molecules, and the molecule is the formula unit. Noble gas molecules are monatomic, so their formulas are their elemental symbols. Seven elements form diatomic molecules. Their formulas are their elemental symbols with a superscript 2 ②.

   Only four binary molecular compounds are included in this chapter ③. The five molecular compounds whose water solutions are called acids, all shown in Figure 6.5 ④, are the basis of the nomenclature scheme in this book. Their names and formulas should be memorized. They lead directly to the names and formulas of the ions that come from the acids ⑤.

   You must also know the names and formulas of a few other polyatomic ions, such as NH_4^+ and OH^-, and many monatomic ions ⑤ and ⑥. The monatomic ions can be picked right off the periodic table. Once you have the ions, you combine them in whatever numbers that are necessary to make the charge zero. This gives you the formula of an ionic compound ⑦. The name of the compound is simply the names of the ions—cation first, anion second.

   In the next several chapters you will have many opportunities to practice this nomenclature system. Use them to perfect your formula writing skill. When other substances are added in Chapter 12, you will find that they fit neatly into the system you have learned here.

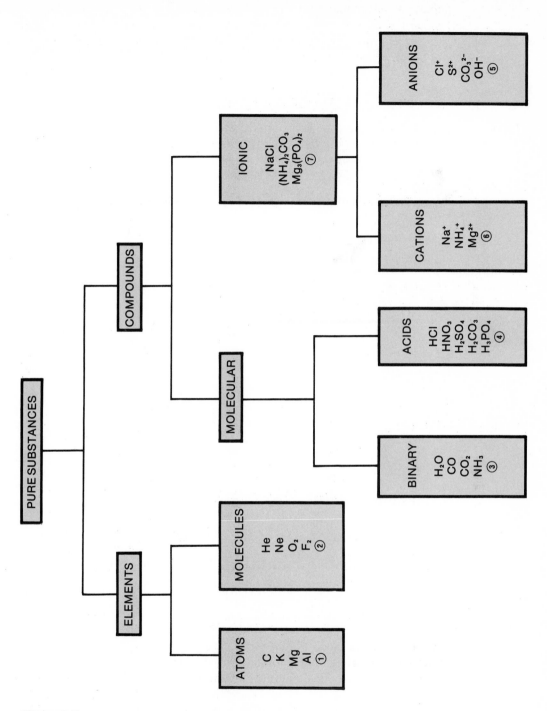

Figure 6.5
Classifications and examples of substances whose names and formulas are studied in the partial nomenclature system of this chapter.

CHAPTER 6 IN REVIEW

6.1 SYMBOLS OF THE ELEMENTS

6 A Given the name (or symbol) of an element in Figure 6.1, write the symbol (or name). (102)

6.2 THE MEANING OF A CHEMICAL FORMULA

6 B Given a chemical formula, state the number of atoms of each element in the formula unit. (104)

6 C Given the number of atoms of each element in a formula unit of a pure substance, write the chemical formula. (104)

6.3 THE FORMULAS OF THE ELEMENTS

6 D Given the name (or formula) of an element listed in Figure 6.1, write its formula (or name). (105)

6.4 COMMON BINARY MOLECULAR COMPOUNDS

6 E Given the name (or formula) of water, ammonia, carbon monoxide, or carbon dioxide, write its formula (or name). (107)

6.5 MONATOMIC IONS

6 F Define or use the word *ion*. Distinguish among a neutral atom, a monatomic ion, and a polyatomic ion; between a cation and an anion. (107)

6 G Given a periodic table and the name and atomic number of a Group A element or a source from which that name and number may be found, write the name and formula of the monatomic ion formed by that element.

6.6 THE HYDROGEN ION (110)

6.7 ACIDS AND THE IONS DERIVED FROM THEM

6 H Given the name (or formula) of an acid or ion in Table 6.1, write its formula (or name). (111)

6.8 OTHER POLYATOMIC IONS

6 I Given the name (or formula) of the ammonium or hydroxide ions, write the formula (or name). (113)

6.9 SUMMARY OF NAMES AND FORMULAS OF IONS (113)

6.10 NAMES AND FORMULAS OF IONIC COMPOUNDS

6 J Given a periodic table and the name (or formula) any ionic compound whose ions are from the following, write its formula (or name):

Monatomic cation from Group 1A, 2A, or 3A of the periodic table, or the ammonium ion;

Monatomic anion from Group 6A or 7A of the periodic table, or the nitrate, sulfate, carbonate, phosphate, or hydroxide ion. (114)

6.11   HYDRATES

6 K Given the formula of a hydrate, state the number of water molecules associated with each formula unit of the anhydrous compound. (119)

6 L Given the name (or formula) of a hydrate, write its formula (or name). (This performance goal is limited to hydrates of ionic compounds covered in this chapter.) (119)

6.12   SUMMARY (120)

TERMS AND CONCEPTS

Chemical formula (104)
Formula unit (104)
Molecule (105)
Monatomic (105)
Diatomic (105)
Halogen (105)
Molecular compound (107)
Binary (107)
Ion (107)
Polyatomic (107)
Cation (107)
Anion (107)
Chemical equation (107)

Acid (110)
Hydronium ion (111)
Ionize (111)
Hydrated (111)
Aqueous (111)
Total ionization (111)
Oxyacid (112)
Oxyanion (112)
Ionic compound (114)
Water of crystallization (hydration) (119)
Hydrate (119)
Anhydrous compound (119)

Most of these terms and many others appear in the Glossary.

QUESTIONS AND PROBLEMS

Section 6.1

6.1) The names, atomic numbers, or symbols of the elements in Figure 6.1 are entered in Table 6.3 (124). Fill in the open spaces, referring only to a periodic table for any information you require.

Section 6.2

For Questions 6.2 to 6.5 and 6.28 to 6.31, for each formula, state the identity of each element in the compound and the number of atoms of that element in the formula unit; or, from the given number of atoms of each element in the formula unit, write the formula.

6.2) N_2O.

6.3) SO_2Cl_2.

6.4) One silicon, four hydrogen atoms.

6.5) Three carbon, eight hydrogen, one oxygen atoms.

6.28) PCl_3.

6.29) CH_3CHO.

6.30) Two hydrogen, one sulfur, three oxygen atoms.

6.31) Four hydrogen, two phosphorus, seven oxygen atoms.

Table 6.3
Table of Elements

Name of Element	Atomic Number	Symbol of Element	Name of Element	Atomic Number	Symbol of Element
Sodium					Mg
		Pb		8	
Aluminum			Phosphorus		
	26				Ca
		F	Zinc		
Boron					Li
	18		Nitrogen		
Silver				16	
	6			53	
Copper			Barium		
		Be			K
Krypton				10	
Chlorine			Helium		
	1				Br
		Mn			Ni
	24		Tin		
Cobalt				14	
	80				

Section 6.3

6.6) What features distinguish atoms from molecules?

6.7) Distinguish between substances that are monatomic and those that are diatomic.

6.8) When a solid element is vaporized, each gaseous particle consists of two atoms. Is the formula unit for the element a molecule or an atom?

Section 6.4

6.9) What characteristic must a compound have if it is classified as molecular?

6.10) Write the names of H_2O and CO_2.

Section 6.5

6.11) Distinguish between monatomic ions and polyatomic ions.

6.32) An element is a gas at room conditions. Is its formula unit (a) an atom, (b) a molecule, or (c) either an atom or a molecule? If (c), give examples of each.

6.33) Write formulas for two elements whose structural units are monatomic molecules and two that exist as diatomic molecules.

6.34) When a solid metallic element is vaporized, each gaseous particle is monatomic. Is the formula unit of the element an atom or a molecule?

6.35) What is the difference between a molecular compound and a binary molecular compound?

6.36) Write the formulas of carbon monoxide and ammonia.

6.37) Identify the polyatomic ions among the following: K^+, S^{2-}, Br^-, NH_4^+, CO_3^{2-}, Al^{3+}.

Table 6.4
Formula Writing and Nomenclature Exercise Number 1

Instructions: For each box write the chemical formula and name of the compound formed by the cation at the head of the column and the anion at the left of the row. Correct formulas and names are listed on page 615.

IONS	Li^+	Mg^{2+}	NH_4^+	Al^{3+}	Na^+	Ba^{2+}	K^+	Ca^{2+}
Br^-	1	2	3	4	5	6	7	8
SO_4^{2-}	9	10	11	12	13	14	15	16
OH^-	17	18	19	20	21	22	23	24
F^-	25	26	27	28	29	30	31	32
O^{2-}	33	34	35	36	37	38	39	40
NO_3^-	41	42	43	44	45	46	47	48
PO_4^{3-}	49	50	51	52	53	54	55	56
Cl^-	57	58	59	60	61	62	63	64
S^{2-}	65	66	67	68	69	70	71	72
I^-	73	74	75	76	77	78	79	80
CO_3^{2-}	81	82	83	84	85	86	87	88

6.12) How does a monatomic anion differ from a monatomic cation?

6.38) Identify the monatomic cations in Question 6.37.

6.13) Explain how monatomic cations are formed from atoms.

6.39) Explain how monatomic anions are formed from atoms.

6.14) The table below gives the atomic numbers of several elements from the A groups of the periodic table. Fill in the blanks with the ion name and symbol for the monatomic ion formed by each element, and classify that ion as a cation or anion. You may refer to a periodic table.

ATOMIC NUMBER	ION NAME	ION FORMULA	CATION OR ANION
11	Sodium ion	Na^+	Cation
53	Iodide	I^-	Anion
3	Lithium ion	Li^+	Cation
20	Calcium ion	Ca^{2+}	Cation
16	Sulfide	S^{2-}	Anion

6.15)* You are not expected to be familiar with the elements selenium ($Z = 34$) and rubidium ($Z = 37$), but from their positions in the periodic table, you should be able to make intelligent guesses as to the names and formulas of the monatomic ions they will form. Complete the following table.

ELEMENT	ION NAME	ION FORMULA	CATION OR ANION
Selenium	Selenide	Se^{2-}	Anion
Rubidium	Rubidium ion	Rb^+	Cation

Sections 6.6 and 6.7

6.16) What property identifies a substance as an acid?

6.40) How do you recognize the formula of an acid? Distinguish between a binary acid and an oxyacid.

6.17) Write the formula of hydrochloric acid.

$$HCl$$

6.41) Show how the hydrogen and chloride ions are derived from the ionization of hydrochloric acid.

6.18) Write the formula of nitric acid, the formula of the anion derived from its ionization, and the name of the anion. $HNO_3 \rightarrow H^+ + NO_3^-$
Nitrate

6.42) Write the formula of phosphoric acid, the formula of the anion derived from its ionization, and the name of the anion.

6.19) Write the name of H_2CO_3, the formula of the anion derived from its ionization, and the name of the anion.

Carbonic Acid CO_3
Carbonate

6.43) Write the name of H_2SO_4, the formula of the anion derived from its ionization, and the name of the anion.

Section 6.8

6.20) What is the name of the OH^- ion?

Hydroxide

6.44) Write the formula of the ammonium ion.

Section 6.10

The formula writing exercises of Tables 6.4 and 6.5 are the principal questions for Section 6.10. For Questions 6.21 to 6.24 and 6.45 to 6.48, given the name of an ionic compound, write the formula or, given the formula, write the name. Refer only to a periodic table while answering these questions.

Table 6.5
Formula Writing Exercise Number 2

Instructions: For each box write the chemical formula of the compound formed by the cation at the head of the column and the anion at the left of the row. Refer only to the periodic table in completing this exercise. Correct formulas are listed on page 616.

IONS	Aluminum	Ammonium	Barium	Calcium	Lithium	Magnesium	Potassium	Sodium
Bromide	1 $AlBr_3$	12	23	34	45	56	67	78
Carbonate	2	13 $(Na)_2CO_3$	24	35	46	57	68	79
Chloride	3	14	25 $BaCl_2$	36	47	58	69	80
Fluoride	4	15	26	37 CaF_2	48	59	70	81
Hydroxide	5	16	27	38	49 $LiOH$	60	71	82
Iodide	6	17	28	39	50	61 MgI_2	72 KI	83
Nitrate	7	18	29	40	51	62	73 K_2NO_3	84
Oxide	8	19	30	41	52	63	74 K_2O	85 Na_2O
Phosphate	9	20	31	42	53	64	75 K_3PO_4	86
Sulfate	10	21	32	43	54	65 $MgSO_4$	76	87
Sulfide	11	22	33	44	55 Li_2S	66	77	88

6.21) Lithium chloride; CaS. *LiCl*
Calcium Sulphte

6.22) Ammonium nitrate; BaCO$_3$.
NH$_4$ NO$_3$ Barium Carbonate

6.23) Barium bromide; K$_3$PO$_4$.
Ba Br$_2$ Potassium Phosphate

6.24) Magnesium phosphate; (NH$_4$)$_2$SO$_4$.
Mg$_3$(PO$_4$)$_2$ A Ammonium Sulphate

Section 6.11

6.25) Distinguish between a hydrate and an anhydrous compound. *Hydrate is molecule w/ H$_2$O in it. Anhydrous is a molecule w/ H$_2$O removed*

6.26) How many water molecules are associated with one anhydrous formula unit of calcium chloride in CaCl$_2$ · 2 H$_2$O? Write the name of the compound.

6.27) One hydrate of barium hydroxide contains eight molecules of water per formula unit of the anhydrous compound. Write the formula and name of the hydrate.

Ba(OH)$_2$ · 8 H$_2$O

Barium Hydroxide Octahydrate

6.45) KF; magnesium oxide.

6.46) NaOH; aluminum phosphate.

6.47) CaI$_2$; sodium sulfate.

6.48) Al$_2$(CO$_3$)$_3$; lithium sulfide.

6.49) Among the following, identify all hydrates and anhydrous compounds: NiSO$_4$ · 6 H$_2$O; KCl; Na$_3$PO$_4$ · 12 H$_2$O.

6.50) How many water molecules are associated with one anhydrous formula unit of magnesium sulfate in MgSO$_4$ · 7 H$_2$O? Write the name of the compound.

6.51) Write the formulas of ammonium phosphate trihydrate and potassium sulfide 5-hydrate.

chemical formula calculations

Chapter 6 in this book is an introduction to chemical nomenclature, the system of naming chemicals and writing their formulas. The introduction is limited to 35 elements and 97 compounds. You will need the formulas of many of these substances to solve examples in the next few chapters. To help you strengthen your formula writing skills within this narrow range of substances, we will identify them by name only and leave it to you to furnish the formulas.

We encourage you to try to write all formulas as they are required, using a periodic table as your only reference. If you need assistance, check Formula Writing Exercise No. 2 on page 127, and the answers to that exercise on page 616.

7.1 SIGNIFICANT FIGURES

Before beginning chemical formula calculations, let's consider very briefly the subject of **significant figures.** This is a method of expressing the precision of measurements and results calculated from measurements. No measurement is absolutely correct. Whatever error, or uncertainty, is present is carried over to the calculated results as well. By the rules of significant figures, the size of the uncertainty is indicated by rounding off the final result to the **doubtful digit,** the last digit in the recorded value. (This topic is discussed in greater detail in Appendix I, Part H. Here we will simply list the practices followed in this book.)

1. Unless there is some reason for doing otherwise, all numbers are expressed in three significant figures. This means that, beginning with the first nonzero digit, there will usually be three digits in any quantity. ("Usually" is in the above sentence because addition sometimes increases the number of significant figures in a result, and subtraction may reduce that number.) Notice that the final digit may be a zero on the right side of the decimal. The location of the decimal point has nothing to do with significant figures.

2. Although atomic weights are given to the second decimal in the periodic table, they will generally be rounded off to the first decimal when used in calculations. When addition of atomic weights results in a sum greater than 100, that sum will be rounded off to the closest whole number.

3. In rounding off, if the first digit to be dropped is in the range 0 to 4, the digit before it will be unchanged. If the first digit to be dropped is from 5 to 9, the digit before it will be increased by 1.

As you compare your calculated results with those in the examples ahead, or those in the answers at the end of the book, do not be concerned if your answer is not identical to the book answer. It is quite normal for you to be plus or minus 1, 2, or 3 in the doubtful digit. That is why it is doubtful. Be more interested in how you get your answer, rather than in duplicating the exact value.

7.2 MOLECULAR WEIGHT; FORMULA WEIGHT

> **PG 7 A** Distinguish among atomic weight, molecular weight, and formula weight.

In Chapter 4 you were introduced to the scale of the relative masses of the atoms, known as atomic weight. Atomic weight is defined as the average mass of the atoms of an element compared to the mass of an atom of carbon-12 at exactly 12 atomic mass units (amu). But what about compounds? Is there such a thing as a "compound weight?" The answer is yes.

When speaking of the "compound weight" of a molecular compound, chemists use a term that is precisely parallel to atomic weight: **molecular weight is the average mass of molecules of a molecular substance compared to the mass of an atom of carbon-12 at exactly 12 atomic mass units.** Because ionic compounds do not exist as individual molecules, it is not appropriate to refer to the *molecular weight* of such a compound. Instead, the term *formula weight* is used. **Formula weight is the average mass of one formula unit of a compound compared to the mass of an atom of carbon-12 at exactly 12 atomic mass units.** For an ionic compound the "formula unit" is an imaginary structural unit consisting of the number of atoms of each element that appears in the chemical formula of the compound.

7.3 THE MOLE CONCEPT

PG 7 B Define *mole*. Identify and state the number that corresponds to one mole.

Dalton's atomic theory says that compounds consist of atoms of different elements combined in a ratio of small whole numbers. To prepare a compound from its elements we must therefore "count out" the proper number of atoms of the different elements. But, if we wish to count out equal numbers of oxygen and carbon atoms so that they can react to produce carbon monoxide, CO, it will be a time-consuming process. This is because of the tremendous number of atoms of each element in the tiniest of samples. There must be a better way—and chemists have found it.

It is a fact that, as atomic weights compare the masses of individual atoms of different elements, so they also compare the masses of *equal numbers* of atoms of those elements. If one atom of carbon has a mass of 12.0 amu, and one atom of oxygen has a mass of 16.0 amu, their mass ratio is

$$\frac{1 \text{ atom of O}}{1 \text{ atom of C}} = \frac{16.0 \text{ amu}}{12.0 \text{ amu}} = \frac{16.0}{12.0}$$

Similarly Y atoms of carbon have a mass of $12.0 \times Y$, and Y atoms of oxygen have a mass of $16.0 \times Y$, where Y can be any number. Their mass ratio is

$$\frac{Y \text{ atoms of O}}{Y \text{ atoms of C}} = \frac{16.0 \times Y \text{ amu}}{12.0 \times Y \text{ amu}} = \frac{16.0}{12.0}$$

The ratios are identical. Therefore, if you wish to "count out" an equal number of atoms of any two elements, all you have to do is to weigh out samples having a mass ratio that is the same as the ratio of their atomic weights.

Chemists have organized the "counting by weighing" idea into what is called the mole concept. A **mole** (abbreviated **mol**) is defined as **that amount of any species that contains the same number of units as the number of atoms in exactly 12 grams of carbon-12.** This number is called **Avogadro's number, N,** in honor of the man whose interpretation of experiments with gases led to an early method of estimating atomic weights.

Notice that the definition of the mole does not identify the number of atoms in exactly 12 grams of carbon-12. That number must be found by experiment. Since a carbon atom is extremely small we would expect the number of atoms in exactly 12 grams of carbon-12 to be huge. This is the case; the number has been estimated to be 602 000 000 000 000 000 000 000. In exponential notation this is 6.02×10^{23}. The following equivalence results from the definition of a mole:

One mole of any species = 6.02×10^{23} units of that species   (7.1)

To get a better idea of what is meant by a mole, let us consider another counting unit with which we are more familiar, the dozen. Suppose we define a dozen as the number of eggs in the egg carton in which they are sold at the local supermarket. By experiment (walking down to the store, opening a box of eggs, and counting them) you can determine that this number is 12. In buying eggs, you would rarely use the number 12 directly; rather, you would think in terms of dozens. When is the last time, for example, you saw a shopping list with "24 eggs" written on it? Would it not be written simply as two dozen eggs? Just as two dozen eggs expresses a quantity of eggs, two moles of carbon expresses a quantity of carbon, a quantity that can be thought of as containing a certain number of carbon atoms. If you ever have difficulty understanding *mole* in a sentence, substitute the word *dozen* and you will see what it means.

Let's think for a moment about this number—Avogadro's number. Its size is tremendous; it staggers the imagination. Many attempts have been made to create a word picture of it, but they can only hint at its vast size by leaving the reader with another huge number that likewise defies the imagination. For example, if you were to try to count the atoms in exactly 12 grams of carbon-12, and proceeded at the rate of 100 atoms per minute without interruption, 24 hours per day, 365 days per year, the task would require 11 500 000 000 000 000 years. This is about two million times as long as the earth has been in existence! If we enlisted the aid of the entire population of the earth, about 4.5 billion people, all counting at the same rate, the time required could be brought down to a somewhat more reasonable 2 540 000 years.

Is it any wonder the chemist prefers to think in terms of moles rather than dealing with actual numbers of molecules?

7.4 MOLAR WEIGHT

PG 7 C Define molar weight.

Molar weight is the weight in grams of one mole of any species. As defined, the units of molar weight are grams per mole. The concept is particularly useful in finding the number of grams in a specified number of moles, or vice versa, as we will see in Section 7.6. *Molar weight* is a general term that may be applied to any species, atoms, molecules, or formula units.

Many chemists and many textbooks use separate terms to identify the molar weights of atoms, molecules, or formula units. For example, the *gram atomic weight* of an element is the weight in grams that contains the same number of atoms as the number of atoms in exactly 12 grams of carbon-12. If you substitute the word "units" for the first "atoms," the back part of that definition becomes ". . . that contains the same number of [units] as the number of atoms in exactly 12 grams of carbon-12." This is the definition of a mole given on page 131. Gram atomic weight, then, is the weight in grams of one mole of atoms of an element, which is what we have called *molar weight*. Identical comments may be made about the terms *gram molecular weight* and *gram formula weight*, so the term *molar weight* does indeed include all three.

There is one situation in which the distinction between *gram atomic weight* and *gram molecular weight* has an advantage over *molar weight*. It is when it is applied to the gaseous elements that form diatomic molecules. For example, the phrase "molar weight of oxygen" is ambiguous. Does it mean the gram atomic weight, 16.0 grams, or the gram molecular weight, 32.0 grams for O_2? The difficulty is removed if you state clearly by word or formula the species whose molar weight is intended, as molar weight of O, oxygen atoms, or molar weight of O_2, oxygen molecules.

Another term widely used by chemists and in textbooks is *gram atom*. It refers to the quantity of an element whose weight is equal to the gram atomic weight. This quantity is the same as one mole, which has been identified as the official unit of *amount of substance* in the SI system of units. *Gram molecule* and *gram formula unit*, the latter term not widely used, are comparable. In this text the word *mole* will be used for the amount of a substance.

There is an important relationship you should catch at this point. Because the definitions of atomic weight, mole and molar weight are all tied to "exactly 12 grams of carbon-12," *atomic weight and molar weight of atoms are numerically equal*. The atomic weight of sodium is 23.0 amu; the molar weight of sodium atoms is 23.0 grams per mole. The periodic table is therefore a source of molar weights of atoms in grams per mole, as well as atomic weights in atomic mass units. Similarly, the molecular weight of molecular compounds and the formula weight of ionic compounds are numerically equal to the molar weights of those compounds.

7.5   CALCULATION OF MOLAR WEIGHT

PG 7 D   Calculate the molar weight of any substance whose chemical formula is known.

What is the molar weight of CO? Based on the time-honored truth that the whole is equal to the sum of its parts, the weight of one mole of CO is equal to the sum of the weights of all the atoms it contains. Each molecule of CO contains one carbon atom and one oxygen atom. Therefore one mole of CO molecules contains one mole of carbon atoms (12.0 grams) and one mole of oxygen atoms (16.0 grams). The molar weight of CO is 12.0 g C + 16.0 g O = 28.0 grams per mole. Similarly . . . .

EXAMPLE 7.1   Calculate the molar weight of sodium fluoride, NaF.

The procedure is straightforward. Solve the problem, remembering that the periodic table is your best source for atomic weights (molar weights of atoms).

23.0 g Na + 19.0 g F = 42.0 g NaF/mol

Now that the procedure is established, try another that is a bit more complicated. . . .

EXAMPLE 7.2 Calculate the molar weight of calcium fluoride.

No formula this time. This you must furnish. What is it? Refer to periodic table for help, if necessary.

CaF_2

Now you should be asking yourself how many moles of calcium and how many moles of fluorine are in one mole of CaF_2 units. Answer, please.

One mole of calcium and two moles of fluorine.

Now find the grams of calcium fluoride per mole.

Ca: $1 \times 40.1 = 40.1$ g Ca
F: $2 \times 19.0 = \underline{38.0}$ g F
                78.1 g CaF_2/mol

The calculation setup in Example 7.2 is for the convenience of ordinary addition. Assuming that you are using a calculator, the addition is just as easy when numbers to be added are in a line, as in Example 7.1. The line setup for Example 7.2 is

$$40.1 \text{ g Ca} + 2(19.0) \text{ g F} = 78.1 \text{ g } CaF_2/\text{mol}$$

We will use the line setup hereafter.

EXAMPLE 7.3   Find the molar weight of magnesium hydroxide.

First you need the formula of magnesium hydroxide. It is. . . .

$Mg(OH)_2$

You know the procedure. Although this problem does introduce a variation not met previously, you probably will be able to think it through. All the way to the answer, please.

$$24.3 \text{ g Mg} + 2(16.0) \text{ g O} + 2(1.0) \text{ g H} = 58.3 \text{ g Mg(OH)}_2/\text{mol}$$

This problem requires you to recognize that one mole of $Mg(OH_2$ contains one mole of magnesium ions and two moles of hydroxide ions. Each hydroxide ion contains one oxygen atom and one hydrogen atom. Hence, there are two moles of oxygen and two moles of hydrogen in the calculations.

EXAMPLE 7.4   Calculate the molar weight of aluminum sulfate.

The formula, please. . . .

$Al_2(SO_4)_3$

Now to be sure the formula is correctly interpreted, how many moles of aluminum, sulfur, and oxygen atoms are in one mole of $Al_2(SO_4)_3$?

2 mol Al, 3 mol S, 12 mol O

Each sulfate ion contains one sulfur atom and four oxygen atoms. There are three sulfate ions in the formula unit. This makes three sulfur atoms and $3 \times 4 = 12$ oxygen atoms in the formula unit.
Now calculate the molar weight.

$$2(27.0) \text{ g Al} + 3(32.1) \text{ g S} + 12(16.0) \text{ g O} = 342 \text{ g Al}_2(SO_4)_3/\text{mol}$$

Let's try a hydrate.

EXAMPLE 7.5   Find the molar weight of $CuSO_4 \cdot 5 H_2O$. (Cu is copper, $Z = 29$.)

The formula of a hydrate shows the number of water molecules locked into a crystal with each formula unit of the compound. Another way of writing the hydrate formula that may make its interpretation easier is $CuSO_4(H_2O)_5$. Now list the number of moles of atoms of each element in one mole of formula units: copper, sulfur, oxygen (careful!) and hydrogen.

------

1 mol Cu, 1 mol S, 9 mol O (4 from $CuSO_4$ + 5 from H_2O), 10 mol H.

Complete the calculation of molar weight.

63.5 g Cu + 32.1 g S + 9(16.0) g O + 10(1.0) g H = 250 g $CuSO_4 \cdot 5\ H_2O$/mol

---

EXAMPLE 7.6   Find the molar weight of bromine vapor, Br_2.

This time you are looking for the molar weight of an element. The problem is easy, but you may wonder about something as you proceed. Write the answer.

2(79.9) g Br = 160 g Br_2/mol

If you wrote 159.8 g Br_2/mol you were correct. The answer above has been rounded off to three significant figures. We thought you might wonder about whether or not you should multiply the atomic weight of bromine by 2. Bromine is not stable as an individual atom but forms diatomic molecules, molecules consisting of two atoms. This is why the formula is Br_2. In one mole of Br_2 there are two moles of Br atoms.

Example 7.6 teaches an important lesson: always perform your molar weight calculation for the formula unit exactly as that formula is written. This also emphasizes the importance of having the correct formula. Keep this in mind when you meet a slightly more complicated example a few pages ahead.

7.6 MASS RELATIONSHIPS BETWEEN ELEMENTS IN A COMPOUND; PERCENTAGE COMPOSITION

PG 7 E   For any compound whose formula is known, given the mass of a sample, calculate the mass of any element in the sample. Or, given the mass of any element in the sample, calculate the mass of the sample or the mass of any other element in the sample.

7 F   Calculate the percentage composition of any compound whose formula is known.

The numbers used in calculating the molar weight of a compound furnish a one-step conversion between the mass of any sample of the compound and the mass of any element in the sample. In Example 7.4 you found that one mole of aluminum sulfate, $Al_2(SO_4)_3$, is made up of $2 \times 27.0 = 54.0$ g Al, plus $3 \times 32.1 = 96.3$ g S, plus $12 \times 16.0 = 192$ g O, a total of 342 grams per mole. For aluminum sulfate these numbers give us equivalent relationships between the masses of the elements and the compound:

$$54.0 \text{ g Al} \simeq 96.3 \text{ g S} \simeq 192 \text{ g O} \simeq 342 \text{ g Al}_2(SO_4)_3 \qquad (7.2)$$

If the number of grams of any species in Equation 7.2 is known, the mass of any other species may be calculated.

EXAMPLE 7.7   How many grams of aluminum are in 138 grams of $Al_2(SO_4)_3$.

The unit path is grams of given quantity $\rightarrow$ grams of wanted quantity. Set up the problem and solve.

------   ------

$$138 \text{ g } Al_2(SO_4)_3 \times \frac{54.0 \text{ g Al}}{342 \text{ g } Al_2(SO_4)_3} = 21.8 \text{ g Al}$$

The mass relationships between elements in a compound are often expressed in its percentage composition. (If percentage problems are troublesome to you, check page 583 for assistance.) The percentage of one element in a compound is found by dividing the mass of that element in a sample by the mass of the whole sample, and then multiplying by 100:

$$\% \text{ of element} = \frac{\text{grams of element}}{\text{grams of sample}} \times 100 \qquad (7.3)$$

The most convenient "sample" to use is one mole. Again, using the numbers for aluminum sulfate, its percentage composition is

$$\text{Al:} \quad \frac{54.0 \text{ g Al}}{342 \text{ g Al}_2(SO_4)_3} \times 100 = 15.8\% \text{ Al}$$

$$\text{S:} \quad \frac{96.3 \text{ g S}}{342 \text{ g Al}_2(SO_4)_3} \times 100 = 28.2\% \text{ S}$$

$$\text{O:} \quad \frac{192 \text{ g O}}{342 \text{ g Al}_2(SO_4)_3} \times 100 = 56.1\% \text{ O}$$

$$\text{total percent} = 100.1\%$$

The sum of the percentages must be 100.0. A difference of 0.1 or 0.2 may be due to rounding off.

The same procedure may be used for any compound.

EXAMPLE 7.8   Calculate the percentage composition of calcium nitrate. First you need the formula of calcium nitrate. It is. . . .

$Ca(NO_3)_2$

Now calculate the molar weight of calcium nitrate. You will have to find the grams of each element to determine percentage. Carry it all the way, and check to see if your percentages total 100.

Element	Grams	Percent
Ca	$1 \times 40.1 = 40.1$ g Ca	$\frac{40.1}{164} \times 100 = 24.4\%$ Ca
N	$2 \times 14.0 = 28.0$ g N	$\frac{28.0}{164} \times 100 = 17.1\%$ N
O	$6 \times 16.0 = 96.0$ g O	$\frac{96.0}{164} \times 100 = 58.5\%$ O
	$\overline{164.1 \text{ g } Ca(NO_3)_2}$	$\overline{100.0\%}$

7.7 CONVERSION BETWEEN MASS AND NUMBER OF MOLES

PG 7 G   Given the number of grams (or moles) of a chemical species of known or calculable molar weight, find the number of moles (or grams).

The "counting by weighing" value of the mole concept is based on the ready conversion of the mass of a chemical species into a number of moles, and vice versa. Using dimensional analysis, this is a simple operation. It is a part of many chemical problems, and is one of the most important skills to be learned in the introductory course. The operation is based upon the equivalence between grams and moles of a chemical that comes from the mole concept. If MW is the number of grams of X in 1 mole, the molar weight of X, then

$$MW \text{ grams of X} \simeq 1 \text{ mole} \qquad (7.4)$$

The following examples illustrate the two possible conversions.

EXAMPLE 7.9 How many moles of aluminum sulfate, $Al_2(SO_4)_3$ are in 150 grams? The molar weight, from Example 7.4, is 342 grams/mole.

Applying Equation 7.4 to the problem at hand yields the equivalence 342 grams $Al_2(SO_4)_3 \simeq 1$ mole $Al_2(SO_4)_3$. Your unit path is from the given quantity, grams, to the desired quantity, moles: g → mol. If one mole is 342 grams, then the given quantity of 150 grams is what number of moles? Set up and solve.

$$150 \text{ g } Al_2(SO_4)_3 \times \frac{1 \text{ mol } Al_2(SO_4)_3}{342 \text{ g } Al_2(SO_4)_3} = 0.439 \text{ mole } Al_2(SO_4)_3$$

EXAMPLE 7.10 You are carrying out a laboratory reaction that requires 0.0250 mole of $NiCl_2 \cdot 6\,H_2O$. How many grams of the compound do you weigh out?

First, find the molar weight of $NiCl_2 \cdot 6\,H_2O$. (Ni is nickel, Z = 28.)

238 g $NiCl_2 \cdot 6\,H_2O$/mol

Use molar weight to convert from the given quantity of moles to grams. If 1 mole is 238 grams, how many grams is 0.0250 mole?

$$0.0250 \text{ mol } NiCl_2 \cdot 6\,H_2O \times \frac{238 \text{ g } NiCl_2 \cdot 6\,H_2O}{1 \text{ mol } NiCl_2 \cdot 6\,H_2O} = 5.95 \text{ g } NiCl_2 \cdot 6\,H_2O$$

7.8 NUMBER OF ATOMS, MOLECULES, OR FORMULA UNITS IN A SAMPLE

PG 7 H Given the mass or number of moles of a pure substance whose formula is known, find the number of atoms, molecules, or formula units. Or, given the number of atoms, molecules, or formula units, find the number of moles or mass.

Equation 7.1 (1 mole = 6.02×10^{23} units) gives us a simple one-step conversion between moles and units. In principle the following example is no more complicated than telling how many eggs are in four dozen.

EXAMPLE 7.11   How many sodium atoms are in 4.00 moles of sodium?

- - - - - -   - - - - - -

$$4.00 \ \cancel{mol\ Na} \ \times \ \frac{6.02 \times 10^{23} \ Na \ atoms}{1 \ \cancel{mol\ Na}} = 24.1 \times 10^{23} \ Na \ atoms = 2.41 \times 10^{24} \ Na \ atoms$$

To find the number of units of a substance in a given mass, you must begin the setup one step earlier with a grams → moles conversion.

EXAMPLE 7.12   How many molecules are in 500 grams of water (about one pint)?

Solution: Using Equation 7.4 first, and then Equation 7.1, the unit path begins with the given quantity: grams → moles → molecules.

$$500 \ \cancel{g\ H_2O} \ \times \ \frac{1 \ \cancel{mol\ H_2O}}{18.0 \ \cancel{g\ H_2O}} \times \ \frac{6.02 \times 10^{23} \ H_2O \ molecules}{1 \ \cancel{mol\ H_2O}} = 1.67 \times 10^{25} \ H_2O \ molecules$$

In the reverse problem, finding the mass of a given number of units, the procedure is logically reversed. . . .

EXAMPLE 7.13   What is the mass of one billion billion (1.00×10^{18}) molecules of ammonia, NH_3?

At one point the molar weight of ammonia will be required. Find it first.

- - - - - -   - - - - - -

17.0 g NH_3/mol

Now solve the problem.

- - - - - -   - - - - - -

$$1.00 \times 10^{18} \text{ NH}_3 \text{ molecules } \times \frac{1 \text{ mol NH}_3}{6.02 \times 10^{23} \text{ NH}_3 \text{ molecules}} \times \frac{17.0 \text{ g NH}_3}{1 \text{ mol NH}_3} = 2.82 \times 10^{-5} \text{ g NH}_3$$

This very small mass, about $\dfrac{6}{100\ 000\ 000}$ of a pound, suggests again the enormous number of molecules in a mole.

The next example shows how important it is to identify chemical substances by formula.

EXAMPLE 7.14   (a) How many atoms of chlorine are in 48.5 grams of chlorine atoms, Cl? (b) How many atoms of chlorine are in 48.5 grams of chlorine gas, Cl_2?

Part (a) is straightforward. Set up and solve.

$$48.5 \text{ g Cl } \times \frac{1 \text{ mol Cl}}{35.5 \text{ g Cl}} \times \frac{6.02 \times 10^{23} \text{ Cl atoms}}{1 \text{ mol Cl}} = 8.22 \times 10^{23} \text{ Cl atoms}$$

In Part (b), the idea is the same, but there is an important difference. You might find it helpful to write the unit path. Then set up and solve the problem.

unit path: $g\ Cl_2 \rightarrow mol\ Cl_2 \rightarrow Cl_2$ molecules $\rightarrow Cl$ atoms

$$48.5 \text{ g Cl}_2 \times \frac{1 \text{ mol Cl}_2}{71.0 \text{ g Cl}_2} \times \frac{6.02 \times 10^{23} \text{ Cl}_2 \text{ molecules}}{1 \text{ mol Cl}_2} \times \frac{2 \text{ Cl atoms}}{1 \text{ Cl}_2 \text{ molecule}} = 8.22 \times 10^{23} \text{ Cl atoms}$$

Answers (a) and (b) must be the same. 48.5 grams of chlorine must have the same number of atoms no matter how they are "packaged." Notice in (a), where you work with chlorine atoms, formula Cl, you use the mass of one mole of chlorine atoms—the atomic weight, 35.5 grams. In (b) you work with chlorine molecules, formula Cl_2. You therefore use the molar weight of chlorine molecules—the molecular weight, 71.0 grams. In each case the weight used matches the formula of the substance.

If you write the chemical formula for the species whose molar weight you are using—and calculate the molar weight straight from the chemical formula—your setups will be correct. *Always include the chemical formula in the dimensional analysis setup.*

7.9 EMPIRICAL FORMULA OF A COMPOUND

EMPIRICAL FORMULAS AND MOLECULAR FORMULAS

> **PG 7 I**   Distinguish between an empirical formula and a molecular formula.

The percentage composition of the compound ethene (also called ethylene) is 85.7% carbon and 14.3% hydrogen. Its chemical formula is C_2H_4. The percentage composition of the compound propene (also called propylene) is likewise 85.7% carbon and 14.3% hydrogen. Its formula is C_3H_6. These are, in fact, two of a whole series of compounds having the general formula C_nH_{2n}, where n is an integer. In ethene and propene, n = 2 and 3, respectively. All compounds in this series have the same percentage composition.

C_2H_4 and C_3H_6 are typical **molecular formulas.** They show the number of carbon and hydrogen atoms actually present in a molecule of ethene and propene. They are formulas of real chemical substances. If, in the general formula C_nH_{2n}, we let n be 1, the resulting formula is CH_2. This formula is the **empirical formula** for all compounds having the general formula C_nH_{2n}. The **empirical formula shows the simplest ratio of atoms of the elements in the compound.** All subscripts are reduced to their lowest terms: they have no common divisors. Empirical formulas can be determined from percentage composition data—as we will shortly see—which are found by chemical analysis. Hence the name *empirical*, which means something that is based on experience or experiment.

Empirical formulas may or may not be molecular formulas of real chemical compounds. There happens to be no known stable compound that has the formula CH_2—and there is good reason to believe that no such compound can exist. On the other hand, the molecular formula of dinitrogen tetroxide is N_2O_4. The subscript numbers have a common divisor, 2. Dividing by 2 we reach the empirical formula, NO_2. This empirical formula is also the molecular formula of a real chemical compound, nitrogen dioxide. In other words, NO_2 is *both* the empirical formula and the molecular formula for nitrogen dioxide, as well as the empirical formula for dinitrogen tetroxide.

EXAMPLE 7.15   For each formula below that could be an empirical formula, write EF; for each formula that could not be an empirical formula, write the empirical formula corresponding to the formula shown: C_4H_{10}; C_2H_6O; Hg_2Cl_2; $(CH)_6$.

_____   _____

C_4H_{10}: C_2H_5; C_2H_6O: EF; Hg_2Cl_2: HgCl; $(CH)_6$: CH

DETERMINATION OF AN EMPIRICAL FORMULA

PG 7 J Given data from which the ratio of relative masses of the elements in a compound can be determined, write the empirical formula of the compound.

To determine the empirical formula of a compound from its percentage composition we must find the whole number ratio of atoms of the elements in the compound. The numbers in this ratio become the subscripts in the empirical formula of the compound. The procedure by which this is done is as follows:

1. Determine the relative weights of different elements in the compound.

2. Convert these relative weights to the relative numbers of moles of atoms of the elements.

3. Express the number of moles of atoms as the smallest possible ratio of integers.

The integers found in the last step are the subscripts in the empirical formula.

It is sometimes helpful in an empirical formula problem to organize the calculations in a table. The headings in the table will be

ELEMENT	GRAMS	MOLES	MOLE RATIO	FORMULA RATIO	EMPIRICAL FORMULA

We will illustrate the procedure for finding empirical formulas with the compound ethene. As noted, ethene is 85.7% carbon and 14.3% hydrogen. The first step in the above procedure calls for converting the percentage figures into relative weights of the elements. Thinking of percent as the number of parts of one element per 100 parts of the compound—or number of *grams* of one element per 100 *grams* of the compound—it follows that a 100-gram sample of the unknown contains 85.7 grams of carbon and 14.3 grams of hydrogen. From this we see that *percentage composition figures may always be interpreted directly as the relative weights of different elements* in satisfying the first step of the procedure. The elemental symbol and the relative grams of each element are entered into the table:

ELEMENT	GRAMS	MOLES	MOLE RATIO	FORMULA RATIO	EMPIRICAL FORMULA
C	85.7				
H	14.3				

We are now ready to find the number of moles of atoms of each element, Step 2 in the procedure. This is a direct one-step conversion from grams to moles, as in Section 7.6.

ELEMENT	GRAMS	MOLES	MOLE RATIO	FORMULA RATIO	EMPIRICAL FORMULA
C	85.7	$\dfrac{85.7\ \text{g C}}{12.0\ \text{g/mol}} = 7.14$			
H	14.3	$\dfrac{14.3\ \text{g H}}{1.0\ \text{g/mol}} = 14.3$			

Students sometimes question the conversion of grams of hydrogen, or any elemental gas that forms diatomic molecules, to moles of atoms. They tend to divide by the molar weight of *molecules*—the *gram molecular* weight—rather than the molar weight of *atoms*—the *gram atomic* weight. But a chemical formula expresses the ratio of moles of *atoms* of the different elements, so the molar weight of atoms, or atomic weight, must be used.

It is the ratio of these moles of atoms that must now be expressed in the smallest possible ratio of integers, Step 3 in the procedure. This is most easily done by *dividing each number of moles by the smallest number of moles*. In this problem the smallest number of moles is 7.14. Thus,

ELEMENT	GRAMS	MOLES	MOLE RATIO	FORMULA RATIO	EMPIRICAL FORMULA
C	85.7	7.14	$\dfrac{7.14}{7.14} = 1.00$		
H	14.3	14.2	$\dfrac{14.2}{7.14} = 1.99$		

The ratio of *atoms* of the elements in a compound is the same as the ratio of *moles* of atoms in the compound. To see this in a more familiar setting, the ratio of handlebars to wheels in bicycles is $\frac{1}{2}$. In four dozen bicycles there are four dozen handlebars and eight dozen wheels. The ratio of dozens is $\frac{4}{8}$, which is the same as $\frac{1}{2}$. Thus the numbers in the Mole Ratio column are in the same ratio as the subscripts in the empirical formula.

The empirical formula subscripts must be whole numbers. The numbers in the Mole Ratio column must therefore be expressed as a ratio of integers. In this case the ratio 1.00/1.99 becomes $\frac{1}{2}$, and the empirical formula is CH_2.

ELEMENT	GRAMS	MOLES	MOLE RATIO	FORMULA RATIO	EMPIRICAL FORMULA
C	85.7	7.14	1.00	1	CH_2
H	14.2	14.2	1.99	2	

It is often necessary to adjust a number in the Mole Ratio column by a few hundredths to get an integer, as 1.99 is adjusted to 2 in this example. These minor changes correct for experimental errors or significant figure roundoffs.

If either quotient in the Mole Ratio column is not very close to a whole number, the Formula Ratio may be found by multiplying both quotients by a small integer. This is shown in the example that follows.

EXAMPLE 7.16 A piece of iron (Z = 26) weighs 1.34 grams. Exposed to oxygen under conditions in which oxygen combines with all of the iron to form a pure oxide of iron, the final weight increases to 1.92 grams. Find the empirical formula of the compound.

As before, relative weights of the elements contained in the compound are required—but this time they are not obtained from percentage composition values. The number of grams of iron in the final compound is given in the problem. How many grams of oxygen combine with 1.34 grams of iron if the iron oxide produced weighs 1.92 grams?

grams oxygen = grams iron oxide − grams iron

= 1.92 grams   − 1.34 grams = 0.58 gram

The table is started below, and the symbols and masses of elements are entered. Step 2 is to compute the number of moles of atoms of each element. Do so, and put the results into the table.

ELEMENT	GRAMS	MOLES	MOLE RATIO	FORMULA RATIO	EMPIRICAL FORMULA
Fe	1.34				
O	0.58				

ELEMENT	GRAMS	MOLES	MOLE RATIO	FORMULA RATIO	EMPIRICAL FORMULA
Fe	1.34	$\dfrac{1.34 \text{ g Fe}}{55.8 \text{ g/mol}} = 0.0240$			
O	0.58	$\dfrac{0.58 \text{ g O}}{16.0 \text{ g/mol}} = 0.036$			

Recalling that the mole ratio figures are obtained by dividing each number of moles by the smallest, find those numbers and place them in the table.

ELEMENT	GRAMS	MOLES	MOLE RATIO	FORMULA RATIO	EMPIRICAL FORMULA
Fe	1.34	0.0240	$\dfrac{0.0240}{0.0240} = 1.00$		
O	0.58	0.036	$\dfrac{0.036}{0.0240} = 1.5$		

This time the numbers in the mole ratio column are not both integers or very close to integers. But they can be changed to integers and kept in the same ratio by multiplying both of them by the same small integer. Find the smallest whole number that will yield integers when used as a multiplier for 1.00 and 1.5, use it to obtain the formula ratio figures, complete the table, and write the empirical formula of the compound.

Element	Grams	Moles	Mole Ratio	Formula Ratio	Empirical Formula
Fe	1.34	0.0240	1.00	2	
O	0.58	0.036	1.5	3	Fe_2O_3

Multiplication of the mole ratio numbers by 2 yields $1.00 \times 2 = 2$ and $1.5 \times 2 = 3$, both whole numbers.

The decimal part of a mole ratio determines what multiplier to use. If the mole ratio is about 1.33 or 1.67, multiplication by 3 yields 4 or 5:

$$1.33 \times 3 = 4 \quad or \quad 1.67 \times 3 = 5$$

A mole ratio of 1.25 or 1.75 should be multiplied by 4 to get 5 or 7:

$$1.25 \times 4 = 5 \quad or \quad 1.75 \times 4 = 7$$

You are not likely to find more complicated ratios in the beginning course.

The procedure is the same for compounds containing more than two elements.

EXAMPLE 7.17   An organic compound is found to contain 20.0% carbon, 2.2% hydrogen, and 77.8% chlorine. Determine the empirical formula of the compound.

Element	Grams	Moles	Mole Ratio	Formula Ratio	Empirical Formula

Element	Grams	Moles	Mole Ratio	Formula Ratio	Empirical Formula
C	20.0	1.67	1.00	3	
H	2.2	2.2	1.3	4	$C_3H_4Cl_4$
Cl	77.8	2.19	1.31	4	

Notice that the experimental results in this problem yield mole ratio figures 1.3 and 1.31, both close to .33 in the decimal part of the number. Multiplication of the mole ratio figures by 3 yields integers for the formula ratio column: $1.00 \times 3 = 3$; $1.3 \times 3 = 3.9$ or 4; $1.31 \times 3 = 3.93$ or 4.

DETERMINATION OF A MOLECULAR FORMULA

PG 7 K Given the molar weight of a compound and data from which the empirical formula can be determined, write the molecular formula of the compound.

The determination of an empirical formula is an important step in finding the molecular formula of an unknown compound if the molar weight of the unknown can be established. For example, suppose we know the empirical formula of a compound is CH_2, as in ethene. Suppose also we can determine that the molar weight of the compound is 70.0 grams per mole. What value of n in C_nH_{2n}—or $(CH_2)_n$, which is a somewhat more convenient form for this purpose—would yield a formula with a molar weight of 70.0? The molar weight of the empirical formula, CH_2, is 14 grams per mole. Dividing 70 by 14 gives five empirical formula units in one mole of unknown:

$$\frac{70.0 \text{ g } (CH_2)_n}{1 \text{ mol } (CH_2)_n} \times \frac{1 \text{ mol empirical formula units}}{14.0 \text{ g } (CH_2)_n}$$

$$= 5 \text{ mol empirical formula units/mol } (CH_2)_n$$

Therefore $(CH_2)_5$ or, more correctly, C_5H_{10}, is the molecular formula for the compound with molar weight 70.0 g/mol and empirical formula CH_2. The compound is called pentene.

EXAMPLE 7.18 An unknown compound is found in the laboratory to be composed of 91.8% silicon and 8.2% hydrogen. Another experiment indicates the molar weight of the compound to be 122 g/mol. Find the empirical formula and the molecular formula.

Start by finding the empirical formula.

ELEMENT	GRAMS	MOLES	MOLE RATIO	FORMULA RATIO	EMPIRICAL FORMULA

-------   -------

ELEMENT	GRAMS	MOLES	MOLE RATIO	FORMULA RATIO	EMPIRICAL FORMULA
Si	91.8	3.27	1.00	2	
					Si_2H_5
H	8.2	8.2	2.5	5	

You now know that the molecular formula can be expressed as $(Si_2H_5)_n$, where n is the number of empirical formula units in the molecule. The molar weight of 122 g/mol is based on the molecular formula. If you can find the molar weight of the empirical formula unit, you can figure out how many times that quantity fits into 122. Calculate the molar weight of the empirical formula, Si_2H_5.

-------   -------

61.2 g Si_2H_5/mol

The mass of one mole of molecules is 122 grams. The mass of one mole of empirical formula units is 61.2 grams. How many moles of empirical formula units are in one mole of molecules? Calculate that number and write the molecular formula.

-------   -------

$$\frac{122 \text{ g } (Si_2H_5)_n}{1 \text{ mol } (Si_2H_5)_n} \times \frac{1 \text{ mol empirical formula units}}{61.2 \text{ g } (Si_2H_5)_n} = 2 \text{ mol empirical formula units/mol } (Si_2H_5)_n$$

molecular formula = 2 empirical formula units = $(Si_2H_5)_2 = Si_4H_{10}$

In Chapter 13 you will learn one way to find the molar weight of an unknown compound from experimental data. You will then have a complete picture of how a chemical formula may be determined.

CHAPTER 7 IN REVIEW

7 A Distinguish among atomic weight, molecular weight, and formula weight. (130)

7.3 THE MOLE CONCEPT

7 B Define *mole*. Identify and state the number that corresponds to one mole. (131)

7.4 MOLAR WEIGHT

7 C Define molar weight. (132)

7.5 CALCULATION OF MOLAR WEIGHT

7 D Calculate the molar weight of any substance whose chemical formula is known. (133)

7.6 MASS RELATIONSHIP BETWEEN ELEMENTS IN A COMPOUND; PERCENTAGE COMPOSITION

7 E For any compound whose formula is known, given the mass of a sample, calculate the mass of any element in the sample. Or, given the mass of any element in the sample, calculate the mass of the sample or the mass of any other element in the sample. (136)

7 F Calculate the percentage composition of any compound whose formula is known. (136)

7.7 CONVERSION BETWEEN MASS AND NUMBER OF MOLES

7 G Given the number of grams (or moles) of a chemical species of known or calculable molar weight, find the number of moles (or grams). (138)

7.8 NUMBER OF ATOMS, MOLECULES OR FORMULA UNITS IN A SAMPLE

7 H Given the mass or number of moles of a pure substance whose formula is known, find the number of atoms, molecules, or formula units. Or, given the number of atoms, molecules, or formula units, find the number or moles or mass. (139)

7.9 EMPIRICAL FORMULA OF A COMPOUND

7 I Distinguish between an empirical formula and a molecular formula. (142)

7 J Given data from which the ratio of relative masses of elements in a compound can be determined, write the empirical formula of the compound. (143)

7 K Given the molar weight of a compound and data from which the empirical formula can be determined, write the molecular formula of the compound. (147)

TERMS AND CONCEPTS

Significant figures (129)   Mole (131)
Doubtful digit (129)   Avogadro's number (131)
Uncertainty (129)   Molar weight (132)
Molecular weight (130)   Percentage composition (137)
Formula weight (130)   Molecular formula (142)
Formula unit (130)   Empirical formula (142)

Most of these terms and many more are defined in the Glossary.

QUESTIONS AND PROBLEMS

In the following questions, only names are given for those compounds whose formulas were introduced in Chapter 6. For reference, you may find all of these formulas on page 598. You are encouraged to write formulas with no assistance other than a periodic table, using the reference page only when absolutely necessary. This is the way to build and improve your formula writing skills.

Section 7.2

7.1) Why is it incorrect to refer to the atomic weight of sodium chloride?

7.2) It may be said that because atomic, molecular, and formula weights are all comparative weights, they are conceptually alike. What, then, is their difference?

7.3) State which of the three terms, *atomic weight, molecular weight* or *formula weight*, is most appropriate for each of the following substances: Mg; Br_2; Al_2O_3; CO_2; NH_4Cl.

7.4) In what units are atomic, molecular, and formula weights expressed? Define that unit.

7.51) Why is it proper to speak of the molecular weight of water, but not the molecular weight of potassium nitrate?

7.52) Other than sodium chloride, potassium nitrate, and water, list two substances to which *atomic weight* applies, two that have a *molecular weight*, and two that have a *formula weight*.

7.53) State which of the three terms, *atomic weight, molecular weight,* or *formula weight*, is most appropriate for each of the following substances: NH_3; CaO; Ba; N_2; Na_2CO_3.

7.54)* One of the three terms, *atomic weight, molecular weight,* or *formula weight*, may logically be applied to any of the substances in Questions 7.3 and 7.53. Which term is it, and why?

Section 7.3

7.5) Explain the meaning of the term *mole*. Why is it used in chemistry?

7.6) Give the name and value of the number associated with the mole.

7.55) What do quantities representing one mole of sodium, one mole of water, and one mole of sodium chloride have in common?

7.56) Is the mole a number? Explain.

7.7) How many grams of sulfur have the same number of atoms as 40.4 grams of neon ($Z = 10$)? Explain how you got your answer.

7.57)* Give a definition of Avogadro's number, N. Explain why the size of that number has changed over the years.

Section 7.4

7.8) Why is the expression "molar weight of hydrogen" uncertain in its meaning? How may that uncertainty be removed?

7.58)* Explain how *molar weight* is related to the older and more specific terms, *gram atomic weight*, *gram molecular weight*, and *gram formula weight*.

Section 7.5

For Problems 7.9 to 7.14 and 7.59 to 7.64, calculate the molar weight of each substance listed.

7.9) Lithium bromide.

7.59) Sodium chloride.

7.10) Chlorine gas.

7.60) Fluorine gas.

7.11) Calcium sulfate.

7.61) Potassium nitrate.

7.12) Ammonium nitrate.

7.62) Ammonium phosphate.

7.13) Benzene, C_6H_6.

7.63) Ethanol, C_2H_5OH.

7.14) Sodium carbonate decahydrate.

7.64) Cobalt sulfate hexahydrate, $CoSO_4 \cdot 6\,H_2O$.

Section 7.6

7.15) The chromium deposited in chromium plating comes from chromium trioxide, CrO_3. How many grams of this compound will be used in a laboratory setup that plates 65.4 grams of chromium?

7.65) When copper sulfide, CuS, is treated with nitric acid, elemental sulfur is produced. How many grams of sulfur will be obtained from 396 grams of CuS?

7.16) Metallic sodium and chlorine gas are produced by running an electric current through liquid sodium chloride. If 426 grams of sodium are collected, how many grams of chlorine will also be produced?

7.66) Carbon reacts with sulfur to form carbon disulfide, CS_2. If a sample of the product contains 71.9 grams of carbon, how many grams of sulfur are present?

7.17) Pyrite, FeS_2, is an iron ore that is called *fool's gold* because it looks like gold. How many grams of FeS_2 must be processed to produce 1.43×10^6 grams of iron?

7.67) Zinc carbonate, $ZnCO_3$, is an ore of zinc. Calculate the mass of zinc that can be recovered from 3.82×10^7 grams of zinc carbonate.

7.18)* A ton of earth is 1.85% galena, PbS, a commercial ore of lead. How many grams of lead can be taken from one ton (9.08×10^5 grams) of this earth?

7.68)* How many grams of dirt that is 0.89% braunite, Mn_2O_3, a manganese ore, must be processed to obtain 6.85×10^4 grams of manganese?

For Problems 7.19 to 7.21 and 7.69 to 7.71, calculate the percentage composition of each of the following compounds, selected from Problems 7.9 to 7.14 and 7.59 to 7.64.

7.19) Lithium bromide.

7.69) Sodium chloride.

7.20) Calcium sulfate.

7.70) Potassium nitrate.

7.21) Benzene, C_6H_6.

7.71) Ethanol, C_2H_5OH.

7.22) Calculate the percentage of water in sodium carbonate decahydrate (see Problem 7.14).

7.72) Find the percentage of anhydrous cobalt sulfate in $CoSO_4 \cdot 6\,H_2O$ (see Problem 7.64).

Section 7.7

For Problems 7.23 to 7.25 and 7.73 to 7.75, calculate the number of moles for each mass of substance selected from Problems 7.9 to 7.14 and 7.59 to 7.64.

7.23) 68.4 g LiBr.

7.24) 17.2 g Cl_2.

7.25) 34.1 g $CaSO_4$

7.73) 19.7 g $(NH_4)_3PO_4$.

7.74) 28.9 g C_2H_5OH.

7.75) 108 g $CoSO_4 \cdot 6 H_2O$.

For Problems 7.26 to 7.28 and 7.76 to 7.78, calculate the mass for each given number of moles of substances selected from Problems 7.9 to 7.14 and 7.59 to 7.64.

7.26) 0.345 mol NH_4NO_3.

7.27) 1.82 mol C_6H_6.

7.28) 0.791 mol $Na_2CO_3 \cdot 10 H_2O$.

7.76) 1.40 mol NaCl.

7.77) 2.19 mol F_2.

7.78) 0.108 mol $(NH_4)_3PO_4$.

Section 7.8

For Problems 7.29 to 7.33 and 7.79 to 7.83, determine how many atoms are in each of the following:

7.29) 1.24 mol Mg.

7.30) 0.713 mol Br_2.

7.31) 29.6 g Na.

7.32) 3.40 g Ca.

7.33) 0.004 61 g I_2.

7.79) 0.436 mol P.

7.80) 1.41×10^{-5} mol Rb (Z = 37).

7.81) 41.2 g Al.

7.82) 0.000 162 g K.

7.83) 38.1 g N_2.

For Problems 7.34 to 7.38 and 7.84 to 7.88, determine how many molecules or formula units are in each of the following:

7.34) 0.521 mol H_2S.

7.35) 0.0626 mol Br_2.

7.36) 12.4 g N_2.

7.37) 6.45 g CO.

7.38) 13.6 g LiBr.

7.84) 0.003 31 mol Ne.

7.85) 6.53 mol N_2O_5.

7.86) 0.106 g Cl_2.

7.87) 40.2 g CO_2.

7.88) 30.9 g NaCl.

For Problems 7.39, 7.40, 7.89, and 7.90, calculate the number of moles in each of the following:

7.39) 2.35×10^{21} Ba atoms.

7.40) 1.09×10^{23} Br_2 molecules.

7.89) 1.84×10^{24} Pb atoms.

7.90) 6.91×10^{20} formula units of $CuSO_4$.

For Problems 7.41 to 7.43 and 7.91 to 7.93, calculate the mass of:

7.41) 7.06×10^{23} He atoms.

7.42) 4.06×10^{22} CO_2 molecules.

7.43) 1.19×10^{20} formula units of KI.

7.91) 1.07×10^{23} Mg atoms.

7.92) 8.19×10^{24} CO molecules.

7.93) 7.23×10^{23} formula units of NaF.

Section 7.9

7.44) Explain why C_6H_{10} must be a molecular formula, while C_7H_{10} could be either a molecular formula or an empirical formula.

7.94) From the following list, identify each formula that could be an empirical formula. Write the empirical formulas of the other compounds. C_2H_5OH; Na_2O_2; CH_3COOH; N_2O_5.

7.45) Analysis of an organic compound shows it to be 40.0% carbon, 6.7% hydrogen, and 53.3% oxygen. Calculate the empirical formula of the compound.

7.46) 28.8 grams of a certain hydrocarbon are found to contain 25.3 grams of carbon and the balance hydrogen. Find the empirical formula of the compound.

7.47) A compound is found to contain 32.4% sodium, 22.6% sulfur, and 45.0% oxygen. Find the empirical formula of the compound.

7.48) The white rock dolomite, used in gardens, is 13.2% magnesium, 21.8% calcium, 13.0% carbon, and 52.1% oxygen. Find the empirical formula of dolomite.

7.49) A compound is, by analysis, 5.00% hydrogen and 95.00% fluorine. Its molar weight is 40.0 grams per mole. Find both the empirical and molecular formulas of the compound.

7.50) The mass of phosphorus is 10.3 grams in an 18.3-gram sample of a compound containing only phosphorus and oxygen. The molar weight of the compound is 220 g/mol. Write the empirical and molecular formulas of the compound.

7.95) The percentage composition of an organic compound is 48.6% carbon, 8.1% hydrogen, and 43.2% oxygen. Determine the empirical formula of the compound.

7.96) Analysis of a sample of a hydrocarbon shows it consists of 17.9 grams of carbon and 4.5 grams of hydrogen. Determine the empirical formula of the compound.

7.97) The percentage composition of a compound is 26.1% nitrogen, 7.5% hydrogen, and 66.4% chlorine. Write the empirical formula of the compound.

7.98) When 7.30 grams of iron powder react and combine with 6.30 grams of powdered sulfur, a sulfide of iron is formed. Determine the empirical formula of the compound.

7.99) A 3.80-gram sample of a pure compound of sulfur and fluorine is found by analysis to contain 1.37 grams of sulfur. Another experiment establishes that the molar weight of the compound is 178 grams/mole. Find the empirical and molecular formulas of this fluoride of sulfur.

7.100) The molar weight of a compound is 549 grams per mole. If the compound is 73.0% mercury, 8.74% carbon, 0.729% hydrogen, and 17.5% oxygen, find the empirical and molecular formulas of the compound.

8

chemical reactions and equations

In Chapter 8 we continue to encourage the development of your formula writing skills by furnishing only the names of compounds whose formulas were introduced in Chapter 6. If you need assistance you may find all these formulas on pages 615 and 616.

8.1 EVOLUTION OF A CHEMICAL EQUATION

Sodium reacts vigorously with water, producing hydrogen gas and a solution of sodium hydroxide, and releasing heat (Fig. 8.1). In that sentence 17 words were used to describe a chemical reaction. With many reactions to consider, the chemist needs a shorter description. A **chemical equation** fulfills that need. You will study chemical equations and learn how to write them in this chapter.

The minimum "word equation" for the reaction referred to above is "sodium plus water yields hydrogen plus sodium hydroxide solution plus heat." The corresponding chemical equation reads exactly the same way:

$$\text{Na} + \text{H}_2\text{O} \rightarrow \text{H}_2 + \text{NaOH(aq)} + \text{heat} \tag{8.1}$$

Nearly all chemical reactions involve some transfer of energy, usually in the form of heat, as shown here. Energy terms are generally omitted from equations unless there is a specific reason for including them. Equations that include the gain or loss of heat will be considered in Chapter 19.

Figure 8.1
Sodium reacting with water. A, Small piece of sodium dropped into beaker of water. B, Sodium forms "ball" that dashes erratically over water surface, releasing hydrogen as it reacts. C, Solution of sodium hydroxide, NaOH (aq), which is hot because of heat released in the reaction. WARNING: Do not "try" this experiment, as it is dangerous, potentially splattering hot alkali into eyes and onto skin and clothing.

The word equation states that a product of the reaction is "sodium hydroxide solution." The (aq) following NaOH indicates that the sodium hydroxide is in *aqueous* (water) solution. There are many times when the state or form of a species in a reaction is important. It may be shown by an appropriate "state" symbol in parentheses, following the formula of the species. Although they are usually omitted, we will include states in all chemical equations in this book. We will call them to your attention when they are important. In addition to the aqueous solution form of a reaction species, the three normal states of matter are solid (s), liquid (ℓ), and gas (g). In this reaction, all four states appear:

$$Na(s) + H_2O(\ell) \rightarrow H_2(g) + NaOH(aq) \qquad (8.2)$$

Sodium at room conditions is a solid, and water is a liquid. Water vapor also exists at room temperature, and sometimes the distinction between liquid and gaseous water is important. Hydrogen is a gas.

Two quantitative interpretations of a chemical equation will be described in the next section, but one must be introduced here. From a "molecular" point of view, the reaction might be considered as sodium atoms reacting with water molecules to produce molecules of hydrogen and "formula units" of sodium hydroxide in aqueous solution. But the equation does not say *how many* of each species are involved. It cannot be one of each. This violates the Law of Conservation of Mass, as well as that part of Dalton's atomic theory which states that atoms are neither created nor destroyed in chemical reactions. Count up the atoms of each element and you will see that this is the case. There is a single sodium atom on each side of the equation. There is also one oxygen atom. But the left side of the equation has two hydrogen atoms, whereas the right side has three.

How do we account for this seeming discrepancy? The answer is that the chemical species do not take part in the reaction in a simple $1:1:1:1$ ratio. The problem can be resolved by taking one or more of the chemicals in multiple— done by adjusting coefficients of the four species—until the atoms of each element are "balanced" on the two sides of the equation. It is the hydrogen that is unbalanced; the left side is short by one. Therefore let's try two water molecules:

$$Na(s) + 2\,H_2O(\ell) \rightarrow H_2(g) + NaOH(aq) \tag{8.3}$$

At first glance, this hasn't helped; indeed it seems to have made matters worse. The hydrogen is still out of balance (four on the left, three on the right) and, furthermore, oxygen is now unbalanced (two on the left, one on the right). We are short one oxygen and one hydrogen atom on the right side. But look closely. Oxygen and hydrogen are part of the same compound on the right, and there is one atom of each in that compound. If we take two NaOH units

$$Na(s) + 2\,H_2O(\ell) \rightarrow H_2(g) + 2\,NaOH(aq) \tag{8.4}$$

there are four hydrogens and two oxygens on both sides of the equation. These elements are now in balance. But, alas, the sodium has been *unbalanced*. Correction of this condition, however, should be obvious:

$$2\,Na(s) + 2\,H_2O(\ell) \rightarrow H_2(g) + 2\,NaOH(aq) \tag{8.5}$$

The equation is now balanced; it satisfies the Law of Conservation of Mass, reflected in Dalton's statement that atoms are neither created nor destroyed in chemical reactions. Note that, in the absence of a numerical coefficient, as with H_2, the coefficient is assumed to be 1.

Balancing an equation involves some important do's and don'ts that are apparent in this example:

DO: *Balance the equation entirely by use of coefficients placed before the different chemical formulas.*
DON'T: *Change a correct chemical formula in order to make an element balance.*
DON'T: *Add some real or imaginary chemical species to either side of the equation just to make an element balance.*

There is a great temptation toward either of the "don'ts." A moment's thought should show why they are improper. The original equation expresses the *correct* formula for each species present. If you change or add a formula, even if the new formula is for a real chemical, it is *not a species in the reaction* and does not belong in the equation. A chemical equation must correspond to reality. It must describe as accurately as possible what actually happens when we carry out a reaction. Coefficients alone must be used in balancing equations.

8.2 MEANING OF A CHEMICAL EQUATION

PG 8 A Given a chemical equation, interpret it in terms of (a) atoms, molecules, and/or formula units and (b) moles.

The "molecular" meaning of a chemical equation guided the balancing procedure used in Section 8.1. A complete interpretation of the balanced equation is as follows: two atoms of sodium react with two molecules of water to produce one molecule of hydrogen and two formula units of sodium hydroxide (i.e., 2 Na^+ ions and 2 OH^- ions). But it is equally true that *four* atoms of sodium react with *four* molecules of water to produce *two* molecules of hydrogen and *four* formula units of sodium hydroxide:

$$4 \, Na(s) + 4 \, H_2O(\ell) \rightarrow 2 \, H_2(g) + 4 \, NaOH(aq) \qquad (8.6)$$

Equation 8.6, which is simply Equation 8.5 multiplied by 2, conforms to the Law of Conservation of Mass and to the atomic theory. As in algebra, both sides of an equation may be multiplied by any number.

Using other multipliers on Equation 8.5 we find that

$$2(12) \, Na(s) + 2(12) \, H_2O(\ell) \rightarrow 1(12) \, H_2(g) + 2(12) \, NaOH(aq)$$

This could be read or written correctly as

$$2 \text{ dozen } Na(s) + 2 \text{ dozen } H_2O(\ell) \rightarrow 1 \text{ dozen } H_2(g) + 2 \text{ dozen } NaOH(aq)$$

Similarly,

$$2(6.02 \times 10^{23}) \, Na(s) + 2(6.02 \times 10^{23}) \, H_2O(\ell) \rightarrow$$
$$1(6.02 \times 10^{23}) \, H_2(g) + 2(6.02 \times 10^{23}) \, NaOH(aq)$$

may be thought of as

$$2 \text{ moles } Na(s) + 2 \text{ moles } H_2O(\ell) \rightarrow 1 \text{ mole } H_2(g) + 2 \text{ moles } NaOH(aq)$$

We may refer to this last equation as a "molar" interpretation of a chemical equation. With the understanding that the coefficients refer to *moles* rather than atoms, molecules, or formula units, the interpretation of the equation may be applied directly to Equation 8.5.

The molar interpretation provides another way to balance chemical equations. Returning to the unbalanced equation,

$$Na(s) + H_2O(\ell) \rightarrow H_2(g) + NaOH(aq) \qquad (8.2)$$

we find that the left side of the equation has two moles of hydrogen atoms, whereas the right side has three moles of hydrogen atoms. A single coefficient, ½, applied to hydrogen balances the equation:

$$Na(s) + H_2O(\ell) \rightarrow \tfrac{1}{2} H_2(g) + NaOH(aq) \qquad (8.7)$$

This says that one mole of sodium reacts with one mole of water to form half a mole of hydrogen plus one mole of sodium hydroxide. There are now two moles of hydrogen atoms on each side of the equation, which is balanced in other atoms as well. From a molecular standpoint you cannot properly think of half of a hydrogen molecule or any fractional part of any atom or molecule any more than you can think of half an egg—but half a *mole* of molecules is fine, just as acceptable as half a dozen eggs.

In chemistry, as in algebra, equations are *usually* written with the lowest set of whole number (integer) coefficients possible, and you should conform to this practice. There are occasions, however, when fractional coefficients or coefficients all divisible by the same integer are required.

8.3 WRITING CHEMICAL EQUATIONS

The procedure for writing chemical equations may be expressed in two steps:

1. Write the correct chemical formula for each reactant on the left and each product on the right.
2. *Using coefficients only* (do NOT change a formula, or add another), balance the number of atoms of each element on each side of the equation.

There are numerous "techniques" that may be used in writing equations, and certain types of reactions are readily recognized. Both techniques and types will be introduced through a series of examples.

8.4 COMBINATION REACTIONS

PG 8 B Given the identity of a compound that is formed from two or more simpler substances, write the equation for the reaction.

Reactions in which two or more substances, often elements, combine to form a compound are called **combination**, or **synthesis**, **reactions**. A general equation for a combination reaction is

$$A + B \rightarrow C \qquad (8.8)$$

Figure 8.2
Charcoal (carbon) burning in air. "Burning in air" is interpreted chemically as "reacting or combining with oxygen." Carbon combines with oxygen to produce carbon dioxide, CO_2.

EXAMPLE 8.1   When charcoal (carbon) is burned completely in air (Fig 8.2), carbon dioxide is formed. Write the equation for the reaction.

Step 1 requires the correct formulas for each reactant and product. The formula of carbon, as an element, is simply C. Identifying the second reactant requires knowledge of what happens chemically in the process of "burning in air." Oxygen in air is capable of combining chemically with many things, sometimes slowly and unnoticed, sometimes rapidly, giving off heat and light. When a substance combines with oxygen, we say **oxidation** has occurred; the substance has been **oxidized.*** If oxidation is accompanied by heat and light, we call the process burning.

Carbon and oxygen, then, are the two reactants in this burning equation. The single product of the reaction is carbon dioxide, CO_2. With this information you should be able to complete the first step by writing the unbalanced equation. (As noted earlier, we will include state symbols in writing equations. You may omit them, unless your instructor requires them.)

------   ------

$$C(s) + \quad O_2(g) \rightarrow \quad CO_2(g)$$

The second step is to balance the number of atoms of each element *by use of coefficients only.* Complete the equation.

------   ------

$$C(s) + O_2(g) \rightarrow CO_2(g)$$

Sometimes balancing is easy, as when all the coefficients are 1!

---

*In Chapter 18 you will learn a broader and more useful meaning for oxidation and oxidized.

Limiting the quantity of air (oxygen) in the burning of carbon brings about a different result. . . .

EXAMPLE 8.2 When carbon is burned in a limited quantity of air, poisonous carbon monoxide, $CO(g)$, is produced. Write the equation.

First, write the unbalanced equation.

$$C(s) + O_2(g) \rightarrow CO(g) \quad \text{(unbalanced)}$$

Now count up the atoms of each element on each side of the equation and balance by inspection, using coefficients only.

$$2\,C(s) + O_2(g) \rightarrow 2\,CO(g) \quad or \quad C(s) + \tfrac{1}{2}\,O_2(g) \rightarrow CO(g)$$

As noted previously, unless there is some specific reason for doing otherwise, the smallest whole number coefficients are usually used. The first equation is preferred.

If, as in this example, you have an equation involving fractional coefficients and wish to convert them to integers, simply multiply by the lowest common denominator. Thus, for $C(s) + \tfrac{1}{2}\,O_2(g) \rightarrow CO(g)$, multiplication by two yields $2 \times \tfrac{1}{2} = 1$ for the coefficient of oxygen, and $2 \times 1 = 2$ for the coefficients of carbon and carbon monoxide.

EXAMPLE 8.3 Write the equation for the formation of sodium chloride by direct combination of its elements.

$$2\,Na(s) + Cl_2(g) \rightarrow 2\,NaCl(s)$$

You did recall, did you not, that chlorine is one of those gaseous elements that form diatomic molecules?

8.5 DECOMPOSITION REACTIONS

PG 8 C Given the identity of a compound that is decomposed into simpler substances, either compounds or elements, write the equation for the reaction.

Decomposition reactions are the opposite of combination reactions, in that chemical compounds break down into simpler substances, usually elements. A general decomposition equation is

$$D \rightarrow E + F \qquad (8.9)$$

EXAMPLE 8.4 When water is electrolyzed it decomposes into its elements (Fig. 2.4, p. 17). Write the equation.

"Electrolyzing" a substance involves running an electric current through it, a process used in many chemical reactions. You know the formulas for water and the elements in it, so you should be able to produce the unbalanced equation readily. Balancing is straightforward. Go all the way.

$$2\,H_2O(\ell) \rightarrow 2\,H_2(g) + O_2(g)$$

Decomposition reactions do not always go all the way to the elements, as the following example illustrates.

EXAMPLE 8.5 Heating potassium chlorate, $KClO_3(s)$, releases oxygen, leaving solid potassium chloride (Fig. 8.3). Write the equation.

From the description of the reaction you must identify reactants and products. Write their formulas in the unbalanced equation.

$$KClO_3(s) \rightarrow \quad KCl(s) + \quad O_2(g) \quad \text{(unbalanced)}$$

Inspection balancing is a bit more complicated this time. A quick glance shows the potassium and chlorine already balanced; only the oxygen remains. Oxygen comes three to a "package" in $KClO_3$ on the left, and two to a "package" in O_2 on the right. What is the smallest number of packages of three that can be repackaged two at a time and have none left over? How many packages of two will there be? It's like tennis balls, packed three in a can. What is the smallest number of cans that must be opened to furnish two balls to each of what number of courts? With that hint, see if you can balance the oxygen. Disregard potassium and chlorine.

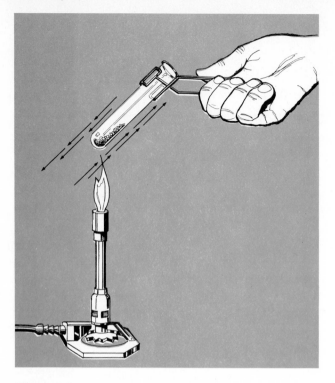

Figure 8.3
*Decomposition of potassium chlorate, KClO₃, into potassium
chloride, KCl, and oxygen in the presence of a catalyst.*

$$2\ KClO_3(s) \rightarrow \quad KCl(s) + 3\ O_2(g) \quad \text{(unbalanced)}$$

Two potassium chlorates, each with 3 oxygens, give six atoms of oxygen; three oxygen molecules, with atoms two at a time, also give six atoms of oxygen. The oxygen is in balance.

Completing the equation follows.

$$2\ KClO_3(s) \rightarrow 2\ KCl(s) + 3\ O_2(g)$$

Now use a slightly different approach. Going back to the unbalanced equation,

$$KClO_3(s) \rightarrow \quad KCl(s) + \quad O_2(g) \quad \text{(unbalanced)}$$

and noting again that oxygen is the only unbalanced element, observe also that oxygen is in elemental form on the right side of the equation. Any time all elements are in balance except one that is in elemental form, the only necessary step is to apply the proper coefficient to that element to complete the balancing. With three oxygen atoms on the left, how many oxygen *molecules* must be used to obtain three oxygen *atoms* on the right?

$1\frac{1}{2}$, or $\frac{3}{2}$

The improper fraction form, $\frac{3}{2}$, is more useful than the mixed form, $1\frac{1}{2}$, in seeing the implied multiplication, $\frac{3}{2} \times 2 = 3$:

$$\frac{3}{2} \text{ molecules} \times \frac{2 \text{ atoms}}{1 \text{ molecule}} = 3 \text{ atoms}$$

Insert the fractional coefficient in the proper space to produce a balanced equation.

$$KClO_3(s) \rightarrow KCl(s) + \tfrac{3}{2} O_2(g)$$

Now multiplication of the entire equation by 2 will reproduce the balanced equation with whole number coefficients:

$$2 \, KClO_3(s) \rightarrow 2 \, KCl(s) + 3 \, O_2(g)$$

EXAMPLE 8.6   Lime, CaO(s), and carbon dioxide gas, CO_2, are the products of the thermal decomposition of limestone, $CaCO_3(s)$ (Fig. 8.4). Write the equation.

$$CaCO_3(s) \rightarrow CaO(s) + CO_2(g)$$

8.6 COMPLETE OXIDATION OR BURNING OF ORGANIC COMPOUNDS

PG 8 D Write the equation for the complete oxidation or burning of any compound consisting only of carbon, hydrogen, and possibly oxygen.

A large number of organic compounds consist of two or three elements: carbon and hydrogen, or carbon, hydrogen, and oxygen. When such compounds are burned in an excess of air, or otherwise oxidized completely, the end products are always the same: carbon dioxide, $CO_2(g)$, and steam, $H_2O(g)$.* When the steam condenses it becomes the "white smoke" so commonly seen rising from chimneys. In writing the following equations, recall the chemical interpretation of "burning in air"—reacting with oxygen. Remember also that oxygen is

*In some cases you may wish to consider *liquid* water, $H_2O(\ell)$, as the product. We will use either in the coming pages, while recognizing that the distinction is important when writing thermochemical equations (p. 457).

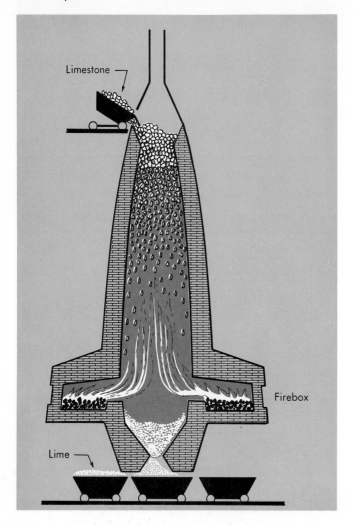

Figure 8.4
*Lime kiln in which limestone, cal-
cium carbonate, CaCO₃, is decomposed into lime, CaO, and carbon
dioxide, CO₂, by heating.*

diatomic; its formula is O_2. The general equation for a complete oxidation
reaction is

$$C_xH_yO_z + O_2 \rightarrow CO_2 + H_2O \qquad (8.10)$$

plus the coefficients required for balancing.

EXAMPLE 8.7   Write the equation for the complete burning of the liquid hydrocarbon
pentane, C_5H_{12}, in air.

Reactants and products are both known, so the unbalanced equation should be
written readily.

$C_5H_{12}(\ell) +\quad O_2(g) \rightarrow\quad CO_2(g) +\quad H_2O(g)$ (unbalanced)

Quite often in balancing equations it is advisable to balance first the elements other than hydrogen and oxygen, then balance hydrogen, and finally oxygen. This is particularly true when elemental oxygen is in the equation, as it is here. Working in the order carbon, hydrogen, oxygen, you should be able to balance this equation.

$C_5H_{12}(\ell) + 8\,O_2(g) \rightarrow 5\,CO_2(g) + 6\,H_2O(g)$

Five carbons in C_5H_{12} require 5 carbon dioxides. Twelve hydrogens in C_5H_{12} call for 6 waters. This adds up to a total of 10 oxygens in CO_2 + 6 in H_2O, or 16 atoms of oxygen, all to come from the oxygen of the air. Thus 8 O_2 molecules are required.

EXAMPLE 8.8   Write the equation for the complete burning of ethane, $C_2H_6(g)$.

$2\,C_2H_6(g) + 7\,O_2(g) \rightarrow 4\,CO_2(g) + 6\,H_2O(g)$

Working with one molecule of ethane produces

$C_2H_6(g) + O_2(g) \rightarrow 2\,CO_2(g) + 3\,H_2O(g)$ (unbalanced)

with the oxygen yet to balance. There are 4 + 3 or 7 oxygen atoms on the right, which takes $\frac{7}{2}$ O_2 molecules on the left ($\frac{7}{2} \times 2 = 7$). Therefore

$C_2H_6(g) + \frac{7}{2}\,O_2(g) \rightarrow 2\,CO_2(g) + 3\,H_2O(g)$

Multiplying by 2 to clear the fractional coefficient yields the answer given.

Try this one, but be careful.

EXAMPLE 8.9   Write the equation for the complete burning of butanol, $C_4H_9OH(\ell)$, in air.

$C_4H_9OH(\ell) + 6\,O_2(g) \rightarrow 4\,CO_2(g) + 5\,H_2O(g)$

If you did not get this answer you probably counted only nine hydrogen atoms in C_4H_9OH (there are ten), or overlooked the fact that the one oxygen in C_4H_9OH takes care of one of the thirteen oxygens on the right, leaving only twelve to come from O_2.

8.7 REACTIONS OCCURRING IN WATER SOLUTIONS

The reactions we have considered in Examples 8.1 through 8.9 have all dealt with pure substances in the gas, liquid, or solid state. Most of the reactions you will study in the laboratory, and many that occur in the world around us, take place in aqueous solutions. Such reactions may be described by equations similar to those we have been writing for reactions between pure substances. They may also be described in the form of *net ionic equations*, which we will consider in Chapter 16. Each type of equation has its unique value; later you will choose between them, depending upon your particular need at the time. In this chapter we will confine ourselves to the conventional equation for describing reactions in aqueous solutions. In so doing, the symbol (aq) will be used to identify dissolved substances.

REDOX REACTIONS—"SINGLE REPLACEMENT" TYPE

> **PG 8 E** Given the reactants of a redox reaction ("single replacement" type only), write the equation.

Many elements are capable of replacing other elements from aqueous solutions. This is one kind of **oxidation-reduction reaction,** or **"redox" reaction,** that will be studied in greater detail in Chapter 18. The equation for such a reaction makes it appear as if one element is replacing another in a compound. It is a **"single replacement" equation;** in fact, the reactions are sometimes called **single replacement reactions.** The general equation is

$$M(s, g, \text{ or } \ell) + NX(aq) \rightarrow MX(aq) + N(s, g, \text{ or } \ell) \qquad (8.11)$$

The next three examples illustrate this type of redox reaction.

EXAMPLE 8.10   Elemental calcium reacts with hydrochloric acid to produce hydrogen gas and a solution of calcium chloride (Fig. 8.5).

To begin, write the unbalanced equation, showing the formula of each reactant and each product.

$$Ca(s) + HCl(aq) \rightarrow H_2(g) + CaCl_2(aq) \qquad \text{(unbalanced)}$$

Balancing the equation is easy.

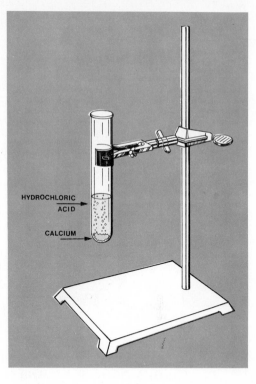

Figure 8.5
Calcium reacts with hydrochloric acid to release hydrogen gas. WARNING: Do not "try" this experiment, as it is potentially dangerous.

HYDROCHLORIC
ACID

CALCIUM

$$Ca(s) + 2\,HCl(aq) \rightarrow H_2(g) + CaCl_2(aq)$$

EXAMPLE 8.11 Copper reacts with a solution of silver nitrate, $AgNO_3$, to produce silver and a solution of copper(II) nitrate,* $Cu(NO_3)_2$ (Fig. 8.6).

The unbalanced equation, please. . . .

$$Cu(s) + \quad AgNO_3(aq \rightarrow \quad Ag(s) + \quad Cu(NO_3)_2(aq) \qquad \text{(unbalanced)}$$

An important technique appears in this equation. Notice that the nitrate ion, NO_3^-, appears on both sides of the equation: as a part of $AgNO_3$, on the left, and as a part of $Cu(NO_3)_2$, on the right. When a polyatomic ion is unchanged in a reaction, it may be balanced as a distinct unit, just as atoms of an element are balanced. The equation is balanced for that matter, except for the nitrate ion; there is one nitrate on the left, and there are two on the right. Balance them as the next step.

*Copper and some other elements identified in Chapter 12 vary in the manner in which they form compounds. These variations are indicated by Roman numerals in the name.

Figure 8.6
Reaction between copper wire and solution of silver nitrate, AgNO₃, in which silver metal grows on wire in thin needles.

$$\text{Cu(s)} + 2\,\text{AgNO}_3\text{(aq)} \rightarrow \quad \text{Ag(s)} + \quad \text{Cu(NO}_3)_2\text{(aq)} \qquad \text{(unbalanced)}$$

The equation now has the nitrates balanced, but it is unbalanced in another respect. Completion of the balancing, however, is easy.

$$\text{Cu(s)} + 2\,\text{AgNO}_3\text{(aq)} \rightarrow 2\,\text{Ag(s)} + \text{Cu(NO}_3)_2\text{(aq)}$$

EXAMPLE 8.12   If chlorine gas is bubbled through a solution of sodium bromide, an aqueous solution of bromine and sodium chloride results.

First write the formulas in the unbalanced equation.

$$\text{Cl}_2\text{(g)} + \quad \text{NaBr(aq)} \rightarrow \quad \text{Br}_2\text{(aq)} + \quad \text{NaCl(aq)}$$

Now balance the equation.

$$\text{Cl}_2\text{(g)} + 2\,\text{NaBr(aq)} \rightarrow \text{Br}_2\text{(aq)} + 2\,\text{NaCl(aq)}$$

PRECIPITATION REACTIONS

PG 8 F Given the reactants in a precipitation reaction, write the equation.

When solutions of two compounds are mixed, a positive ion from one compound may combine with a negative ion from the other to form a solid compound that settles to the bottom. The solid produced is called a **precipitate;** the reaction is called a **precipitation reaction.** In the equation for a precipitation reaction, ions of the two reactants appear to change partners. The equation, and sometimes the reaction itself, is called a **double replacement equation** or **double replacement reaction.** The general equation has the form

$$MY(aq) + NX(aq) \rightarrow MX(aq) + NY(s) \tag{8.12}$$

The next two examples illustrate chemical precipitations.

EXAMPLE 8.13 If solutions of sodium chloride and silver nitrate, $AgNO_3$, are combined, silver chloride, $AgCl$, precipitates (Fig. 8.7). Write the equation.

The word description of the reaction may appear incomplete; in fact, it *is* incomplete. What happens to the sodium and nitrate ions on the product side of the equation? These ions remain in solution, effectively as a solution of sodium nitrate. In solution reactions, any ions not otherwise accounted for may be considered as compounds in solution.

With this explanation you should be able to write the unbalanced equation.

$$NaCl(aq) + AgNO_3(aq) \rightarrow AgCl(s) + NaNO_3(aq)$$

Now for the balancing.

$$NaCl(aq) + AgNO_3(aq) \rightarrow AgCl(s) + NaNO_3(aq)$$

The equation is balanced, with all coefficients equal to 1. In checking this balance the nitrate ion may again be regarded as a distinct unit. There is one nitrate on each side of the equation.

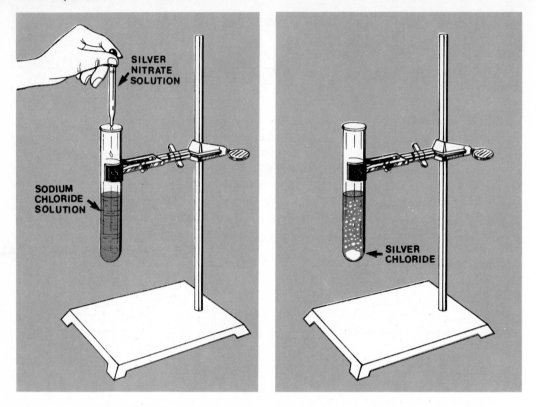

Figure 8.7
Silver chloride, AgCl, precipitates when solutions of silver nitrate, AgNO₃, and sodium chloride, NaCl, are combined.

EXAMPLE 8.14 Aluminum hydroxide precipitates on combining solutions of potassium hydroxide and aluminum nitrate. Write the equation.

Recalling what was said above about ions unaccounted for in precipitation reactions, write the unbalanced equation.

$$KOH(aq) + Al(NO_3)_3(aq) \rightarrow Al(OH)_3(s) + KNO_3(aq) \qquad \text{(unbalanced)}$$

In this equation there are two polyatomic (many atom) ions, OH^- and NO_3^-, which appear unchanged in the reaction. They may be treated as units in balancing the equation. Doing so makes the remainder of the balancing routine.

$$3\ KOH(aq) + Al(NO_3)_3(aq) \rightarrow Al(OH)_3(s) + 3\ KNO_3(aq)$$

The three hydroxides in $Al(OH)_3$ on the right side of the unbalanced equation require three KOH on the left, and the three nitrates in $Al(NO_3)_3$ on the left require three KNO_3 on the right. Balancing the polyatomic ions leaves the potassium and aluminum in balance.

NEUTRALIZATION

PG 8 G   Given the reactants in a neutralization reaction, write the equation.

An acid has been identified as a solution that contains hydrogen ions, H^+ (p. 110). A solution that contains hydroxide ions, OH^-, is called a **base**. When an acid is added to an equal amount of a base, each hydrogen ion forms a chemical bond with a hydroxide ion to produce a molecule of water. The acid and base **neutralize** each other; the process is called **neutralization**.* An ionic compound called a **salt** is also formed; it usually remains in solution. Neutralization reactions are described by double replacement equations. In general,

$$HX(aq) + MOH(aq) \rightarrow HOH(\ell) + MX(aq) \qquad (8.13)$$
$$\text{acid} \qquad \text{base} \qquad \text{water} \qquad \text{salt}$$

The formula of water has been written as HOH in Equation 8.13, rather than the usual H_2O. We use the HOH formula because it makes it a little easier to treat the hydroxide ion as a polyatomic unit in balancing neutralization equations. Whichever water formula you use, remember that the compound is *molecular*, not ionic. It is *not correct* to think of water as "hydrogen hydroxide."

EXAMPLE 8.15   Write the equation for the reaction between hydrochloric acid and sodium hydroxide solutions.

Begin with the unbalanced equation.

$HCl(aq) + \quad NaOH(aq) \rightarrow \quad HOH(\ell) + \quad NaCl(aq)$
Now balance the equation.

$HCl(aq) + NaOH(aq) \rightarrow HOH(\ell) + NaCl(aq)$
The equation is balanced, with all coefficients equal to 1.

---

*There are other kinds of acids that do not contain hydrogen ions, and bases that do not contain hydroxide ions, and the reactions between them are also neutralization reactions. The H^+ and OH^- neutralization is the most common, and the only one we will consider here.

EXAMPLE 8.16 Write the equation for the reaction between sulfuric acid and solid alumi-num hydroxide.

The products here are similar to the earlier ones. Write the unbalanced equation.

- - - - - -   - - - - - -

$H_2SO_4(aq) + Al(OH)_3(s) \rightarrow HOH(\ell) + Al_2(SO_4)_3(aq)$ (unbalanced)

Balancing this time is a bit trickier than before, but everything falls into place if you treat the polyatomic ions, SO_4^{2-} and OH^-, as distinct units in balancing.

- - - - - -   - - - - - -

$3\,H_2SO_4(aq) + 2\,Al(OH)_3(s) \rightarrow 6\,HOH(\ell) + Al_2(SO_4)_3(aq)$

Starting from the unbalanced equation and balancing first the sulfate ions, three on the right in $Al_2(SO_4)_3$ require three H_2SO_4:

$3\,H_2SO_4(aq) + Al(OH)_3(s) \rightarrow Al_2(SO_4)_3(aq) + HOH(\ell)$ (unbalanced)

Next, two aluminums in $Al_2(SO_4)_3$ on the right require two $Al(OH)_3$ on the left:

$3\,H_2SO_4(aq) + 2\,Al(OH)_3(s) \rightarrow Al_2(SO_4)_3(aq) + HOH(\ell)$ (unbalanced)

Two aluminum hydroxides make six hydroxide ions available for the formation of six water molecules:

$3\,H_2SO_4(aq) + 2\,Al(OH)_3(s) \rightarrow Al_2(SO_4)_3(aq) + 6\,HOH(\ell)$

The hydroxide step also balances the hydrogens other than those in the hydroxide ions, six from 3 molecules of H_2SO_4 on the left, and six from six water molecules on the right. All oxygen is in polyatomic ions, so it too should be in balance. It is, with 18 atoms on each side.

Your sequence of steps in balancing this equation may have been different, but the final result should be the same.

8.8 OTHER REACTIONS

PG 8 H   Given a word description of a chemical reaction in which all reac-tants and products are identified, write the chemical equation for the reaction.

In addition to the reactions already considered, there are many others that do not fit into any of the classifications above. We will use a few of these to provide further practice in balancing equations.

EXAMPLE 8.17   When rust, Fe_2O_3, is treated with hydrogen gas, it is reduced to metallic iron, with water vapor as a second product. Write and balance the equation.

$$Fe_2O_3(s) + 3\,H_2(g) \rightarrow 2\,Fe(s) + 3\,H_2O(g)$$

From the unbalanced equation, $Fe_2O_3(s) + H_2(g) \rightarrow Fe(s) + H_2O(g)$, the iron is readily balanced by a coefficient of 2 before Fe on the right side. This leaves only oxygen unbalanced. Three waters take care of the oxygen, but unbalance the hydrogen. Hydrogen, being in elemental form on the left, is readily restored to balance by a coefficient of 3. Whenever an equation can be balanced completely except for some elemental species, balancing that element last is readily accomplished by whatever coefficient is necessary. Note that this is a single replacement redox reaction that does not occur in water solution.

EXAMPLE 8.18   Carbon disulfide, $CS_2(\ell)$, reacts with oxygen gas to produce two gaseous products, carbon dioxide, CO_2, and sulfur dioxide, SO_2. Write the equation.

$$CS_2(\ell) + 3\,O_2(g) \rightarrow CO_2(g) + 2\,SO_2(g)$$

There is one carbon on each side of the unbalanced equation. Sulfur is readily balanced by a coefficient of 2 for the sulfur dioxide. Elemental oxygen is balanced last.

EXAMPLE 8.19   Balance the following equation:
$$PCl_5(s) + \quad H_2O(\ell) \rightarrow \quad H_3PO_4(aq) + \quad HCl(aq)$$

$$PCl_5(s) + 4\,H_2O(\ell) \rightarrow H_3PO_4(aq) + 5\,HCl(aq)$$

Taking chlorine first, one PCl_5 on the left requires 5 HCl on the right. Next, phosphorus is already balanced. Then hydrogen: three from H_3PO_4 and five from HCl is a total of eight on the right, requiring 4 H_2O on the left. Oxygen, as frequently happens when not present in elemental form, is balanced by the time we get to it.

8.9 SUMMARY OF TYPES OF REACTIONS AND EQUATIONS

In Sections 8.4 through 8.7 you were introduced to different types of reactions and the equations that describe them. One of the most valuable skills you can

TABLE 8.1
Summary of Types of Reactions and Equations

Reactants	Reaction Type	Products	Reference
Element + element Element + compound Compound + compound	Combination	One compound	Equation 8.8
One compound	Decomposition	Element + element Element + compound Compound + compound	Equation 8.9
O_2* + compound of C, H, and O	Complete Oxidation or Burning	CO_2* + H_2O*	Equation 8.10
Element + ionic compound or acid	Oxidation- Reduction†	Element + ionic compound	Equation 8.11
Two ionic compounds	Precipitation§	Two ionic compounds	Equation 8.12
Acid + base	Neutralization§	Ionic compound (salt) + H_2O	Equation 8.13

*The reactant oxygen and the products carbon dioxide and water are usually not mentioned in the description of a reaction of this kind.
†There are several kinds of oxidation-reduction reactions. The one described in this chapter occurs in water solution. Its equation is called a single replacement equation.
§Precipitation and neutralization reactions are usually described by a double replacement equation.

develop for equation writing is looking at formulas of one or more given reactants and predicting the kind of reaction possible to them. Writing the equation follows much more easily when the reaction has been "classified." Table 8.1 has been prepared to help you develop this skill. It summarizes all of the reaction types in this chapter, the nature of the reactants that are likely to engage in each type, and the products.

8.10 MISCELLANEOUS SYMBOLS USED IN CHEMICAL EQUATIONS

When it is necessary to show more information about a chemical reaction than the basic equation provides, other symbols are sometimes used. State symbols (s), (ℓ), and (g), and for solutions, (aq), which are used throughout this book, are examples. An alternate way of indicating the formation of a precipitate is to place an arrow pointing down next to its formula, as in the equation of Example 8.13:

$$NaCl + AgNO_3 \rightarrow AgCl\downarrow + NaNO_3$$

Formation of a gas may be shown by an arrow pointing up, as in the equation of Example 8.10:

$$Ca + 2\,HCl \rightarrow H_2\uparrow + CaCl_2$$

If a reaction is brought about by the application of heat, as in Example 8.6, it may be indicated by placing a Δ above the arrow:

$$CaCO_3(s) \xrightarrow{\Delta} CaO(s) + CO_2(g)$$

The use of other reaction conditions is also shown above, and sometimes above and below, the arrow in the equation. The electrolysis of water (Example 8.4) may be designated as

$$2\ H_2O(\ell) \xrightarrow{\text{electrolysis}} 2\ H_2(g) + O_2(g)$$

Many reactions are carried out in the presence of a **catalyst**, a substance that speeds up the rate of a reaction without being consumed itself. Manganese dioxide, MnO_2, is a catalyst for the decomposition of potassium chlorate (Example 8.5):

$$2\ KClO_3(s) \xrightarrow{MnO_2} 2\ KCl(s) + 3\ O_2(g)$$

Both heat and the catalyst may be shown:

$$2\ KClO_3(s) \xrightarrow[MnO_2]{\Delta} 2\ KCl(s) + 3\ O_2(g)$$

CHAPTER 8 IN REVIEW

8.1 EVOLUTION OF A CHEMICAL EQUATION (154)

8.2 MEANING OF A CHEMICAL EQUATION

8 A Given a chemical equation, interpret it in terms of (a) atoms, molecules, and/or formula units, and (b) moles. (157)

8.3 WRITING CHEMICAL EQUATIONS (158)

8.4 COMBINATION REACTIONS

8 B Given the identity of a compound that is formed from two or more simpler substances, write the equation for the reaction. (158)

8.5 DECOMPOSITION REACTIONS

8 C Given the identity of a compound that is decomposed into simpler substances, either compounds or elements, write the equation for the reaction. (160)

8.6 COMPLETE OXIDATION OR BURNING OF ORGANIC COMPOUNDS

8 D Write the equation for the complete oxidation or burning of any compound consisting only of carbon, hydrogen, and possibly oxygen. (163)

8.7   REACTIONS OCCURRING IN WATER SOLUTIONS

8 E Given the reactants of a redox reaction ("single replacement" type only), write the equation. (166)

8 F Given the reactants in a precipitation reaction, write the equation. (169)

8 G Given the reactants in a neutralization reaction, write the equation. (171)

8.8   OTHER REACTIONS

8 H Given a word description of a chemical reaction in which all reactants and products are identified, write the chemical equation for the reaction. (172)

8.9   SUMMARY OF TYPES OF REACTIONS AND EQUATIONS (173)

8.10   MISCELLANEOUS SYMBOLS USED IN CHEMICAL EQUATIONS (174)

TERMS AND CONCEPTS

Chemical equation (154)
Aqueous (155)
"State symbol" (155)
Balanced equation (156)
Combination reaction (158)
Oxidized-oxidation (159)
Burning (159)
Decomposition reaction (160)
Oxidation-reduction reaction (166)

Redox reaction (166)
"Single replacement" equation, reaction (166)
Precipitate; precipitation reaction (169)
"Double replacement" equation, reaction (169)
Neutralize, neutralization (171)
Base (171)
Salt (171)
Catalyst (175)

Most of these terms and many more are defined in Glossary.

EQUATION BALANCING EXERCISE

Instructions: Balance the following equations, for which correct chemical formulas are already written. Balanced equations are on pages 619 and 620.

1) $Na + O_2 \rightarrow Na_2O$
2) $H_2 + Cl_2 \rightarrow HCl$
3) $P + O_2 \rightarrow P_2O_3$
4) $KClO_4 \rightarrow KCl + O_2$
5) $Sb_2S_3 + HCl \rightarrow SbCl_3 + H_2S$
6) $NH_3 + H_2SO_4 \rightarrow (NH_4)_2SO_4$
7) $CuO + HCl \rightarrow CuCl_2 + H_2O$
8) $Zn + Pb(NO_3)_2 \rightarrow Zn(NO_3)_2 + Pb$
9) $AgNO_3 + H_2S \rightarrow Ag_2S + HNO_3$

10) $Cu + S \rightarrow Cu_2S$

11) $Al + H_3PO_4 \rightarrow H_2 + AlPO_4$

12) $NaNO_3 \rightarrow NaNO_2 + O_2$

13) $Mg(ClO_3)_2 \rightarrow MgCl_2 + O_2$

14) $H_2O_2 \rightarrow H_2O + O_2$

15) $BaO_2 \rightarrow BaO + O_2$

16) $H_2CO_3 \rightarrow H_2O + CO_2$

17) $Pb(NO_3)_2 + KCl \rightarrow PbCl_2 + KNO_3$

18) $Al + Cl_2 \rightarrow AlCl_3$

19) $P + O_2 \rightarrow P_2O_5$

20) $NH_4NO_2 \rightarrow N_2 + H_2O$

21) $H_2 + N_2 \rightarrow NH_3$

22) $Cl_2 + KBr \rightarrow Br_2 + KCl$

23) $BaCl_2 + (NH_4)_2CO_3 \rightarrow BaCO_3 + NH_4Cl$

24) $MgCO_3 + HCl \rightarrow MgCl_2 + CO_2 + H_2O$

25) $P + I_2 \rightarrow PI_3$

26) $PbO_2 \rightarrow PbO + O_2$

27) $Al + HCl \rightarrow AlCl_3 + H_2$

28) $Fe_2(SO_4)_3 + Ba(OH)_2 \rightarrow BaSO_4 + Fe(OH)_3$

29) $Al + CuSO_4 \rightarrow Al_2(SO_4)_3 + Cu$

30) $KClO_3 \rightarrow KCl + O_2$

31) $Mg + N_2 \rightarrow Mg_3N_2$

32) $C_6H_{14} + O_2 \rightarrow CO_2 + H_2O$

33) $FeCl_2 + Na_3PO_4 \rightarrow Fe_3(PO_4)_2 + NaCl$

34) $Li_2O + HOH \rightarrow LiOH$

35) $HgO \rightarrow Hg + O_2$

36) $CaSO_4 \cdot 2\,H_2O \rightarrow CaSO_4 + H_2O$

37) $C_3H_7CHO + O_2 \rightarrow CO_2 + H_2O$

38) $NaHCO_3 + HCl \rightarrow NaCl + H_2O + CO_2$

39) $Bi(NO_3)_3 + NaOH \rightarrow Bi(OH)_3 + NaNO_3$

40) $FeS + HBr \rightarrow FeBr_2 + H_2S$

41) $Zn(OH)_2 + H_2SO_4 \rightarrow ZnSO_4 + HOH$

42) $P_4O_{10} + H_2O \rightarrow H_3PO_4$

43) $C_4H_9OH + O_2 \rightarrow CO_2 + H_2O$

44) $CaC_2 + H_2O \rightarrow C_2H_2 + Ca(OH)_2$

45) $CaCO_3 + H_3PO_4 \rightarrow Ca_3(PO_4)_2 + CO_2 + H_2O$

46) $PCl_5 + H_2O \rightarrow H_3PO_4 + HCl$

47) $CaI_2 + H_2SO_4 \rightarrow HI + CaSO_4$

48) $C_3H_7COOH + O_2 \rightarrow CO_2 + H_2O$

49) $Mg(CN)_2 + HCl \rightarrow HCN + MgCl_2$

50) $(NH_4)_2S + HgBr_2 \rightarrow NH_4Br + HgS$

QUESTIONS AND PROBLEMS

As in Chapter 7, the following questions give only the names of compounds whose formulas were introduced in Chapter 6. For reference, you may find all of these formulas on page 615. You are encouraged to write formulas with no assistance other than a periodic table, using the reference page only if necessary. This is the way you will build and improve your formula writing skills.

Section 8.2

8.1) Write a sentence giving the "molecular" meaning of the equation for the complete oxidation of benzene:

$$2\,C_6H_6 + 15\,O_2 \rightarrow 12\,CO_2 + 6\,H_2O$$

8.31) State the "molar" interpretation of the equation in 8.1. Explain the difference in the two interpretations.

Section 8.4

Write balanced chemical equations for the combination reactions described in Problems 8.2 to 8.6 and 8.32 to 8.36.

8.2) Calcium forms calcium oxide by reaction with the oxygen in the air.

8.3) P_4O_{10} is the product of the direct combination of phosphorus and oxygen.

8.4) When potassium contacts fluorine gas—two highly reactive elements—potassium fluoride is produced.

8.5) Silicon reacts with chlorine to form silicon tetrachloride, $SiCl_4$.

8.6) Magnesium nitride is one of the few metallic nitrides. Its formula is what you would expect from the positions of the elements in the periodic table, and it is formed by direct combination of the elements.

8.32) Lithium combines with oxygen to form lithium oxide.

8.33) Boron $(Z = 5)$ combines with oxygen to form B_2O_3.

8.34) Calcium combines with bromine to make calcium bromide.

8.35) Phosphorus tribromide, PBr_3, is produced by direct combination of phosphorus and bromine.

8.36) There are only a few carbides—compounds made up of a metal and carbon. Although aluminum carbide is not ionic, its formula is as would be expected from the periodic table positions of the elements, counting carbon at -4 to balance aluminum's $+3$. Write the equation for its formation by direct combination of the elements.

Section 8.5

Write the equations for the following decomposition reactions in Problems 8.7 to 8.10 and 8.37 to 8.40:

8.7) Melted crystals of table salt, sodium chloride, may be decomposed to its elements by electrolysis.

8.8) Mercury $(Z = 80)$ and oxygen gas are produced by heating mercury(II) oxide, HgO.

8.9) Carbonic acid is an unstable compound, that decomposes into carbon dioxide and water.

8.10) Hydrogen peroxide, H_2O_2, the familiar bleaching compound, decomposes slowly into water and oxygen.

8.37) Pure hydrogen iodide, HI, decomposes spontaneously to its elements.

8.38) Silver oxide, Ag_2O, may be decomposed to its elements by heating.

8.39) Sulfurous acid, H_2SO_3, decomposes spontaneously to water and sulfur dioxide.

8.40) When calcium hydroxide—sometimes called slaked lime—is heated, it decomposes to calcium oxide—or lime—and water vapor.

Section 8.6

Write equations for the complete oxidation of the following organic compounds in Problems 8.11 to 8.14 and 8.41 to 8.44:

8.11) Propane, C_3H_8.

8.12) Acetylene, C_2H_2.

8.13) Acetaldehyde, CH_3CHO.

8.41) Butane, C_4H_{10}.

8.42) Ethyl alcohol, C_2H_5OH.

8.43) Acetic acid, CH_3COOH.

8.14) Sugar, $C_{12}H_{22}O_{11}$.

8.44) Glycerin, $C_3H_8O_3$.

Section 8.7

Write chemical equations for the reactions in Problems 8.15 to 8.26 and 8.45 to 8.56 that occur in water solution:

8.15) Hydrogen gas and a solution of magnesium sulfate are the result of the reaction that occurs when metallic magnesium is placed into sulfuric acid.

8.16) When barium metal is placed into water, hydrogen gas bubbles off and a solution of barium hydroxide remains.

8.17) A strip of zinc ($Z = 30$) replaces the silver from a solution of silver nitrate, $AgNO_3$, leaving needle-like crystals of silver ($Z = 47$) and a solution of zinc nitrate, $Zn(NO_3)_2$.

8.18) Magnesium metal will replace the nickel from a solution of nickel chloride, $NiCl_2$, yielding metallic nickel ($Z = 28$) and a solution of magnesium chloride.

8.19) Bromine liberates iodine from a solution of sodium iodide and leaves a solution of sodium bromide.

8.20) If silver nitrate, $AgNO_3$, and potassium bromide solutions are combined, solid silver bromide, $AgBr$, precipitates.

8.21) Combination of solutions of lead nitrate, $Pb(NO_3)_2$, and copper(II) sulfate, $CuSO_4$, yields lead sulfate, $PbSO_4$, as a precipitate, and a solution of copper(II) nitrate, $Cu(NO_3)_2$.

8.22) Magnesium chloride solution combined with a solution of sodium fluoride yields a precipitate of magnesium fluoride.

8.23) If sodium sulfide solution is poured into silver nitrate solution, $AgNO_3(aq)$, silver sulfide, Ag_2S, precipitates.

8.24) Sodium hydroxide solution added to magnesium bromide solution yields a precipitate of magnesium hydroxide.

8.25) Potassium hydroxide solution neutralizes nitric acid.

8.26) Solid magnesium hydroxide reacts with hydrochloric acid.

8.45) Calcium metal reacts with hydrobromic acid, HBr, to form hydrogen gas and a solution of calcium bromide.

8.46) Potassium and water react to produce a solution of potassium hydroxide and hydrogen gas.

8.47) If zinc ($Z = 30$) is placed into a solution of copper(II) nitrate, $Cu(NO_3)_2$, metallic copper ($Z = 29$) and a solution of zinc nitrate, $Zn(NO_3)_2$, result.

8.48) If metallic manganese ($Z = 25$) is placed in a solution of chromium(III) chloride, $CrCl_3$, solid chromium ($Z = 24$) and manganese(II) chloride solution, $MnCl_2(aq)$, result.

8.49) If chlorine gas is bubbled through a solution of potassium iodide, a solution of potassium chloride and elemental iodine is the product.

8.50) Lead nitrate solution, $Pb(NO_3)_2(aq)$, reacts with sodium iodide solution to precipitate lead iodide, PbI_2.

8.51) Combining solutions of barium chloride and sodium sulfate produces a precipitate of barium sulfate.

8.52) Calcium fluoride precipitates from combining solutions of calcium nitrate and potassium fluoride.

8.53) Ammonium sulfide solution added to a solution of copper(II) nitrate, $Cu(NO_3)_2$, yields a precipitate of copper(II) sulfide, CuS.

8.54) Zinc chloride solution, $ZnCl_2(aq)$, produces a precipitate of zinc hydroxide, $Zn(OH)_2$, when added to a solution of potassium hydroxide.

8.55) Sulfuric acid neutralizes barium hydroxide solution.

8.56) Solid nickel hydroxide, $Ni(OH)_2$, reacts with sulfuric acid. Nickel sulfate solution, $NiSO_4$ (aq), is one product.

Section 8.8

For each general reaction described in Problems 8.27 to 8.30 and 8.57 to 8.60, write the chemical equation:

8.27) Metallic zinc (Z = 30) reacts with steam at high temperatures to produce solid zinc oxide, ZnO, and hydrogen gas.

8.28) When solid barium oxide is placed into water, a solution of barium hydroxide is produced.

8.29) Igniting a mixture of powdered iron(II) oxide, FeO, and aluminum produces a vigorous reaction in which aluminum oxide and iron are the products.

8.30) Solid iron(III) oxide, Fe_2O_3, reacts with gaseous carbon monoxide to produce iron and carbon dioxide.

8.57) A solid oxide of iron, $Fe_3O_4(s)$, and hydrogen are the products of the reaction between iron (Z = 26) and steam.

8.58) Sulfuric acid is produced when gaseous sulfur trioxide, $SO_3(g)$, reacts with water.

8.59) At high temperatures carbon monoxide and steam react to produce carbon dioxide and hydrogen.

8.60) Magnesium nitride, Mg_3N_2, and hydrogen are the products of the reaction between magnesium and ammonia.

quantity relationships in chemical reactions

9.1 THE QUANTITATIVE MEANING OF A CHEMICAL EQUATION

> **PG 9 A** Given a chemical equation, or a reaction for which the equation can be written, and the number of moles of one species in the reaction, calculate the number of moles of any other species.

A balanced chemical equation tells how many moles of individual reactants are required for the reaction, as well as the number of moles of different species that will be produced. Using the equation

$$PCl_5(s) + 4\,H_2O(\ell) \rightarrow H_3PO_4(aq) + 5\,HCl(aq),$$

we see that four moles of water are required to react with each mole of PCl_5. *For this reaction* there is, in effect, an equivalence between PCl_5 and H_2O: 1 mole $PCl_5 \simeq 4$ moles H_2O. Furthermore, for this reaction, each mole of PCl_5 that reacts will produce one mole of H_3PO_4 and five moles of HCl. The equivalence may thus be extended to the products:

$$1 \text{ mole } PCl_5 \simeq 4 \text{ moles } H_2O \simeq 1 \text{ mole } H_3PO_4 \simeq 5 \text{ moles HCl} \quad (9.1)$$

Notice that *the numbers in the series of mole relationships derived from an equation are the same as the coefficients in the equation.*

If the number of moles of one species in a reaction, either reactant or product, is known, the equivalences establish a unit path to the moles of any other species. A single-step conversion is involved. If, for example, 3.20 moles of PCl_5 react according to the above equation, how many moles of HCl will be produced? If 5 moles of HCl are produced by one mole of PCl_5 (1 mole $PCl_5 \simeq 5$ moles HCl), then the given quantity of PCl_5 (3.20 moles) will yield

$$3.20 \ \text{mol } PCl_5 \times \frac{5 \ \text{mol HCl}}{1 \ \text{mol } PCl_5} = 16.0 \ \text{mol HCl}$$

The procedure may be applied to any equation.

EXAMPLE 9.1   How many moles of oxygen are required to burn 2.40 moles of ethane, C_2H_6, according to the equation

$$2 \ C_2H_6(g) + 7 \ O_2(g) \rightarrow 4 \ CO_2(g) + 6 \ H_2O(\ell)$$

From the equivalences available in the equation, what unit path and what specific equivalence are required for this problem?

---

mol $C_2H_6 \rightarrow$ mol O_2; 2 mol $C_2H_6 \simeq 7$ mol O_2

The problem requires conversion from moles of ethane to moles of oxygen, so only those two species are needed. The numbers are the coefficients in the equation. For every 2 moles of C_2H_6, 7 moles of O_2 are required.

Complete the problem.

$$2.40 \ \text{mol } C_2H_6 \times \frac{7 \ \text{mol } O_2}{2 \ \text{mol } C_2H_6} = 8.40 \ \text{mol } O_2$$

EXAMPLE 9.2   In the reaction $N_2(g) + 3 \ H_2(g) \rightarrow 2 \ NH_3(g)$, how many moles of hydrogen are required to produce 4.20 moles of ammonia (NH_3)?

$$4.20 \text{ mol } \cancel{\text{NH}_3} \times \frac{3 \text{ mol } H_2}{2 \text{ mol } \cancel{\text{NH}_3}} = 6.30 \text{ mol } H_2$$

The conversion of moles of one substance to moles of another, moles given → moles wanted, is a step that appears in every chemical reaction problem having to do with amounts of reactants or products.

9.2 MASS CALCULATIONS: STOICHIOMETRY

PG 9 B Given a chemical equation, or a reaction for which the equation can be written, and the number of grams or moles of one species in the reaction, find the number of grams or moles of any other species.

Section 7.7 (p. 138) presented the method by which the mass of a chemical of known formula can be converted to moles, or the number of moles converted to grams. Unit paths were *grams → moles* or *moles → grams*. The conversion ratio was molar weight, grams/mole, in both directions. In Section 9.1 we have just seen how to follow a unit path of *moles given species → moles wanted species*, using coefficients from the equation for the conversion ratio. By combining these techniques we have a unit path from grams of one substance to grams of another:

$$\begin{matrix} \text{grams} \\ \text{given species} \end{matrix} \rightarrow \begin{matrix} \text{moles} \\ \text{given species} \end{matrix} \rightarrow \begin{matrix} \text{moles} \\ \text{wanted species} \end{matrix} \rightarrow \begin{matrix} \text{grams} \\ \text{wanted species} \end{matrix} \quad (9.2)$$

If you know the grams of any one substance in a reaction you can, in three steps, calculate the grams of any other.

Finding the amount of one substance in a reaction from a known quantity of another is called **stoichiometry**. "Quantity" may be expressed in a number of ways. Officially, according to the SI system of units, quantity is the number of moles. We more commonly think of amount in terms of measurable units, such as grams for mass and liters for volume. Regardless of the measurements in which we choose to express quantity, there is a single "stoichiometry pattern" for the solution of *all* such problems. It consists of these three steps:

1. Convert the quantity of the given species to moles (as in grams → moles, Section 7.7, p. 138).
2. Convert the moles of given species to moles of wanted species (moles given → moles wanted, Section 9.1).
3. Convert the moles of wanted species to the quantity units required (as in moles → grams, Section 7.7, p. 138).

"Equation" 9.2 is the unit path of the stoichiometry pattern.

If you learn how to use the stoichiometry pattern now, with grams as the only quantity unit, you will find later that you already know how to solve problems in which other units appear.

Occasionally you may be given the quantity of one substance in grams and asked to find the number of moles of a second species. In this case the first two steps of the stoichiometry pattern complete the problem. Or, you may be given the moles of one substance and asked to find the grams of another. Steps 2 and 3 accomplish this objective.

The burning of ethane may be used to illustrate the method of stoichiometry.

EXAMPLE 9.3 Calculate the number of grams of oxygen that are required to burn 150 grams of ethane, C_2H_6, in the reaction $2\,C_2H_6(g) + 7\,O_2\,(g) \rightarrow 4\,CO_2(g) + 6\,H_2O(\ell)$.

Solution: Step 1 calls for the conversion of the given quantity to moles—in this case, the conversion of 150 grams of ethane to moles of ethane as in Section 7.7, p. 138. The setup begins

$$150\,g\,C_2H_6 \times \frac{1\ mol\ C_2H_6}{30.0\,g\,C_2H_6}$$

In Step 2 the setup is extended to convert moles of C_2H_6 to moles of O_2, as in Section 9.1. The conversion equivalence is 7 moles $O_2 \simeq 2$ moles C_2H_6, from the equation:

$$150\,g\,C_2H_6 \times \frac{1\ mol\ C_2H_6}{30.0\,g\,C_2H_6} \times \frac{7\ mol\ O_2}{2\ mol\ C_2H_6}$$

Finally, in Step 3, the moles of oxygen are converted to grams, as in Section 7.7, p. 138.

$$150\,g\,C_2H_6 \times \frac{1\ mol\ C_2H_6}{30.0\,g\,C_2H_6} \times \frac{7\ mol\ O_2}{2\ mol\ C_2H_6} \times \frac{32.0\ g\ O_2}{1\ mol\ O_2} = 560\ g\ O_2$$

Now you try one . . . .

EXAMPLE 9.4 The equation for burning heptane, C_7H_{16}, is

$$C_7H_{16}\,(\ell) + 11\,O_2(g) \rightarrow 7\,CO_2(g) + 8\,H_2O(\ell)$$

How many grams of oxygen are required to burn 3.50 moles of C_7H_{16}?

The given quantity is *moles* of heptane. In other words, we already have moles of the given substance, so the first step in the stoichiometry pattern is completed. The equation gives us the molar equivalence between heptane and oxygen. Start with the second step, and set up, but do not calculate, the number of moles of the wanted substance, oxygen.

$$3.50 \; \text{mol} \; \cancel{C_7H_{16}} \; \times \; \frac{11 \; \text{mol} \; O_2}{1 \; \text{mol} \; \cancel{C_7H_{16}}} \times \underline{\hspace{3cm}}$$

From the equation, 11 moles $O_2 \approx 1$ mole C_7H_{16}. This establishes the conversion factor.

Changing from moles of O_2 to grams of O_2 via molar weight is Step 3 in the stoichiometric sequence, and finishes the problem. Complete the setup by inserting the required numerator and denominator. Calculate for the answer.

$$3.50 \; \text{mol} \; \cancel{C_7H_{16}} \; \times \; \frac{11 \; \cancel{\text{mol} \; O_2}}{1 \; \text{mol} \; \cancel{C_7H_{16}}} \times \frac{32.0 \; \text{g} \; O_2}{1 \; \cancel{\text{mol} \; O_2}} = 1.23 \times 10^3 \; \text{g} \; O_2$$

EXAMPLE 9.5 How many moles of H_2O will be produced in the heptane burning reaction that also yields 115 grams of CO_2? (See equation in Example 9.4.)

This time we are given grams of one species and must find moles of another. The first two steps in the stoichiometric pattern are all that is necessary. Set up, but do not solve, the first step, beginning with the given quantity.

$$115 \; \text{g} \; \cancel{CO_2} \; \times \; \frac{1 \; \text{mol} \; CO_2}{44.0 \; \text{g} \; \cancel{CO_2}} \times \underline{\hspace{3cm}}$$

By Step 2, and by drawing the proper molar equivalence from the equation, extend the setup and solve the problem.

$$115 \; \text{g} \; \cancel{CO_2} \; \times \; \frac{1 \; \cancel{\text{mol} \; CO_2}}{44.0 \; \text{g} \; \cancel{CO_2}} \times \frac{8 \; \text{mol} \; H_2O}{7 \; \cancel{\text{mol} \; CO_2}} = 2.99 \; \text{mol} \; H_2O$$

Now we'll go all the way from grams to grams.

EXAMPLE 9.6   How many grams of CO_2 will be produced by burning 66.0 grams of heptane? The equation is

$$C_7H_{16}(\ell) + 11\,O_2(g) \rightarrow 7\,CO_2(g) + 8\,H_2O(\ell)$$

Beginning with the given quantity, set up, but do not solve, the first step, the conversion of 66.0 grams of heptane to moles.

$$66.0\,g\,\cancel{C_7H_{16}} \times \frac{1\text{ mol }C_7H_{16}}{100\,g\,\cancel{C_7H_{16}}} \times \underline{\hspace{3cm}}$$

Step 2 converts the moles of given species (heptane) to moles of wanted species (carbon dioxide). Extend the setup that far.

$$66.0\,g\,\cancel{C_7H_{16}} \times \frac{1\text{ }\cancel{\text{mol }C_7H_{16}}}{100\,g\,\cancel{C_7H_{16}}} \times \frac{7\text{ mol }CO_2}{1\text{ }\cancel{\text{mol }C_7H_{16}}} \times \underline{\hspace{3cm}}$$

Finally, convert moles of CO_2 to grams, Step 3. Calculate the numerical answer.

$$66.0\,g\,\cancel{C_7H_{16}} \times \frac{1\text{ }\cancel{\text{mol }C_7H_{16}}}{100\,g\,\cancel{C_7H_{16}}} \times \frac{7\text{ }\cancel{\text{mol }CO_2}}{1\text{ }\cancel{\text{mol }C_7H_{16}}} \times \frac{44.0\text{ g }CO_2}{1\text{ }\cancel{\text{mol }CO_2}} = 203\text{ g }CO_2$$

9.3   PERCENTAGE YIELD

PG 9 C   Given two of the following, or information from which two of the following may be determined, calculate the third: theoretical yield, actual yield, and percentage yield.

Example 9.6 indicated that the burning of 66.0 grams of heptane will produce 203 grams of CO_2. This is called a **theoretical yield**, where *yield* is the amount produced. In actual practice, factors such as impure reactants, incomplete reactions, and side reactions cause the **actual yield** to be less than the calculated yield. Knowing the actual yield found in the laboratory and the theoretical yield calculated from the equation for the reaction, we can find the **percentage yield**:

$$\frac{\text{actual yield}}{\text{theoretical yield}} \times 100 = \text{percentage yield} \qquad (9.3)$$

If only 180 grams of carbon dioxide resulted in Example 9.6 instead of the theoretical 203 grams, the percentage yield would be

$$\frac{\text{actual yield}}{\text{theoretical yield}} \times 100 = \frac{180\,g}{203\,g} \times 100 = 88.7\%$$

EXAMPLE 9.7   A solution containing excess* sodium sulfate is added to a second solution containing 3.18 grams of barium nitrate. Barium sulfate precipitates. (a) Calculate the theoretical yield of barium sulfate. (b) If the actual yield is 2.69 grams, calculate the percentage yield. The equation is

$$Ba(NO_3)_2(aq) + Na_2SO_4(aq) \rightarrow BaSO_4(s) + 2\,NaNO_3(aq)$$

Calculation of the theoretical yield is a typical stoichiometry problem. Solve part (a).

$$3.18\,g\,Ba(NO_3)_2 \times \frac{1\,mol\,Ba(NO_3)_2}{261\,g\,Ba(NO_3)_2} \times \frac{1\,mol\,BaSO_4}{1\,mol\,Ba(NO_3)_2} \times \frac{233\,g\,BaSO_4}{1\,mol\,BaSO_4} = 2.84\,g\,BaSO_4$$

Now, if the actual yield is 2.69 grams, find the percentage yield by substituting into Equation 9.3.

$$\frac{2.69\,g\,BaSO_4\ (\text{actual})}{2.84\,g\,BaSO_4\ (\text{theoretical})} \times 100 = 94.7\%$$

---

*The word "excess" as used here means "more than enough" . . . more than enough sodium sulfate than is required to precipitate all the barium in 3.18 grams of barium nitrate.

The basic meaning of per cent is parts per hundred—the number of parts of one kind of thing in 100 parts of all kinds of things (see Appendix I, Part E). Applied to percentage yield, it means grams of actual product per 100 grams of theoretical product. This leads to an equivalence that can be used in a dimensional analysis approach to percentage yield problems. An X% yield may be interpreted as

$$X \text{ grams actual product} \simeq 100 \text{ grams theoretical product} \qquad (9.4)$$

Notice that percentage yield is concerned with actual and theoretical grams of *product*, not reactant.

EXAMPLE 9.8   A procedure for preparing sodium sulfate is summarized in the equation

$$2 \, S(s) + 3 \, O_2(g) + 4 \, NaOH(aq) \rightarrow 2 \, Na_2SO_4(aq) + 2 \, H_2O(\ell)$$

The percentage yield of the process is 82.2%. Find the number of grams of sodium sulfate that will be recovered from the reaction of 36.9 grams of sodium hydroxide.

Using the stoichiometry procedures already explained, you can calculate the grams of Na_2SO_4 that can be produced by 36.9 grams of NaOH. That quantity is the theoretical yield. But you will not recover that much; you will recover only 82.2% of that amount. This is the strategy for solving the problem. Begin with the given quantity and set up the three steps to the theoretical grams of Na_2SO_4. Do not calculate the answer.

--------   --------

$$36.9 \text{ g NaOH} \times \frac{1 \text{ mol NaOH}}{40.0 \text{ g NaOH}} \times \frac{2 \text{ mol Na}_2\text{SO}_4}{4 \text{ mol NaOH}} \times \frac{142 \text{ g Na}_2\text{SO}_4 \text{ (theoretical)}}{1 \text{ mol Na}_2\text{SO}_4} \times \underline{}$$

We asked you not to calculate the answer to this point, but we did. The theoretical yield from the above setup is 65.5 g Na_2SO_4. Now forget that number for a moment; we'll return to it shortly.

The above setup represents the unit path g NaOH → mol NaOH → mol Na_2SO_4 → g Na_2SO_4 (theoretical). Can you extend that path one more step to the actual yield: g Na_2SO_4 (theoretical) → g Na_2SO_4 (actual)? Equation 9.4 provides the needed conversion factor. Insert above the necessary numbers and units to complete the setup. Then calculate the answer.

--------   --------

$$36.9 \text{ g NaOH} \times \frac{1 \text{ mol NaOH}}{40.0 \text{ g NaOH}} \times \frac{2 \text{ mol Na}_2\text{SO}_4}{4 \text{ mol NaOH}} \times \frac{142 \text{ g Na}_2\text{SO}_4 \text{ (theoretical)}}{1 \text{ mol Na}_2\text{SO}_4}$$

$$\times \frac{82.2 \text{ g Na}_2\text{SO}_4 \text{ (actual)}}{100 \text{ g Na}_2\text{SO}_4 \text{ (theoretical)}} = 53.8 \text{ g Na}_2\text{SO}_4 \text{ (actual)}$$

The above setup represents the overall unit path $g\ NaOH \rightarrow mol\ NaOH \rightarrow$ $mol\ Na_2SO_4 \rightarrow g\ Na_2SO_4$ (theoretical) $\rightarrow g\ Na_2SO_4$ (actual). It also illustrates the dimensional analysis "parts per 100" approach to solving percentage problems.

Before you started to solve the problem the strategy was laid out: find the theoretical yield, and then take 82.2% of that quantity. As noted above, the theoretical yield is 65.5 g Na_2SO_4. Thus,

$$82.2\%\ of\ 65.5\ g\ Na_2SO_4 = 0.822 \times 65.5\ g\ Na_2SO_4 = 53.8\ g\ Na_2SO_4$$

The decimal factor 0.822, coming from 82.2%, is the same as the dimensional analysis factor, $\dfrac{82.2\ g\ Na_2SO_4\ (actual)}{100\ g\ Na_2SO_4\ (theoretical)}$.

In Example 9.8 a percentage yield conversion appears at the end of the stoichiometric pattern, Equation 9.2. Sometimes it must be used at the beginning. Suppose a chemical manufacturer wants to make X grams of a product. Suppose also that he knows his actual yield is only 50% of the theoretical yield. If he is going to *get* X grams of actual product, he must use enough reactant to produce 2X grams of theoretical product because he will recover only half that amount. The next example shows how this is figured.

EXAMPLE 9.9 Sodium nitrite, $NaNO_2$, is produced from sodium nitrate, $NaNO_3$, by the reaction $2\ NaNO_3 \rightarrow 2\ NaNO_2 + O_2$. How many grams of $NaNO_3$ must be used to produce an actual yield of 60.0 g $NaNO_2$ if the percentage yield is 76.3%?

In this problem, 60.0 g $NaNO_2$ is only 76.3% of the theoretical yield from which the amount of reactant must be calculated. 76.3% of what theoretical yield is 60.0 g $NaNO_2$ (actual)? The easiest way to solve this is by dimensional analysis. The unit path is g $NaNO_2$ (actual) $\rightarrow NaNO_2$ (theoretical), and Equation 9.4 furnishes the conversion relationship. Set up the conversion, but do not calculate the answer.

$$60.0\ \cancel{g\ NaNO_2\ (actual)} \times \frac{100\ g\ NaNO_2\ (theoretical)}{76.3\ \cancel{g\ NaNO_2\ (actual)}} \times \underline{\qquad}$$

The conventional arithmetic approach to this percentage problem is

$$76.3\%\ of\ X = 0.763X = 60.0$$

$$X = \frac{60.0}{0.763} = \ldots .$$

The setups are the same.

The setup so far represents the theoretical yield. Extend it with the usual three steps of the stoichiometry pattern and calculate the answer.

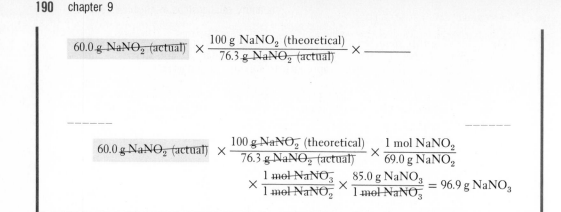

$$60.0 \text{ g NaNO}_2 \text{ (actual)} \times \frac{100 \text{ g NaNO}_2 \text{ (theoretical)}}{76.3 \text{ g NaNO}_2 \text{ (actual)}} \times \underline{\hspace{2cm}}$$

$$60.0 \text{ g NaNO}_2 \text{ (actual)} \times \frac{100 \text{ g NaNO}_2 \text{ (theoretical)}}{76.3 \text{ g NaNO}_2 \text{ (actual)}} \times \frac{1 \text{ mol NaNO}_2}{69.0 \text{ g NaNO}_2}$$
$$\times \frac{1 \text{ mol NaNO}_3}{1 \text{ mol NaNO}_2} \times \frac{85.0 \text{ g NaNO}_3}{1 \text{ mol NaNO}_3} = 96.9 \text{ g NaNO}_3$$

In Example 9.9 the percentage conversion appears before the stoichiometry pattern, and in Example 9.8 it is after. How do you know where to put it? It is *always* applied to the product, never to a reactant. *Yield* always refers to a product, so that is where percentage yield conversions belong. You can get some right answers by using it on reactants, but the thought process is much more complicated and much less logical—and errors are much more likely.

9.4 LIMITING REAGENT PROBLEMS

PG 9 D   Given a chemical equation or information from which it may be determined, and initial quantities of two or more reactants, (a) identify the limiting reagent; (b) calculate the theoretical yield of a specified product, assuming complete use of the limiting reagent; and (c) calculate the quantity of the reactant initially in excess that remains unreacted.

Zinc reacts with sulfur to form zinc sulfide: $Zn(s) + S(s) \rightarrow ZnS(s)$. Suppose you put three moles of zinc and two moles of sulfur into a reaction vessel and caused them to react until one is totally used up. How many moles of zinc sulfide will result? Also, how many moles of which element will remain unreacted?

This question is something like asking: "How many pairs of gloves can you assemble out of 20 left gloves and 30 right gloves, and how many unmatched gloves, and for which hand, will be left over?" The answer, of course, is 20 pairs of gloves. After you have assembled 20 pairs you run out of left gloves, even though you have 10 right gloves remaining. Similarly, in the zinc sulfide question where the elements combine on a 1-to-1 mole ratio, you will run out of sulfur after two moles of zinc sulfide form. One mole of zinc, the excess reactant, will remain unreacted. Sulfur, **the reactant that is completely used up by** the reaction, is called the **limiting reagent.** This analysis is summarized below:

Equation	Zn	+	S	→	ZnS
Moles at start	3		2		0
Moles used (−) or produced (+)	−2		−2		+2
Moles left	1		0		2

There is another way to think about the zinc sulfide reaction. Start again with three moles of zinc and two moles of sulfur. How many moles of ZnS can be produced by 3 moles of zinc, assuming there is an excess of sulfur? Three, of course: 1 mol Zn ≏ 1 mol ZnS, so 3 moles of zinc can produce 3 moles of zinc sulfide. Similarly, how many moles of zinc sulfide can be formed from 2 moles of sulfur if there is an excess of zinc? By the same reasoning, 2 can be formed. Now combine these. The available zinc can produce 3 moles of zinc sulfide if there is enough sulfur, but the sulfur available can produce only 2 moles of zinc sulfide. The smaller product quantity, 2 moles of zinc sulfide, is the most the reaction can yield before running out of sulfur, the limiting reagent.

Now for the moles of zinc that remain unchanged: How many moles of zinc are used by all of the limiting reagent, sulfur? From the equation, 1 mol Zn ≏ 1 mol S. Therefore, 2 moles of sulfur will use 2 moles of zinc. If you start with 3 moles of zinc and use 2, then 1 mole of zinc (3 − 2) will be left.

You now have the idea behind solving limiting reagent problems. The reasoning is not difficult in terms of moles, but calculations become a bit more complicated when grams are introduced. To illustrate, consider the next example.

EXAMPLE 9.10 How many grams of ZnS (97.5 g/mol) can be produced by 8.50 g Zn and 6.84 g S in Zn + S → ZnS? How many grams of which element will remain unreacted?

Begin by finding the number of grams of ZnS that can be formed by 8.50 g Zn, assuming that excess sulfur is present.

$$8.50 \text{ g Zn} \times \frac{1 \text{ mol Zn}}{65.4 \text{ g Zn}} \times \frac{1 \text{ mol ZnS}}{1 \text{ mol Zn}} \times \frac{97.5 \text{ g ZnS}}{1 \text{ mol ZnS}} = 12.7 \text{ g ZnS}$$

Now find the grams of ZnS that can come from 6.84 g S, assuming excess zinc.

$$6.84 \, g\,S \times \frac{1 \, mol\,S}{32.1 \, g\,S} \times \frac{1 \, mol\,ZnS}{1 \, mol\,S} \times \frac{97.5 \, g\,ZnS}{1 \, mol\,ZnS} = 20.8 \, g \, ZnS$$

These two results show that there is enough zinc to produce 12.7 grams of zinc sulfide, and enough sulfur to form 20.8 grams of zinc sulfide. Which element is the limiting reagent, and how many grams of zinc sulfide are possible?

Zinc is the limiting reagent. It will yield 12.7 grams of zinc sulfide.

Notice that, in this problem, the substance present in greater mass was the limiting reagent. You cannot draw any conclusions about the limiting reagent by the relative masses of reactants.

Now you must find the number of grams of sulfur that are left over. How many grams of sulfur will react with all of the limiting reagent, 8.50 g Zn?

$$8.50 \, g\,Zn \times \frac{1 \, mol\,Zn}{65.4 \, g\,Zn} \times \frac{1 \, mol\,S}{1 \, mol\,Zn} \times \frac{32.1 \, g\,S}{1 \, mol\,S} = 4.17 \, g \, S$$

If you start with 6.84 grams of sulfur and use 4.17 grams, how many will be left?

6.84 g S at start − 4.17 g S used = 2.67 g S unreacted

The procedure for solving limiting reagent problems may be summarized as follows:

1. For each reactant, calculate the amount of product that can result from the given quantity of that reactant. The *smallest* amount calculated in this way is the quantity of product that will form. The reactant that produces the smallest amount of product is the limiting reagent.

2. For each reactant other than the limiting reagent, calculate the amount that will react with all of the limiting reagent.

3. For each reactant other than the limiting reagent, subtract the amount that reacts (Step 2, above) from the quantity initially present. This is the amount of that substance which remains unreacted.

Follow these steps in solving this example:

EXAMPLE 9.11   A solution containing 29.0 grams of calcium nitrate is added to a solution containing 33.0 grams of sodium fluoride. Calcium fluoride precipitates, as shown by the equation

$$Ca(NO_3)_2(aq) + 2\,NaF(aq) \rightarrow CaF_2(s) + 2\,NaNO_3(aq)$$

How many grams of calcium fluoride will precipitate? How many grams of which reactant are in excess?

Begin by calculating the grams of CaF_2 that can be precipitated by 29.0 g $Ca(NO_3)_2$, assuming that calcium nitrate is the limiting reagent.

------   ------

$$29.0\;g\;Ca(NO_3)_2 \times \frac{1\;mol\;Ca(NO_3)_2}{164\;g\;Ca(NO_3)_2} \times \frac{1\;mol\;CaF_2}{1\;mol\;Ca(NO_3)_2} \times \frac{78.1\;g\;CaF_2}{1\;mol\;CaF_2} = 13.8\;g\;CaF_2$$

Now do the same for 33.0 g NaF.

------   ------

$$33.0\;g\;NaF \times \frac{1\;mol\;NaF}{42.0\;g\;NaF} \times \frac{1\;mol\;CaF_2}{2\;mol\;NaF} \times \frac{78.1\;g\;CaF_2}{1\;mol\;CaF_2} = 30.7\;g\;CaF_2$$

Which reactant, $Ca(NO_3)_2$ or NaF, is the limiting reagent? How many grams of CaF_2 will it form?

------   ------

$Ca(NO_3)_2$ is the limiting reagent. It will form 13.8 g CaF_2.

The next step (Step 2, above) is to calculate the grams of NaF that will react with the limiting quantity of $Ca(NO_3)_2$.

------   ------

$$29.0 \text{ g } \cancel{\text{Ca(NO}_3)_2} \times \frac{1 \text{ mol } \cancel{\text{Ca(NO}_3)_2}}{164 \text{ g } \cancel{\text{Ca(NO}_3)_2}} \times \frac{2 \text{ mol } \cancel{\text{NaF}}}{1 \text{ mol } \cancel{\text{Ca(NO}_3)_2}} \times \frac{42.0 \text{ g NaF}}{1 \text{ mol } \cancel{\text{NaF}}} = 14.9 \text{ g NaF}$$

Finally, how many grams of NaF will remain unreacted (Step 3, above)?

33.0 g NaF at start -14.9 g NaF used $= 18.1$ g NaF unreacted

CHAPTER 9 IN REVIEW

9.1   QUANTITATIVE INTERPRETATION OF A CHEMICAL EQUATION

9 A Given a chemical equation, or a reaction for which the equation can be written, and the number of moles of one species in the reaction, calculate the number of moles of any other species. (181)

9.2   MASS CALCULATIONS: STOICHIOMETRY

9 B Given a chemical equation, or a reaction for which the equation can be written, and the number of grams or moles of one species in the reaction, find the number of grams or moles of any other species. (183)

9.3   PERCENTAGE YIELD

9 C Given two of the following, or information from which two of the following may be determined, calculate the third: theoretical yield, actual yield, and percentage yield. (186)

9.4   LIMITING REAGENT PROBLEMS

9 D Given a chemical equation, or information from which it may be determined, and initial quantities of two or more reactants, (a) identify the limiting reagent; (b) calculate the theoretical yield of a specified product, assuming complete use of the limiting reagent; and (c) calculate the quantity of the reactant initially in excess that remains unreacted. (190)

TERMS AND CONCEPTS

Stoichiometry (183)
"Stoichiometry pattern" (183)
Theoretical yield (187)

Actual yield (187)
Percentage yield (187)
Limiting reagent (191)

QUESTIONS AND PROBLEMS

An asterisk (*) identifies a question that is relatively difficult, or that extends beyond the performance goals in the chapter.

Section 9.1

9.1) If carbon dioxide is bubbled through lime water, $Ca(OH)_2(aq)$, the solution becomes cloudy with finely divided calcium carbonate: $Ca(OH)_2(aq) + CO_2(g) \rightarrow CaCO_3(s) + H_2O(\ell)$. If 3.91 moles of carbon dioxide pass through the solution, how many moles of calcium carbonate will form?

9.2) The possibility of a rock sample being limestone (calcium carbonate) may be tested with hydrochloric acid. The formation of bubbles indicates carbon dioxide: $CaCO_3(s) + 2 HCl(aq) \rightarrow CaCl_2(aq) + CO_2(g) + H_2O(\ell)$. How many moles of carbon dioxide will be released for the reaction of 0.284 mole of HCl?

Section 9.2

9.3) In the reaction of Problem 9.1, how many grams of CO_2 are required to produce 0.462 mole of $CaCO_3$?

9.4) Calculate the number of moles of $CaCl_2$ formed from 54.1 grams of $CaCO_3$ in the reaction of Problem 9.2.

9.5) How many grams of sodium sulfate can be crystallized from the reaction of 95.2 grams of sodium chloride with excess sulfuric acid in the reaction $2 NaCl(s) + H_2SO_4(aq) \rightarrow Na_2SO_4(s) + 2 HCl(g)$?

9.6) Silver bromide that remains on photographic film after development is removed by reaction with "hypo," known chemically as sodium thiosulfate, $Na_2S_2O_3$. How many grams of sodium thiosulfate must react to remove 1.09 grams of silver bromide? The equation for the reaction is $2 Na_2S_2O_3(aq) + AgBr(s) \rightarrow Na_3Ag(S_2O_3)_2(aq) + NaBr(aq)$.

9.27) When acetylene, C_2H_2, burns in an oxyacetylene torch, which is used in cutting steel, the equation is $2 C_2H_2(g) + 5 O_2(g) \rightarrow 4 CO_2(g) + 2 H_2O(g)$. How many moles of oxygen are required for each mole of C_2H_2 used?

9.28) The fermentation of molasses, $C_6H_{12}O_6$, in the presence of zymase, a catalyst derived from yeast, may be represented by the equation

$C_6H_{12}O_6(\ell) \xrightarrow{\text{zymase}} 2 C_2H_5OH(\ell) + 2 CO_2(g)$. How many moles of alcohol, C_2H_5OH, are produced from the decomposition of 2.80 moles of molasses?

9.29) Determine the number of grams of C_2H_2 that burned in producing 1.06 moles of CO_2 in the reaction of Problem 9.27.

9.30) If 29.8 grams of C_2H_5OH are formed in the reaction of Problem 9.28, how many moles of $C_6H_{12}O_6$ fermented?

9.31) Stannous fluoride, SnF_2, the "fluoride" in many toothpastes, may be made by the reaction $Sn(s) + 2 HF(g) \rightarrow SnF_2(s) + H_2(g)$. Calculate the grams of HF that are required to prepare 0.750 gram of stannous fluoride.

9.32) A soda acid (sodium hydrogen carbonate) fire extinguisher makes carbon dioxide by the reaction $2 NaHCO_3(s) + H_2SO_4(aq) \rightarrow Na_2SO_4(aq) + 2 H_2O(\ell) + CO_2(g)$. How many grams of carbon dioxide will be produced by an extinguisher charged with 55.0 grams of $NaHCO_3$?

9.7) Baking soda, $NaHCO_3$, reacts with cream of tartar, $KHC_4H_4O_6$, in the baking of a cake. A French recipe calls for 12 grams of cream of tartar. (A standard item in a French kitchen is a small scale, as quantities of solids appear in European recipes in grams rather than cups, teaspoons, or other volume units.) How many grams of baking soda should be called for in the same recipe if all of both materials are to react completely? The equation is $KHC_4H_4O_6(s) + NaHCO_3(s) \rightarrow KNaC_4H_4O_6(s) + H_2O(\ell) + CO_2(g)$.

9.8) How many grams of chlorine will be produced by the reaction of 2.69 grams of manganese dioxide, MnO_2, in the typical laboratory method of preparing chlorine: $MnO_2(s) + 4\,HCl(aq) \rightarrow MnCl_2(aq) + 2\,H_2O(\ell) + Cl_2(g)$?

9.9) Chloroform, $CHCl_3$, may be prepared in the laboratory by the reaction between chlorine and methane:

$$3\,Cl_2(g) + CH_4(g) \rightarrow CHCl_3(\ell) + 3\,HCl(g)$$

Calculate the number of grams of chlorine that are required to produce 45.0 grams of $CHCl_3$.

9.10) One form of soap has the formula $C_{17}H_{35}COONa$. It is prepared by the reaction $C_3H_5(C_{17}H_{35}COO)_3(s) + 3\,NaOH(aq) \rightarrow 3\,C_{17}H_{35}COONa(s) + C_3H_5(OH)_3(\ell)$. Calculate the number of grams of sodium hydroxide (lye) required to prepare 175 grams of soap.

9.11) The purification ability and bleaching action of chlorine both come from the release of oxygen in a reaction which may be summarized as $2\,H_2O(\ell) + 2\,Cl_2(aq) \rightarrow 4\,HCl(aq) + O_2(g)$. How many grams of oxygen will be liberated by 255 grams of chlorine?

9.12) Photosynthesis is the process by which carbon dioxide is converted into sugar, $C_6H_{12}O_6$, with the help of sunlight and the chlorophyll of green plants acting as a catalyst. It may be summarized in the equation $6\,CO_2(g) + 6\,H_2O(g) \xrightarrow[\text{light energy}]{\text{chlorophyll}} C_6H_{12}O_6(aq) + 6\,O_2(g)$. How many grams of sugar may be produced in the reaction that consumes 6.90 grams of carbon dioxide?

9.13) Plaster of Paris is a compound whose formula may be written $CaSO_4 \cdot \frac{1}{2}\,H_2O$ or $(CaSO_4)_2 \cdot H_2O$. Casts made from this material are a higher hydrate of calcium sulfate, $CaSO_4 \cdot 2\,H_2O$. The equation is $3\,H_2O(\ell) + (CaSO_4)_2 \cdot H_2O(s) \rightarrow 2\,CaSO_4 \cdot 2\,H_2O(s)$. What mass of $CaSO_4 \cdot 2\,H_2O$ will result from the reaction of 48.3 grams of $(CaSO_4)_2 \cdot H_2O$?

9.33) Hydrogen peroxide, H_2O_2, is capable of "cleaning" oil paintings of the black lead sulfide, PbS, formed by reaction of white lead in the paint with hydrogen sulfide in the air. The cleaning equation is $PbS(s) + 4\,H_2O_2(aq) \rightarrow PbSO_4(s) + 4\,H_2O(\ell)$. How many grams of PbS can be removed by a solution containing 1.43 grams of H_2O_2, assuming all of the peroxide reacts?

9.34) The hydrate of copper sulfate, $CuSO_4 \cdot 5\,H_2O$, is deep blue. When it is heated, water is driven off, leaving a nearly white anhydrous salt: $CuSO_4 \cdot 5\,H_2O(s) \rightarrow CuSO_4(s) + 5\,H_2O(g)$. How many moles of the hydrate are required to release 48.6 grams of water?

9.35) Dichlorodifluoromethane is the chemical name for the common refrigerant called Freon. How many grams of Freon, CCl_2F_2, can be prepared from 26.8 grams of carbon tetrachloride, CCl_4, by the following catalytic reaction?

$$CCl_4(\ell) + 2\,HF(g) \rightarrow CCl_2F_2(g) + 2\,HCl(g)$$

9.36) The well-known preservative from the biology laboratory, formaldehyde, HCHO, is prepared commercially in solution form by heating methanol, CH_3OH, with oxygen in the presence of copper: $2\,CH_3OH(\ell) + O_2(g) \rightarrow 2\,HCHO(aq) + 2\,H_2O(\ell)$. What mass of formaldehyde can be prepared from 14.8 grams of CH_3OH?

9.37) One part of the blast furnace operation in which iron is extracted from oxide ore is expressed in the equation $Fe_2O_3(s) + 3\,CO(g) \rightarrow 2\,Fe(s) + 3\,CO_2(g)$. How many kg of CO are necessary to produce 1250 kg of iron?

9.38) Electrical energy is derived from chemical energy in starting an automobile engine with a storage battery. The chemical equation for the reaction that occurs is $Pb(s) + PbO_2(s) + 2\,H_2SO_4(aq) \rightarrow 2\,PbSO_4(s) + 2\,H_2O(\ell)$. Calculate the number of grams of H_2SO_4 used in the reaction of 39.4 grams of lead dioxide, PbO_2.

9.39) Limestone caves, such as Mammoth Cave and Carlsbad Caverns, are made when carbon dioxide and water combine to dissolve the limestone, $CaCO_3$:

$$CaCO_3(s) + H_2O(\ell) + CO_2(g) \rightarrow Ca(HCO_3)_2(aq)$$

How many grams of carbon dioxide are required to dissolve 454 grams (1 pound) of limestone?

9.14) Dinitrogen oxide, N_2O—known also as nitrous oxide and laughing gas—was one of the early anesthetics and has recently regained popularity in this role in the dental field. It is made by the decomposition of ammonium nitrate: $NH_4NO_3(s) \rightarrow N_2O(g) + 2 H_2O (\ell)$. How many grams of NH_4NO_3 are required to prepare 12 800 grams of N_2O?

9.15)* One of the early steps in the manufacture of sulfuric acid is to prepare sulfur dioxide from sulfide ores. How many tons of an ore that is effectively 12.1% FeS_2 must be processed to produce 5.00 tons of sulfur dioxide in the reaction $4 FeS_2(s) + 11 O_2(g) \rightarrow 2 Fe_2O_3(s) + 8 SO_2(g)$? (Hint: You can work with a *ton-mole*—the quantity of a substance containing the same number of units as there are atoms in exactly 12 tons of carbon-12—in the same manner as you have been using the "gram-mole." See also page 583 on the use of dimensional analysis for problems involving percent.)

9.16)* Calculate the volume of hydrogen that will result from the reaction of 3.94 grams of zinc in the reaction $Zn(s) + 2 HCl(aq) \rightarrow H_2(g) + ZnCl_2(aq)$, when conditions are such that 1 mole of hydrogen occupies 24.5 liters.

9.40) A chemical reaction that changed the history of the world is $N_2(g) + 3 H_2(g) \rightarrow 2 NH_3(g)$. Conducted at 500°C, up to 1000 atmospheres pressure, and in the presence of a catalyst, the so-called Haber process enabled Germany to manufacture explosives from the nitrogen of the air and thereby wage World War I. How many grams of ammonia, NH_3, can be made from 525 grams of nitrogen?

9.41)* One liter of air at normal atmospheric conditions contains 0.89 gram of nitrogen. What volume of air must be processed to make 100 kilograms (220 pounds) of ammonia by the Haber process if the reaction is $N_2(g) + 3 H_2(g) \rightarrow 2 NH_3(g)$?

9.42)* Excess sulfuric acid is added to 125 milliliters of a solution in which the concentration of barium chloride is 0.750 mole per liter. Calculate the grams of barium sulfate that precipitate by the reaction $H_2SO_4(aq) + BaCl_2(aq) \rightarrow BaSO_4(s) + 2 HCl(aq)$.

Section 9.3

9.17) 3.81×10^4 grams of acetylene, C_2H_2, are produced when 5.19×10^4 grams of methane, CH_4, are treated in an electric arc in the commercial production of acetylene: $2 CH_4(g) \rightarrow C_2H_2(g) + 3 H_2(g)$. Find the percentage yield.

9.18) In the extraction of copper from its sulfide ore, the overall process may be summarized by the equation $Cu_2S(s) + O_2(g) \rightarrow 2 Cu(s) + SO_2(g)$. If the percentage yield is 61.2%, how much copper will result from treatment of 7.00×10^6 grams of Cu_2S?

9.19) In one commercial method for preparing sodium hydroxide, sodium carbonate reacts with calcium hydroxide in boiling water: $Na_2CO_3(aq) + Ca(OH)_2(aq) \rightarrow 2 NaOH(aq) + CaCO_3(s)$. If the process is known to have a 92.0% yield, how many kilograms of sodium carbonate should be used to produce 325 kilograms of sodium hydroxide?

9.43) When 9.08×10^5 grams (one ton) of sodium chloride are electrolyzed in the commercial production of chlorine, 5.24×10^5 grams of chlorine are produced: $2 NaCl(aq) + 2 H_2O(\ell) \rightarrow H_2(g) + Cl_2(g) + 2 NaOH(aq)$. Calculate the percentage yield.

9.44) One step of the Ostwald process for manufacturing nitric acid involves making nitrogen monoxide by oxidizing ammonia in the presence of a platinum catalyst: $4 NH_3(g) + 5 O_2(g) \rightarrow 4 NO(g) + 6 H_2O(g)$. If the percentage yield is 80.3%, how many grams of NO can be produced from 4.00×10^3 grams of NH_3?

9.45) The reaction $2 Al(s) + 2 H_3PO_4(aq) \rightarrow 3 H_2(g) + 2 AlPO_4(s)$ is known to have a yield of 77.8%. If the actual yield in an experiment is 39.6 grams of $AlPO_4$, how many grams of H_3PO_4 reacted?

9.20) In a laboratory preparation of chloroform, $CHCl_3$, a student introduced 3.85 grams of methane, CH_4, and excess chlorine into a reaction chamber. On completion of the reaction, $CH_4(g) + 3\,Cl_2(g) \rightarrow CHCl_3(\ell) + 3\,HCl(g)$, he recovered 23.0 grams of chloroform. What does this represent in percent?

9.21) A common laboratory experiment is making aspirin, $C_9H_8O_4$, from salicylic acid, $C_7H_6O_3$, and acetic anhydride, $C_4H_6O_3$ by the reaction $C_7H_6O_3 + C_4H_6O_3 \rightarrow C_9H_8O_4 + HC_2H_3O_2$. If the yield for this reaction is 76.1%, how many grams of aspirin will be produced from 15.1 grams of salicylic acid and excess acetic anhydride?

9.22) Calcium cyanamide, $CaCN_2$, is an important fertilizer chemical. It is made from calcium carbide, CaC_2, and nitrogen from the air: $CaC_2(s) + N_2(g) \rightarrow CaCN_2(s) + C(s)$. How many grams of CaC_2 must be processed to produce 5.00×10^4 g $CaCN_2$ if the percentage yield is 72.6%?

Section 9.4

9.23) 40.0 grams of ammonia and 40.0 grams of oxygen are made to react until all of the limiting reagent is used. Calculate the grams of water vapor released if the reaction equation is $4\,NH_3(g) + 3\,O_2(g) \rightarrow 2\,N_2(g) + 6\,H_2O(g)$. How many grams of which reactant will remain unchanged at the end of the reaction?

9.24) Zinc may be extracted from its sulfide ore by first roasting to produce zinc oxide, and then heating with powdered coal. The second step is $ZnO(s) + C(s) \rightarrow Zn(s) + CO(g)$. In a laboratory run for the same reaction a chemist mixed 19.6 grams of carbon with 135 grams of zinc oxide. Calculate the grams of zinc metal that will be produced by complete conversion of the limiting species and the grams of the other solid reactant that will remain.

9.25)* The solvent carbon tetrachloride—once widely used in dry cleaning and for extinguishing fires, but now considered too toxic (poisonous) to be used in the home, even as a spot remover— is prepared commercially by the reaction $CS_2(\ell) + 2\,S_2Cl_2(\ell) \rightarrow CCl_4(\ell) + 6\,S(s)$. How

9.46) The sea is a major industrial source of magnesium, where it exists as a very dilute solution of magnesium chloride. The hydroxide is precipitated from sea water by treatment with slaked lime (calcium hydroxide): $MgCl_2(aq) + Ca(OH)_2(s) \rightarrow Mg(OH)_2(s) + CaCl_2(aq)$. Calculate the percentage yield in the process if 489 kilograms of magnesium hydroxide are recovered from the addition of 891 kilograms of calcium hydroxide to an excess of sea water.

9.47) The metal chromium may be extracted from its oxide ore by reaction with carbon: $3\,C_{(s)} + Cr_2O_3(s) \rightarrow 2\,Cr(s) + 3\,CO(g)$. How many kilograms of chromium can be obtained from 402 kilograms of Cr_2O_3 if the percentage yield for the process is 86.4%?

9.48) When iron pyrite, FeS_2, an impurity in coal, reacts with oxygen, harmful SO_2 is produced: $4\,FeS_2(s) + 11\,O_2(g) \rightarrow 2\,Fe_2O_3(s) + 8\,SO_2(g)$. In a laboratory test in which the percentage yield is 89.3%, how many grams of FeS_2 reacted if the mass of SO_2 collected is 21.8 grams?

9.49) A laboratory experiment calls for the use of 6.00 grams of manganese dioxide, MnO_2, and a solution containing 25.8 grams of HCl in the preparation of chlorine by the reaction $MnO_2(s) + 4\,HCl(aq) \rightarrow MnCl_2(aq) + 2\,H_2O(\ell) + Cl_2(g)$. Assuming that the limiting reagent reacts completely, how many grams of chlorine will result? Also, how many grams of the excess reagent will be unreacted at the end of the experiment?

9.50) Although nearly 80% of the air is elemental nitrogen, it is very difficult to change this nitrogen into useful compounds. One successful process uses the reaction $Na_2CO_3(s) + 4\,C(s) + N_2(g) \rightarrow 2\,NaCN(s) + 3\,CO(g)$. In a laboratory experiment with this reaction, 26.8 grams of carbon and 122 grams of sodium carbonate reacted with an abundance of air until one of the solids was completely consumed. Calculate the yield of sodium cyanide, NaCN, a highly poisonous salt used in electroplating. Also find the number of grams of the solid reactant that was left.

9.51)* One step in the extraction of copper from its ore involves the reaction of copper(I) sulfide with copper(I) oxide: $2\,Cu_2O(s) + Cu_2S(s) \rightarrow 6\,Cu(s) + SO_2(g)$. To maximize this yield of copper it is evident that the producer must use two copper compounds in the proper proportion. Sup-

many tons of carbon tetrachloride will result from the reaction of all of the limiting reagent in a mixture of 3.00 tons of S_2Cl_2 and 1.00 ton of CS_2? Which reactant is in excess, and by how many tons? (The hint for Problem 9.15 applies here too.)

pose he uses 1.00 ton of each compound in a production run. How many tons of elemental copper will he produce? How many *additional* tons of which compound should be added to balance the 1.00 ton of the limiting reagent? What then will be the total yield of copper in tons?

At this point we close the chapter with a pair of 'challenge' questions that do not specify a chemical equation. The molar relationships between given and wanted species can be found, however, in the wording of the problems. With those relationships, the problems can be solved, using the principles developed in this chapter.

9.26)* A mixed industrial product contains magnesium as a hexahydrate of magnesium chloride, $MgCl_2 \cdot 6 H_2O$. A 5.25-gram sample of the material is placed in water, dissolving the soluble components of the mixture, including the magnesium chloride. The chloride ion in the solution is then precipitated as silver chloride, $AgCl$, which is filtered, dried, and weighed at 4.01 grams. Assuming magnesium chloride hexahydrate to be the only source of chloride ion, calculate (a) the grams of magnesium and (b) the percentage of $MgCl_2 \cdot 6 H_2O$ in the sample.

9.52)* One of the methods of analyzing steel for its manganese content involves a series of chemical changes resulting in the complete conversion of the manganese to Mn_3O_4. From prior experience, however, it is known that the percentage yield is 96%. One such analysis of an 11.2-gram sample of steel resulted in the separation of 0.337 gram of Mn_3O_4. Find the percentage of manganese in the steel.

10

chemical bonding

For several chapters we have accepted Dalton's idea that atoms of different elements combine to form compounds. We've given them names and written their formulas. But never have we considered how these combinations come about—what causes atoms to stick together in molecular or ionic species. This is the kind of thing we will consider in this chapter on chemical bonding. Specifically we will examine two types of bonds, ionic bonds and covalent bonds. But first, let's introduce a tool that makes it easier to picture the role played by electrons in chemical bonding: the Lewis symbol.

10.1 LEWIS SYMBOLS OF THE ELEMENTS

PG 10 A Write the Lewis symbol for any atom in one of the A groups of the periodic table.

In Chapter 5 (p. 90) it was stated that the similar chemical properties of members of the same family of elements can be attributed to their **valence electrons.** These are the s and p electrons of the highest occupied energy level when the atom is in its ground state. The ground state valence electron configuration of the alkali metals, for example, is ns^1, where n is the principal energy level of the valence electrons. For lithium this configuration is $2s^1$, and $n = 2$; for sodium, $3s^1$, and $n = 3$; and so forth. The ground state valence electron

Table 10.1
Lewis Symbols of the Elements

Group	A	2A	3A	4A	5A	6A	7A	0
Highest energy electron configuration	ns^1	ns^2	ns^2np^1	ns^2np^2	ns^2np^3	ns^2np^4	ns^2np^5	ns^2np^6
Number of valence electrons	1	2	3	4	5	6	7	8
Lewis symbol, third period element	Na·	Mg·	·Al·	·Si·	·P:	·S:	:Cl:	:Ar:

configurations of the representative elements—the elements in the A groups of the periodic table—are summarized in Table 10.1.

The last line of Table 10.1 shows the **Lewis symbols,** or the **electron dot symbols,** of the elements of the third period. The Lewis symbol of an element is the chemical symbol with one or more dots distributed around it. The number of dots corresponds to the number of valence electrons in an atom of the element. Lewis symbols for alkali metals, represented by sodium in Table 10.1, have one dot for one valence electron, the ns^1 electron. The symbol for lithium is Li·; for potassium, K·; and so forth.

Lewis symbols for the alkaline earth metals have two dots, matching the ns^2 configuration of the valence electrons, as shown for magnesium. For calcium the Lewis symbol is Ca: and for barium, Ba:. Note that the exact location of the dots is not important. For magnesium, for example, the symbol may be Mg: to suggest that the electrons occupy the same atomic orbital. If the purpose of the symbol is to suggest that magnesium can form two bonds, ·Mg· may be more useful. Lewis symbols for the halogens have seven dots, as shown for chlorine, to match the ns^2np^5 (two s electrons + five p electrons = seven valence electrons) of their highest energy level. The noble gas family, Group 0, is characterized by a full set of 8 valence electrons, or an **octet,** as it is often called.

Lewis symbols are a great aid in accounting for what we believe to be taking place at the atomic level during the making and breaking of chemical bonds. We will use them for that purpose throughout this chapter.

10.2 MONATOMIC IONS WITH NOBLE GAS ELECTRON CONFIGURATIONS

PG 10 B Identify by name and symbol the monatomic ions that are isoelectronic with a given noble gas atom, and write the electron configurations of those ions.

In Figure 5.11 (p. 89) we illustrated the formation of a sodium ion by the removal of an electron from a neutral sodium atom. This concept was presented again in Equations 6.2 to 6.4. Table 10.2 offers the same idea, this time ex-

Table 10.2

Formation of Monatomic Cations

Element	Atom	Monatomic Ion		Electron(s)
Sodium	11 p$^+$ 11 e$^-$ →	11 p$^+$ 10 e$^-$	+	e$^-$
	Na· →	Na$^+$	+	·(e$^-$)
	$1s^22s^22p^63s^1$ →	$1s^22s^22p^6$	+	e$^-$
Magnesium	12 p$^+$ 12 e$^-$ →	12 p$^+$ 10 e$^-$	+	e$^-$ + e$^-$
	Mg· →	Mg^{2+}	+	·(e$^-$) + ·(e$^-$)
	$1s^22s^22p^63s^2$ →	$1s^22s^22p^6$	+	2 e$^-$
Aluminum	13 p$^+$ 13 e$^-$ →	13 p$^+$ 10 e$^-$	+	e$^-$ + e$^-$ + e$^-$
	·Al· →	Al^{3+}	+	·(e$^-$) + ·(e$^-$) + ·(e$^-$)
	$1s^22s^22p^63s^23p^1$ →	$1s^22s^22p^6$	+	3 e$^-$

panded to show the Lewis symbol for sodium and the electron configurations for the different species. A neutral sodium atom contains 11 protons and 11 electrons. Its electron configuration is $1s^22s^22p^63s^1$, and it has Na· as its Lewis symbol. If you remove the highest energy electron, the 3s electron, from the neutral sodium atom, the sodium ion that remains has the configuration $1s^22s^22p^6$. There are still 11 protons, leaving a net charge of +1 for the ion, which is represented by the symbol Na$^+$.

The formation of a magnesium ion is similar. The magnesium atom begins with 12 protons and 12 electrons, and produces an ion with a +2 charge by losing two electrons. Notice the electron configuration of the magnesium ion, Mg^{2+}; it is identical to that of the sodium ion. Two species having identical electron configurations are said to be **isoelectronic** with each other. The prefix *iso-* suggests *sameness*, as in *isotope.*

What do you suppose is the electron configuration of the aluminum ion, Al^{3+}? The neutral atom must lose three electrons, two 3s and one 3p, to produce an ion with a +3 charge. Do you see that this will reduce the total electron population to 10, which will again yield the electron configuration $1s^22s^22p^6$? This is shown in Table 10.2.

Sodium, magnesium, and aluminum are "active" metals. They do not occur in nature as pure elements, nor are they completely stable in their elemental states when exposed to air or water. In air, freshly cut sodium is promptly covered with an oxide coating in which the sodium exists as the more stable Na$^+$ ion. Magnesium and aluminum are both used as structural metals, but it is possible to do so only because the elements cover themselves with a thin protective film when exposed to air. The film, a carbonate for magnesium and an oxide for aluminum, prevents further corrosion. The magnesium and aluminum are present as ions, Mg^{2+} and Al^{3+}, in these films. It appears that sodium, magnesium, and aluminum are more stable in nature as ions than as neutral atoms, and that these ions are isoelectronic with each other.

At this point ask yourself: Is there an *element* whose neutral atoms have this same $1s^2 2s^2 2p^6$ electron configuration? The answer, of course, is yes. Its neutral atom must have ten protons to match the ten electrons, and that identifies the element with atomic number 10, neon. And what are the chemical properties of neon? It is a noble gas. The word *noble* means that it is unreactive, or stable, in its elemental state. The valence electron configuration of neon—of all noble gases except helium—is the complete octet of highest energy s and p electrons, an $ns^2 np^6$ configuration, where n = 2 for neon. It appears that the noble gas configuration represents stability not only for the noble gases, but for many metal cations too.

An examination of nonmetal anions (ions with negative charges) leads to the same conclusion. What must be done with neutral atoms of fluorine, oxygen, and nitrogen to change them into negatively charged ions, F^-, O^{2-}, and N^{3-}?* To make negatively charged ions from neutral atoms you must add one or more electrons to each atom. Lewis symbols and electron configurations again illustrate the process, which is summarized in Table 10.3. The fluoride, oxide, and nitride anions are isoelectronic with each other—and with the stable noble gas neon.

Table 10.3
Formation of Monatomic Anions

Element	Atom		Electron(s)		Monatomic ion
Fluorine	9 p⁺ 9 e⁻	+	e⁻	→	9 p⁺ 10 e⁻
	:F̈:	+	·(e⁻)	→	[:F̈:]⁻ F⁻
	$1s^2 2s^2 2p^5$	+	e⁻	→	$1s^2 2s^2 2p^6$
Oxygen	8 p⁺ 8 e⁻	+	e⁻ + e⁻	→	8 p⁺ 10 e⁻
	·Ö:	+	·(e⁻) + ·(e⁻)	→	[:Ö:]²⁻ O²⁻
	$1s^2 2s^2 2p^4$	+	2 e⁻	→	$1s^2 2s^2 2p^6$
Nitrogen	7 p⁺ 7 e⁻	+	e⁻ + e⁻ + e⁻	→	7 p⁺ 10 e⁻
	·N̈:	+	·(e⁻) + ·(e⁻) + ·(e⁻)	→	[:N̈:]³⁻ N³⁻
	$1s^2 2s^2 2p^3$	+	3 e⁻	→	$1s^2 2s^2 2p^6$

*Nitride ion, N^{3-}, is not common, but does exist in a small number of compounds.

The fact that all Group 1A monatomic ions have a +1 charge, all Group 2A monatomic ions have a +2 charge, and all Group 7A monatomic ions have a −1 charge, to name just three groups, suggests that the pattern built around the neon configuration is duplicated around other noble gas atoms. With a slight modification for lithium and hydrogen (see below), this is the case.

EXAMPLE 10.1   Write the electron configurations for the calcium and chloride ions, Ca^{2+} and Cl^-. With what noble gases are these ions isoelectronic?

To begin, note the locations of calcium and chlorine in the periodic table. Write their Lewis symbols and from them state the number of electrons that must be gained or lost to achieve complete octets at the highest energy level.

Ca: must lose two electrons to achieve an octet;

:Cl: must gain one electron to achieve an octet.

You are now ready to answer the main questions: the electron configuration for Ca^{2+} and Cl^-, please, and the noble gases with which they are isoelectronic.

Both ions are isoelectronic with argon, $1s^2 2s^2 2p^6 3s^2 3p^6$

The calcium atom starts with the configuration $1s^2 2s^2 2p^6 3s^2 3p^6 4s^2$. In losing two electrons to yield Ca^{2+} it reaches the electron configuration of argon. Chlorine, with configuration $1s^2 2s^2 2p^6 3s^2 3p^5$, must gain one electron to become Cl^-, which is isoelectronic with argon.

EXAMPLE 10.2   Identify at least one more cation and one more anion that are isoelectronic with an atom of argon.

K^+, Sc^{3+}, S^{2-}, and P^{3-}

The formation of K^+, S^{2-}, and P^{3-} are identical to the formation of Na^+, O^{2-}, and N^{3-}, respectively, the ions immediately above them in the periodic table. You have not yet had information that would lead you to include the scandium ion, Sc^{3+}, in your answer to the question. If given the electron configuration of scandium (Z = 21) $1s^22s^22p^63s^23p^64s^23d^1$, you might have guessed the charge on the ion would be +3 because removal of the three highest energy electrons leaves the configuration of the noble gas, argon. Your guess would have been correct.

It was noted earlier that these ideas must be modified slightly when applied to ions formed by lithium and hydrogen. Both elements form ions that are isoelectronic with a noble gas atom, but they do not have a complete octet of electrons. The noble gas they duplicate is helium, which has the electron configuration of $1s^2$. Lithium, Li, with electron configuration $1s^22s^1$, loses its 2s electron to form the lithium ion, Li^+. Hydrogen, H, with configuration $1s^1$, gains an electron to reach the helium configuration and form the hydride ion, H^-.

You may wonder about the hydrogen ion, H^+. This ion does not normally exist by itself (p. 111), but rather as a "hydrated" hydrogen ion, $H \cdot H_2O^+$, commonly called the hydronium ion and written H_3O^+. This polyatomic ion exists in aqueous acid solutions, and is not properly a part of a consideration of monatomic ions.

Figure 10.1 is a periodic table showing most of the monatomic ions that are isoelectronic with noble gases, as well as the noble gases themselves. Refer-

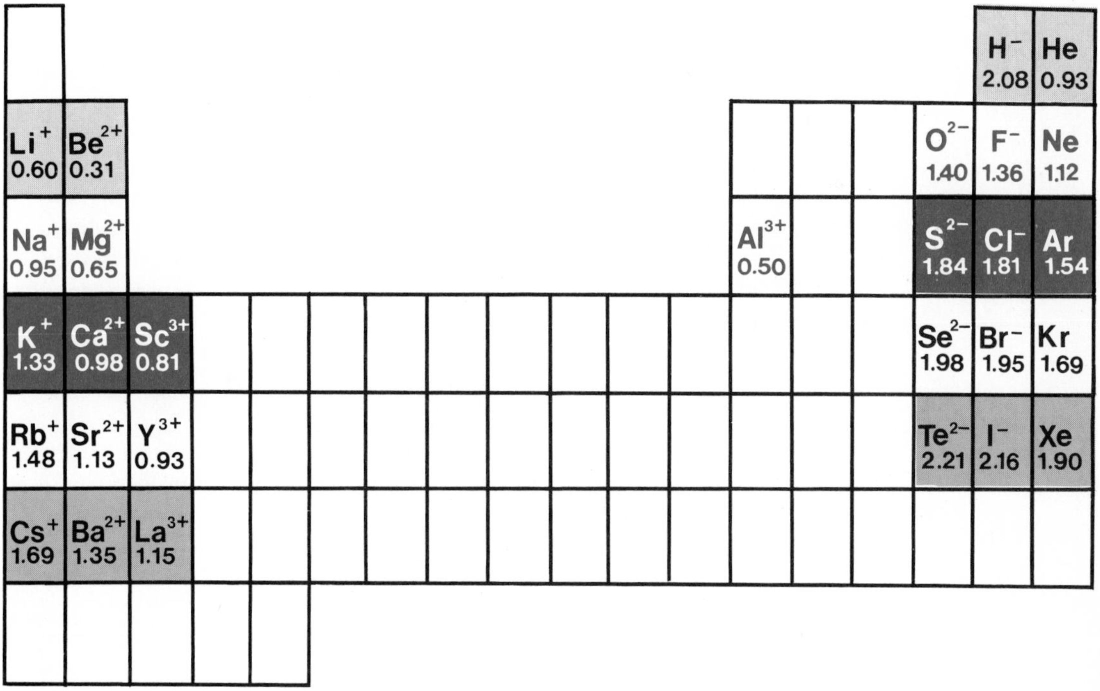

Figure 10.1
Monatomic ions having noble gas electron configurations. Each color group includes one noble gas atom and the monatomic ions that are isoelectronic with it. The numbers show the ionic radius in angstroms. (An angstrom is a unit of length equal to 0.1 nanometer.) For all isoelectronic species, size decreases as nuclear charge increases.

ring to the table, this section may be summarized in the following generalization: *Metal atoms with 1, 2, or 3 electrons more than the preceding noble gas tend to acquire noble gas configurations by losing these electrons, thereby forming cations with charges of +1, +2, and +3, respectively. Nonmetal atoms with 1, 2, or 3 electrons fewer than the following noble gas reach noble gas configurations by gaining these electrons, thereby forming anions with charges of −1, −2, and −3, respectively.*

Not all monatomic ions are isoelectronic with atoms of noble gases. Cations of the elements to the right of Group 3B in the periodic table generally contain d electrons that are not present in the preceding noble gas atom. We will not attempt to describe these ions in terms of electron configuration, except to note that they usually differ from the parent atom by the absence of the highest energy s electrons and sometimes a d electron.

10.3 IONIC BONDS

We've been considering the formation of monatomic ions by atoms gaining or losing electrons as if this is something they did quite freely. In fact, not many atoms exist in nature uncombined, as pure elements rather than as part of a compound. Nowhere, for example, are sodium or chlorine atoms to be found, but there are large natural deposits of sodium chloride made up of positively charged sodium ions and negatively charged chloride ions. The compound may also be obtained by evaporating sea water, which contains sodium and chloride ions in solution. The ions arrange themselves in crystals with definite geometric patterns, as shown in Figure 10.2. They are held in these fixed positions by strong electrostatic forces called **ionic bonds.** Compounds made up of ions held in place by ionic bonds are called **ionic compounds.**

In the laboratory we can produce sodium chloride by direct combination of the elements. Chlorine molecules are diatomic. In the presence of sodium the chlorine molecules are broken into individual atoms. It is then that a sodium atom can lose an electron to become a sodium ion, and a chlorine atom gains that electron to become a chloride ion. Using Lewis diagrams,

$$Na \cdot \overset{\frown}{+} \cdot \overset{..}{\underset{..}{Cl}} : \rightarrow Na^+ + \left[: \overset{..}{\underset{..}{Cl}} : \right]^- \rightarrow NaCl \text{ crystal}$$

The ionic crystal (Fig. 10.2) forms from the ions produced in this way.

There are many kinds of ionic crystals. The ions present do not have to be in a 1:1 ratio, as with sodium chloride. If the compound is calcium chloride, in which each calcium atom has two valence electrons to lose, there must be two chlorine atoms to receive them:

$$Ca \vdots + \begin{matrix} \cdot \overset{..}{\underset{..}{Cl}} : \\ \\ \cdot \overset{..}{\underset{..}{Cl}} : \end{matrix} \rightarrow Ca^{2+} + 2 \left[: \overset{..}{\underset{..}{Cl}} : \right]^- \rightarrow CaCl_2 \text{ crystal}$$

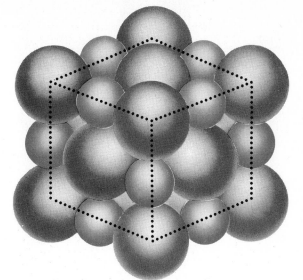

Figure 10.2
Arrangement of ions in a sodium chloride type crystal. The small spheres represent the sodium ions, the larger spheres the chloride ions.

The 1:2 ratio of calcium ions to chloride ions in calcium chloride is reflected in the formula of the compound, $CaCl_2$. Several combinations of numbers and charges enter into ionic crystals, but always in such proportions as to yield a compound that is electrically neutral. Crystals are not restricted to monatomic ions; polyatomic ions do exactly the same thing in compounds such as $(NH_4)_2SO_4$, although their structures are more complicated. One example is calcium carbonate, $CaCO_3$ (Fig. 10.3).

Figure 10.3
Model of calcium carbonate. Each carbon atom is bonded to three oxygen atoms, making up the carbonate ion. There are equal numbers of calcium ions, Ca^{2+}, and carbonate ions, CO_3^{2-}, yielding a compound that is electrically neutral, $CaCO_3$.

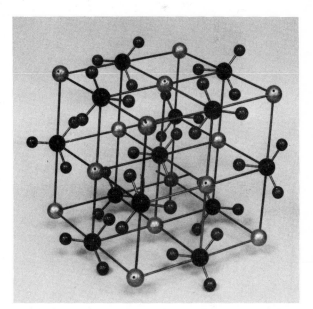

Ionic bonds are very strong. Because of this, nearly all ionic compounds are solids at room temperature. To melt an ionic compound there must be enough kinetic energy to break the ionic bonds and free the ions from each other. This takes a high temperature, often 800°C or more. Solid ionic compounds are poor conductors of electricity because the ions are locked in place in the crystal. When the substance is melted or dissolved, the crystal is destroyed and the ions are free to move. These mobile charged particles can carry electric current, so liquid ionic compounds and their solutions are good conductors.

10.4 SIZES OF IONS

PG 10 C   Compare the radii of several given isoelectronic monatomic ions and suggest an explanation for the trend observed.

10 D   Compare the radii of several given ions formed by elements in the same group in the periodic table, and suggest an explanation for the trend observed.

Returning to Figure 10.2 for a moment, notice the relative sizes of the two kinds of ions in the crystal. The smaller spheres are the Na^+ ions, and the larger spheres are the Cl^- ions. This situation applies in most ionic compounds; the negative ion is usually the larger of the two. Occasionally, the effect is so great that anions in the crystal are actually "in contact" with each other, with the cations fitting into crevices between them. This happens with lithium iodide (Fig. 10.4). Ionic size apparently plays a significant role in determining the properties of some species, particularly in regard to some of the smaller ions such as Li^+.

In Chapter 5 (p. 94), three influences on atomic size were identified: (1) number of occupied principal energy levels, (2) nuclear charge, and (3) the "shielding" of part of the nuclear charge by inner electrons. These same influences determine ionic sizes. Ionic radii, as well as the radii of the noble gases with which the ions are isoelectronic, are given in Figure 10.1. As with atoms in the same group in the periodic table, the sizes of monatomic ions increase as you go down the group because the number of occupied principal energy levels increases, in accord with (1) above.

The difference in size between Na^+ and F^- is explained in terms of nuclear charge, (2) above. Recall that F^- and Na^+ are isoelectronic. The greater nuclear charge in Na^+ (11 protons vs. 9 in F^-) pulls the outer electrons in more tightly and hence makes the ion smaller. From Figure 10.1 we see that this effect is general. Compare, for example, the radii of the series of ions O^{2-}, F^-, Na^+, Mg^{2+}, and Al^{3+}. All of these ions have the neon structure with a total of 10 electrons. As the nuclear charge increases from 8 with O^{2-} to 13 for Al^{3+}, the radius decreases steadily, from 0.140 nm for O^{2-} to 0.050 nm for Al^{3+}. We note further that when the accepted size values of noble gas atoms are included, as they are in Figure 10.1, they fit perfectly into the size trend among isoelectronic

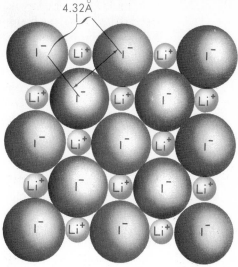

4.32Å

Figure 10.4
Iodide ions in lithium iodide appear to be in contact. Ionic radius is considered to be half the distance between adjacent iodide ions.

species. While the regularity in sizes of isoelectronic species encourages us to believe the two measuring techniques—one for ions, the other for noble gas molecules—are compatible, we must remain cautious in reaching this conclusion.

10.5 COVALENT BONDS

PG 10 E Distinguish between ionic and covalent bonds.

The type of bonding described in Section 10.3, where a transfer of electrons produces oppositely charged ions that are held together by electrostatic forces, explains quite satisfactorily the properties of many compounds such as sodium fluoride, NaF. There are, however, a great many compounds whose properties are inconsistent with an ionic model. Consider, for example, hydrogen fluoride, HF, and methane, CH_4. Both HF and CH_4 are gases at room temperature and atmospheric pressure, while NaF is a solid. When condensed to liquids, both hydrogen fluoride and methane are nonconductors, while liquid sodium fluoride is a good conductor. Such striking contrasts argue strongly against the existence of ions in HF and CH_4.

HF and CH_4 are **molecular compounds;** they consist of individual particles, or **molecules** (p. 109). What holds atoms together in molecules? The forces appear to be electrostatic, and to that extent similar to the forces in ionic bonds. But they are also different—and there is a large "in between" area as well.

In 1916, G. N. Lewis proposed that two atoms in a molecule are held together by a **covalent bond** in which they share a pair of electrons. The idea is that when the bonding electrons can spend most of their time between two atoms, they attract *both* positively charged nuclei and "couple" the atoms to each other, much as two railroad cars are held together by the coupler between them. The result is a bond that is permanent until broken by a chemical change.

The simplest molecule and the simplest covalent bond appear in hydrogen, H_2. The formation of a molecule of H_2 can be represented as

$$H\cdot + \cdot H \rightarrow H:H \quad or \quad H—H$$

The two dots or the straight line drawn between the two atoms represents the covalent bond that holds the atoms together. In modern terms we say that the *electron cloud* or *charge cloud* formed by the two electrons is concentrated in the region between the two nuclei. This is where there is the greatest probability of locating the bonding electrons. The atomic orbitals of the separated atoms are said to *overlap* (Fig. 10.5).

A similar approach shows the formation of the covalent bond between two fluorine atoms to form a molecule of F_2, and between one hydrogen atom and one fluorine atom to form an HF molecule:

$$:\ddot{F}\cdot + \cdot\ddot{F}: \rightarrow :\ddot{F}:\ddot{F}: \quad or \quad :\ddot{F}—\ddot{F}: \quad or \quad F—F$$

$$H\cdot + \cdot\ddot{F}: \rightarrow H:\ddot{F}: \quad or \quad H—\ddot{F}: \quad or \quad H—F$$

Fluorine has seven valence electrons. The 2s orbital and two of the 2p orbitals are filled, but the remaining 2p orbital has only one electron. The F_2 bond is considered to be formed by the overlap of the half-filled 2p orbitals of two

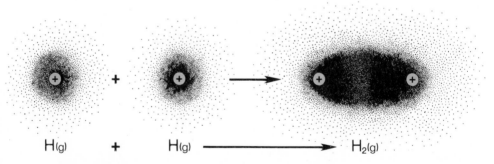

Figure 10.5
The "electron cloud" picture of the hydrogen atom and the hydrogen molecule. Because nobody has ever seen an atom or the electrons that constitute chemical bonds between atoms, it is difficult to describe them accurately or to draw a satisfactory picture of them. The quantum mechanical concept, however, suggests that the nucleus of a hydrogen atom is, over a period of time, "surrounded" by an "electron cloud," or "charge cloud," which is represented as dots showing many instantaneous positions of the single electron. The uniform distribution of dots around the nucleus shows the spherical shape of the 1s orbital. When two hydrogen atoms form a covalent bond and become a molecule, their 1s orbitals are said to overlap. The bonding electrons are believed to spend more time between the two nuclei, as suggested by the heavier density of electron position dots between nuclei in the illustration, and lower dot density at the outer edges of the molecule.

fluorine atoms. In the HF molecule, the bond forms from the overlap of the half-filled 1s orbital of a hydrogen atom with the half-filled 2p orbital of a fluorine atom.

When used to show the bonding arrangement between atoms in a molecule, electron dot diagrams are commonly called **Lewis diagrams, Lewis formulas,** or **Lewis structures.** Notice that the unshared electron pairs of fluorine are shown for two of the three Lewis diagrams for F_2 and HF above, but not for the third. Technically they should always be shown, but they are frequently omitted when not required by the context in which the diagrams appear. Unshared electron pairs are often called **lone pairs.**

The valence electron population around each atom in the Lewis diagrams of H_2, F_2, and HF is significant. When two bonding electrons are shared by two atoms, the electrons effectively "belong" to both atoms. They are therefore counted as valence electrons for each bonded atom. Thus each hydrogen atom in H_2 and the hydrogen atom in HF have two electrons, the same number as an atom of the noble gas helium. Each fluorine atom in F_2 and the fluorine atom in HF have eight valence electrons, matching neon and the other noble gas atoms. These and many similar observations lead us to believe that the stability of a noble gas electron configuration contributes to the formation of covalent bonds, just as it contributes to the formation of ions. This generalization is often referred to as the **octet rule,** or **rule of eight,** because of the eight electrons that characterize the highest occupied energy level of a noble gas atom.

The tendency toward a completed octet of electrons in a bonded atom reflects one of the driving forces for all physical or chemical change—a natural tendency for a system to move to the lowest energy state possible. Every time you drop something, it falls to the floor because its energy is lower on the floor than when you hold it in your hand. The formation of hydrogen molecules from hydrogen atoms illustrates this point. It may be represented by the equation

$$H(g) + H(g) \rightarrow H_2(g) + 435 \text{ kJ}$$

in which kJ is the abbreviation for kilojoule, the unit in which reaction energies are measured (see Chapter 19). This equation indicates that the energy of a system involving two moles of hydrogen atoms is lower by 435 kilojoules when those atoms are bonded into one mole of hydrogen molecules than when they are separate atoms.

10.6 POLAR AND NONPOLAR COVALENT BONDS

PG 10 F   Distinguish between polar and nonpolar covalent bonds.

10 G   Given the electronegativities of two elements, classify the bond between them as nonpolar covalent, polar covalent, or primarily ionic. Identify the positive and negative ends of the bonds, if any.

As we might expect, the two electrons joining the atoms in the H_2 molecule are shared equally by the two nuclei. Another way of saying this is that the electron cloud distribution is balanced in the molecule, with the greatest probability of locating the bonding electrons in the overlap region between the bonded atoms, as shown in Figure 10.5. **A bond in which the distribution of bonding electron charge is symmetrical, or balanced, is said to be nonpolar.** A bond between identical atoms, as in H_2 or F_2, is always nonpolar.

In the HF molecule, the distribution of the bonding electrons is somewhat different from that in H_2 or F_2. Here the density of the electron cloud favors the fluorine atom; the bonding electrons, on the average, are shifted towards fluorine and away from hydrogen (Fig. 10.6). **A bond with an unsymmetrical electron distribution is a polar bond.** In the hydrogen fluoride molecule, the fluorine atom acts as a negative pole and the hydrogen atom as a positive pole.

Bond polarity may be described in terms of the **electronegativities** of the bonded atoms. **Electronegativity is a measure of the relative ability of two atoms to attract the pair of electrons forming a single covalent bond between them.** High electronegativity identifies an element with a strong attraction for bonding electrons.

Electronegativity values of Group A elements are shown in Figure 10.7. Notice that electronegativities tend to be greater at the top of any column. This is because the bonding electrons are closer to the nucleus in a smaller atom, and are therefore attracted by it more strongly. Electronegativities also increase from left to right across any row of the periodic table. This matches the increase in nuclear charge among atoms whose bonding electrons are in the same principal energy level. Perhaps you recognize these two explanations. They are identical with those given for atomic and ionic sizes (pp. 94 and 208). In general, electronegativities are highest at the upper right region of the periodic table, and lowest in the lower left region.

You can estimate the polarity of a bond by calculating the difference between the electronegativity values for the two elements: the greater the difference, the more polar the bond. In nonpolar H_2 and F_2 molecules, where two atoms of the same element are bonded, the electronegativity difference is zero. In the polar HF molecule the electronegativity difference is 4.0 for fluorine minus 2.1 for hydrogen, or 1.9. A bond between carbon and chlorine, for example, with an electronegativity difference of $3.0 - 2.5 = 0.5$, is more polar than an H—H bond, but less polar than an H—F bond.

The more electronegative element toward which the bonding electrons are displaced acts as the "negative pole" in a polar bond. This is sometimes indicated by using an arrow rather than a simple dash, with the arrow pointing to the negative pole. In a bond between hydrogen and fluorine this is $H \rightarrow F$.

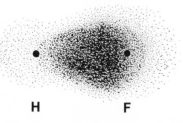

Figure 10.6
A polar bond. Fluorine in a molecule of HF has a higher electronegativity than hydrogen. The bonding electron pair is therefore shifted toward fluorine. The nonsymmetrical distribution of charge yields a polar bond.

H   F

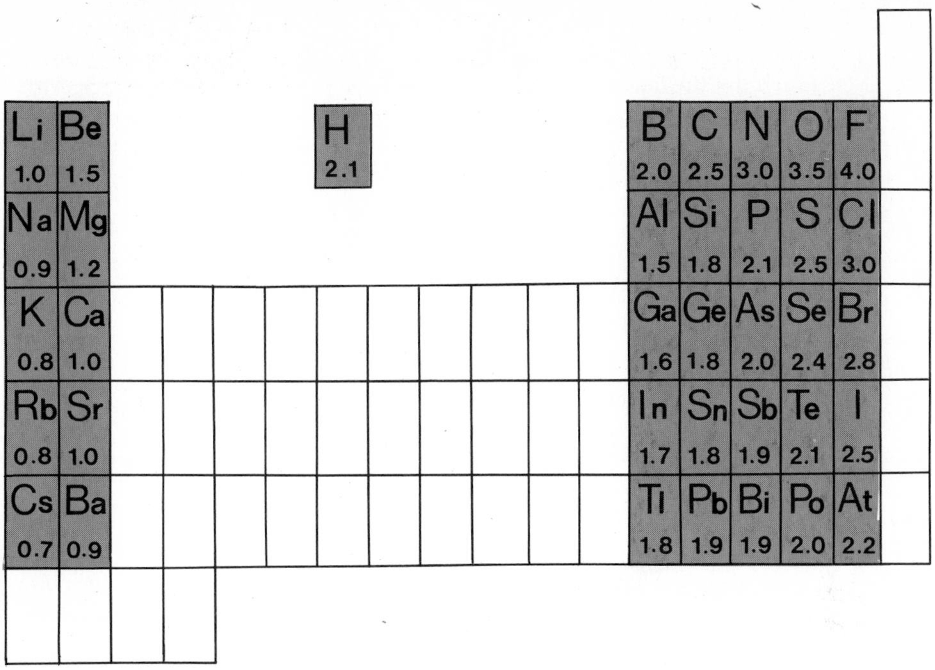

Li	Be											B	C	N	O	F	
1.0	1.5											2.0	2.5	3.0	3.5	4.0	
Na	Mg											Al	Si	P	S	Cl	
0.9	1.2											1.5	1.8	2.1	2.5	3.0	
K	Ca											Ga	Ge	As	Se	Br	
0.8	1.0											1.6	1.8	2.0	2.4	2.8	
Rb	Sr											In	Sn	Sb	Te	I	
0.8	1.0											1.7	1.8	1.9	2.1	2.5	
Cs	Ba											Tl	Pb	Bi	Po	At	
0.7	0.9											1.8	1.9	1.9	2.0	2.2	

(H 2.1)

Figure 10.7
Electronegativities of the representative elements.

EXAMPLE 10.3 Using data from Figure 10.7, arrange the following bonds in order of increasing polarity, and circle the element that will act as the negative pole.

H—O   H—S

P—H   H—C

Locate the elements in the table, calculate the differences in electronegativity, and enter those differences.

_ _ _ _ _ _   _ _ _ _ _ _

H—O: $3.5 - 2.1 = 1.4$ H—S: $2.5 - 2.1 = 0.4$

P—H: $2.1 - 2.1 = 0.0$ H—C: $2.5 - 2.1 = 0.4$

Now arrange the bonds in order from the least polar to the most polar.

_ _ _ _ _ _   _ _ _ _ _ _

P—H, H—S, H—C, H—O   *or*   P—H, H—C, H—S, H—O

The P—H bond is nonpolar (electronegativity difference = 0). H—O is the most polar bond because it has the largest electronegativity difference (1.4). The other bonds, with equal electronegativity differences, have about the same polarity.

Now draw a circle around the atom in each bond that will act as the negative pole.

————   ——————

P—H   H—Ⓢ   H—©   H—Ⓞ

Since sulfur, carbon, and oxygen are all more electronegative than hydrogen, the electron density in these three bonds is shifted away from hydrogen toward the other element. There is no electronegativity difference in the P—H bond, so neither atom will act as a negative pole.

Electronegativity numbers can also be used to *predict* whether a bond will be nonpolar covalent, polar covalent, or ionic. We must be cautious, however, not to give these predictions more credit than they deserve. In the first place, there is no sharp difference between the three bond classifications. The whole range of polarities passes *gradually* from a pure nonpolar covalent bond between identical atoms to the most ionic bond in cesium fluoride, CsF. Various authors use electronegativity differences anywhere from 1.5 to 2.0 as the cross-over point between covalent and ionic bonds, which shows how arbitrary the classifications are. Only laboratory measurements can really determine the polarity of a bond. It is probably safe to say that:

1. The only truly nonpolar bond is between identical atoms.
2. The bond between atoms with an electronegativity difference between 0.0 and 1.5 is polar covalent.
3. An electronegativity difference greater than 2.0 identifies a bond that is primarily ionic.
4. If the electronegativity difference is from 1.5 to 2.0, the bond may be considered as very strongly polar or slightly ionic.

EXAMPLE 10.4   Using Figure 10.7, classify each of the following bonds as nonpolar covalent, polar covalent, or ionic:

Li—F   C—I   P—Cl

First determine each electronegativity difference.

————   ——————

Li—F, 3.0   C—I, 0.0   P—Cl, 0.9

Now the classification: nonpolar covalent, polar covalent, or ionic.

Li—F, ionic (electronegativity difference greater than 2.0)
C—I, nonpolar covalent (electronegativity difference, 0.0)
P—Cl, polar covalent (electronegativity difference between 0.0 and 1.5)

Although the atoms are not identical, the electronegativity difference between carbon and iodine is negligible, so the bond may be considered nonpolar.

In Chapter 11 you will use bond polarities to predict the polarities of molecules. In Chapter 14 you will find that molecular polarity is largely responsible for the physical properties of many compounds.

10.7 MULTIPLE BONDS

So far our consideration of covalent bonds has been limited to the sharing of one pair of electrons by two bonded atoms. Such a bond is called a **single bond.** In many molecules we find two atoms bonded by two pairs of electrons; this is a **double bond.** When two atoms are bonded by three pairs of electrons it is called a **triple bond.** All four electrons in a double bond and all six electrons in a triple bond are counted as valence electrons for each of the bonded atoms.

Probably the most abundant substance containing a triple bond is in the nitrogen molecule, N_2. Its Lewis diagram may be thought of as the combination of two nitrogen atoms, each with three unpaired electrons:

$$:\!\overset{\cdot}{N}\!\cdot \;+\; \cdot\overset{\cdot}{N}\!: \;\rightarrow\; :N\!:\!:\!N: \quad or \quad :N\!\equiv\!N:$$

Counting the bonding electrons for both atoms, each nitrogen atom is satisfied with a full octet of electrons.

There is experimental evidence to support the idea of **multiple bonds,** a general term that includes both double and triple bonds. A triple bond is stronger and the distance between bonded atoms is shorter than the same measurements for a double bond between the same atoms, and a double bond is shorter and stronger than a single bond. Bond strength is measured as the energy required to break a bond. The triple bond in N_2 is among the strongest bonds known. This is why elemental nitrogen is so stable and unreactive in the earth's atmosphere.

In Chapter 11 you will examine multiple bonds that are found in more complex molecules.

CHAPTER 10 IN REVIEW

10.1 LEWIS SYMBOLS OF THE ELEMENTS

10 A Write the Lewis symbol of any atom in one of the A groups of the periodic table. (200)

10.2 MONATOMIC IONS WITH NOBLE GAS ELECTRON CONFIGURATIONS

10 B Identify by name and symbol the monatomic ions that are isoelectronic with a given noble gas atom, and write the electron configurations of those ions. (201)

10.3 IONIC BONDS (206)

10.4 SIZES OF IONS

10 C Compare the radii of several given isoelectronic monatomic ions and suggest an explanation for the trend observed. (208)

10 D Compare the radii of several given ions formed by elements in the same group in the periodic table, and suggest an explanation for the trend observed. (208)

10.5 COVALENT BONDS

10 E Distinguish between ionic and covalent bonds. (209)

10.6 POLAR AND NONPOLAR COVALENT BONDS

10 F Distinguish between polar and nonpolar covalent bonds. (211)

10 G Given the electronegativities of two elements, classify the bond between them as nonpolar covalent, polar covalent, or primarily ionic. Identify the positive and negative ends of the bonds, if any. (211)

10.7 MULTIPLE BONDS (215)

TERMS AND CONCEPTS

Valence electrons (200)
Lewis symbol (201)
Electron dot symbol (201)
Octet (201)
Isoelectronic (202)
Ionic bond (206)
Ionic compound (206)

Molecular compound (209)
Molecule (209)
Covalent bond (210)
Electron (charge) cloud (210)
Overlap (of atomic orbitals) (210)
Lewis diagram, formula, or structure (211)
Lone pair (211)

Octet rule; rule of eight (211)   Single bond (215)
Nonpolar bond (212)   Double bond (215)
Polar bond (212)   Triple bond (215)
Electronegativity (212)   Multiple bond (215)

Most of these terms and many others appear in the Glossary.

QUESTIONS AND PROBLEMS

An asterisk (*) identifies a question that is relatively difficult, or that extends beyond the performance goals in the chapter.

Section 10.1

10.1) Write the Lewis symbols for atoms of potassium, phosphorus, and bromine.

10.2) Draw the electron dot symbols for the elements whose atomic numbers are 31 and 82.

10.3) Write the elemental symbols that might be represented by X if the Lewis symbol is $\cdot \ddot{\text{X}} :$.

10.22) Write the electron dot symbols for oxygen, calcium, and boron.

10.23) Write the Lewis symbols for the elements whose atomic numbers are 52 and 55.

10.24) The electron dot symbol M⦂ represents a group of elements. Write the symbols of the elements in that group.

Section 10.2

10.4) Identify those elements in the third period of the periodic table that form monatomic ions which are isoelectronic with a noble gas atom. Write the symbol for each such ion (example: Ca^{2+} in the fourth period).

10.5) Identify two negatively charged monatomic ions that are isoelectronic with neon.

10.6) Write the symbols of two ions that are isoelectronic with the chloride ion.

10.7)* A monatomic ion with a -2 charge has the electron configuration $1s^2 2s^2 2p^6 3s^2 3p^6 4s^2 3d^{10} 4p^6$. (a) What neutral noble gas atom has the same electron configuration? (b) What is the monatomic ion with a -2 charge and this configuration? (c) Write the symbol of an ion with a $+1$ charge that is isoelectronic with the two above species.

10.25) Write the electron configuration of each third period monatomic ion identified in Question 10.4. Also identify the noble gas atoms having the same configurations.

10.26) Identify by symbol two positively charged monatomic ions that are isoelectronic with krypton ($Z = 36$).

10.27) Write the symbols of two ions that are isoelectronic with the barium ion.

10.28)* If the monatomic ions in Question 10.7(b) and (c) combine to form an ionic compound, what will be the formula of that compound? Which atom will be at the negative end of each bond in the compound?

Section 10.3

10.8)* Using Lewis symbols, show how ionic bonds are formed by atoms of sulfur and potassium, leading to the correct formula of potassium sulfide.

10.29)* Aluminum oxide is an ionic compound. Sketch the transfer of electrons from aluminum atoms to oxygen atoms that accounts for the chemical formula of the compound.

Section 10.4

10.9) Identify the largest ion among the following: Li^+; F^-; S^{2-}.

10.30) Which among the following is the smallest ion: Se^{2-} ($Z = 34$); Br^-; Te^{2-} ($Z = 52$)?

10.10) Arrange the following ions in order of their increasing size (smallest ion first): Br⁻; Ca²⁺; Cl⁻; K⁺.

10.11) State the size trend among ions and noble gas atoms that have the same electron configuration. Suggest an explanation for this trend.

Section 10.5

10.12) Explain why ionic bonds are called electron *transfer* bonds, and covalent bonds are known as electron *sharing* bonds.

10.13) Compare the bond between potassium and chlorine in potassium chloride with the bond between two chlorine atoms in chlorine gas. Which bond is ionic, and which is covalent? Describe how each bond is formed.

10.14) Show how a covalent bond forms between an atom of iodine and an atom of chlorine, yielding a molecule of ICl.

10.15)* "The bond between a metal atom and a nonmetal atom is most apt to be ionic, whereas the bond between two nonmetal atoms is most apt to be covalent." Explain why this statement is true.

10.16)* Does the energy of a system tend to increase, decrease, or remain unchanged as two atoms form a covalent bond?

Section 10.6

10.17) What is meant by saying that a bond is *polar* or *nonpolar*? What bonds are completely nonpolar?

10.18) Consider the following bonds: F—Cl; Cl—Cl; Br—Cl; I—Cl. Arrange these bonds in order of increasing polarity (lowest polarity first). Based on Figure 10.7, classify each bond as (a) nonpolar or (b) polar covalent.

10.19) For each polar bond in Question 10.18, identify the atom that acts as the negative pole.

10.20) What is electronegativity? Why are the noble gases not included in the electronegativity table?

Section 10.7

10.21)* What is a multiple bond? Distinguish between single, double, and triple bonds.

10.31) List the following ions in order of their decreasing size (largest ion first): Ba²⁺; Cs⁺ (Z = 55); I⁻; Rb⁺ (Z = 37).

10.32) State the size trend among monatomic ions derived from elements in the same group in the periodic table. Suggest an explanation for this trend.

10.33) Show how atoms achieve the stability of noble gas atoms in forming covalent bonds.

10.34) Considering bonds between the following pairs of elements, which are most apt to be ionic and which are most apt to be covalent: sodium and sulfur; fluorine and chlorine; oxygen and sulfur? Explain your choice in each case.

10.35)* Sketch the formation of two covalent bonds by an atom of sulfur in making a molecule of hydrogen sulfide, H_2S.

10.36)* The bond between two metal atoms is neither ionic nor covalent. Explain, according to the octet rule, why this is so.

10.37)* How do the energy and stability of bonded atoms and noble gas electron configurations appear to be related in forming covalent and ionic bonds?

10.38) Compare the electron cloud formed by the bonding electron pair in a polar bond with that in a nonpolar bond.

10.39) List the following bonds in order of decreasing polarity: K—Br; S—O; N—Cl; Li—F; C—C. Classify each bond as (a) nonpolar, (b) polar covalent, or (c) primarily ionic.

10.40) For each bond in Question 10.39, identify the positive pole, if any.

10.41) Identify the trends in electronegativities that may be observed in the periodic table.

10.42)* Double bonds and triple bonds conform to the octet rule. Could a quadruple (4) bond obey that rule? a quintuple (5) bond?

11

the structure of molecules

In our introduction to covalent bonds in Chapter 10 we concentrated on the *bond* between two atoms, rather than on the molecule in which the bond was to be found. In Section 10.5 we considered the bond between hydrogen atoms in H_2, between fluorine atoms in F_2, and between a hydrogen atom and a fluorine atom in HF. These particular substances happen to form two-atom molecules, and their Lewis diagrams describe the molecules as well as the bonds, but that was not the point of the presentation at that time. Later we considered bonds between atoms that are parts of larger molecules. Example 10.3 included the bond between a hydrogen atom and a phosphorus atom. There is no known species that has the formula PH or HP, made up of one atom of each element. Bonds between hydrogen and phosphorus atoms are present in larger molecules. Three such bonds appear in the compound phosphine, PH_3. It is natural to wonder how bonds such as these are arranged in molecules, and what effect this arrangement has on the physical and chemical properties of the substance. In this chapter we will address the first of these questions; in Chapter 14 we will study the second.

11.1 DRAWING LEWIS DIAGRAMS BY INSPECTION

PG 11 A Draw the Lewis diagram for any molecule or polyatomic ion made up of elements in the A groups of the periodic table (Sections 11.1 to 11.3).

Because each atom in a covalent bond usually has the electron configuration of a noble gas, we can often draw Lewis diagrams by inspection. With the exception of hydrogen, a nonmetal atom tends to surround itself with eight valence electrons, corresponding with all noble gases except helium. On page 211 this was called the **octet rule.** Hydrogen is satisfied with two valence electrons, matching the configuration of helium.

The bonds in the molecules formed between hydrogen, on the one hand, and fluorine, oxygen, nitrogen, or carbon, on the other, illustrate the inspection method of drawing Lewis diagrams. A hydrogen atom, H·, has one electron. It needs another to reach the $1s^2$ configuration of helium. A fluorine atom has seven valence electrons, one of which is unpaired. It needs one to reach the eight of a noble gas, or to "complete its octet." By being shared in a covalent bond, both unpaired electrons "belong to" and are counted for both atoms, and each atom thereby reaches the noble gas configuration. Using electron dot diagrams, as on page 210,

$$\text{H} \cdot + \cdot \ddot{\underset{..}{\text{F}}} : \rightarrow \text{H} : \ddot{\underset{..}{\text{F}}} : \quad or \quad \text{H}—\ddot{\underset{..}{\text{F}}} :$$

In general, unpaired electrons from two different atoms are capable of pairing to form covalent bonds, and do so until all atoms reach the electron configuration of a noble gas.

The Lewis symbols for oxygen, nitrogen, and carbon are

$$\cdot \ddot{\underset{.}{\text{O}}} : \qquad \cdot \ddot{\underset{.}{\text{N}}} \cdot \qquad \cdot \underset{.}{\text{C}} \cdot$$

Oxygen has six valence electrons, two of which are unpaired. It needs two more electrons to complete its octet. It can reach this point if each unpaired oxygen electron forms a bond with the electron from one hydrogen atom. In other words, it does twice what the fluorine atom did once in forming a molecule of HF. The result is a molecule of water, H_2O. With nitrogen there are three unpaired electrons, so it can form three bonds, one with each of three different hydrogen atoms. This yields a molecule of ammonia, NH_3. Carbon has four valence electrons, so it forms covalent bonds with four hydrogen atoms to produce a molecule of methane, CH_4.* The Lewis diagrams for these compounds are shown:

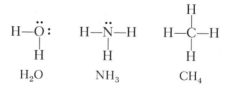

*A carbon atom, with its electron configuration $1s^2 2s^2 2p^2$, actually has only two unpaired electrons, the two 2p electrons. It is a fact, however, that carbon atoms form covalent bonds with four hydrogen atoms. One explanation for this involves "hybridized" orbitals, a topic beyond the scope of this text.

EXAMPLE 11.1   Draw Lewis diagrams for carbon tetrachloride, CCl_4, and phosphorus tribromide, PBr_3.

The Lewis symbol for carbon is given above. It has been shown that a carbon atom requires four additional electrons to complete its octet and does so by forming four bonds. Write the Lewis symbol for chlorine. From that symbol state the number of electrons each chlorine atom must gain to complete its octet.

------   ------

$\cdot\ddot{C}l:$ The atom has seven valence electrons and needs
          one to reach eight.

Assuming that new bonds will be formed when each atom contributes one electron to the bond, draw the Lewis diagram of CCl_4.

------   ------

$$:\ddot{C}l:$$
$$\ |\ $$
$$:\ddot{C}l-\underset{|}{C}-\ddot{C}l:$$
$$:\ddot{C}l:$$

Each of four electrons from a carbon atom forms a bond with the unpaired electron from a separate chlorine atom.

Now draw the Lewis symbols of a phosphorus atom and a bromine atom. For each, state the number of electrons required to complete the octet.

------   ------

$\cdot\ddot{P}\cdot$ Three electrons required. $\cdot\ddot{B}r:$ One electron required.

Finally, the Lewis diagram for PBr_3. . . .

------   ------

$$:\ddot{B}r-\ddot{P}-\ddot{B}r:$$
$$:\ddot{B}r:$$

Notice the similarity between this diagram and that for NH_3. The central elements are both in Group 5A and have five valence electrons. The other element in each case requires one additional electron to reach a noble gas configuration. The inspection methods for drawing the diagrams are the same.

In all examples so far, covalent bonds have been formed when each atom contributes one electron to the bonding pair. This is not always the case. Many bonds are formed where one atom contributes both electrons and the other atom offers only an empty orbital. To illustrate, an ammonium ion is produced when a hydrogen ion is bonded to the unshared electron pair of the nitrogen atom in an ammonia molecule:

$$H^+ + \;:\!\overset{\displaystyle H}{\underset{\displaystyle H}{N}}\!-\!H \rightarrow \left[\overset{\displaystyle H}{\underset{\displaystyle H}{H\!-\!N\!-\!H}}\right]^+$$

The four bonds in an ammonium ion are identical. This shows that a bond formed when one atom contributes both electrons is the same as any other bond. But Lewis diagrams of species produced in this way are not readily drawn by inspection. They are found instead by the method given in Section 11.3.

Notice that in drawing the Lewis diagram of an ion, the diagram is enclosed in brackets and the charge is shown as a superscript.

11.2 MULTIPLE BONDS

Sometimes the number of electrons and the number of atoms in a molecule do not permit a Lewis diagram in which every bond is formed by a single pair of electrons, and each atom has the configuration of a noble gas. The development of the diagram for ethene, C_2H_4, illustrates both the problem and the solution. To emphasize the contrast, this will be shown side by side with the Lewis diagram of ethane, C_2H_6, a compound where the problem is not present. In both molecules the carbon atoms are bonded to each other, leaving six electrons—three on each carbon—still available for bonding:

$$\cdot\overset{\displaystyle ..}{\underset{\displaystyle ..}{C}}\!-\!\overset{\displaystyle ..}{\underset{\displaystyle ..}{C}}\cdot$$

With ethane there are six hydrogen atoms to form covalent bonds with the six available electrons, but with ethene there are only four hydrogen atoms:

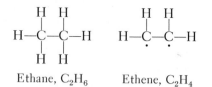

Ethane, C_2H_6 Ethene, C_2H_4

At this point each carbon atom in ethane is surrounded by a full octet of electrons, but the carbon atoms in ethene have only seven electrons around them. The problem is resolved, however, if the unpaired electrons in ethene combine to form a *second* bond between the two carbon atoms.

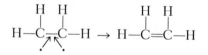

With two pairs of electrons between the carbon atoms, each carbon atom is surrounded by eight electrons, satisfying the octet rule. When two atoms are bonded by *two* pairs of electrons, they are held together by a double bond.

The idea of multiple bonds was introduced in Chapter 10 (p. 215). The N_2 molecule was used to illustrate a triple bond between two nitrogen atoms. Multiple bonding is not limited to bonds between atoms of the same element. For example, the Lewis diagrams of formaldehyde and the cyanide ion are,

respectively, $\text{H—C} \overset{\overset{\displaystyle \ddot{O}:}{\diagup}}{\underset{\diagdown}{}}$ and $[\,:\text{C}\equiv\text{N}:\,]^-$.

$$\quad\quad\quad\quad\quad\quad\quad\quad\quad\quad\quad\quad \text{H}$$

11.3 DRAWING COMPLEX LEWIS DIAGRAMS

Lewis diagrams are not readily drawn by inspection for some of the larger or more complex molecules and polyatomic ions. The procedure that follows may be used to sketch a Lewis structure for any species that obeys the octet rule.

1. **Count the number of valence electrons available.** For a molecule this can be done simply by adding the valence electrons contributed by each atom in the molecule. For a polyatomic ion this total must be adjusted to account for the charge on the ion. An ion with a -1 charge would have one more electron than the number of valence electrons in the neutral atoms; a -2 ion, two more; a $+1$ ion, one less; and so forth.

2. **Draw a tentative structure for the molecule or ion, joining atoms by single bonds. Place electron dots around each symbol so that the total number of electrons for each atom is eight.** Count both bonding electron pairs and unshared electrons in placing electron dots. In some cases, only one arrangement of atoms is possible. In others, two or more structures may be drawn. Ultimately chemical or physical evidence must be used to decide which of the possible structures is correct. A few general rules will help you in making diagrams that are most likely to be correct:

   a. A hydrogen atom always forms one bond; a carbon atom normally forms four bonds.

   b. When several carbon atoms appear in the same molecule, they are often bonded to each other. In some compounds they are arranged in a closed loop; however, we will avoid such so-called cyclic compounds in this text.

   c. In compounds or ions having two or more oxygen atoms and one atom of another nonmetal, the oxygen atoms are usually arranged *around* the central nonmetal atom.

   d. In an oxyacid (hydrogen + oxygen + a nonmetallic element, such as H_2SO_4 and HNO_3), hydrogen is usually bonded to an oxygen atom, which is then bonded to the nonmetallic atom: H—O—X, where X is a nonmetal.

3. **Count the total number of electrons in your structure and compare it with the number available from Step 1.** If they are the same, your diagram is complete. If your count yields two electrons more than you have available, the molecule must have a double bond. It is usually between an oxygen and some other element, or between two carbon atoms. If your count is four more than there are available, a triple bond or two double bonds are indicated. Multiple bonds should be used only when necessary, and then use as few as possible.

Although some of the examples that follow might be done simply and quickly by inspection, we will follow the procedure outlined above in order that you may become familiar with it, as well as see similarities and regularities that might otherwise be overlooked.

EXAMPLE 11.2   Write Lewis structures for the ClF molecule and the ClO⁻ ion.

Following the rules given above, first determine the number of valence electrons in each species.

ClF   ClO⁻

_ _ _ _ _ _   _ _ _ _ _ _

ClF, 14; ClO⁻, 14

In the ClF molecule, we simply add the valence electrons of the neutral atoms. Since both Cl and F are in Group 7A and therefore have seven valence electrons (number of valence electrons equal to group number for A groups), we have: $7 + 7 = 14$. For the ClO⁻ ion, we add the number of valence electrons in Cl (7) to the number contributed by O in Group 6A (6), and then add 1 to take care of the -1 charge: $7 + 6 + 1 = 14$.

There is only one possible tentative structure in each case. Write it, and surround each atom with eight electrons.

_ _ _ _ _ _   _ _ _ _ _ _

:Cl̈—F̈:   [:Cl̈—Ö:]⁻

The shared electron pair constituting the bond is counted for both atoms, so each atom requires six additional electrons.

Now count the electrons in the above diagram. Compare the result with the number of electrons available, determined in Step 1. Modify the above structures with multiple bonds to correct any differences in the two counts.

_ _ _ _ _ _   _ _ _ _ _ _

Both structures have 14 electrons, the same as the number available. The Lewis diagrams are therefore correct as shown above.

The above example illustrates the fact that two species having the same number of electrons and the same number of atoms have similar Lewis diagrams, whether they are molecules or polyatomic ions.

EXAMPLE 11.3 Draw the Lewis diagram for $SO_3{}^{2-}$, the sulfite ion.

Begin by counting up the number of valence electrons. Don't forget the ionic charge.

- - - - - -   - - - - - -

26 electrons

Sulfur has six valence electrons; each oxygen atom has six electrons; and there are two extra electrons for the -2 charge: $6 + 3(6) + 2 = 26$.

Now proceed to the tentative structure, with each atom surrounded by an octet of electrons. Remember, when there are two or more oxygen atoms and another nonmetal atom, the oxygens are usually arranged around the central nonmetal atom.

- - - - - -   - - - - - -

$$:\ddot{O}\!-\!\overset{\displaystyle ..}{S}\!-\!\ddot{O}: \quad {}^{2-}$$
$$\underset{..}{\overset{|}{:}O}:$$

Now count up the electrons in the tentative structure and compare the result with the 26 valence electrons available. Modify the structure with multiple bonds as necessary.

- - - - - -   - - - - - -

The tentative structure has 26 electrons, equal to the valence electrons available. No modification is necessary.

EXAMPLE 11.4 Derive the Lewis structure of SO_2.

First determine the number of valence electrons.

- - - - - -   - - - - - -

Sulfur has six valence electrons, and oxygen six for each atom: $6 + 2(6) = 18$.

Now sketch the tentative structure, complete with an electron octet around each atom.

$$:\ddot{O}-\ddot{S}-\ddot{O}:$$

As usual, the oxygen atoms are arranged around the central nonmetal atom.

Now count the electrons and compare to the 18 valence electrons available. Modify the structure with multiple bonds, if necessary, to correct for an unequal number.

Tentative diagram has 20 electrons; only 18 electrons available.

$$:\ddot{O}-\ddot{S}=\ddot{O}:$$

This time the tentative structure has two electrons more than the number available. The total in the tentative structure must therefore be reduced by two. This is accomplished by replacing two electron pairs—the lone pair from the sulfur atom and one unshared pair from one of the oxygen atoms—with one electron pair that is shared between the atoms as a second bond. This removes four electrons and returns two, a net reduction of two electrons:

$$:\ddot{O}-\ddot{S}-\ddot{O}: \qquad :\ddot{O}-\ddot{S}=\ddot{O}:$$

$$\text{Replace} \qquad\qquad \text{New}$$

Each atom is now surrounded by eight electrons, and the total number of electrons in the structure matches the number available.

You may wonder if it makes any difference on which side of the sulfur atom the double bond is placed. It does not, for the limited purpose of learning how to draw Lewis diagrams. It is a fact, however, that experimentally the bonds are identical, and have strengths and lengths that are between those normally associated with single bonds and double bonds connecting sulfur and oxygen atoms. This condition is known as *resonance,* and is frequently shown as

$$\ddot{O}=\ddot{S}-\ddot{O}: \leftrightarrow :\ddot{O}-\ddot{S}=\ddot{O}$$

In this text we will not be concerned with resonance structures beyond this point of information.

The rules we are following are readily applied to simple organic* molecules, which always contain carbon atoms, usually include hydrogen atoms, and may contain atoms of other elements, notably oxygen. If oxygen is present in an organic compound, it usually forms two bonds. If you remember that carbon forms four bonds and hydrogen forms one, and that two or more carbon atoms

*Organic chemistry is the chemistry of compounds containing carbon, other than certain "inorganic" carbon compounds such as carbonates, CO and CO_2. A knowledge of bonding, including Lewis diagrams, is important in organic chemistry.

usually bond to each other (Rules 2a and 2b, page 223), your tentative structures will probably be those that are found to be correct in the laboratory.

EXAMPLE 11.5   Write the Lewis structure for propane, C_3H_8.

First determine the electron count.

20

Four for each carbon atom, and one for each hydrogen: $3(4) + 8(1) = 20$.

The tentative structure follows readily from
the rules mentioned just before this example.

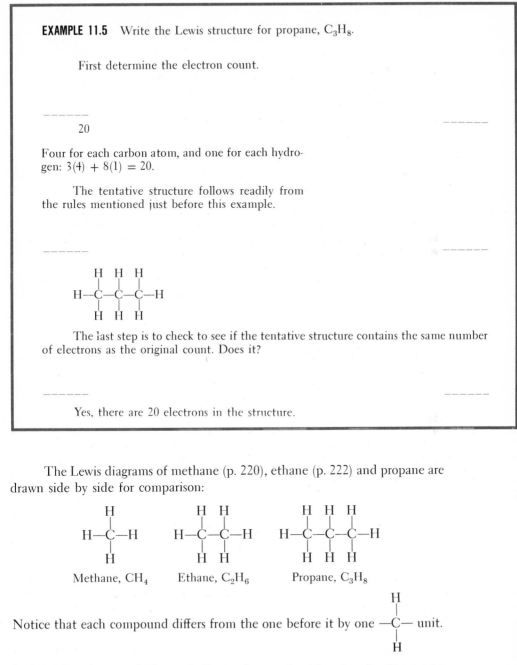

The last step is to check to see if the tentative structure contains the same number
of electrons as the original count. Does it?

Yes, there are 20 electrons in the structure.

The Lewis diagrams of methane (p. 220), ethane (p. 222) and propane are
drawn side by side for comparison:

Methane, CH_4 Ethane, C_2H_6 Propane, C_3H_8

Notice that each compound differs from the one before it by one —C— unit.

In fact, there is no end, theoretically, to the number of such units that might
be inserted in the chain. The next member of the *alkane series*, as this is

called, is C_4H_{10}, butane. It is possible to draw two structures of butane, and both are real but different compounds, each with its own set of physical properties. The diagrams are

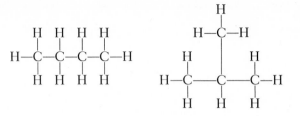

Compounds having the same molecular formula but different structures are called **isomers** of each other. Among the alkanes the number of isomers increases dramatically as the number of carbon atoms increases. There are three pentanes, C_5H_{12}, five hexanes, C_6H_{14}, 35 nonanes, C_9H_{20}, over 300 000 with the formula $C_{20}H_{42}$, and over 100 000 000 having the formula $C_{30}H_{62}$! Needless to say, they have not all been studied. This gives you some idea why there are so many organic compounds—and we haven't even mentioned compounds that contain elements other than carbon and hydrogen. One kind of soap, for example, is $C_{18}H_{35}O_2Na$.

EXAMPLE 11.6   Prepare the Lewis diagram of acetylene, C_2H_2.

As usual, start with the electron count.

$2(4) + 2(1) = 10$

Now draw the tentative structure.

H—C̈—C̈—H

The electron count and structure modification, if necessary. . . .

H—C≡C—H

The tentative structure has 14 electrons, four more than the number available. This requires a triple bond.

EXAMPLE 11.7 Draw a Lewis diagram for C_2H_6O.

State the electron count, please.

- - - - - -   - - - - - -

$2(4) + 6(1) + 6 = 20$

It was mentioned earlier that oxygen atoms usually form two bonds in organic molecules. Try for the tentative structure, complete with unshared electrons. Bond the carbons to each other.

- - - - - -   - - - - - -

$$\begin{array}{ccc} & H & H \\ & | & | \\ H- & C- & C-\overset{\cdot\cdot}{\underset{\cdot\cdot}{O}}-H \\ & | & | \\ & H & H \end{array}$$

Now check the electrons in the diagram against the valence electron count, adjusting the structure with multiple bonds, if necessary.

- - - - - -   - - - - - -

20 electrons in both places. The Lewis diagram above is correct.

The compound in Example 11.7 is ethyl alcohol. If we had not insisted that the carbon atoms be bonded together you might have produced the diagram for an isomer of alcohol, another well-known compound, dimethyl ether:

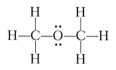

11.4 EXCEPTIONS TO THE OCTET RULE

Lewis diagrams for many substances may be drawn with the help of the octet rule, but some substances do not "obey" the rule. Two common oxides of nitrogen, NO and NO_2, have an odd number of electrons. It is therefore impossible to write Lewis diagrams for these compounds in which each atom is surrounded by eight electrons. Phosphorus pentafluoride, PF_5, places five electron pair bonds around the phosphorus atom, and six pairs surround sulfur in SF_6.

Certain molecules whose Lewis diagrams obey the octet rule do not have the properties that would be predicted. Oxygen, O_2, was not used to introduce the double bond on page 215 for that reason. On paper, O_2 appears to have an ideal double bond:

$$:\overset{..}{O}=\overset{..}{O}:$$

But liquid oxygen is *paramagnetic*, meaning that it is attracted by a magnetic field. This is characteristic of molecules that have unpaired electrons in their structures. This might suggest a Lewis diagram that has each oxygen surrounded by seven electrons:

$$:\overset{..}{\underset{.}{O}}—\overset{..}{\underset{.}{O}}:$$

But this is in conflict with other evidence that the oxygen atoms are connected by something other than a single bond. In essence, it is impossible to write a single Lewis diagram that satisfactorily explains all the properties of molecular oxygen.

Two other substances for which satisfactory octet rule diagrams can be drawn, but are contradicted experimentally, are the fluorides of beryllium and boron. We might even expect BeF_2 and BF_3 to be ionic compounds, but laboratory evidence strongly supports covalent structures having the Lewis diagrams:

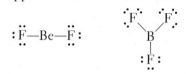

In these compounds the beryllium and boron atoms are surrounded by two and three pairs of electrons, respectively, rather than four.

11.5 MOLECULAR GEOMETRY

A Lewis diagram shows the order in which atoms are bonded to each other in a molecule, and by how many electron pairs they are bonded, but the diagram tells little about the geometry of the molecule. **Molecular geometry** identifies the *bond angles* in the molecule and its *shape*. A **bond angle** is the angle between any two bonds formed by the same atom (Fig. 11.1). It turns out that the shape of a molecule plays an important role in determining the physical and chemical properties of a substance, as you will see in Chapter 14.

To illustrate the shortcomings of a Lewis diagram, glance at the three shown for H_2O, NH_3, and CH_4 (p. 220). All three diagrams suggest that the bond angle between the central atom and any two hydrogen atoms is a right angle. This is not so. All of the bond angles are near 109°. In two dimensions it is impossible to draw four 109° angles around the central carbon atom in CH_4, or three such angles around a central nitrogen atom in NH_3. There is no way a two-dimensional Lewis diagram can give an accurate description of a three-dimensional shape.

Water is a two-dimensional molecule, since its three atoms lie in one plane. The Lewis diagram for water can be drawn with the proper bond angle. Even so, the diagram is not completely satisfactory because it provides no space

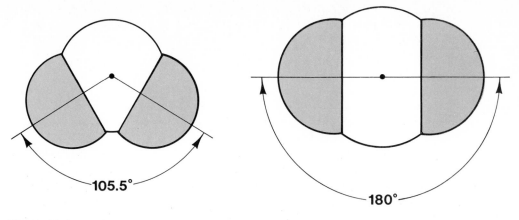

Figure 11.1
Bond angle. In a water molecule the bonds between the oxygen atom and each of two hydrogen atoms form an angle of 105.5°. In a carbon dioxide molecule the bonds between a carbon atom and each of two oxygen atoms lie in a straight line. The bond angle is 180°.

for the unshared electron pairs, which should be placed above and below the plane of the paper.

If the atoms in a three-atom molecule are not in a straight line, as with water, the geometry of the molecule is said to be **bent** (Fig. 11.1). If the three atoms are in a straight line, as with carbon dioxide, it is a **linear** molecule, and the bond angle is 180°.

When a molecule has four or more atoms they may or may not lie in the same plane. If they are in the same plane, the molecule is **planar.** Among molecules that are *not* planar, several geometric shapes are known. We will study only two, leaving the more complex forms to advanced courses. Each of these two looks like a pyramid with a triangular base. If the base and all three sides of the pyramid are identical equilateral triangles, the molecule has the shape of a **tetrahedron*** (Fig. 11.2A). The other pyramidal structure is called a **trigonal pyramid** to indicate a triangular base. It looks like a tetrahedron that has been "squashed down" (Fig. 11.2B).

11.6 ELECTRON PAIR REPULSION; ELECTRON PAIR GEOMETRY

No single theory or model yet developed succeeds in explaining all the molecular shapes that have been observed and measured in the laboratory. A theory that explains one group of molecules fails when applied to another group. Each approach has advantages—and limitations. Chemists therefore use them all within the areas to which they apply, fully recognizing that there is still much to learn about how atoms are assembled in molecules.

---

* In solid geometry, a tetrahedron is the simplest regular solid. A *regular* solid is a solid figure with identical equilateral faces. A cube is a regular solid, having six identical squares for its faces.

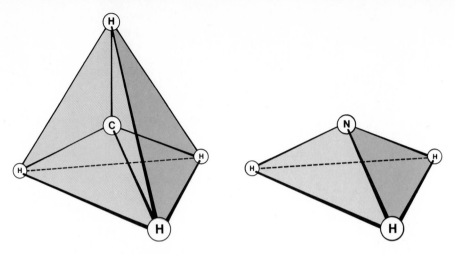

Figure 11.2
A, Methane, CH₄, is a typical five-atom molecule. The four hydrogen atoms are at the corners of a tetrahedron, and the carbon atom is at its center. The molecule is three dimensional. If the top hydrogen atom and the carbon atom are in the plane of the paper, the large hydrogen atom in the base is closer to you than the paper, and the other hydrogen atoms are behind the page. B, Ammonia, NH₃, is a four-atom molecule having the shape of a pyramid with a triangular base. It is like the CH₄ molecule without the top hydrogen atom. The nitrogen atom is in the plane of the paper, the large hydrogen atom is in front of the paper, and the smaller hydrogen atoms are behind the page.

In this text we will explore only one of the models used to explain molecular geometry. It is called the **electron pair repulsion principle.** According to this theory, the pairs of electrons that surround any atom in a molecule will, because of repulsion between similarly charged species, arrange themselves as far from each other as possible. The electron pairs may be shared in a covalent bond, or they may be unshared; it makes no difference, as far as electron pair repulsion is concerned. The **electron pair geometry** tells how electron pairs are arranged around a central atom.

Let us pause to distinguish clearly between two kinds of geometry. Electron pair geometry describes the arrangement of all *electron pairs*, shared or unshared, around a central atom. Molecular geometry describes the arrangement of all *atoms* around the central atom to which they are bonded. The two are related: molecular geometry is the direct effect of electron pair geometry.

Earlier in this chapter we drew Lewis diagrams for molecules in which central atoms were surrounded by an octet of electrons, four electron pairs. Some pairs make covalent bonds with other atoms, and some are lone pairs. In Section 11.4 you saw that not all molecules obey the octet rule. Sometimes a central atom is surrounded by only two or three pairs of electrons. We must therefore determine how two, three, or four electron pairs will distribute themselves around a central atom if they are to be as far from each other as possible. It is a geometric fact that, when this condition is met, all angles formed between any two pairs of electrons and the central atom are equal. The three electron pair geometries that result are shown in Figure 11.3. The reasoning

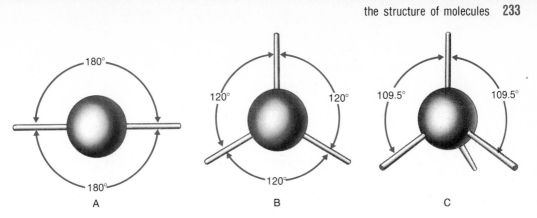

Figure 11.3
Electron pair geometry. "Tinker-toy" models show the arrangement of two, three and four electron pairs around a central atom, according to the electron pair repulsion principle. The ball represents the central atom, and the sticks represent the electron pairs. A, Two sticks are as far from each other as possible when they are at opposite ends of a diameter of the ball, the same as the North and South Poles are at opposite ends of a diameter of the Earth. The sticks and the center of the ball are on a straight line. The geometry is linear, and the angle formed is 180°. B, Three sticks are as far from each other as possible when equally spaced on a circumference of the ball. The sticks and the center of the ball are in the same plane. The geometry is trigonal (triangular) planar, and the angles are 120°. C, Four electron pairs are as far from each other as possible when arranged at the corners of a tetrahedron, a regular solid whose faces are four identical equilateral triangles. The center of the ball is the center of the tetrahedron. The geometry is tetrahedral, and the angles are 109.5°. This is sometimes called a "tetrahedral angle."

behind those geometries is explained in the caption. The geometries are summarized below and in Table 11.1.

1. If the central atom is surrounded by two electron pairs, the atom is on the line between the pairs, the geometry is linear, and the "electron pair" angle is 180°;

2. If the central atom is surrounded by three electron pairs, the atom is at the center of an equilateral triangle formed by the pairs, the geometry is **trigonal (triangular) planar,** and the "electron pair" angle is 120°;

3. If the central atom is surrounded by four electron pairs, the atom is at the center of a tetrahedron formed by the pairs, the geometry is **tetrahedral,** and the "electron pair" angle is 109.5°, often called the **tetrahedral angle.**

11.7 MOLECULAR GEOMETRY BASED ON ELECTRON PAIR REPULSION

PG 11 B Given or having derived the Lewis diagram of a molecule in which a second period central atom is surrounded by two, three, or four pairs of electrons, predict the electron pair geometry and the molecular geometry around that atom.

According to the electron pair repulsion theory, bond angles between atoms are roughly the same as the "electron pair" angles stated above. Minor changes appear if one or two electron pairs are unshared. We will give the molecular geometries for all combinations of electron pairs, including those in which some pairs are unshared. The descriptions that follow are summarized in Table 11.1. Line references below are to that table.

Two electron pairs, two bonded atoms. Two electron pairs yield a linear electron pair geometry (BeF_2, Line 1). Each electron pair bonds a fluorine atom to the central beryllium atom, creating a linear molecular geometry. The bond angle is $180°$.

Three electron pairs, three bonded atoms. Three electron pairs yield a trigonal planar electron pair geometry (BF_3, Line 2). Each electron pair bonds a fluorine atom to the central boron atom, creating a trigonal planar molecular geometry. Each F—B—F angle is $120°$.

Four electron pairs, four bonded atoms. Four electron pairs produce a tetrahedral electron pair geometry (CH_4, Line 3). Each electron pair bonds a hydrogen to the central carbon atom. The molecular geometry is the same as the electron pair geometry, tetrahedral. Each H—C—H bond angle is the tetrahedral angle, $109.5°$.

Four electron pairs, three bonded atoms. As before, four electron pairs produce a tetrahedral electron pair geometry (NH_3, Line 4). Only three of the electron pairs are bonded to hydrogen atoms, however. The four atoms make a molecule that has the shape of a low pyramid with the nitrogen atom at the top and the three hydrogen atoms forming an equilateral triangular base. This molecular geometry is a trigonal pyramid.

The bond angles in NH_3 do not conform to the "ideal" tetrahedral angle predicted by the four electron pairs. The absence of a hydrogen atom to compete with the nitrogen atom in attracting the unshared pair evidently allows the lone pair to be drawn closer to the nitrogen nucleus than the other three pairs. The lone pair appears to "push" the other three pairs closer to each other. This makes each H—N—H bond angle slightly less than the tetrahedral angle, $107.5°$ instead of $109.5°$.

Four electron pairs, two bonded atoms. Again, a tetrahedral electron pair geometry is predicted (H_2O, Line 5). Only two hydrogen atoms are bonded to the central oxygen atom. The two lone pairs push the bonding pairs closer to each other than the tetrahedral angle, producing an actual bond angle of $105.5°$. The three atoms, not in a straight line, have a bent molecular geometry.

Table 11.1

Electron Pair and Molecular Geometries

Line	Electron Pairs	Bonded Atoms	Electron Pair Geometry	Ball and Stick Model	Ideal Bond Angle	Molecular Geometry	Lewis Diagram	Ball and Stick model	Space Filling Model	Example	Actual Bond Angle
1	2	2	Linear		180°	Linear	A—B—A		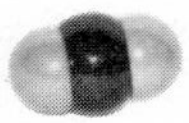	BeF_2	180°
2	3	3	Trigonal (triangular) planar		120°	Trigonal (triangular) planar	A—B(—A)—A			BF_3	120°
3	4	4	Tetrahedral		109.5°	Tetrahedral	A—B(—A)(—A)—A			CH_4	109.5°
4	4	3	Tetrahedral		109.5°	Trigonal (triangular) pyramid	A—$\ddot{B}$(—A)—A			NH_3	107.5°
5	4	2	Tetrahedral		109.5°	Bent	A—$\ddot{B}$:—A or A—$\ddot{B}$:(—A)			H_2O	105.5°

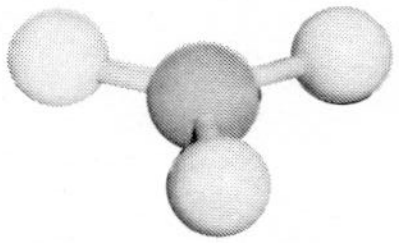

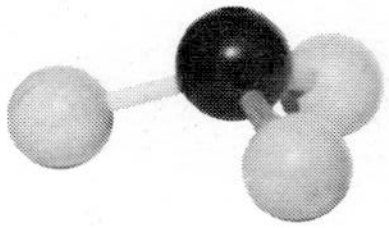

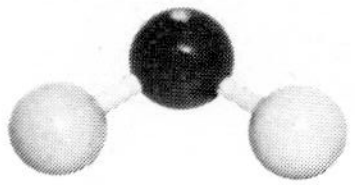

235

EXAMPLE 11.8   Predict the electron pair and molecular geometries of carbon tetrachloride, CCl_4.

It is always helpful in predicting electron pair and molecular geometries to have the Lewis diagram. Draw this for CCl_4.

------   ------

$$:\overset{..}{\underset{}{\text{Cl}}}:$$
$$:\overset{..}{\underset{..}{\text{Cl}}}-\overset{|}{\underset{|}{\text{C}}}-\overset{..}{\underset{..}{\text{Cl}}}:$$
$$:\overset{..}{\underset{..}{\text{Cl}}}:$$

From the Lewis diagram you should establish the number of electron pairs around the central atom and the number of atoms bonded to the central atom. Both geometries follow.

------   ------

With four electron pairs around carbon, all bonded to other atoms, both geometries are tetrahedral.

EXAMPLE 11.9   Describe the shape of a molecule of boron trihydride, BH_3.

First draw the Lewis diagram. Remember that boron has only three valence electrons to contribute to covalent bonds. From the structure answer the question.

------   ------

$$\begin{array}{c} \text{H} \\ \diagdown \\ \text{B}-\text{H} \\ \diagup \\ \text{H} \end{array}$$ Trigonal planar

Three electron pairs yield both an electron pair geometry and a molecular geometry that are trigonal planar with 120° bond angles.

EXAMPLE 11.10 Predict the electron pair geometry and shape of a molecule of dichlorine oxide, Cl_2O.

:C̈l—Ö: Electron pair geometry: tetrahedral; molecular geometry: bent
 :C̈l:

Oxygen has four electron pairs around it, yielding an electron pair geometry that is approximately tetrahedral. Only two of the electron pairs are bonded to other atoms, so the molecule is bent. The structure is similar to that of water.

11.8 THE GEOMETRY OF THE MULTIPLE BOND

Experimental evidence shows that bond angles involving atoms connected by multiple bonds are the same as those that would be predicted for atoms connected only by single bonds. For example, in carbon dioxide the Lewis structure is

$$\ddot{O}=C=\ddot{O}$$

The carbon atom is connected to two oxygen atoms by two double bonds. Where might these bonds best locate themselves to be as far apart as possible? The answer is the same as it was for two pairs of electrons in beryllium fluoride on page 235: at 180° from each other. The carbon dioxide molecule is therefore linear.

The reasoning is not quite so clear when we consider mixtures of single and multiple bonds around the central atom. The Lewis diagram for formaldehyde was given on page 223 as

As in BF_3, there are three atoms surrounding the central atom. Two are bonded by electron pairs and the third by two pairs of electrons. Finding the maximum distance from each other again suggests a planar molecule, but the bond angles might be questionable. Laboratory data, however, show that the bond angles surrounding the carbon atom are 120°—just like the three bond angles surrounding the boron atom in BF_3.

11.9 POLARITY OF MOLECULES

We previously considered the polarity of covalent bonds. Now that we have some idea about how atoms are arranged in molecules, we are ready to discuss the polarity of molecules themselves. **A polar molecule is one in which there is an unsymmetrical distribution of charge,** resulting in + and − poles. A simple example is the HF molecule. The fact that the bonding electrons are closer to the fluorine atom gives the fluorine end of the molecule a partial negative charge, while the hydrogen end acts as a positive pole. In general, any diatomic molecule in which the two atoms differ from each other will be at least slightly polar. Other examples are HCl and BrCl. In both of these molecules the chlorine atom acts as a negative pole.

When a molecule has more than two atoms we must know something about the bond angles in order to decide whether the molecule is polar or nonpolar. Consider, for example, the two triatomic molecules, BeF_2 and H_2O. Despite the presence of two strongly polar bonds, the linear BeF_2 *molecule* is nonpolar. Since the fluorine atoms are symmetrically arranged around Be, the two polar Be—F bonds cancel each other. This may be shown as

$$: \ddot{F} \leftarrow Be \rightarrow \ddot{F} :$$

in which the arrows point to the more electronegative atoms.

In contrast, the bent water molecule is polar; the two polar bonds do not cancel each other because the molecule is not symmetrical around a horizontal axis.

$$\overset{(-)}{\underset{H \ (+) \ H}{\ddot{O}}}$$

The negative pole is located at the more electronegative oxygen atom; the positive pole is midway between the two hydrogen atoms. In an electric field, water molecules tend to line up with the hydrogen atoms pointing toward the plate with the negative charge and the oxygen atoms toward the plate with the positive charge (Fig. 11.4).

Another molecule which is nonpolar despite the presence of polar bonds is CCl_4. The four C—Cl bonds are themselves polar, with the bonding electrons displaced toward the chlorine atoms. But because the four chlorines are symmetrically distributed about the central carbon atom (Fig. 11.5), the polar bonds cancel each other. If one of the chlorine atoms in CCl_4 is replaced by hydrogen, the symmetry of the molecule is destroyed. The chloroform molecule, $CHCl_3$, is polar.

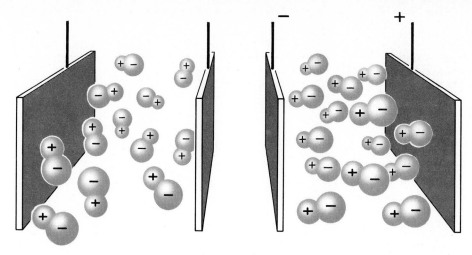

Field off   Field on

Figure 11.4
Orientation of polar molecules in an electric field. Two plates, immersed in a liquid whose molecules are polar, are connected through a switch to a source of an electric field. With the switch open the orientation of the molecules is random. When the switch is closed the molecules tend to line up with the positive end toward the negative plate, and the negative pole toward the positive plate.

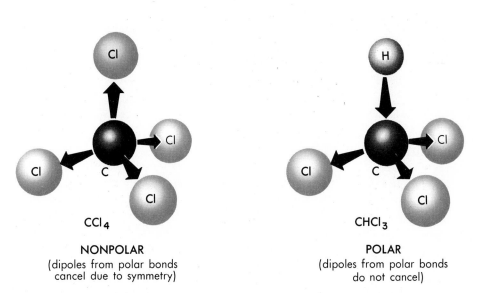

CCl₄   CHCl₃

NONPOLAR   POLAR
(dipoles from polar bonds   (dipoles from polar bonds
cancel due to symmetry)   do not cancel)

Figure 11.5
Polar and nonpolar molecules. CCl₄ is nonpolar because polar bonds cancel because of symmetry. CHCl₃ is polar because polar bonds do not cancel because of lack of symmetry.

EXAMPLE 11.11   Is the BF_3 moleculer polar? Is the NH_3 molecule polar?

The geometries of both of these molecules are described in Table 11.1. Consider BF_3 first. Sketch the Lewis diagram, with arrows pointing to the more electronegative element. Is the molecule polar?

—————   —————

BF_3 is nonpolar.

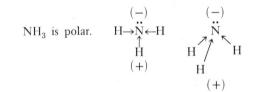

Even though fluorine is more electronegative than boron, the three fluorine atoms are arranged symmetrically around the boron atom. The polar bonds cancel.

Now sketch the trigonal pyramidal structure of NH_3, with arrows pointing to the more electronegative element. Is it polar or nonpolar?

—————   —————

NH_3 is polar.

$$
\begin{array}{cc}
(-) & (-) \\
H \rightarrow \ddot{N} \leftarrow H & \ddot{N} \\
\uparrow & H \nearrow\quad\nwarrow H \\
H & H \\
(+) & (+)
\end{array}
$$

The bonding electrons in ammonia are displaced toward the more electronegative nitrogen atom. The bonds do not cancel in the unsymmetrical pyramidal shape, so the molecule is polar. The diagram at the right, which attempts to show the molecular shape, indicates the charge displacement toward the nitrogen atom more accurately.

CHAPTER 11 IN REVIEW

11.1   DRAWING LEWIS DIAGRAMS BY INSPECTION

11 A   Draw the Lewis diagram for any molecule or polyatomic ion made up of elements in the A groups of the periodic table. (219)

11.2   MULTIPLE BONDS (222)

11.3   DRAWING COMPLEX LEWIS DIAGRAMS (223)

11.4   EXCEPTIONS TO THE OCTET RULE (229)

11.5   MOLECULAR GEOMETRY (230)

11.6 ELECTRON PAIR REPULSION; ELECTRON PAIR GEOMETRY (231)

11.7 MOLECULAR GEOMETRY BASED ON ELECTRON PAIR RE-PULSION

11 B Given or having derived the Lewis diagram of a molecule in which a second period central atom is surrounded by two, three, or four pairs of electrons, predict the electron pair geometry and the molecular geometry around that atom. (233)

11.8 THE GEOMETRY OF THE MULTIPLE BOND (237)

11.9 POLARITY OF MOLECULES

11 C Given or having determined the Lewis diagram of a molecule, predict whether the molecule is polar or nonpolar. (237)

TERMS AND CONCEPTS

Octet rule (220)
Isomer (228)
Molecular geometry (230)
Bond angle (230)
Bent (geometry) (231)
Linear (geometry) (231)
Planar (geometry) (231)
Tetrahedron (231)

Trigonal pyramid (geometry) (231)
Electron pair repulsion principle (232)
Electron pair geometry (232)
Trigonal planar (geometry) (233)
Tetrahedral (geometry) (233)
Tetrahedral angle (233)
Polar, nonpolar molecule (238)

Most of these terms and many others appear in the Glossary.

QUESTIONS AND PROBLEMS

An asterisk (*) identifies a question that is relatively difficult, or that extends beyond the performance goals in this chapter.

Sections 11.1 to 11.3

Write Lewis diagrams for each of the following sets of molecules:

11.1) HBr; H_2S; PH_3.

11.2) OF_2; CO; SO_4^{2-}.

11.3) ClO^-; BrO_4^-; H_2SO_4.

11.4) C_4H_{10}; C_4H_8; C_4H_6.

11.5)* CH_3F; CH_2F_2; CF_4.

11.6)* All possible isomers of C_5H_{12}.

11.7)* Two isomers of C_5H_{10} in which all carbon atoms are in a continuous chain.

11.23) BrF; SF_2; PF_3.

11.24) OH^-; ClO_3^-; NO_3^-.

11.25) IO_2^-; H_3PO_4; HSO_4^-.

11.26)* CH_4O; C_2H_4O; $C_2H_6O_2$.

11.27) CH_2ClF; CBr_2F_2; $ClBrClF$.

11.28)* All possible isomers of C_6H_{14}.

11.29)* Two isomers of C_3H_8O, one of which, an ether, does not have all of the carbon atoms bonded to each other.

242 chapter 11

11.8)* Two isomers of C_3H_6O.

11.9)* Formic acid, HCOOH, a compound produced by ants.

11.10) HS^-; H_2S; H_2S_2; H_2S_3; H_2S_4; H_2S_5.

11.30)* Two isomers of $C_2H_2Cl_2$.

11.31)* Propionic acid, C_2H_5COOH.

11.32) Hydrogen peroxide, H_2O_2, and the peroxide ion, O_2^{2-}.

Section 11.4

11.11)* Why is it not possible to draw a Lewis diagram that obeys the octet rule if the species has an odd number of electrons?

11.12)* Two iodides of arsenic (Z = 33) are AsI_3 and AsI_5. One of these iodides has a Lewis diagram that conforms to the octet rule, and one does not. Draw the diagram that is possible, and explain why the other cannot be drawn.

11.33)* Because of its five valence electrons, it is sometimes difficult to fit a nitrogen atom into a Lewis diagram that obeys the octet rule. Why is this so? Without actually drawing them, can you tell for which of the following species it is impossible to draw such a diagram? N_2O; NO_2; NO_2^-; NF_3; NO; N_2O_3; N_2O_4; N_2O_5; NOCl; NO_2Cl.

11.34)* On page 229 it says that five electron pairs surround the phosphorus atom in PF_5, and six surround sulfur in SF_6. How can this be when there is a total of four s and p orbitals?

Section 11.7

For each molecule, or for the atom specified in a molecule, describe (a) the electron pair geometry, and (b) the molecular geometry predicted by the electron pair repulsion principle.

11.13) BeH_2; CF_4; OF_2.

11.14) IO_4^-; ClO_2^-; CO_3^{2-}.

11.15) Each carbon atom in C_2H_5OH.

11.16) Nitrogen atom in CH_3NH_2.

11.35) BH_3; NF_3; HF.

11.36) ClO^-; IO_3^-; NO_3^-.

11.37) Oxygen atom in C_2H_5OH.

11.38) Silicon atom in SiF_4.

Section 11.8

For the atom specified in each of the following molecules, describe (a) the electron pair geometry, and (b) the molecular geometry predicted by the electron pair repulsion principle.

11.17)* Each carbon atom in C_2H_4.

11.18)* Carbon atom in HCHO.

11.39)* Each carbon atom in C_2H_2.

11.40)* Carbon atom in SCN^-.

Section 11.9

11.19) Explain how the carbon tetrafluoride molecule, CF_4, which contains four polar bonds (electronegativity difference 1.5), can be nonpolar.

11.20) Compare the polarities of the HCl and HI molecules. In each case identify the end of the molecule that is more negative.

11.41) The nitrogen-fluorine bond has an electronegativity difference of 1.0—less than the electronegativity difference between carbon and fluorine in CF_4. Yet the NF_3 molecule is polar, while CF_4 is nonpolar. How can this be?

11.42) Compare the polarities of the following molecules: ClF; Cl_2; BrCl; ICl. In each case identify the end of the molecule that is more negative.

11.21) Compare the polarities of the H_2O and H_2S molecules. Which molecule is more polar? What would you predict about the polarity of the H_2Te molecule ($Z = 52$ for Te)?

11.22) Sketch the water molecule, paying particular attention to the bond angle and using arrows to indicate the polarity of the individual bonds. Then sketch the methanol molecule, $HOCH_3$, again using arrows to show bond polarity and predicting the approximate shape of the molecule around the oxygen atom. Estimate the relative polarities of the water and methanol molecules, and explain your prediction.

11.43) Draw Lewis diagrams of the CF_4 and CH_2F_2 molecules, using arrows pointing to the more electronegative element in each bond. From these diagrams show that CH_2F_2 is polar and CF_4 is not.

11.44) As noted on page 229, there are two possible Lewis structures for the compound with the molecular formula C_2H_6O, and both are real compounds. C_2H_5OH is ethanol, or ethyl alcohol, and CH_3OCH_3 is diethyl ether, the anesthetic. Sketch these molecules with arrows to indicate the direction of bond polarity around the oxygen atom. Predict the relative polarities of these molecules. Predict also the relative polarity between CH_3OH and C_2H_5OH. What would you expect of the polarity of $C_5H_{11}OH$?

12

inorganic nomenclature

In Chapter 6 we presented a brief introduction to a formal system of chemical nomenclature. In this chapter we will review that introduction and enlarge it to include a broader range of acids and ionic compounds. Binary molecular compounds will be added. We will stop short of the interesting nomenclature that is used for the vast number of organic compounds, saving that system for Chapter 22.

The language of chemistry is like most languages in having many variations—many dialects, if you wish. Few people who work with chemicals speak a pure dialect. The standard for such a dialect, if one really exists, has been established by the International Union of Pure and Applied Chemistry (IUPAC), whose function it is to unify chemical terminology as it develops in laboratories throughout the world. The system they use is called the Stock system. At the other extreme we have common names, names given by craftsmen who identified a substance by its use or by some obvious physical or chemical property (Appendix III, p. 598). In between are various levels of custom and formality.

In this chapter we will strive to develop familiarity with the "professional" language of chemistry, the language used by chemists today. In areas where both old and new terms are employed we will mention both, but lean toward the more modern terminology thereafter. This is, after all, the way a language grows.

12.1 REVIEW OF INTRODUCTION TO NOMENCLATURE, CHAPTER 6

This section summarizes the nomenclature we have used for the past five chapters. Before you begin to expand your name-and-formula skill, you may wish to read about some items in detail.

1. Names and symbols of 35 elements are related to the periodic table in Figure 6.1 (p. 103). Seven of these elements form stable diatomic molecules. Their names and formulas are nitrogen, N_2, oxygen, O_2, hydrogen, H_2, fluorine, F_2, chlorine, Cl_2, bromine, Br_2, and iodine, I_2. The formulas of all other elements are simply their elemental symbols.

2. The name of a monatomic cation (positively charged ion) formed by a representative element (an element from an A group of the periodic table) is the name of the element, followed by *ion*. The name of a monatomic anion (negatively charged ion) is the name of the element changed to end in *-ide*, followed by *ion* (p. 107).

3. The formula of a monatomic ion is the elemental symbol with the ion charge as a superscript. Charges on monatomic ions formed by representative elements are the same for all elements in a group. They are (Fig. 6.3):

Group	1A	2A	3A	5A	6A	7A
Charge	+1	+2	+3	−3	−2	−1

4. Names and formulas of five acids and the anions derived from their "total ionization"—the removal of all ionizable hydrogen from the molecule—were given on page 112 in relation to Table 6.1. This table is basic to the nomenclature system in this chapter, so it will be reprinted here as Table 12.1 for your convenience. Binary (two-element) acids are named by placing the stem of the name of the second element between *hydro-* and *-ic*. Thus, HCl is *hydro*chloric acid. The anion from a binary acid is monatomic, named according to Item 2, above. Oxyacids whose names end in *-ic* produce anions with names that end in *-ate*. The formula of each anion is the formula of the

Table 12.1
Acids and Anions

Acid	Ionization Equation	Ion Name
Hydrochloric acid	$HCl \rightarrow H^+ + Cl^-$	Chloride
Nitric acid	$HNO_3 \rightarrow H^+ + NO_3^-$	Nitrate
Sulfuric acid	$H_2SO_4 \rightarrow 2\,H^+ + SO_4^{2-}$	Sulfate
Carbonic acid*	$H_2CO_3 \rightarrow 2\,H^+ + CO_3^{2-}$	Carbonate
Phosphoric acid*	$H_3PO_4 \rightarrow 3\,H^+ + PO_4^{3-}$	Phosphate

*The carbonic and phosphoric acid ionizations occur only slightly in water solutions. They are used here to illustrate the derivation of the formulas and names of the carbonate and phosphate ions, both of which are quite abundant from sources other than their parent acids.

acid minus the hydrogen(s), followed by a negative charge equal to the number of hydrogens taken from the acid.

5. NH_4^+ is the ammonium ion; OH^- is the hydroxide ion (p. 113).

6. The name of an ionic compound is the name of the cation followed by the name of the anion (p. 115).

7. The formula of an ionic compound is the formula of the cation followed by the formula of the anion, each taken the number of times required to bring the total charge on the formula unit to zero. Polyatomic ions are enclosed in parentheses when used more than once in a formula (p. 115).

The above list summarizes material included in Performance Goals D, G, H, I, and J in Chapter 6, p. 122. These performance goals should be considered as part of Chapter 12, too.

12.2 OXIDATION STATE; OXIDATION NUMBER

Names for many ionic compounds include the **oxidation state** or **oxidation number** of an element in the compound. The terms are synonymous and are used interchangeably. Oxidation numbers are a sort of electron bookkeeping system used to keep track of electrons in oxidation-reduction reactions. These reactions will be considered in some detail in Chapter 18. Right now we are interested in the rules by which oxidation numbers are assigned. They are:

1. The oxidation number of any elemental substance is 0 (zero).

2. The oxidation number of a monatomic ion is the same as the charge on the ion.

3. The oxidation number of combined oxygen is -2, except in peroxides (-1) and superoxides ($-\frac{1}{2}$). (We will not emphasize peroxides or superoxides in this text.)

4. The oxidation number of combined hydrogen is $+1$, except in hydrides (-1).

5. In any molecular or ionic species, the sum of the oxidation numbers of all atoms in the formula unit is equal to the charge on the species.

12.3 NAMES AND FORMULAS OF CATIONS AND MONATOMIC ANIONS

PG 12 A Given the name (or formula) of any ion in Figure 12.1, write the formula (or name) of that ion.

Oxidation number Rule 2 furnishes another way to look at the charge on a monatomic ion. Ions formed by Group 1A elements have a $+1$ charge. As ions, the elements are said to be in the +1 oxidation state. Similarly, ions formed by Group 2A elements are in the +2 oxidation state, and the oxidation number of Group 3A monatomic ions is +3. These are the only oxidation states known to Group 1A, 2A, and 3A elements as monatomic ions.

Some transition elements—elements in the B groups or Group 8 of the periodic table—are capable of forming monatomic cations with different charges. Iron is one example. If a neutral iron atom loses two electrons, the ion has a +2 charge, Fe^{2+}. A monatomic ion of iron may also be in the +3 oxidation state, Fe^{3+}. In this case the neutral atom has lost three electrons. Copper is another common element that is capable of more than one oxidation state as a monatomic ion. Its oxidation state may be +1, as in Cu^+, or it may be +2, as in Cu^{2+}.

For the purpose of naming chemical compounds it is necessary to distinguish between the two monatomic ions of iron. This is done in two ways. The older system applies either the -ic or the -ous suffix to the stem of the Latin name of the element. The Latin name for iron is *ferrum*. The -ic suffix is always applied to the higher oxidation state, and -ous to the lower. Hence the name of the Fe^{3+} ion is *ferric*; the -ous suffix applied to the Fe^{2+} ion gives us *ferrous*. By this system the name of $FeCl_2$, made up of an Fe^{2+} ion and two Cl^- ions, is ferrous chloride. When Fe^{3+} combines with three Cl^- ions the resulting compound is ferric chloride, $FeCl_3$.

In this text we emphasize the newer Stock system whereby an ion is identified by its English name, followed immediately by the oxidation state, written in Roman numerals and enclosed in parentheses. Thus, Fe^{2+} is written iron(II), and Fe^{3+} is written iron(III). In speaking, iron(II) becomes "iron two," and iron(III) is "iron three." Applied to the two chlorides, $FeCl_2$ is iron(II) chloride and $FeCl_3$ is iron(III) chloride.

As a general rule, if an element forms only one monatomic cation, its oxidation number is not included in the name. If an element forms two monatomic cations, the oxidation number is used in the name to distinguish between them.

Figure 12.1 is a partial periodic table showing most of the cations you are apt to find in an introductory course—perhaps all of them. The ammonium ion, NH_4^+, has been added beneath the alkali metal ions it so closely resembles. Notice that the mercury(I) ion, Hg_2^{2+}, is a *diatomic elemental* ion. It is the only polyatomic elemental cation of importance. Its +2 charge may be thought of as arising from each mercury atom losing one electron. Therefore the name is mercury(I) ion, or *mercurous* by the older system.

Monatomic anions form when neutral atoms gain one or more electrons (p. 109). Their names end in -ide. Both names and formulas are reviewed in Items 2 and 3 (p. 245).

EXAMPLE 12.1   Referring only to a periodic table (not Fig. 12.1), write the formula for each ion whose name is given, and the name where the formula is given.

barium ion Ba^{2+} I^- Iodide

zinc ion Zn^{2+}

Ag⁺ Silver ion

cobalt(II) ion Co^{2+}

Cr³⁺ Chromium(III)

rubidium ion Rb^{1+}
(Z = 37)

Se²⁻
(Se is selenium, Z = 34)

The last item in each column involves an element that is not among those you are expected to recognize. The location of the element in the periodic table should give you all the information you need to write the formula or name requested.

barium ion, Ba^{2+}
zinc ion, Zn^{2+}
cobalt(II) ion, Co^{2+}
rubidium ion, Rb^+

I^-, iodide ion
Ag^+, silver ion
Cr^{3+}, chromium(III) ion
Se^{2-}, selenide ion

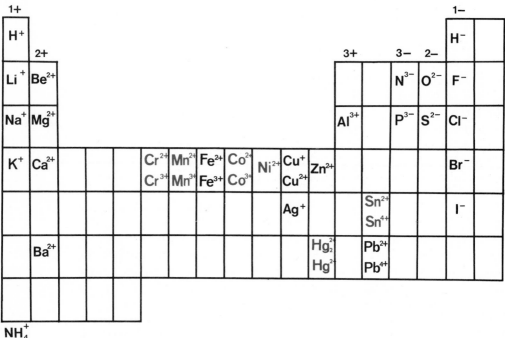

Figure 12.1
Partial periodic table of common ions. Notes: (1) Tin (Sn) and lead (Pb) form monatomic ions in a +2 oxidation state. In their +4 oxidation states they are more accurately described as being covalently bonded, but such compounds are frequently named as if they were ionic compounds. (2) Hg_2^{2+} is a diatomic elemental ion. Its name is mercury (I), indicating a +1 charge from each atom in the diatomic ion. (3) Ammonium ion, NH_4^+, is included as the only other common polyatomic cation, thereby completing this table as a minimum list of the cations you should be able to recall simply by referring to a full periodic table.

The (II) in cobalt(II) identifies the oxidation state and charge as +2.

Chromium is capable of more than one oxidation state, so that oxidation state must appear in the name of the ion. It is the same as the ionic charge, so chromium(III).

There is only one ionic charge for each of the first two elements in each column, so those oxidation states do not appear in the names of the ions.

Atomic number 37 in the periodic table shows that Rb is the symbol for rubidium. Its appearance in Group 1A establishes the ionic charge as +1.

Given that selenium is the name of an element whose monatomic ion has a negative charge, the name of the ion is found by applying an *-ide* suffix to the elemental name: selenide.

12.4 THE ACIDS OF CHLORINE AND THEIR ANIONS

Chlorine forms five acids that furnish quite a complete picture of acid nomenclature. You are already familiar with one: hydrochloric acid, HCl, a binary acid. There is also a chloric acid, $HClO_3$. From what you know about four Chapter 6 acids whose names end in *-ic*, can you tell the formula and name of the anion that comes from $HClO_3$? If you can, you are already catching on to the system of nomenclature we are about to develop.

All five chlorine acids are shown in Table 12.2. Each acid name, formula, and ionization equation, and the name of the anion from that ionization, are given. The table illustrates the nomenclature system as it applies to chlorine. Table 12.3 isolates the *system*, the prefixes (beginnings) and suffixes (endings) of acid and anion names, as the number of oxygen atoms changes. Our immediate purpose is to learn the system so that it may be applied to elements other than chlorine.

Begin with chloric acid on the second line of Table 12.2. Its formula, $HClO_3$, must be memorized. Do this now. The rest of this section and much of the next begin with that name and formula.

The name and formula of the anion from chloric acid are found by the rules in Item 4 (p. 245). An anion formula is the acid formula minus hydrogen(s), with a negative charge equal to the number of hydrogens removed from the acid. From $HClO_3$ you take *one* hydrogen, leaving ClO_3^-. Oxyacids ending in *-ic* produce anions with names ending in *-ate*. Hence, chlor*ic* acid yields the chlor*ate* ion. These endings are set apart on Line 2 of Table 12.3.

If an acid has one hydrogen more than the number in the *-ic* acid, the prefix *per-* is added before the name. Line 1 of Table 12.2 shows the formula $HClO_4$, with one oxygen more than $HClO_3$, chloric acid. Adding *per-* to the *-ic* name, $HClO_4$ becomes *per*chloric acid. The *per-* prefix goes to the anion name, too. As ClO_3^- is the chlorate ion, ClO_4^- is *per*chlorate ion. These prefixes are shown in Line 1 of Table 12.3, the line representing one more oxygen atom than the number in the *-ic* acid and the *-ate* anion.

If the number of oxygen atoms is one fewer than the number in an *-ic* acid and an *-ate* anion, it is the ending that changes. The *-ic* suffix becomes *-ous;* with one fewer oxygen than chlor*ic* acid, $HClO_2$ is chlor*ous* acid. The anion ending

Table 12.2
Acids of Chlorine

Line*	Acid	Ionization Equation	Anion Name
1	Perchloric acid	$HClO_4 \rightarrow H^+ + ClO_4^-$	Perchlorate
2	Chloric acid	$HClO_3 \rightarrow H^+ + ClO_3^-$	Chlorate
3	Chlorous acid	$HClO_2 \rightarrow H^+ + ClO_2^-$	Chlorite
4	Hypochlorous acid	$HClO \rightarrow H^+ + ClO^-$	Hypochlorite
5	Hydrochloric acid	$HCl \rightarrow H^+ + Cl^-$	Chloride

*Line numbers are references from text.

Table 12.3
Prefixes and Suffixes in Acid and Anion Nomenclature

Line	Oxygen Atoms Compared to -ic Acid and -ate Anion	Acid Prefix and/or Suffix (Example)	Anion Prefix and/or Suffix (Example)
1	One more	per-ic (perchloric)	per-ate (perchlorate)
2	Same	-ic (chloric)	-ate (chlorate)
3	One fewer	-ous (chlorous)	-ite (chlorite)
4	Two fewer	hypo-ous (hypochlorous)	hypo-ite (hypochlorite)
5	No oxygen	hydro-ic (hydrochloric)	-ide (chloride)

changes from *-ate* to *-ite:* ClO_2^- is the chlor*ite* ion. Line 3 of both tables shows these suffixes.

If the acid has two fewer oxygens than the *-ic* acid—or one fewer than the *-ous* acid—the prefix *hypo-* is applied to the name of the *-ous* acid. The name of HClO is therefore *hypo*chlorous acid. Similarly, the name of the anion is formed by attaching *hypo-* to the name of the *-ite* ion. ClO^- is the *hypo*chlorite ion. These prefixes appear on Line 4 of both tables.

Line 5 of Tables 12.2 and 12.3 shows the rule for binary acids, in which the prefix *hydro-* and suffix *-ic* surround the stem of the name of the second element.

You may wonder how chlorine can form acids with different numbers of oxygen atoms. Actually, this is quite reasonable from the standpoint of Lewis diagrams, discussed in Chapter 11. Table 12.4 shows that molecules of all five acids satisfy the octet rule. Notice that the ionizable hydrogen of an oxyacid is always bonded to an oxygen atom, not directly to the central element. This is characteristic of oxyacids. Notice also the oxidation states of chlorine in each molecule. Assigning oxidation states of -2 to oxygen and $+1$ to hydrogen (Rules 3 and 4, p. 246), and then applying Rule 5 to find the oxidation state of chlorine, yields $+7$, $+5$, $+3$, $+1$, and -1, as the number of oxygens decreases from four to zero.

Table 12.4
Acids of Chlorine

Acid		Lewis Diagram	Oxidation States
Perchloric	$HClO_4$	:Ö: | H—Ö—Cl—Ö: | :Ö:	H Cl 4 O +1 +7 4(−2)
Chloric	$HClO_3$	H—Ö—Cl—Ö: | :Ö:	H Cl 3 O +1 +5 3(−2)
Chlorous	$HClO_2$	H—Ö—Cl: | :Ö:	H Cl 2 O +1 +3 2(−2)
Hypochlorous	HClO HOCl	H—Ö—Cl:	H Cl O +1 +1 −2
Hydrochloric	HCl	H—Cl:	H Cl +1 −1

12.5 ACIDS AND ANIONS OF GROUP 5A, 6A, AND 7A ELEMENTS

PG 12 B   Given the name (or formula) of an acid of a Group 5A, 6A, or 7A element, or an anion derived from such an acid, write its formula (or name).

One of the characteristics of chemical families—elements in the same group in the periodic table—is that their members usually form similar compounds. This is true of the binary acids of the halogens. Rather than simply supplying the names and formulas of these acids, we'll give you the opportunity to find them yourself, and thereby learn the system. We'll even include a binary acid that does not contain a halogen.

EXAMPLE 12.2   For each of the following names, write the formula; for each formula, write the name.

hydrofluoric acid   HI

hydrobromic acid H_2S

Chlorine, fluorine, bromine, and iodine are all from the same family. If you know the formula of hydrochloric acid, you should be able to find the formulas of hydrofluoric and hydrobromic acids by substitution of elemental symbols. Then, if you reverse the thought process, you should be able to write the names of HI and H_2S.

------

hydrofluoric acid, HF
hydrobromic acid, HBr

HI, hydroiodic acid
H_2S, hydrosulfuric acid

The names are established by the rule for naming binary acids: the prefix *hydro-* followed by the elemental name changed to end in *-ic*.

Bromine and iodine form oxyacids comparable to oxyacids of chlorine. In name and formula they may be substituted for chlorine in chloric acid and the anion derived from it. These ideas do not extend to fluorine because that element doesn't happen to form oxyacids.

EXAMPLE 12.3   Complete the name and formula blanks in the following table:

ACID NAME	ACID FORMULA	ANION FORMULA	ANION NAME
bromic acid			
		IO_3^-	

ACID NAME	ACID FORMULA	ANION FORMULA	ANION NAME
bromic acid	$HBrO_3$	BrO_3^-	bromate
iodic acid	HIO_3	IO_3^-	iodate

Thought process: Bromic acid corresponds with chloric acid, $HClO_3$. The acid and ion formulas come from substituting Br for Cl in the corresponding chlorine formulas. The anion name for an *-ic* acid is the name of the central element changed to end in *-ate*.

IO_3^- corresponds to ClO_3^-, the chlorate ion that comes from $HClO_3$, chloric acid. The acid and ion names and formulas are the same, except that iodine replaces chlorine. Hence, chlorate becomes iodate as the name of IO_3^-, chloric becomes iodic as the name of the acid, and $HClO_3$ becomes HIO_3 as the acid formula. The acid formula can also be derived directly from the anion formula. The ion has a -1 charge, so the acid must have one hydrogen: HIO_3.

Now that you've seen how halogen acids and anions can be related to chloric acid, try your skill on the following.

EXAMPLE 12.4 Fill in the name and formula blanks in the following table. Try to do it without referring to Tables 12.2 and 12.3, but use them if absolutely necessary.

ACID NAME	ACID FORMULA	ANION FORMULA	ANION NAME
periodic	*HIO₄*	*IO₄⁻*	*periodate*
Hypo Bromous	HBrO	*Hypo Brom BrO⁻*	*Hypobromite*
Iodous	*HIO₂*	IO₂⁻	*Iodite*

ACID NAME	ACID FORMULA	ANION FORMULA	ANION NAME
periodic	HIO_4	IO_4^-	periodate

Let's think about the top line only. The prefix *per-*, applied to the memorized formula of chloric acid, $HClO_3$, means one more oxygen atom. Therefore, perchloric acid is $HClO_4$. Substituting iodine for chlorine we have periodic acid as HIO_4. Remove one hydrogen ion to get the anion formula, IO_4^-, with a negative charge equal to the number of hydrogens removed from the neutral molecule. The prefix *per-* is applied to the anion name as it is to the acid name. As perchloric acid → perchlorate ion, so periodic acid → periodate ion.

If you now wish to reconsider any of your entries on the other two lines, make any changes you wish. The answers follow.

ACID NAME	ACID FORMULA	ANION FORMULA	ANION NAME
periodic	HIO_4	IO_4^-	periodate
hypobromous	HBrO	BrO^-	hypobromite
iodous	HIO_2	IO_2^-	iodite

Reasoning processes are similar for all lines in the table. In the second line there are two fewer oxygen atoms than in chloric acid, $HClO_3$, and in the third line one fewer. Prefixes and suffixes match those for the corresponding chlorine substances in Tables 12.2 and 12.3.

Nitric, sulfuric, and phosphoric acids have important variations with different numbers of oxygen atoms. We will limit ourselves to two of these. The acid and anion nomenclature system in Table 12.3 remains the same. See if you can apply it to this new situation.

HeSo₄   N₃⁻¹
        S₄⁻²
        C₃⁻²

        P₄ ⁻3

EXAMPLE 12.5   Fill in the name and formula blanks in the following table.

ACID NAME	ACID FORMULA	ANION FORMULA	ANION NAME
Nitrous	HNO_2	NO_2^-	nitrite
Sulfurous	H_2SO_3	SO_3^{--}	sulfite

ACID NAME	ACID FORMULA	ANION FORMULA	ANION NAME
nitrous	HNO_2	NO_2^-	nitrite
sulfurous	H_2SO_3	SO_3^{2-}	sulfite

HNO_2 has one fewer oxygen atoms than nit*ric* acid, HNO_3, so its name must be nit*rous* acid. One hydrogen ion must be removed from the acid formula to produce the ion, NO_2^-. The anion from an *-ous* acid has an *-ite* suffix: nitrous acid → nitrite ion. From the memorized sulfuric acid, H_2SO_4, you have the sulfate ion, SO_4^{2-}. The sulfite ion has one fewer oxygen, SO_3^{2-}. If the anion has a −2 charge, the acid must have two hydrogens: H_2SO_3. The name of the acid with one fewer oxygens than sulfuric acid is sulfurous acid.

$HClO_3$, $HBrO_3$, HIO_3

HNO_3

H_2SO_4

H_2CO_3

H_3PO_4

The following example gives you the opportunity to practice what you have learned about acid and anion nomenclature. If you have memorized what you must, and know how to apply the rules that have been given, you will be able to write the required names and formulas with reference to nothing other than a periodic table. If you have really mastered the system, you will be able to extend it to the last substance in each column. They have not been mentioned anywhere in this chapter.

EXAMPLE 12.6   For each of the following names, write the formula; for each formula, write the name.

phosphoric acid H_3PO_4 CO_3^{2-} Carbonate

sulfate ion SO_4^{-2} HF   Hydro Flouric Acid.

✗ bromous acid $HBrO_2$ ✗ NO_2^- Nitrate ion

periodate ion IO_4^- ✓ H_2SO_3 Sulfuric Acid

nitric acid HNO_3 PO_4^{3-} Phosfate phosphate ion

telluric acid $HTeO_3$ SeO_3^{2-} Selenate ion
(tellurium, Z = 52)   (Se, selenium Z = 34)

phosphoric acid, H_3PO_4 (Table 12.1)

sulfate ion, SO_4^{2-} (Table 12.1)

bromous acid, HBrO (HClO and Tables 12.2 and 12.3)

periodate ion, IO_4^- (Example 12.4)

nitric acid, HNO_3 (Table 12.1)

telluric acid, H_2TeO_4

CO_3^{2-}, carbonate ion (Table 12.1)

HF, hydrofluoric acid, (Example 12.2)

NO_2^-, nitrite ion (Example 12.5)

H_2SO_3, sulfurous acid (Example 12.5)

PO_4^{3-}, phosphate ion (Table 12.1)

SeO_3^{2-}, selenite ion

References in parentheses tell where each name or formula may be found, or a starting point from which it may be figured out. The last items in each column include elements from Group 6A, the same chemical family as sulfur. The formula of telluric acid matches that of sulfuric acid, H_2SO_4. From sulfuric acid the name of SO_4^{2-} is sulfate ion. One fewer oxygen atom makes it SO_3^{2-}, sulfite ion. Substitution of selenium for sulfur in name and formula gives SeO_3^{2-}, selenite ion.

12.6 NAMES AND FORMULAS OF ANIONS DERIVED FROM THE STEPWISE IONIZATION OF POLYPROTIC ACIDS

PG 12 C Given the name (or formula) of an ion formed by the stepwise ionization of hydrosulfuric, sulfuric, sulfurous, phosphoric, or carbonic acid, write its formula (or name).

A hydrogen atom consists of one proton and one electron. To form what we have called a hydrogen ion, H^+, the neutral atom must lose its electron. That leaves only the proton; a hydrogen ion is simply a proton. (In aqueous solution, the only place hydrogen ions commonly exist, the proton is hydrated. Check what was said about the hydronium ion on p. 111.) Acids are sometimes classified by the number of hydrogen ions, or protons, that can be released by a single molecule. **Acids like hydrochloric, HCl, and nitric, HNO_3, that can yield only one proton per molecule, are called monoprotic acids. Acids that can yield more than one proton per molecule are called polyprotic acids.** Sulfuric acid, H_2SO_4, and phosphoric acid, H_3PO_4, are polyprotic acids capable of releasing two and three hydrogen ions per molecule, respectively.

When polyprotic acids ionize, they do so by steps, releasing one hydrogen ion with each step. The intermediate anions produced are stable chemical species that are the negative ions in many ionic compounds. The ionization of carbonic acid, H_2CO_3, is an example:

$$H_2CO_3 \xrightarrow{-H^+} HCO_3^- \xrightarrow{-H^+} CO_3^{2-} \qquad (12.1)$$

Brain Power (super nutrition)

Biotin Cream

Table 12.5
Names and Formulas of Anions Derived from the Stepwise Ionization of Acids

Acid	Ion	Names of Ions	
		Preferred	*Other*
H_2CO_3	HCO_3^-	Hydrogen carbonate	Bicarbonate Acid carbonate
H_2S	HS^-	Hydrogen sulfide	Bisulfide Acid sulfide
H_2SO_4	HSO_4^-	Hydrogen sulfate	Bisulfate Acid sulfate
H_2SO_3	HSO_3^-	Hydrogen sulfite	Bisulfite Acid sulfite
H_3PO_4	$H_2PO_4^-$	Dihydrogen phosphate	Monobasic phosphate
$H_2PO_4^-$	HPO_4^{2-}	Hydrogen phosphate	Dibasic phosphate

The intermediate ion, HCO_3^-, is the hydrogen carbonate ion—a logical name, since the ion is literally a hydrogen ion bonded to a carbonate ion. The ion is also called the bicarbonate ion.

Phosphoric acid, H_3PO_4, has three steps in its ionization process:

$$H_3PO_4 \xrightarrow{-H^+} H_2PO_4^- \xrightarrow{-H^+} HPO_4^{2-} \xrightarrow{-H^+} PO_4^{3-} \qquad (12.2)$$

$H_2PO_4^-$ is the dihydrogen phosphate ion, signifying two hydrogen ions attached to a phosphate ion, and HPO_4^{2-} is the monohydrogen phosphate ion, or simply hydrogen phosphate ion. It is essential that the prefix *di-* be used in naming the $H_2PO_4^-$ ion to distinguish it from HPO_4^{2-}. The prefix *mono-* in monohydrogen phosphate is optional; absence of a prefix in a case like this is understood to mean *one*.

If you recognize the logic of this part of the nomenclature system you will be able to extend it to intermediate ions from the stepwise ionization of hydrosulfuric, sulfuric, and sulfurous acids. All of these are shown in Table 12.5.

12.7 NAMES AND FORMULAS OF OTHER ACIDS AND IONS

Organic acids. Generally, organic acids ionize only slightly in water, but the anions produced are often abundant from other sources. Acetic acid, $HC_2H_3O_2$, the component of vinegar that is responsible for its odor and taste, is a good example. The ionization equation is

$$HC_2H_3O_2 \rightarrow H^+ + C_2H_3O_2^- \qquad (12.3)$$

Notice that only the hydrogen written first in the formula ionizes; the others do not (see below). $C_2H_3O_2^-$ is the acetate ion.

An organic chemist is more apt to write CH_3COOH for the formula of acetic acid. This "line formula," as it is called, shows the presence of the *carboxyl group*, —COOH, which is the source of ionizable hydrogen in almost all organic acids. The uniqueness of that hydrogen, compared to the other three, appears if Equation 12.3 is written with line formulas and Lewis diagrams:

$$CH_3COOH \rightarrow H^+ + CH_3COO^- \qquad (12.4)$$

$$(12.5)$$

In discussing the structure of the oxyacids of chlorine (p. 250), it was mentioned that ionizable hydrogens are bonded to oxygen which, in turn, is bonded to the central element. This is true for organic acids, too. The hydrogens linked to carbon through oxygen can ionize, whereas hydrogens bonded directly to carbon do not.

Hydrocyanic acid, HCN. When hydrocyanic acid ionizes it produces the cyanide ion, CN^-. This ion and the hydroxide ion are the only two common polyatomic anions that have an *-ide* ending, a suffix otherwise reserved for monatomic anions and binary molecular compounds.

Polyatomic anions from transition elements. Chromium and manganese form some polyatomic anions that theoretically may be traced to acids, but only the ions are important. Their names and formulas are the chromate ion, CrO_4^{2-}, the dichromate ion, $Cr_2O_7^{2-}$, and the permanganate ion, MnO_4^-.

Other ions. The performance goals have identified the important ions whose names and formulas you should recognize and be able to write. There are, of course, many others. Some of these, plus the ions already discussed, are listed in Tables 12.6 and 12.7. We recommend that you use these tables as a reference for the less common ions, and only as a last resort if you happen to forget one of the ions you should know.

12.8 FORMULAS OF IONIC COMPOUNDS

PG 12 D Given the name of any ionic compound made up of ions that are included in Performance Goals 12A, 12B, or 12C, or the hydroxide ion, write the formula of that compound.

You have been writing the formulas of a limited number of ionic compounds through the last five chapters. Now you know the names and formulas of many more ions, and therefore you can write the formulas of many more compounds such as those in the following example.

Table 12.6
Cations

Ionic Charge: +1		Ionic Charge: +2		Ionic Charge: +3	
Alkali Metals: *Group 1A*		*Alkaline Earths:* *Group 2A*		*Group 3A*	
Li^+	Lithium	Be^{2+}	Beryllium	Al^{3+}	Aluminum
Na^+	Sodium	Mg^{2+}	Magnesium	Ga^{3+}	Gallium
K^+	Potassium	Ca^{2+}	Calcium	*Transition Elements*	
Rb^+	Rubidium	Sr^{2+}	Strontium	Cr^{3+}	Chromium(III)
Cs^+	Cesium	Ba^{2+}	Barium	Mn^{3+}	Manganese(III)
Transition Elements		*Transition Elements*		Fe^{3+}	Iron(III)
Cu^+	Copper(I)	Cr^{2+}	Chromium(II)	Co^{3+}	Cobalt(III)
Ag^+	Silver	Mn^{2+}	Manganese(II)		
Polyatomic Ions		Fe^{2+}	Iron(II)		
NH_4^+	Ammonium	Co^{2+}	Cobalt(II)		
		Ni^{2+}	Nickel		
Others		Cu^{2+}	Copper(II)		
H^+	Hydrogen	Zn^{2+}	Zinc		
	or	Cd^{2+}	Cadmium		
H_3O^+	Hydronium	Hg_2^{2+}	Mercury(I)		
		Hg^{2+}	Mercury(II)		
		Others			
		Sn^{2+}	Tin(II)		
		Pb^{2+}	Lead(II)		

Table 12.7
Anions

Ionic Charge: −1				Ionic Charge: −2		Ionic Charge: −3	
Halogens: *Group 7A*		*Oxyanions*		*Group 5A*		*Group 5A*	
F^-	Fluoride	ClO_4^-	Perchlorate	O^{2-}	Oxide	N^{3-}	Nitride
Cl^-	Chloride	ClO_3^-	Chlorate	S^{2-}	Sulfide	P^{3-}	Phosphide
Br^-	Bromide	ClO_2^-	Chlorite	*Oxyanions*		*Oxyanion*	
I^-	Iodide	ClO^-	Hypochlorite	CO_3^{2-}	Carbonate	PO_4^{3-}	Phosphate
Acidic Anions		BrO_3^-	Bromate	SO_4^{2-}	Sulfate		
HCO_3^-	Hydrogen carbonate	BrO_2^-	Bromite	SO_3^{2-}	Sulfite		
		BrO^-	Hypobromite	$C_2O_4^{2-}$	Oxalate		
HS^-	Hydrogen sulfide			CrO_4^{2-}	Chromate		
		IO_4^-	Periodate	$Cr_2O_7^{2-}$	Dichromate		
HSO_4^-	Hydrogen sulfate	IO_3^-	Iodate	*Acidic Anion*			
HSO_3^-	Hydrogen sulfite	NO_3^-	Nitrate	HPO_4^{2-}	Hydrogen phosphate		
		NO_2^-	Nitrite				
$H_2PO_4^-$	Dihydrogen phosphate			*Diatomic Elemental*			
		OH^-	Hydroxide	O_2^{2-}	Peroxide		
Other Anions		$C_2H_3O_2^-$	Acetate				
SCN^-	Thiocyanate	MnO_4^-	Permanganate				
CN^-	Cyanide						
H^-	Hydride						

EXAMPLE 12.7 Write the formula for each compound listed below.

barium hydroxide | ammonium nitrate

potassium chlorate | sodium hydrogen carbonate

copper(II) chloride | calcium hypobromite

magnesium hydrogen sulfate | iron(III) sulfite

sodium hydrogen telluride (tellurium, Z = 52) | strontium dihydrogen phosphate (strontium, Z = 38)

----

barium hydroxide, $Ba(OH)_2$
potassium chlorate, $KClO_3$
copper(II) chloride, $CuCl_2$
magnesium hydrogen sulfate,
    $Mg(HSO_4)_2$
sodium hydrogen telluride, $NaHTe$

ammonium nitrate, NH_4NO_3
sodium hydrogen carbonate, $NaHCO_3$
calcium hypobromite, $Ca(BrO)_2$
iron(III) sulfite, $Fe_2(SO_3)_3$
strontium dihydrogen phosphate,
    $Sr(H_2PO_4)_2$

The ions of all compounds except the last in each column are among those you should know and recognize. Tellurium is in the same family as sulfur, so the hydrogen telluride ion should correspond with the hydrogen sulfide ion, HS^-. It is the intermediate step in the ionization of hydrotelluric acid, H_2Te, giving HTe^-. Strontium ion has a charge of $+2$, according to its position in Group 2A. It therefore requires two dihydrogen phosphate ions, $H_2PO_4^-$, to reach a total charge of zero.

12.9 NAMES OF IONIC COMPOUNDS

PG 12 E Given the formula of an ionic compound made up of identifiable ions, write the name of the compound.

As you know, the name of an ionic compound is the name of the cation followed by the name of the anion. If you recognize the two ions, you have the name of the compound.

There is one place where you are apt to be uncertain about the name of a "familiar" ion. For example, what is the name of Fe_2O_3? Iron oxide is not an adequate answer; it fails to distinguish between the two possible oxidation states of iron. Is it iron(II) oxide or iron(III) oxide? To decide, you must use a combination of oxidation number Rules 3 and 5. Rule 3 states that oxygen has an oxidation number of -2. Fe_2O_3 has three oxygen atoms in the formula unit, so oxygen contributes $3(-2) = -6$ to the total oxidation number for the formula unit.

Oxidation number Rule 5 requires that the total oxidation number of the formula unit be equal to the charge on the unit. In a compound this charge is zero. This means a total of $+6$ must come from the two atoms of iron in the formula, or $+3$ from each atom. The compound is therefore iron(III) oxide. FeO is iron(II) oxide. That conclusion would be reached by recognizing that the -2 of a single oxide ion is balanced by the $+2$ of a single iron(II) ion.

In writing or speaking the name of an ionic compound containing a metal that is capable of more than one possible oxidation state, it is essential that the compound name include the oxidation state of that metal.

EXAMPLE 12.8   Write the name of each compound below.

LiBr   *Lithium Bromide*

$Mg(IO_4)_2$ *Magnesium Periodate*

$AgNO_3$ *Silver Nitrate*

$MnCl_3$ *Manganese chloride*

Hg_2Br_2 *Mercury I Bromide*

$NaHSO_3$ *Sodium Hydrogen Sulfite*

K_2HPO_4 *Potassium Hydrogen Phosphate*

$ZnCO_3$ *Zinc (I) Carbonate*

HgS   *Mercury II Sulfide*

$(NH_4)_2SeO_4$ *Selenate*
(selenium, $Z = 34$)

LiBr, lithium bromide
$Mg(IO_4)_2$, magnesium periodate
$AgNO_3$, silver nitrate
$MnCl_3$, manganese(III) chloride
Hg_2Br_2, mercury(I) bromide

$NaHSO_3$, sodium hydrogen sulfite
K_2HPO_4, potassium hydrogen phosphate
$ZnCO_3$, zinc carbonate
HgS, mercury(II) sulfide
$(NH_4)_2SeO_4$, ammonium selenate

In $MnCl_3$, three -1 charges from three Cl^- ions require $+3$ from the manganese ion, so it is the manganese(III) ion. Similarly, one -2 charge from the sulfide ion in HgS must be balanced by $+2$ from a mercury(II) ion. In Hg_2Br_2 the two -1 charges from two Br^- ions are balanced by the $+2$ charge from the diatomic mercury(I) ion. In $(NH_4)_2SeO_4$, selenium substitutes for its family member, sulfur, in sulfate ion, SO_4^{2-}, so SeO_4^{2-} is the selenate ion.

Before moving to the next section, notice that only once so far have you used a prefix that suggests a number. The $H_2PO_4^-$ ion has been called the dihydrogen phosphate ion, and HPO_4^{2-} may be called the monohydrogen phosphate ion. The dichromate ion, $Cr_2O_7^{2-}$, was mentioned, and it appears in Table 12.7. Number prefixes are technically acceptable in naming compounds containing other ions, but they are not commonly found. With the above exceptions, we recommend that you do not use them for ionic compounds. Number prefixes are required for molecular compounds. This is covered in the next section.

12.10 BINARY MOLECULAR COMPOUNDS

PG 12 F   Given the name (or formula) of a binary molecular compound, write its formula (or name).

Atoms in molecular compounds are held together by covalent bonds, in which two atoms share one or more pairs of electrons. Molecular compounds are sometimes called covalent compounds. A binary compound, as noted earlier, has two elements. The elements are generally *both nonmetals*. You may use this rule to distinguish between binary molecular compounds and binary ionic compounds, for which you already know the nomenclature rules.

Names of binary molecular compounds consist of two words:

1. The first word is the name of the element appearing first in the chemical formula, including a prefix to indicate the number of atoms of that element in the molecule.

2. The second word is the name of the element appearing second in the chemical formula, changed to end in -*ide*, and also including a prefix to indicate the number of atoms of that element in the molecule.

Elemental symbols usually appear in the order of their increasing electronegativities (p. 212). The electronegativity of oxygen is less than that of fluorine, but more than that of chlorine. Therefore, oxygen appears before fluorine in OF_2, but after chlorine in Cl_2O.

The same two nonmetals often form more than one binary compound. Their names are distinguished by the prefixes mentioned in the rules above. Silicon and chlorine form silicon tetrachloride, $SiCl_4$, and disilicon hexachloride, Si_2Cl_6. The prefix *tetra-* identifies four chlorine atoms in a molecule of $SiCl_4$. In Si_2Cl_6 *di-* indicates two silicon atoms and *hexa-* shows six chlorine atoms in the molecule. Technically, $SiCl_4$ should be monosilicon tetrachloride, but the prefix *mono-* for one is usually omitted. If an element has no prefix in the name of a binary molecular compound, you may assume that there is only one atom of that element in the molecule.

Table 12.8 gives the first ten number prefixes. The letter "o" in *mono*, and the letter "a" in prefixes 4 to 10, are omitted if the resulting word "sounds better." This usually occurs when the next letter is a vowel. For example, a compound whose formula ends in O_5 is a *pentoxide* rather than a *pentaoxide*.

The oxides of nitrogen are ideal for practicing the nomenclature of binary molecular compounds.

Table 12.8
Numerical Prefixes Used in Chemical Names

Number	Prefix	Number	Prefix
1	mono-	6	hex-
2	di-	7	hepta-
3	tri-	8	octa-
4	tetra-	9	nona-
5	penta-	10	deca-

EXAMPLE 12.9   For each name below, write the formula; for each formula, write the name.

nitrogen monoxide NO_2

dinitrogen oxide N_2O_3

dinitrogen pentoxide N_2O_4

— — — — —   — — — — —

nitrogen monoxide, NO NO_2, nitrogen dioxide
dinitrogen oxide, N_2O N_2O_3, dinitrogen trioxide
dinitrogen pentoxide, N_2O_5 N_2O_4, dinitrogen tetroxide

Nitrogen dioxide could be correctly identified as mononitrogen dioxide. Two of the above compounds continue to be called by their older names: N_2O is nitrous oxide and NO is nitric oxide.

In Chapter 6, p. 110, it was shown that the water solution of hydrogen chloride, a gaseous binary molecular compound, is called hydrochloric acid. Both the acid and the compound have the same formula, HCl. When necessary, state symbols may be used to distinguish between them. HCl(g) clearly identifies the compound, hydrogen chloride, while HCl(aq) refers specifically to the water solution, hydrochloric acid. The same distinction may be used for any binary acid and its parent compound. The procedure is not necessary for oxyacids, however. HNO_3 is always nitric acid, never "hydrogen nitrate."

CHAPTER 12 IN REVIEW

12.1   REVIEW OF INTRODUCTION TO NOMENCLATURE, CHAPTER 6 (255)

12.2   OXIDATION STATE; OXIDATION NUMBER (246)

12.3   NAMES AND FORMULAS OF CATIONS AND MONATOMIC ANIONS

TERMS AND CONCEPTS

QUESTIONS AND PROBLEMS

To answer all questions in this nomenclature chapter, write the name of the species for which the formula is given, or the formula if the name is given. In naming ions or ionic compounds, use oxidation numbers when necessary to avoid uncertainty, but not otherwise. For reference materials, try to limit yourself to a periodic table giving no more information than the one inside the front cover of this book. Atomic numbers are included in questions involving elements that are not in Figure 6.1 p. 103.

Section 12.1

12.1) Cu; Kr; Mn; N_2.

12.2) Sodium; hydrogen; silicon; lead.

12.25) Cr; Cl_2; Be; S.

12.26) Boron; silver; argon; iodine.

Section 12.3

12.3) Ca^{2+}; Cr^{3+}; Zn^{2+}; P^{3-}; Br^-.

12.4) Lithium ion; ammonium ion; nitride ion; fluoride ion; mercury(II) ion.

12.27) Cu^+; I^-; K^+; Hg_2^{2+}; S^{2-}.

12.28) Iron(III) ion; hydride ion; oxide ion; aluminum ion; barium ion.

Section 12.5

12.5) HNO_3(aq); H_2SO_3(aq); perchloric acid; selenic acid (selenium: $Z = 34$).

12.6) Sulfate ion; chlorite ion; IO_3^-; BrO^-.

12.29) $HClO$(aq); H_2TeO_3(aq) (Te is tellurium, $Z = 52$); bromic acid; phosphoric acid.

12.30) Selenite ion (selenium: $Z = 34$); periodate ion; BrO_2^-; NO_2^-.

Section 12.6

12.7) Hydrogen carbonate ion; dihydrogen phosphate ion; HSO_4^-.

12.31) HPO_4^{2-}; HS^-; hydrogen sulfite ion.

Section 12.8

12.8) Potassium sulfide; copper(II) nitrate; sodium hydrogen carbonate.

12.32) Barium sulfate; chromium(III) oxide; calcium hydrogen phosphate.

Section 12.9

12.9) $MgSO_3$; AlF_3; $PbCO_3$.

12.33) $CuSO_4$; $Ba(OH)_2$; Hg_2I_2.

Section 12.10

12.10) SO_2; N_2O; phosphorus tribromide; hydrogen iodide.

12.34) Dichlorine oxide; uranium hexafluoride (uranium: $Z = 92$); HBr (g); P_2O_3.

From this point items in the nomenclature exercise are selected at random from any section of the chapter. Unless marked with an asterisk (*), all names and formulas are included in the performance goals, and should be found with reference to no more than the periodic table. Ions in compounds marked with an asterisk are included in Tables 12.6 and 12.7, p. 258; or, if the unfamiliar ion is monatomic, the atomic number of the element is given.

12.11) Hydrogen sulfite ion; potassium nitrate; $MnSO_4$; SO_3.

12.12) BrO_3^-; $Ni(OH)_2$; silver chloride; silicon hexafluoride.

12.35) Perchlorate ion; barium carbonate; NH_4I; PCl_3.

12.36) HS^-; $MgSO_3$; aluminum nitrate; oxygen difluoride.

12.13) Tellurate ion (tellurium: $Z = 52$); iron(III) phosphate; $NaC_2H_3O_2$*; H_2S (g).

12.14) HPO_4^{2-}; CuO; sodium oxalate*; ammonia.

12.15) Hypochlorous acid; chromium(II) bromide; $KHCO_3$; $Na_2Cr_2O_7$.*

12.16) Co_2O_3; Na_2SO_3; mercury(II) iodide; aluminum hydroxide.

12.17) Calcium dihydrogen phosphate; potassium permanganate*; NH_4IO_3; H_2SeO_4 (Se is selenium, $Z = 34$).

12.18) Hg_2Cl_2; HIO_4(aq); cobalt(II) sulfate; lead(II) nitrate.

12.19) Uranium trifluoride (uranium: $Z = 92$); barium peroxide*; $MnCl_2$; $NaClO_2$.

12.20) K_2TeO_4 (Te is tellurium, $Z = 52$); $ZnCO_3$; chromium(II) chloride; acetic acid*.

12.21) Barium chromate*; calcium sulfite; $CuCl$; $AgNO_3$.

12.22) Na_2O_2*; $NiCO_3$; iron(II) oxide; hydrosulfuric acid.

12.23) Zinc phosphide; cesium nitrate (cesium: $Z = 55$); NH_4CN*; S_2F_{10}.

12.24) N_2O_3; $LiMnO_4$*; indium selenide (indium: $Z = 49$; selenium: $Z = 34$); mercury(I) thiocyanate.*

12.37) Mercury(I) ion; cobalt(II) chloride; SiO_2; $LiNO_2$.

12.38) N^{3-}; $Ca(ClO_3)_2$; iron(III) sulfate; phosphorus pentabromide.

12.39) Tin(II) fluoride; potassium chromate*; LiH; $FeCO_3$.

12.40) HNO_2; $Zn(HSO_4)_2$; potassium cyanide*; copper(I) fluoride.

12.41) Magnesium nitride; lithium bromate; $NaHSO_3$; $KSCN$.*

12.42) $Ni(HCO_3)_2$; CuS; chromium(III) iodide; potassium hydrogen phosphate.

12.43) Selenium dioxide (selenium: $Z = 34$); magnesium nitrite; $FeBr_2$; Ag_2O.

12.44) SnO; $(NH_4)_2Cr_2O_7$*; sodium hydride; oxalic acid*.

12.45) Cobalt(III) sulfate; iron(III) iodide; $Cu_3(PO_4)_2$; $Mn(OH)_2$.

12.46) Al_2Se_3 (Se is selenium, $Z = 34$); $MgHPO_4$; potassium perchlorate; bromous acid.

12.47) Strontium iodate (strontium: $Z = 38$); sodium hypochlorite; Rb_2SO_4 (Rb is rubidium, $Z = 37$); P_2O_5.

12.48) ICl; $AgC_2H_3O_2$*; lead(II) dihydrogen phosphate; gallium fluoride (gallium: $Z = 31$).

13

---

the gaseous state

13.1 PROPERTIES OF GASES

The air that surrounds us is a sea of mixed gases, called the atmosphere. It is not necessary, then, to search very far to find a gas whose properties we may study. Some of the familiar characteristics of air—in fact, of all gases—are the following:

1. Gases may be compressed. A fixed quantity of air may be made to occupy a smaller volume by applying pressure. Figure 13.1A shows a quantity of air in a cylinder having a leak-proof piston which can be moved to change the volume occupied by the air. Push the piston down by applying more force and the volume of air is reduced (Fig. 13.1B).

2. Gases expand to fill their containers uniformly. If less force were applied to the piston, as shown in Figure 13.1C, air would respond immediately, pushing the piston upward, expanding to fill the larger volume uniformly. If the piston were pulled up (Fig. 13.1D), air would again expand to fill the additional space.

3. All gases have low density. The density of air is 0.0013 g/cm^3. The air in an empty 1-gallon (3.785-liter) bottle weighs 4.9 grams (about $\frac{1}{6}$ ounce); one gallon of water has a mass of 3785 grams (8.3 pounds), about 770 times greater. All of the air in a typical bedroom weighs about 35 kg (77 pounds).

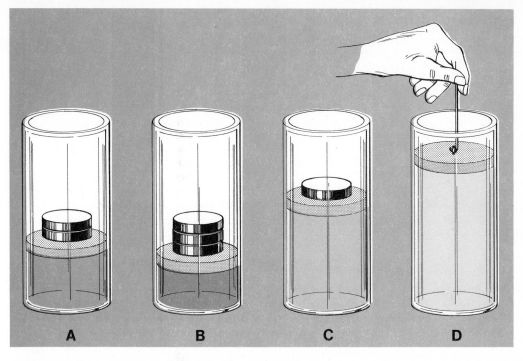

Figure 13.1
Properties of gases. The piston and cylinder show that gases may be compressed, and that they expand to fill uniformly the volume available to them.

4. Gases may be mixed. "There's always room for more," is a phrase that may be applied to gases. You may add the same or a different gas to that gas already occupying a rigid container of fixed volume, provided there is no chemical reaction between them.

5. A confined gas exerts constant pressure on the walls of its container uniformly in all directions. This pressure, illustrated in Figure 13.2, is a unique property of the gas, independent of external factors such as gravitational forces.

Figure 13.2
Gas pressures are exerted uniformly in all directions; liquid pressures depend upon the depth of the liquid.

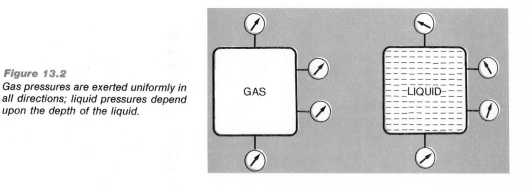

13.2 THE KINETIC THEORY OF GASES AND THE IDEAL GAS MODEL

PG 13 A Explain physical properties of gases, or physical phenomena relating to gases, in terms of the ideal gas model.

In trying to account for the properties of gases, scientists have devised the **kinetic theory of gases.** This theory is actually the best understood portion of the **kinetic molecular theory** referred to in Chapter 2. The theory describes an **ideal gas model*** by which we can visualize the nature of a gas by comparing it with a physical system that can be seen, or at least readily imagined. The main features of the ideal gas model are:

1. Gases consist of molecular particles moving at any given instant in straight lines.
2. Molecules collide with each other and with the container walls without loss of energy.
3. Gas molecules behave as independent particles; attractive forces between them are negligible.
4. Gas molecules are very widely spaced.
5. The actual volume of molecules is negligible compared to the space they occupy.

Particle motion explains why gases fill their containers. It also suggests how they exert pressure. When an individual particle strikes a container wall it exerts a force at the point of collision. When this is added to billions upon billions of similar collisions occurring continuously, the total effect is the steady force that is responsible for gas pressure.

There can be no loss of energy as a result of these collisions. If the particles lost energy, or slowed down, the combined forces would become smaller and the pressure would gradually decrease. Furthermore, because of the relationship between temperature and average molecular speed (p. 277), temperature would drop if energy were lost in collisions. But these things do not happen, so we conclude that energy is not lost in molecular collisions, either with the walls or between molecules.

Gas molecules must be widely spaced; otherwise the density of a gas would not be so low. One gram of liquid water at the boiling point occupies 1.04 cm^3. When changed to steam at the same temperature, the same number of molecules fills 1670 cm^3, an expansion of 1600 times. If the molecules were

---

*An "ideal" gas is one that conforms precisely to the equations that will be developed in this chapter. Most real gases at normal temperatures and pressures approach ideal behavior quite closely, but at certain conditions every real gas departs from this ideal.

touching each other in the liquid state, they must be widely separated in the vapor state. The compressibility and mixing ability of gases are also attributable to the open space between the molecules. Finally, it is because of the large intermolecular distance that attractions between molecules are negligible.

In summary, the ideal gas model pictures a gas as consisting of a large number of independent and widely spaced molecules, moving in random and chaotic fashion at high speed. Individual molecules move in straight lines until they collide with other molecules or the wall of the container without loss of energy.

13.3 GAS MEASUREMENTS

PG 13 B List the measurable properties of a gas.

PG 13 C Given a gas pressure in atmospheres, torr, millimeters of mercury, centimeters of mercury, inches of mercury, pounds per square inch, pascals, or kilopascals, express that pressure in each of the other units.

Experiments involving gases ordinarily require that one or more of the following be measured:

Quantity. The mass of a sample of a gas may be determined by weighing, although the procedure is somewhat more involved than the weighing of a solid or liquid. The amount of gas in a sample is more frequently expressed in moles.

Temperature. Gases expand when heated and contract when cooled, much more than solids or liquids. Temperature is therefore an important variable in experiments with gases. Temperature is generally measured with a thermometer.

Volume. The volume occupied by a gas is the full volume of its container. It may be measured in the usual way.

Pressure. Because the pressure of a gas is closely related to its quantity, temperature, and volume, pressure measurement is particularly important. By definition, pressure is the force exerted on a unit area:

$$\text{pressure} = \frac{\text{force}}{\text{area}} \quad or \quad P = \frac{F}{A} \tag{13.1}$$

Units of pressure come from the definition. In the English system, if force is measured in pounds and area in square inches, the pressure unit is pounds per square inch (psi). The SI unit of pressure is the **pascal**, which is **one newton per square meter.** (The *newton* is the SI unit of force.) One pascal is a very small pressure; the kilopascal is a more practical unit. The **millimeter of mercury,** or its equivalent, the **torr,** and the **atmosphere** are the common units for expressing pressure.

Weather bureaus generally report *barometric* pressure, the pressure exerted by the atmosphere at a given weather station, in inches or centimeters of mercury. It is measured by a device known as a barometer, developed by Evangelista Torricelli in the 17th century. Torricelli found that if a tube, closed at one end and filled with mercury, is inverted in a dish of mercury (Fig. 13.3), the liquid level inside the tube will fall until the pressure of the atmosphere on the surface of the liquid is exactly balanced by the pressure of the mercury column, whose height may be measured. This height thereby becomes a measure of atmospheric pressure. On a day when the mercury column in a barometer is 752 mm high, we say that atmospheric pressure is 752 mm Hg.

At sea level on an average day the height of the mercury column in a barometer is 760.0 mm or 29.92 inches. This pressure is arbitrarily called **one standard atmosphere** of pressure. The atmosphere unit is particularly useful in referring to very high pressures. The English unit that is equal to one atmosphere is 14.69 pounds per square inch. In summary, the different pressure units and their relationships to each other are:

$$1.000 \text{ atm} = 14.69 \text{ lbs/in.}^2 = 29.92 \text{ inches Hg} = 76.00 \text{ cm Hg} =$$

$$\text{(13.2)}$$

$$760.0 \text{ mm Hg} = 760.0 \text{ torr} = 1.013 \times 10^5 \text{ Pa} = 101.3 \text{ kPa}$$

where the Pa and kPa are the pascal and kilopascal, respectively.

The *torr* is the recently introduced substitute for the *millimeter of mercury,* honoring the work of Torricelli. Both terms are widely used, and the choice between them is one of personal preference. The advantage of *millimeter of mercury* is that it has physical meaning; it may be read by direct observation on an open-end **manometer,** the instrument by which pressure is most commonly measured in the laboratory (Fig. 13.4). *Torr,* however, is both easier to say and write. We will use *torr* hereafter in this text.

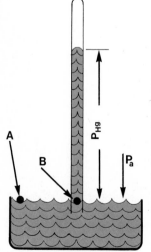

Figure 13.3
Mercury barometer. Two operational principles govern the mercury barometer. (1) The total pressure at any point in a liquid system is the sum of the pressures of each gas or liquid phase above that point. (2) The total pressures at any two points at the same level in a liquid system are always equal. Point A at the liquid surface outside the tube is at the same level as Point B inside the tube. The only thing exerting downward pressure at A is the atmosphere; P_a represents atmospheric pressure. The only thing exerting downward pressure at Point B is the mercury above that point, designated P_{Hg}. A and B being at the same level, the pressures at these points are equal: $P_a = P_{Hg}$.

Outside of the laboratory, mechanical gauges are used to measure gas pressure. A typical tire gauge is probably the most common. Mechanical gauges show the pressure *above* atmospheric pressure, rather than the absolute pressure measured by a manometer. Even a flat tire contains air that exerts pressure. If it did not, the entire tire would collapse, not just the bottom. The pressure of the gas remaining in a flat tire is equal to atmospheric pressure. If a tire gauge shows 25 psi, that is the **gauge pressure** of the gas (air) in the tire. The absolute pressure is nearly 40 psi—the 25 psi shown by the gauge plus about 15 psi from the atmosphere.

A pressure given in one unit is changed to another unit by a one-step conversion, using the required relationship from Equation 13.2.

EXAMPLE 13.1   The pressure inside a steam boiler is 1050 psi. Express this pressure in atmospheres.

Set up and solve the problem.

$$1050 \text{ psi} \times \frac{1 \text{ atm}}{14.7 \text{ psi}} = 71.4 \text{ atm}$$

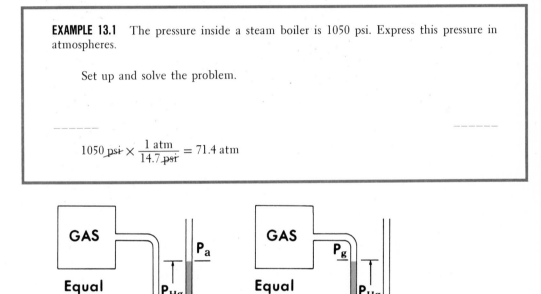

Figure 13.4
Open-end manometers. Open-end manometers are governed by the same principles as mercury barometers (Fig. 11.3). The pressure of the gas, P_g, is exerted on the mercury surface in the closed (left) leg of the manometer. Atmospheric pressure, P_a, is exerted on the mercury surface in the open (right) leg. Using a meter stick, the difference between these two pressures, P_{Hg}, may be measured directly in millimeters of mercury (torr). Gas pressure is determined by equating the total pressures at the lower liquid level. In A, the pressure in the left leg is the gas pressure, P_g. Total pressure at the same level in the right leg is the pressure of the atmosphere, P_a, plus the pressure difference, P_{Hg}. Equating the pressures, $P_g = P_a + P_{Hg}$. In B, total pressure in the closed leg is $P_g + P_{Hg}$, which is equal to the atmospheric pressure, P_a. Equating and solving for P_g yields $P_g = P_a - P_{Hg}$. In effect, the pressure of a gas, as measured by a manometer, may be found by adding the pressure difference to, or subtracting the pressure difference from, atmospheric pressure; $P_g = P_a \pm P_{Hg}$

13.4 BOYLE'S LAW

PG 13 D Given the initial volume (or pressure) and initial and final pressures (or volumes) of a fixed quantity of gas at constant temperature, calculate the final volume (or pressure).

Robert Boyle, in the 17th century, investigated the quantitative relationship between pressure and volume of a fixed amount of gas at constant temperature. A modern laboratory experiment finds this relationship with a mercury-filled manometer such as that shown in Figure 13.5A. The closed and stationary left leg of the manometer is a gas-measuring tube, graduated in milliliters.

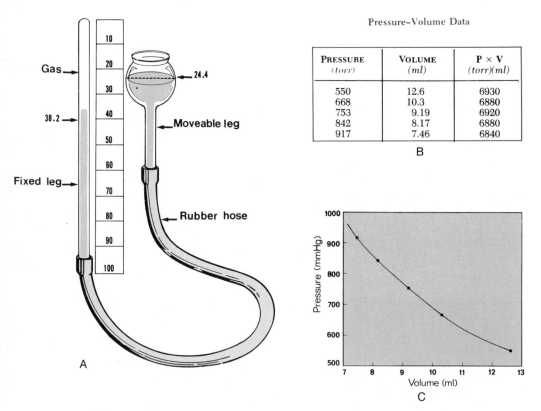

Pressure–Volume Data

PRESSURE *(torr)*	VOLUME *(ml)*	P × V *(torr)(ml)*
550	12.6	6930
668	10.3	6880
753	9.19	6920
842	8.17	6880
917	7.46	6840

B

C

Figure 13.5
Results of a student's experimental determination of the pressure-volume relationship of a fixed quantity of gas at constant temperature. By raising or lowering the moveable leg of the apparatus (A), the pressure and volume of the trapped gas may be determined. The first two columns of the table (B), are the student's data. A plot of pressure vs. volume (C) indicates that the variables are inversely proportional to each other. This is confirmed by the constant (within experimental error) product of pressure × volume, shown in the third column of the table.

Trapped above the mercury is a constant quantity of gas. The open right leg of the manometer is connected to the left leg by a flexible tube. The pressure on the confined gas may be changed simply by moving the right leg up or down. An actual student experiment with this apparatus yielded the data in the table shown in Figure 13.5B. A graph of pressure versus volume for these data is shown in Figure 13.5C. To a mathematician, the shape of the curve suggests an inverse proportionality between pressure and volume. Expressed mathematically,

$$P \propto \frac{1}{V} \tag{13.3}$$

Introducing a proportionality constant (see Appendix I, p. 581), we have

$$P = k_1 \frac{1}{V} \tag{13.4}$$

Multiplying both sides of the equation by V yields

$$PV = k_1 \tag{13.5}$$

Returning to the data of the experiment and multiplying pressure times volume for each setting of the manometer we find that, within experimental error, PV is indeed a constant (Fig. 13.5B).

Boyle's Law, which these data illustrate, states that **for a fixed quantity of gas at constant temperature, pressure is inversely proportional to volume** (Fig. 13.6). Equation 13.5 represents the usual mathematical statement of Boyle's Law. Since the product of P and V is constant, when either factor increases the other must decrease, and vice versa. This is what is meant by an inverse proportionality.

From Equation 13.5, we see that $P_1V_1 = k_1 = P_2V_2$, or

$$P_1V_1 = P_2V_2 \tag{13.6}$$

where subscripts 1 and 2 refer to first and second measurements of pressure and volume of the gas sample at constant temperature. Solving for V_2,

$$V_2 = V_1 \frac{P_1}{P_2} \tag{13.7}$$

Condition: Temperature constant—no gas gained or lost

Figure 13.6
Boyle's Law. A fixed quantity of gas is confined to a cylinder, as in A, at a given pressure and constant temperature. If pressure is doubled, as in B, the volume is reduced to one half its original value. If pressure is doubled again, now four times the original pressure, the volume is reduced to one fourth of its original value (C).

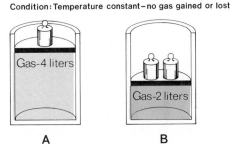

A   B   C

In other words, *if the pressure of a confined gas is changed, the final volume may be calculated by multiplying the initial volume by a ratio of pressures*—a pressure correction. The inverse proportionality between pressure and volume leads to one of two possibilities:

1. If final pressure is greater than initial pressure, then final volume must be less than initial volume. Therefore the pressure correction must be a ratio less than 1—the numerator must be smaller than the denominator.
2. If final pressure is less than initial pressure, then final volume must be more than initial volume. Therefore the pressure correction must be a ratio more than 1—the numerator must be larger than the denominator.

The above statements are the keys to the "reasoning" method of solving pressure-volume problems. By knowing the inverse character of the pressure-volume relationship, you may *reason* whether final volume is more or less than initial volume, and thereby choose a pressure correction greater or less than one. Solve the following example by the reasoning method:

EXAMPLE 13.2 A certain gas sample occupies 3.25 liters at 740 torr. Find the volume of the gas sample if the pressure is changed to 790 torr. Temperature remains constant.

From the statement of the problem, does pressure increase or decrease?

- - - - - -   - - - - - -

It increases—from 740 to 790 torr.

Will this cause an increase or decrease in volume?

- - - - - -   - - - - - -

Decrease. Pressure and volume are inversely related: as one goes up, the other goes down.

The new volume will be found by applying a pressure correction to the initial volume—by multiplying the initial volume by a ratio of initial and final pressures. The ratio will be either $\frac{740 \text{ torr}}{790 \text{ torr}}$ or $\frac{790 \text{ torr}}{740 \text{ torr}}$. Bearing in mind that the final volume must be less than the initial volume, which ratio is correct?

- - - - - -   - - - - - -

$\frac{740 \text{ torr}}{790 \text{ torr}}$

If the final volume, the product of the multiplication, is to be less than the initial volume, the initial volume must be multiplied by a ratio less than 1.

Now complete the problem.

$$3.25 \text{ L} \times \frac{740 \text{ torr}}{790 \text{ torr}} = 3.04 \text{ L}$$

Example 13.2 may also be solved by substitution into Equation 13.6 or 13.7. This "formula" method is equally correct and preferred by some. These alternatives will be mentioned as they arise. Which method is "better" is, of course, a matter of opinion. It is recommended that you use the method presented in your chemistry class. In this text we will generally use the reasoning approach.

EXAMPLE 13.3   1.44 liters of gas at 0.935 atmosphere are compressed to a volume of 0.275 liter. Find the new pressure in atmospheres.

Will the reduction in volume cause an increase or decrease in pressure?

Increase. Pressure and volume are inversely related.

The correction factor this time is a ratio of volumes. Complete the problem.

$$0.935 \text{ atm} \times \frac{1.44 \text{ L}}{0.275 \text{ L}} = 4.90 \text{ atm}$$

This problem might also be solved by substitution into Equation 13.6.

13.5 ABSOLUTE TEMPERATURE

PG 13 E   Given a temperature in degrees Celsius (or Kelvin), convert it to Kelvin (or degrees Celsius).

We know that the temperature of things can be reduced until they become very cold. But how cold? Is there a bottom limit to temperature? Experiments suggest that there is. One such experiment performed by students in college laboratories shows how pressure varies with temperature when the volume of a gas sample is held constant. The experiment is described in Figure 13.7. Another experiment that measures the volume of a sample at different temperatures and constant pressure gives similar results. This includes a graph that is similar to Figure 13.7C, except that pressure is replaced by volume. These experiments, as well as many far more sophisticated investigations of low temperature phenomena, predict an *absolute zero* temperature at −273°C.

The SI temperature scale is the **Kelvin temperature scale** (p. 48), which has its zero at −273°C. All temperatures on the Kelvin scale therefore have positive values. The size of the Kelvin is the same as the size of the Celsius degree. This leads to the equation

$$\text{temperature (K)} = \text{temperature (°C)} + 273 \qquad (13.8)$$

by which temperatures expressed in either scale may be converted to the other. From Equation 13.8 the freezing point of water, 0°C, is 273 K. Under SI, the word "degree" and its symbol are not used, and there is no space between the

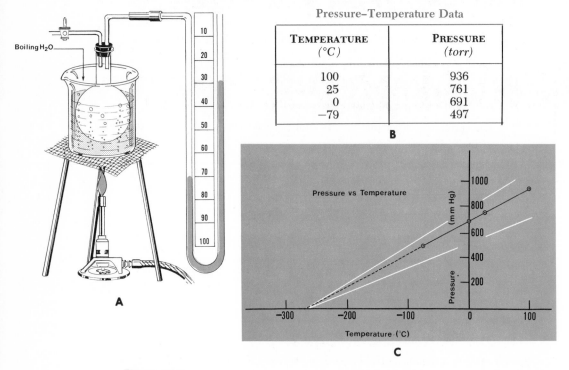

Pressure–Temperature Data

TEMPERATURE (°C)	PRESSURE (torr)
100	936
25	761
0	691
−79	497

B

Figure 13.7

Results of a student's experimental determination of the pressure-temperature relationship of a fixed quantity of gas at constant volume. The flask is immersed in a liquid bath at different temperatures (A). The pressures are measured at each temperature and tabulated (B). A graph of the data (C) is plotted in black. The white lines in C show what the graph would be if the experiment were repeated with larger or smaller quantities of gas. Extrapolation to zero pressure suggests an "absolute zero" temperature at −273°C.

temperature and the symbol K. The freezing point of water is therefore "two-hundred seventy-three Kelvin," or simply "two-hundred seventy-three K."

EXAMPLE 13.4 Tomorrow's weather forecast is for a high temperature of 77°F, or 25°C. Express this temperature in Kelvin.

------   ------

25°C + 273 = 298 K

To understand why there is an absolute zero in temperature it is necessary to recognize what is measured by temperature. Experiments indicate that temperature is a measure of the average translational kinetic energy of the particles. Translational kinetic energy is the energy of motion as a particle goes from one place to another. It is expressed mathematically as $\frac{1}{2} mv^2$, where m is the mass of the particle and v is its velocity. There is no reason to believe that the mass of a particle changes as temperature is reduced, so we conclude that particle velocity is less at lower temperatures. At absolute zero we imagine that all translational molecular movement stops.

Research with gases at absolute zero is impossible because the gases condense to liquids before absolute zero is reached. Other experiments support the concept of absolute zero, however, and fix the temperature somewhat more accurately at −273.16°C. No attempt to penetrate this bottom temperature barrier has ever been successful, although some researchers claim to have reached within 0.0014 K of the theoretical zero.

13.6 CHARLES' LAW

PG 13 F Given the initial volume, and initial and final temperatures of a fixed quantity of gas at constant pressure, calculate the final volume.

A graph of volume or pressure vs. *absolute* temperature may be obtained by shifting the vertical axis of the graph in Figure 13.7 to −273°C. This is shown in Figure 13.8. A straight line passing through the origin is the graph of a direct proportionality. We conclude, therefore, that V ∝ T and P ∝ T, where V is volume, T is absolute temperature, and P is pressure. Applying a proportionality constant to the volume-temperature relationship yields

$$V = k_2 T \qquad\qquad (13.9)$$

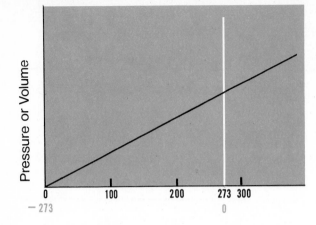

Figure 13.8
Graph of pressure or volume vs. absolute temperature, with other variables held constant. This graph is produced from Figure 13.7, with the vertical axis moved to absolute zero at −273 K. The straight line through the origin shows the pressure to be directly proportional to the absolute temperature at constant volume and quantity, and volume to be directly proportional to absolute temperature at constant pressure and quantity.

Equation 13.9 is the mathematical expression of what is known as Charles' Law: **the volume of a fixed quantity of gas at constant pressure is proportional to absolute temperature.** The physical significance of Charles' Law is illustrated in Figure 13.9.

Dividing both sides of Equation 13.9 by T yields

$$\frac{V}{T} = k_2 \qquad (13.10)$$

From this it follows that $\dfrac{V_1}{T_1} = k_2 = \dfrac{V_2}{T_2}$, or

$$\frac{V_1}{T_1} = \frac{V_2}{T_2} \qquad (13.11)$$

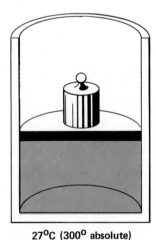

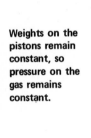

Weights on the pistons remain constant, so pressure on the gas remains constant.

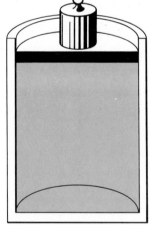

27°C (300° absolute)   327°C (600° absolute)

Figure 13.9
Illustration of Charles' Law. If the absolute temperature is doubled, the volume is doubled at constant pressure and amount of gas.

where the subscripts 1 and 2 again refer to first and second measurements of the two variables. Solving for V_2,

$$V_2 = V_1 \times \frac{T_2}{T_1} \qquad (13.12)$$

Thus, if the temperature of a confined gas is changed at constant pressure, the final volume may be found by multiplying the initial volume by a *ratio of absolute temperatures*—a temperature correction. Again there are two possibilities, this time dictated by the *direct* proportionality between temperature and volume:

1. If final temperature is greater than initial temperature, then final volume must be greater than initial volume. Therefore the temperature correction must be a ratio more than 1—the numerator must be larger than the denominator.
2. If final temperature is less than initial temperature, then final volume must be less than initial volume. Therefore the temperature correction must be a ratio less than 1—the numerator must be smaller than the denominator.

One extremely important fact must be remembered in working gas law problems involving temperature: the proportional relationships apply to *absolute* temperatures rather than to Celsius temperatures, which are frequently given in the statement of the problem. Before solving gas law problems, Celsius temperatures must be converted to K.

EXAMPLE 13.5   2.42 liters of a gas, measured at 22°C, are heated to 45°C at constant pressure. What will be the new volume of the gas?

From the conditions of the problem, will the final volume be more or less than 2.42 liters?

More. Gas volume varies directly with temperature; if temperature increases, volume must also increase.

What temperature ratio will you use as a multiplier to find the new volume?

$$\frac{318\ K}{295\ K}. \qquad \frac{(45 + 273)\ K}{(22 + 273)\ K} = \frac{318\ K}{295\ K}.$$

Always be sure to convert to K for gas law problems.

Complete the problem: find the final volume when the temperature of 2.42 liters of gas is changed from 295 K to 318 K.

-------   -------

$$2.42 \text{ L} \times \frac{318\ \cancel{K}}{295\ \cancel{K}} = 2.61 \text{ L}$$

Direct substitution into Equation 13.12 would yield the same result.

13.7 THE IDEAL GAS EQUATION

PG 13 G Write the equation that shows how the measurable properties of an ideal gas are related to each other (the ideal gas equation).

If we assume that the laws of physics can be applied to the kinetic theory of gases and to the ideal gas model, it is possible to derive an equation that includes all four measurable properties of gases: pressure (P), volume (V), absolute temperature (T), and quantity in moles (n). This equation is

$$PV = nRT \qquad (13.13)$$

R is a constant known as the **universal gas constant.** Its value is the same for any gas or mixture of gases that behaves like an ideal gas. The equation is called the **ideal gas equation** or the **ideal gas law.**

The ideal gas equation can also be derived from measurements of pressure, volume, absolute temperature and quantity of gas. This derivation makes no assumption about the kinetic theory of gases or their ideal behavior. The fact that theoretical calculations and experimental data yield the same equation makes us quite confident that the kinetic theory of gases and the ideal gas model provide a true picture of the nature of a gas.

Boyle's Law and Charles' Law are specific examples of the ideal gas equation. Under the Boyle's Law conditions of constant temperature and quantity of gas, the entire right side of Equation 13.13 becomes a constant. The equation then has exactly the same form as Equation 13.5. If both sides of Equation 13.13 are divided by PT, the result is

$$\frac{V}{T} = \frac{nR}{P} \qquad (13.14)$$

If quantity and pressure are constant, as they are under Charles' Law, the right side of Equation 13.14 is a constant. The equation then has exactly the same form as Equation 13.10.

13.8 PRESSURE, VOLUME, AND TEMPERATURE CHANGES WITH A FIXED SAMPLE OF GAS

PG 13 H   For a fixed quantity of a confined gas, given the initial volume, pressure, and temperature, and the final pressure and temperature, calculate the final volume.

Quite often it is necessary to make calculations involving the pressure, volume, and temperature of a given sample of gas. In this case both n and R in Equation 13.13 are constant. Dividing both sides of the equation by T gives

$$\frac{PV}{T} = nR = K \qquad (13.15)$$

where K is a constant for that gas sample. Using subscripts 1 and 2 for first and second measurements of the three variables, $\dfrac{P_1 V_1}{T_1} = K = \dfrac{P_2 V_2}{T_2}$, or

$$\frac{P_1 V_1}{T_1} = \frac{P_2 V_2}{T_2} \qquad (13.16)$$

If the pressure, volume, and temperature of a fixed quantity of gas (P_1, V_1, and T_1) are known at any time, and any two of the variables (any two of P_2, V_2, and T_2) for the same gas sample are known at another time, the missing variable can be calculated. If the unknown is the final volume, V_2, Equation 13.16 may be solved for that unknown:

$$V_2 = V_1 \times \frac{P_1}{P_2} \times \frac{T_2}{T_1} \qquad (13.17)$$

This shows that the final volume may be found by multiplying the initial volume by a pressure correction and a temperature correction. Both of these ratios may be reasoned out just as they were in earlier examples.

EXAMPLE 13.6   A certain gas occupies 3.40 liters at 65°C and 680 torr. What volume will it occupy if it is cooled to room temperature, 21°C, and compressed to 800 torr?

We will approach this problem by first setting up for the volume change caused by a change in pressure, holding temperature constant. Then we will find the further volume change caused by a change in temperature, holding pressure constant. By steps, what will happen to the volume because of the change in pressure from 680 torr to 800 torr, considering temperature constant? Will volume increase or decrease?

Volume will *decrease* if pressure increases; they are inversely proportional.

Now begin the setup of the problem with the 3.40 L volume multiplied by the proper ratio as pressure increases from 680 torr to 800 torr. Do not solve.

$$3.40 \text{ L} \times \frac{680 \text{ torr}}{800 \text{ torr}} \times \underline{\hspace{3cm}}$$

3.40 liters is the volume at 680 torr and 65°C. Solving the setup as far as it is written would give the volume at 800 torr and 65°C. Now extend the setup by applying the proper temperature correction to get the volume at 800 torr and 21°C. Solve for the answer.

$$3.40 \text{ L} \times \frac{680 \text{ torr}}{800 \text{ torr}} \times \frac{294 \text{ K}}{338 \text{ K}} = 2.51 \text{ L}$$

Reducing temperature reduces volume; they are directly proportional. The temperature correction is therefore less than 1.

Direct substitution into Equation 13.16 or Equation 13.17 would reproduce the above setup and yield the same result.

You now have two ways to solve volume problems by the gas laws: you may *reason* your way through temperature and pressure corrections, or you may solve the problems algebraically by Equation 13.16. If your instructor states a preference, by all means adopt it, at least for the present. If you must memorize an equation, either by your own choice or by teacher direction, Equation 13.16 is recommended. With it you may solve for any variable, given the other five.

13.9 STANDARD TEMPERATURE AND PRESSURE

PG 13 I Given the volume of a gas at one temperature and pressure (or at STP), find the volume it would occupy at STP (or at a stated temperature and pressure).

The volume of a fixed quantity of gas depends upon its temperature and pressure; if either changes, the volume changes. It is therefore not possible to state the amount of gas in volume units without also specifying the temperature and pressure. To meet this need, 0°C (273 K) and 1 atmosphere (760 torr) have been adopted as **standard temperature and pressure (STP)**. Many gas laws problems require changing volume to or from STP. The problems are solved in the same manner as Example 13.6.

EXAMPLE 13.7   What would be the volume at STP of 4.06 liters of nitrogen, measured at 712 torr and 28°C?

Set up the problem in its entirety and solve.

$$4.06 \text{ L} \times \frac{712 \text{ torr}}{760 \text{ torr}} \times \frac{273 \text{ K}}{301 \text{ K}} = 3.45 \text{ L}$$

13.10 APPLICATIONS OF THE IDEAL GAS EQUATION

PG 13 J   Write the variation of the ideal gas equation that includes the mass of the gas sample and its molar weight.

13 K   Given values for all except one of the variables in either form of the ideal gas equation, calculate the value of that remaining variable.

Before the ideal gas equation, Equation 13.13, can be applied, you must have a value for the gas constant, R. This value has been determined in the laboratory. Solving Equation 13.13 for R gives

$$R = \frac{PV}{nT} \tag{13.18}$$

It is an experimental fact that at STP one mole of any ideal gas occupies a volume of 22.4 liters. Substituting these values of pressure, volume, moles, and absolute temperature into Equation 13.18 gives both the values and units of R:

$$R = \frac{1.00 \text{ atm} \times 22.4 \text{ L}}{1.00 \text{ mol} \times 273 \text{ K}} = 0.0821 \frac{\text{L atm}}{\text{mol K}} \tag{13.19}$$

$$R = \frac{760 \text{ torr} \times 22.4 \text{ L}}{1.00 \text{ mol} \times 273 \text{ K}} = 62.4 \frac{\text{L torr}}{\text{mol K}} \tag{13.20}$$

The choice between these values of R for a given problem is dictated by the units in the problem. *It is essential that the measurement units of pressure, volume, temperature, and quantity correspond with the units of R.*

A useful variation of the ideal gas equation may be found as follows. If the weight of any chemical species, g, is divided by the molar weight, MW, the quotient is the number of moles: $\frac{\text{grams}}{\text{grams/mole}}$ = moles. Therefore $\frac{g}{MW}$ may be substituted for its equivalent, n, in Equation 13.13:

$$PV = \frac{g}{MW}RT \qquad (13.21)$$

In using the ideal gas equation, Equation 13.13 or 13.21, the recommended procedure is to solve the equation algebraically for the unknown, substitute known values of the other variables, and calculate the answer. Include and cancel units in the usual way; it is your best check on the correctness of your setup. While the equations may be used to determine the value of any unknown when the others are given, we will limit our examples to the calculation of moles, volume, and molar weight.

EXAMPLE 13.8   What volume will be occupied by 0.393 mole of nitrogen at 738 torr and 24°C?

Solution.   We begin by solving Equation 13.13 for the required volume:

$$V = \frac{nRT}{P}$$

Notice that pressure is given in torr. We therefore use the value of R in which torr is the pressure unit, Equation 13.20. Substituting this and other given data,

$$V = \frac{nRT}{P} = \frac{0.393 \text{ mol} \times \frac{62.4 \text{ L torr}}{\text{mol K}} \times (273 + 24) \text{ K}}{738 \text{ torr}}$$

$$= \frac{0.393 \text{ mol}}{738 \text{ torr}} \times \frac{62.4 \text{ L torr}}{\text{mol K}} \times (273 + 24) \text{ K} = 9.87 \text{ L}$$

EXAMPLE 13.9   How many moles of ammonia are in a 5.00-liter gas cylinder at 18°C if they exert a pressure of 8.65 atmospheres?

The method is again direct. Solve Equation 13.13 for n, substitute, and compute the answer.

$$n = \frac{PV}{RT} = \frac{8.65 \text{ atm} \times 5.00 \text{ L}}{\frac{0.0821 \text{ L atm}}{\text{mol K}} \times (273 + 18) \text{ K}} = 8.65 \text{ atm} \times \frac{\text{mol K}}{0.0821 \text{ L atm}} \times \frac{5.00 \text{ L}}{291 \text{ K}} = 1.81 \text{ mol}$$

Notice that, to divide by R, $\frac{0.0821 \text{ L atm}}{\text{mol K}}$, you multiply by its inverse, 1/R, $\frac{\text{mol K}}{0.0821 \text{ L atm}}$ (p. 579).

One of the most useful applications of the ideal gas equation is determining the molar weight of an unknown substance in the vapor state. The following examples illustrate the method.

EXAMPLE 13.10   1.67 grams of an unknown liquid are vaporized at a temperature of 125°C. Its volume is measured as 0.421 liter at 749 torr. Calculate the molar weight.

Using Equation 13.21 we can solve for the molar weight, substitute, and calculate the answer. Complete the problem.

$$MW = \frac{gRT}{PV} = \frac{1.67 \text{ g} \times \frac{62.4 \text{ L torr}}{\text{mol K}} \times (273 + 125) \text{ K}}{749 \text{ torr} \times 0.421 \text{ L}}$$

$$= \frac{1.67 \text{ g}}{749 \text{ torr}} \times \frac{398 \text{ K}}{0.421 \text{ L}} \times \frac{62.4 \text{ L torr}}{\text{mol K}} = 132 \text{ g/mol}$$

EXAMPLE 13.11   Find the molar weight of an unknown gas if its density is 1.45 grams/liter at 25°C and 756 torr.

Density does not appear in the ideal gas equation as such, but its component units, grams/liter, do. The problem is solved just as the last one, interpreting density as the mass of 1.45 grams and the volume as 1.00 liter. Complete the problem.

$$MW = \frac{gRT}{PV} = \frac{1.45 \text{ g}}{1.00 \text{ L}} \times \frac{62.4 \text{ L torr}}{\text{mol K}} \times \frac{298 \text{ K}}{756 \text{ torr}} = 35.7 \text{ g/mol}$$

Real gases do not always obey the ideal gas equation exactly. Deviations are most likely to occur at relatively high pressure and/or low temperature, which cause a gas to condense to a liquid. Intermolecular attractions become strong, so the gas no longer behaves like an ideal gas. Fortunately, deviations from the equation are negligible for most gases over wide ranges of temperature and pressure, so the equation is generally satisfactory for quantitative work.

13.11 MOLAR VOLUME AT STANDARD TEMPERATURE AND PRESSURE

PG 13 L Define molar volume. State the molar volume of any gas at STP.

13 M Given the volume (or number of moles) of any gas at STP, find the number of moles (or volume).

13 N Given two of the following for any gas at STP, find the third: grams, volume, molar weight.

13 O Given gas density at STP (or molar weight), find molar weight (or gas density at STP).

Molar volume is comparable to molar weight: as molar weight represents grams per mole, molar volume is liters per mole. An expression for molar volume may be found by solving Equation 13.13 for $\dfrac{V}{n}$:

$$\frac{V}{n} = \frac{RT}{P} \tag{13.22}$$

From this it appears that the volume occupied by one mole of *any gas* depends upon the gas pressure and temperature. On page 283 it was noted that the volume of one mole of any ideal gas at standard temperature, $0°C$ (273 K), and 1 atmosphere is 22.4 liters. Thus the equivalence

$$22.4 \text{ L (gas at STP)} \approxeq 1 \text{ mol} \tag{13.23}$$

provides a unit path from L (gas at STP) $\rightarrow$ mol, or mol $\rightarrow$ L (gas at STP).

Because standard temperature and pressure are so frequently used as reference conditions for gases, the molar volume at these conditions is a useful quantity. You must be aware, however, of the restrictions placed on the value 22.4 liters per mole. It is the molar volume of a *gas*—never a solid or a liquid—*if that volume is measured at standard temperature and pressure*, not some other combination of temperature and pressure. Do not use 22.4 liters per mole unless these conditions are satisfied.

Three quantities are usually involved in molar volume problems. They are density, measured in grams/liter; molar weight, grams/mole; and molar volume, 22.4 liters/mole at STP. While density and molar weight are unique for each gas, molar volume at STP is 22.4 liters/mole for all gases. The use of molar volume is illustrated by the following examples.

EXAMPLE 13.12 Find the volume of 0.350 mole of helium at STP.

At 22.4 liters per mole, the calculation for this problem should be apparent. Solve completely.

$$0.350 \text{ mol} \times \frac{22.4 \text{ L}}{1 \text{ mol}} = 7.84 \text{ L}$$

EXAMPLE 13.13 Find the volume of 12.0 grams of oxygen at STP.

The only difference between this example and the one before it is that the quantity is given in grams rather than moles. Conversion of 12.0 grams of oxygen to moles is straightforward, and converting moles to liters as in the previous example completes the problem.

$$12.0 \text{ g } O_2 \times \frac{1 \text{ mol } O_2}{32.0 \text{ g } O_2} \times \frac{22.4 \text{ L } O_2}{1 \text{ mol } O_2} = 8.40 \text{ L } O_2$$

EXAMPLE 13.14 Find the density of ammonia, NH_3, at STP.

This time there is no "given quantity." But we do know, or can find, two things about ammonia. Its molar volume at STP is 22.4 liters/mole, and its molar weight is 17.0 grams per mole. The mole is the connecting link. One mole weighs 17.0 grams, and one mole—the same quantity—occupies 22.4 liters. We seek the density, grams/liter. With that information you can complete the problem.

$$\frac{17.0 \text{ g/mol}}{22.4 \text{ L/mol}} = \frac{17.0 \text{ g}}{22.4 \text{ L}} = 0.759 \text{ g/L}$$

Without the reasoning shown above, the problem may be solved strictly from the units. Knowing that grams are in the numerator of the answer, it is reasonable to assume that grams will be in the numerator of the setup of the problem. Starting with molar weight, grams/mole, by what must we multiply to get grams/liter? Apparently moles in the denominator of grams/mole must be replaced by liters—and we know the relationship between liters and moles of gas at STP. Therefore,

$$\frac{17.0 \text{ g}}{1 \text{ mol}} \times \frac{1 \text{ mol}}{22.4 \text{ L}} = 0.759 \text{ g/L}$$

EXAMPLE 13.15   The density of an unknown gas at STP is 1.25 grams per liter. Estimate the molar weight of the gas.

If 1 liter weighs 1.25 grams, and there are 22.4 liters in 1 mole, what is the weight of 1 mole? Set up and solve.

- - - - - -   - - - - -

$$\frac{1.25 \text{ g}}{1\,\cancel{L}} \times \frac{22.4\,\cancel{L}}{1 \text{ mol}} = 28.0 \text{ g/mol}$$

13.12 GAS STOICHIOMETRY

PG 13 P   For a chemical reaction for which the equation may be written, given the grams of any species *or* the volume of any gaseous species at specified temperature and pressure, calculate the grams of any other species *or* the volume of any gaseous species at specified temperature and pressure.

In Section 9.2, p. 183, you learned a three-step pattern for solving stoichiometry problems. It is repeated here for your ready reference:

1. Convert the quantity of the given species to moles.
2. Convert the moles of given species to moles of wanted species.
3. Convert the moles of wanted species to the quantity units required.

The only measurable quantity unit in Chapter 9 was grams. Conversion between grams and moles in either direction was by molar weight, grams per mole (Equation 7.4, p. 138). In this chapter you have learned that the quantity of a gas can be expressed in liters *if* the temperature and pressure are known. The ideal gas equation ($PV = nRT$) furnishes a way to convert between liters and moles. If the volume is measured at STP, the conversion is an even more direct one by the use of 22.4 L $\simeq$ 1 mol. This gives you two new ways to accomplish Steps 1 and 3 of the stoichiometric pattern. They can be added to the unit path representation of the pattern (Equation 9.2, p. 183) as follows:

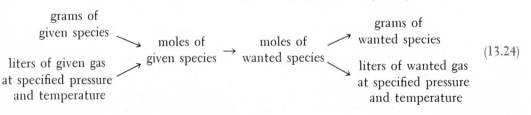

(13.24)

The next two examples involve gas volumes at STP. Equation 13.23 is the conversion relationship.

EXAMPLE 13.16 Hydrogen gas is released when sodium reacts with water: $2 \, Na(s) + 2 \, HOH(\ell) \rightarrow H_2(g) + 2 \, NaOH(aq)$. What volume of H_2, measured at STP, will be given off by 8.62 grams of sodium?

Solution.   Step 1 is to convert the given quantity of moles:

$$8.62 \; g \, Na \; \times \; \frac{1 \; mol \; Na}{23.0 \; g \, Na}$$

Using equation coefficients, we now extend the setup to moles of H_2 (Step 2):

$$8.62 \; g \; Na \; \times \; \frac{1 \; mol \; Na}{23.0 \; g \, Na} \times \frac{1 \; mol \; H_2}{2 \; mol \; Na}$$

Finally we convert the moles of wanted species to liters (Step 3), using molar volume at STP, as in Example 13.12:

$$8.62 \; g \, Na \; \times \; \frac{1 \; mol \; Na}{23.0 \; g \, Na} \times \frac{1 \; mol \; H_2}{2 \; mol \; Na} \times \frac{22.4 \; L \, H_2}{1 \; mol \, H_2} = 4.20 \; L \; H_2$$

EXAMPLE 13.17 Calculate the number of grams of oxygen that are required to react with sulfur dioxide to yield 18.6 liters of sulfur trioxide, measured at STP: $2 \, SO_2(g) + O_2(g) \rightarrow 2 \, SO_3(g)$.

Begin with the given quantity and set up the three conversions of the stoichiometric pattern. Solve completely.

- - - - - -   - - - - - -

$$18.6 \; L \, SO_3 \; \times \; \frac{1 \; mol \; SO_3}{22.4 \; L \, SO_3} \times \frac{1 \; mol \; O_2}{2 \; mol \, SO_3} \times \frac{32.0 \; g \; O_2}{1 \; mol \, O_2} = 13.3 \; g \; O_2$$

In this example your gas conversion was from liters to moles, dividing by the molar volume of 22.4 liters/mole.

Other techniques are required if the gas volume is measured at non-STP conditions. The method we will consider will divide the stoichiometric pattern into two parts. In one part the volume-mole conversion will be accomplished by means of the ideal gas law. (Equation 13.13). The other two steps of the problem will be completed in the usual manner.

EXAMPLE 13.18 How many grams of ammonia can be produced by the reaction of 3.85 liters of hydrogen, measured at 15.0 atmospheres and 85°C? The equation is $N_2(g) + 3 H_2(g) \rightarrow 2 NH_3(g)$.

Solution. Step 1 of the stoichiometric pattern calls for the conversion of the given quantity, a volume of hydrogen at non-STP conditions, to moles. The ideal gas equation may be used for this purpose, as it was in Example 13.9.

$$n = \frac{PV}{RT} = \frac{mol\,K}{0.0821\,L\,atm} \times \frac{15.0\,atm}{(273 + 85)\,K} \times 3.85\,L = 1.96\,mol$$

The remaining steps of the stoichiometric pattern are conversion of moles of hydrogen to moles of ammonia via the equation coefficients, and then moles of ammonia to grams by molar weight:

$$1.96\,mol\,H_2 \times \frac{2\,mol\,NH_3}{3\,mol\,H_2} \times \frac{17.0\,g\,NH_3}{1\,mol\,NH_3} = 22.2\,g\,NH_3$$

EXAMPLE 13.19 How many liters of CO_2, measured at 740 torr and 130°C, will be produced by the complete burning of 16.2 grams of butane, C_4H_{10}? The equation is $2 C_4H_{10}(g) + 13 O_2(g) \rightarrow 8 CO_2(g) + 10 H_2O(g)$.

Two steps of the stoichiometric pattern take you from grams of given species to moles of wanted species. The gas law equation guides you from moles of wanted species to volume. Complete the problem.

$$16.2\,g\,C_4H_{10} \times \frac{1\,mol\,C_4H_{10}}{58.0\,g\,C_4H_{10}} \times \frac{8\,mol\,CO_2}{2\,mol\,C_4H_{10}} = 1.12\,mol\,CO_2$$

$$V = \frac{nRT}{P} = \frac{1.12\,mol\,CO_2}{740\,torr} \times \frac{62.4\,L\,torr}{mol\,K} \times (273 + 130)\,K = 38.1\,L\,CO_2$$

There is an alternative approach to Examples 13.18 and 13.19 that is preferred by some teachers and students. Applied to Example 13.18, instead of using the ideal gas equation to convert a given volume of gas at non-STP conditions to moles, the given volume is first converted to STP by pressure and temperature corrections, as in Section 13.9:

$$\overset{}{\underset{}{3.85\,L\,H_2}} \times \underset{P\ Correction}{\frac{15.0\,atm}{1.00\,atm}} \times \underset{T\ Correction}{\frac{273\,K}{(273 + 85)\,K}}$$

Starting with this volume at STP, the remainder of the problem carries through the three steps of the stoichiometric pattern:

$$\left[\; 3.85 \; \cancel{L \, H_2} \times \frac{15.0 \, \cancel{atm}}{1.00 \, \cancel{atm}} \times \frac{273 \, \cancel{K}}{(273 + 85) \, \cancel{K}} \times \frac{1 \, \cancel{mol \, H_2}}{22.4 \, \cancel{L \, H_2}} \right.$$

$$\underbrace{\qquad\qquad}_{\text{P Correction}} \quad \underbrace{\qquad\qquad}_{\text{T Correction}} \quad \underbrace{\qquad\qquad}_{\text{Step 1}}$$

$$\left. \times \frac{2 \, \cancel{mol \, NH_3}}{3 \, \cancel{mol \, H_2}} \times \frac{17.0 \, g \, NH_3}{1 \, \cancel{mol \, NH_3}} = 22.3 \; g \; NH_3 \right]$$

$$\underbrace{\qquad\qquad}_{\text{Step 2}} \quad \underbrace{\qquad\qquad}_{\text{Step 3}}$$

Applying this method to Example 13.19, the volume of product gas at STP is first set up as in Example 13.16 and then converted to the required pressure and temperature by appropriate corrections:

$$\left[\; 16.2 \; g \, \cancel{C_4 H_{10}} \times \frac{1 \, \cancel{mol \, C_4 H_{10}}}{58.0 \, g \, \cancel{C_4 H_{10}}} \times \frac{8 \, \cancel{mol \, CO_2}}{2 \, \cancel{mol \, C_4 H_{10}}} \times \frac{22.4 \, L \, CO_2}{1 \, \cancel{mol \, CO_2}} \right.$$

$$\underbrace{\qquad\qquad}_{\text{Step 1}} \quad \underbrace{\qquad\qquad}_{\text{Step 2}} \quad \underbrace{\qquad\qquad}_{\text{Step 3}}$$

$$\left. \times \frac{760 \, \cancel{torr}}{740 \, \cancel{torr}} \times \frac{(273 + 130) \, \cancel{K}}{273 \, \cancel{K}} = 37.9 \; L \; CO_2 \right]$$

$$\underbrace{\qquad\qquad}_{\text{P Correction}} \quad \underbrace{\qquad\qquad}_{\text{T Correction}}$$

13.13 AVOGADRO'S HYPOTHESIS: VOLUME-VOLUME STOICHIOMETRY

PG 13 Q State Avogadro's Hypothesis regarding gas volumes and number of molecules.

Solving the ideal gas equation, $PV = nRT$, for n gives

$$n = \frac{PV}{RT} \qquad\qquad (13.25)$$

This equation is true for *all* gases. If you have equal volumes of two gas samples—they may be the same gas or different gases—at the same temperature and pressure, all four factors on the right side of Equation 13.25 are the same for both samples. Therefore the two samples must contain the same number of moles and the same number of molecules. In other words, **equal volumes of all gases at the same temperature and pressure contain the same number of molecules.** This statement is **Avogadro's Hypothesis;** it is illustrated in Figure 13.10.

Avogadro did not derive his hypothesis from the ideal gas equation. Instead, he suggested it as an explanation for the experimental observation that when gases react with each other, if the reacting volumes are measured at the

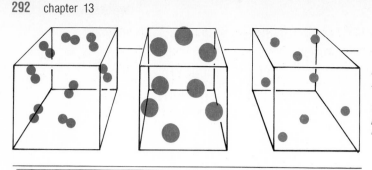

Figure 13.10
Avogadro's hypothesis. Equal volumes of gases, measured at the same temperature and pressure, contain the same number of molecules.

same temperature and pressure, those volumes are in a ratio of small whole numbers. This is known as the **law of combining volumes.** It may be demonstrated in the laboratory by these and other reactions (Fig. 13.11):

$$H_2(g) + Cl_2(g) \rightarrow 2\,HCl(g) \tag{13.26}$$

$$2\,H_2(g) + O_2(g) \rightarrow 2\,H_2O(g) \tag{13.27}$$

$$N_2(g) + 3\,H_2(g) \rightarrow 2\,NH_3(g) \tag{13.28}$$

The interesting fact derived from these observations is that the ratio of volumes in the above reactions is identical to the ratio of moles, represented by

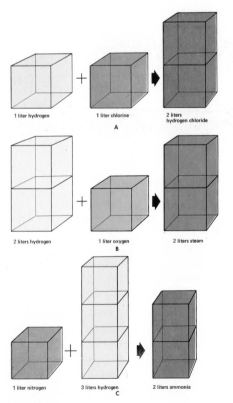

1 liter hydrogen   1 liter chlorine   2 liters hydrogen chloride

A

2 liters hydrogen   1 liter oxygen   2 liters steam

B

1 liter nitrogen   3 liters hydrogen   2 liters ammonia

C

Figure 13.11
A, Volume ratios in the reaction of hydrogen and chlorine to form hydrogen chloride.
B, Volume ratios in the reaction of hydrogen and oxygen to form water (steam).
C, Volume ratios in the reaction of nitrogen and hydrogen to form ammonia.

the coefficients in the equation. Numbers demonstrate that this is necessary if Avogadro's Hypothesis is correct. In Equation 13.28, for example, the ratio of reacting moles of hydrogen to moles of nitrogen is 3/1, as shown by the coefficients. Suppose the reacting gas volumes are measured at STP. What will be the ratio of volumes of hydrogen and nitrogen? At STP there are 22.4 liters per mole. Three moles of hydrogen occupy 3×22.4 liters; 1 mole of nitrogen fills 1×22.4 liters. The ratio of volumes is

$$\frac{3 \times 22.4 \text{ L H}_2}{1 \times 22.4 \text{ L N}_2} = \frac{3 \text{ L H}_2}{1 \text{ L N}_2}$$

the same as the coefficients in the equation. Therefore, if 3 mol $H_2 \simeq 1$ mol N_2 for Equation 13.28, then 3 L $H_2 \simeq 1$ L N_2. This gives a direct volume-given-gas → volume-wanted-gas shortcut in solving stoichiometry problems involving two gases—*provided the gas volumes are measured at the same temperature and pressure.*

EXAMPLE 13.20   1.30 liters of ethene, C_2H_4, are burned completely. What volume of oxygen is required if both gas volumes are measured at STP? The equation is

$$C_2H_4(g) + 3 O_2(g) \rightarrow 2 CO_2(g) + 2 H_2O(\ell)$$

The equivalence 1 liter $C_2H_4 \simeq 3$ liters O_2 tells us that the volume of oxygen is three times the volume of ethene. Set up and solve.

------   ------

$$1.30 \text{ L C}_2\text{H}_4 \times \frac{3 \text{ L O}_2}{1 \text{ L C}_2\text{H}_4} = 3.90 \text{ L O}_2$$

This result is confirmed if the problem is solved by the full three steps of the stoichiometric pattern:

$$1.30 \text{ L C}_2\text{H}_4 \times \frac{1 \text{ mol C}_2\text{H}_4}{22.4 \text{ L C}_2\text{H}_4} \times \frac{3 \text{ mol O}_2}{1 \text{ mol C}_2\text{H}_4} \times \frac{22.4 \text{ L O}_2}{1 \text{ mol O}_2} = 3.90 \text{ L O}_2$$

Be sure to recognize the restriction on this simplified solution process: both gas volumes *must be measured at the same temperature and pressure,* but not necessarily STP as in the above example.

EXAMPLE 13.21   When oxygen comes into contact with nitrogen monoxide, nitrogen dioxide is produced:

$$2 NO(g) + O_2(g) \rightarrow 2 NO_2(g)$$

How many liters of nitrogen dioxide will form by the reaction of 4.30 liters of oxygen if both volumes are measured at the same temperature and pressure? Again, the equation is

$$2\,NO(g) + O_2(g) \rightarrow 2\,NO_2(g)$$

$$4.30\,\cancel{L\,O_2} \times \frac{21\,L\,NO_2}{1\,\cancel{L\,O_2}} = 8.60\,L\,NO_2$$

Sometimes the given and wanted gas volumes are at different temperatures and pressures. The procedure for this kind of problem is to use pressure and temperature correction ratios to convert the volume of the *given* species from its temperature and pressure to the volume it *would occupy* at the temperature and pressure specified for the *wanted* species. From that point the solution is as in the last two examples. To illustrate, consider the following example.

EXAMPLE 13.22   1.75 liters of oxygen, measured at 24°C and 755 torr, are consumed in burning sulfur. At one point in the exhaust hood the sulfur dioxide produced is at 165°C and a pressure of 785 torr. Find the volume of the sulfur dioxide at those conditions. The equation is

$$S(s) + O_2(g) \rightarrow SO_2(g)$$

First, we find the volume that would be occupied by the oxygen at 165°C and 785 torr. Example 13.6, p. 281, is similar, so you may find it helpful to look back. Set up that far, but do not solve.

$$1.75\,L\,O_2 \times \frac{755\,\cancel{torr}}{785\,\cancel{torr}} \times \frac{438\,\cancel{K}}{297\,\cancel{K}} \times \underline{\hspace{3cm}}$$

The pressure change is from 755 torr to 785 torr. An increase in pressure reduces volume, so the pressure correction is less than 1. Temperature increases from 297K to 438K. A temperature rise increases volume, so the temperature correction is more than 1.

From here the problem is of the form, "How many liters of SO₂ are equivalent to the volume of O₂ in the above setup, both gases measured at the same temperature and pressure?" Extend the setup and complete the problem.

$$1.75\,\cancel{L\,O_2} \times \frac{755\,\cancel{torr}}{785\,\cancel{torr}} \times \frac{438\,K}{297\,K} \times \frac{1\,L\,SO_2}{1\,\cancel{L\,O_2}} = 2.48\,L\,SO_2$$

13.14 DALTON'S LAW OF PARTIAL PRESSURES

PG 13 R Given the partial pressure of each component in a mixture of gases, find the total pressure.

13 S Given the total pressure of a gaseous mixture and the partial pressures of all components except one, or information from which those partial pressures can be obtained, find the partial pressure of the remaining component.

Figure 13.12 shows the apparatus for an experiment. Suppose that n_a moles of gas A occupy volume V at absolute temperature T. The pressure exerted by A, p_a, may be found by solving Equation 13.13 for pressure: $p_a = (n_aRT)/V$. Now suppose that a container of equal volume holds n_b moles of gas B at the same temperature. The pressure of the second gas is $p_b = (n_bRT)/V$. Let both gases be forced into one of the containers, all at constant temperature. The mixture of gases consists of $n_a + n_b$ moles. They occupy volume V at temperature T. What is the pressure of the mixture?

The ideal gas equation is valid for a mixture of gases as well as for a pure gas. Therefore, if n is the total number of moles, $n_a + n_b$, the pressure of the mixture, P, is

$$P = \frac{nRT}{V} = \frac{(n_a + n_b)RT}{V} = \frac{n_aRT}{V} + \frac{n_bRT}{V} = p_a + p_b$$

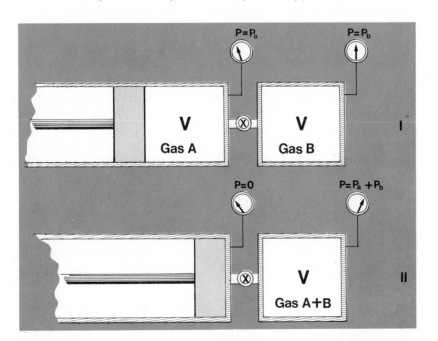

Figure 13.12
Experimental apparatus to demonstrate Dalton's Law of Partial Pressures.

The ideal gas equation thus predicts that the pressure of the mixed gases is equal to the sum of the pressures of the individual gases. The prediction is confirmed in experiments such as that shown in Figure 13.12.

Observations such as these are summed up in **Dalton's Law of Partial Pressures: The total pressure exerted by a mixture of gases is the sum of the partial pressures of the components. The partial pressure of a component is the pressure that component would exert if it alone occupied the total volume at the same temperature.** Figure 13.13 "states" the partial pressure law in pictures. Mathematically it is

$$P = p_1 + p_2 + p_3 + \cdots \qquad (13.29)$$

where P is the total pressure and p_1, p_2, and $p_3 \ldots$ are the partial pressures of components 1, 2, 3. . . .

EXAMPLE 13.23   In a gas mixture the partial pressure of methane is 150 torr, of ethane, 180 torr, and of propane, 450 torr. Find the total pressure exerted by the mixture.

This is a straightforward application of Equation 13.29.

- -

P = 150 torr + 180 torr + 450 torr = 780 torr

A laboratory application of Dalton's Law of Partial Pressures has to do with mixtures of gases with water vapor. Gases may be prepared and collected by an apparatus such as that shown in Figure 13.14. The oxygen formed in the test tube is bubbled through water, at which time it becomes "saturated" with water vapor. The "gas" collected is therefore actually a mixture of the oxygen being generated and water vapor. The pressure exerted by the mixture is the

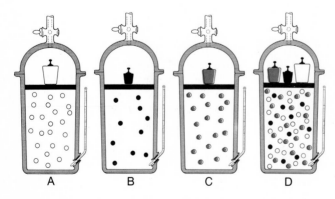

A   B   C   D

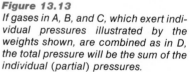

Figure 13.13
If gases in A, B, and C, which exert individual pressures illustrated by the weights shown, are combined as in D, the total pressure will be the sum of the individual (partial) pressures.

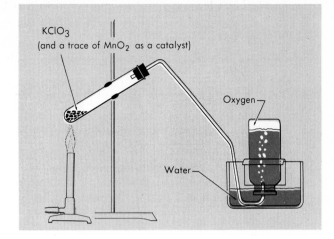

Figure 13.14
Laboratory preparation of oxygen.

sum of the partial pressure of the oxygen and the partial pressure of the water vapor. The latter is the equilibrium vapor pressure of water (Chapter 14, p. 318) which depends only upon temperature. Values for this quantity at various temperatures are given in Appendix V, p. 599.

EXAMPLE 13.24 If the total gas pressure in an oxygen generator such as that shown in Figure 13.14 is 755 torr, and the temperature of the system is 22°C, find the partial pressure of the oxygen.

First, to see clearly where you are headed in this problem, write the partial pressure equation (Equation 13.29) as it applies specifically to this problem.

$$P = p_{O_2} + p_{H_2O}$$

The water vapor pressure, p_{H_2O}, may be found from Appendix V. It is . . .

19.8 torr

From the total pressure given in the problem and the partial pressure of the water vapor, the partial pressure of oxygen follows readily.

$$p_{O_2} = P - p_{H_2O} = 755 \text{ torr} - 19.8 \text{ torr} = 735 \text{ torr}$$

Hydrogen may also be collected over water, as in Figure 13.14. A popular laboratory experiment combines partial pressure and ideal gas law calculations.

EXAMPLE 13.25 85.5 mL of a mixture of hydrogen and water vapor are collected at 29°C. If the total pressure of the mixture is 748 torr, how many moles of hydrogen are present?

Your strategy is to solve the ideal gas equation (Equation 13.13) for moles, n, as in Example 13.9, p. 284. This gives you n in terms of the partial pressure of hydrogen—which you do not know. But you do know the total pressure of hydrogen plus water vapor, and the vapor pressure of water at 29°C is listed in Appendix V. As in Example 13.24, calculate the partial pressure of hydrogen in the mixture.

$$p_{H_2} = P - p_{H_2O} = 748 \text{ torr} - 30 \text{ torr} = 718 \text{ torr}$$

Now you have everything you need to calculate n from Equation 13.19. Complete the problem. (Watch your volume figure.)

$$n = \frac{PV}{RT} = 718 \cancel{\text{torr}} \times \frac{0.0855 \cancel{\text{L}}}{302 \cancel{\text{K}}} \times \frac{\text{mol} \cancel{\text{K}}}{62.4 \text{ L} \cancel{\text{torr}}} = 3.26 \times 10^{-4} \text{ mol } H_2$$

CHAPTER 13 IN REVIEW

13.1 PROPERTIES OF GASES (266)

13.2 THE KINETIC THEORY OF GASES AND THE IDEAL GAS MODEL

13 A Explain physical properties of gases, or physical phenomena related to gases, in terms of the ideal gas model. (268)

13.3 GAS MEASUREMENTS

13 B List the measurable properties of a gas. (269)

13 C Given a gas pressure in atmospheres, torr, millimeters of mercury, centimeters of mercury, inches of mercury, pounds per square inch, pascals, or kilopascals, express that pressure in each of the other units. (271)

13.4 BOYLE'S LAW

13 D Given the initial volume (or pressure) and initial and final pressures (or volumes) of a fixed quantity of gas at constant temperature, calculate the final volume (or pressure). (272)

13.5   ABSOLUTE TEMPERATURE

13 E   Given a temperature in degrees Celsius (or Kelvin), convert it to Kelvin (or degrees Celsius). (275)

13.6   CHARLES' LAW

13 F   Given the initial volume, and initial and final temperatures of a fixed quantity of gas at constant pressure, calculate the final volume. (277)

13.7   THE IDEAL GAS EQUATION

13 G   Write the equation that shows how the measurable properties of an ideal gas are related to each other (the ideal gas equation). (280)

13.8   PRESSURE, VOLUME, AND TEMPERATURE CHANGES WITH A FIXED SAMPLE OF GAS

13 H   For a fixed quantity of a confined gas, given the initial volume, pressure, and temperature, and the final pressure and temperature, calculate the final volume. (281)

13.9   STANDARD TEMPERATURE AND PRESSURE

13 I   Given the volume of a gas at one temperature and pressure (or at STP), find the volume it would occupy at STP (or at a stated temperature and pressure). (282)

13.10   APPLICATIONS OF THE IDEAL GAS EQUATION

13 J   Write the variation of the ideal gas equation that includes the mass of the gas sample and its molar weight. (283)

13 K   Given values for all except one of the variables in either form of the ideal gas equation, calculate the value of that remaining variable. (283)

13.11   MOLAR VOLUME AT STANDARD TEMPERATURE AND PRESSURE

13 L   Define molar volume. State the molar volume of any gas at STP. (286)

13 M   Given the volume (or number of moles) of any gas at STP, find the number of moles (or volume). (286)

13 N   Given two of the following for any gas at STP, find the third: grams, volume, molar weight. (286)

13 O   Given gas density at STP (or molar weight), find molar weight (or gas density at STP). (286)

13.12 GAS STOICHIOMETRY

13 P For a chemical reaction for which the equation may be written, given the grams of any species *or* the volume of any gaseous species at specified temperature and pressure, calculate the grams of any other species *or* the volume of any gaseous species at specified temperature and pressure. (288)

13.13 AVOGADRO'S HYPOTHESIS: VOLUME-VOLUME STOICHIOMETRY

13 Q Explain Avogadro's Hypothesis regarding gas volume and number of molecules. (291)

13.14 DALTON'S LAW OF PARTIAL PRESSURES

13 R Given the partial pressure of each component in a mixture of gases, find the total pressure. (295)

13 S Given the total pressure of a gaseous mixture and the partial pressures of all components except one, or information from which those partial pressures can be obtained, find the partial pressure of the remaining component. (295)

TERMS AND CONCEPTS

Kinetic theory of gases (268)
Kinetic molecular theory (268)
Ideal gas model (268)
Pressure (269)
Pascal (pressure unit) (269)
Millimeter of mercury (pressure unit) (269)
Torr (pressure unit) (270)
Atmosphere (pressure unit) (270)
Barometer (270)
Barometric (atmospheric) pressure (270)
One standard atmosphere (270)
Manometer (270)

Gauge pressure (271)
Boyle's Law (272)
Absolute zero (276)
Kelvin temperature scale (276)
Charles' Law (278)
Universal gas constant, R (280)
Ideal gas equation, ideal gas law (280)
Standard temperature and pressure, STP (282)
Molar volume (286)
Avogadro's Hypothesis (291)
Law of Combining Volumes (292)
Dalton's Law of Partial Pressures (296)

Most of these terms and many others appear in the Glossary.

QUESTIONS AND PROBLEMS

An asterisk (*) identifies a question that is relatively difficult, or that extends beyond the performance goals of the chapter.

Section 13.2

13.1) What is the kinetic theory of gases? How does it describe an "ideal gas?"

13.49) What is kinetic energy? What "properties" of gases are involved in their kinetic energies?

13.2) What causes pressure in a gas?

13.3) List some properties of air that make it suitable for use in automobile tires. Explain how each property is related to the ideal gas model.

For the next five questions in each column, explain how each physical phenomenon described is related to one or more of the features of the ideal gas model.

13.4) Gases with a distinctive odor can be detected some distance from their source.

13.5) Steam bubbles rise to the top in boiling water.

13.6) The density of liquid oxygen is about 1.4 grams/cm³. Vaporized at 0°C and 760 torr, this same 1.4 grams occupies 980 cm³, an expansion of nearly 1000 times.

13.7) Properties of gases become less "ideal"—the substance adopts behavior patterns not typical of gases—when subjected to very high pressures such that the individual molecules are close to each other.

13.8) Any container, regardless of size, will be completely filled by one gram of hydrogen.

Section 13.3

13.9) Four properties of gases may be measured. Name them.

13.10) Define and/or distinguish between the atmosphere, millimeter of mercury, and the torr as units of pressure.

13.11)* An open-end manometer (Fig. 13.15) is used to measure the pressure of a confined gas. Calculate this pressure if $P_a = 747$ torr and the mercury levels in the manometer are as shown.

13.12) A manometer measures a pressure at 64.9 cm Hg. Write this pressure in atmospheres.

13.13) A process that makes ammonia from the nitrogen of the air operates at 445 atm. What is the pressure in pounds per square inch?

13.50) State how and explain why the pressure exerted by a gas is different from the pressure exerted by a liquid or solid.

13.51) What are the desirable properties of air in an air mattress used by a camper? Show which part of the ideal gas model is related to each property.

13.52) Pressure is exerted on the top of a tank holding a gas, as well as on its sides and bottom.

13.53) Balloons expand in all directions when blown up, not just at the bottom as when filled with water.

13.54) Even though an automobile tire is "filled" with air, more air can always be added without increasing the volume of the tire significantly.

13.55) Very small dust particles, seen in a beam of light passing through a darkened room, appear to be moving about erratically.

13.56) Gas bubbles always rise through a liquid.

13.57) What is the meaning of *pressure?*

13.58) Distinguish between a barometer and a manometer. Explain how each measures the pressure of a gas.

13.59)* Figure 13.16 shows an open-end manometer attached to an aspirator, a device commonly used to draw air through laboratory equipment. With mercury levels as shown and $P_a = 758$ torr, calculate the pressure in the hose.

13.60) Canada uses metric units for all weather reports. Atmospheric pressure is recorded at 97.8 kPa on a stormy day in Montreal. What pressure is this in inches of mercury, as pressure is presently expressed in the United States?

13.61) Steam in a main propulsion engine in an ocean-going vessel is kept in the vapor state by drawing a vacuum on the system. The pressure is 50 torr as the steam leaves the turbine. Express this pressure in atmospheres.

Figure 13.15

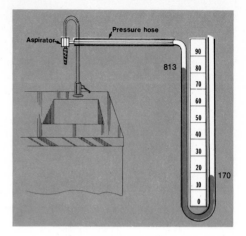

Figure 13.16

Section 13.4

13.14) 3.15 liters of gas at 0.940 atmospheres are compressed to 6.26 atmospheres. Calculate the new volume.

13.15)* An air compressor delivers air to a 59.8-liter storage tank at 9.50 atmospheres. A 3.16-liter portable compressed air tank is filled by connecting it to the larger tank and opening the valve between them. The air distributes itself between the tanks. If the compressor delivers no additional air to the system during the process, what will be the pressure in the portable tank when they are disconnected? in the storage tank? (Disregard the air initially in the smaller tank.)

Section 13.5

13.16) What does the term "absolute zero" mean? What physical condition theoretically exists at absolute zero?

13.17) Lead melts at 328°C. Express this in absolute temperature, K.

13.18) Liquid oxygen "boils" at 90 K. What is its boiling point in °C?

Section 13.6

13.19) An industrial gas storage tank has a "floating top" that maintains constant pressure. If, on a day during which there is no gas consumption, the volume is 85 600 liters at 5 A.M., when the temperature is 15°C, what will be the volume at 3 P.M. when it has warmed to 25°C?

13.62) To what volume must 127 milliliters of a confined gas at 749 torr be expanded to reduce pressure to 506 torr?

13.63)* If a light bulb contained air, the tungsten filament in it would oxidize and break; if the bulb held a vacuum, the filament would vaporize. Bulbs therefore contain argon, an unreactive gas, to prevent both events. 130 mL argon at 1 atm is introduced to an evacuated 150-mL light bulb before it is sealed in the manufacturing process. What is the pressure of the argon in the bulb?

13.64)* Describe an experiment by which the Celsius equivalent of absolute zero may be estimated.

13.65) Find the Celsius boiling point of sulfur, which boils at 718 K.

13.66) Antifreeze may be added to water to reduce the freezing point to −29°C. What is this temperature in K?

13.67) Only 620 milliliters of a collapsible plastic bag are filled with air at a temperature of −4°C. What will the volume be when the temperature rises to 21°C?

13.20) In a laboratory experiment, gas collected at 68°C exerts a pressure of 912 torr. What will the pressure be after cooling to 27°C, assuming no volume change?

13.68)* Beside the ideal gas storage tank of Problem 13.19 is a tank of compressed air, which is also unused. In the morning, when the temperature is 15°C, its pressure gauge records 1650 pounds per square inch. What pressure will it reach in the afternoon when the temperature is 25°C?

Section 13.8

13.21) 73.4 mL of oxygen, measured at 43°C and 824 torr, are collected in a laboratory experiment. What will be the volume when the gas cools to 23°C and the pressure is adjusted to 749 torr?

13.22) A variable volume marine research device is designed to hold a fixed quantity of gas at a pressure that balances the pressure outside the unit. At the surface of the ocean its volume is 1.62 liters at 23°C and 1.02 atm. Find its volume when lowered to a depth where the pressure is 4.86 atm and the temperature is 3°C.

13.69) 7.92 liters of nitrogen, measured at 1.28 atm and 17°C, are in a piston-fitted variable volume cylinder. The nitrogen is warmed to 39°C, and the piston moved to adjust the pressure to 1.39 atm. Calculate the new volume.

13.70) A weather balloon is partly filled with 34.9 liters of helium, measured at 752 torr and 19°C. What volume will the balloon occupy after rising to an altitude where the pressure is 512 torr and temperature is −45°C?

Section 13.9

13.23) What is the meaning of "STP"?

13.24) 47.9 ml hydrogen are collected at 26°C and 718 torr. Find the volume occupied at STP.

13.25) If a gas occupies 46.9 ml at STP, what volume will it fill at 24°C and 738 torr?

13.71) Why have the arbitrary conditions of STP been established?

13.72) Find the STP volume of a certain quantity of air if it occupies 48.6 liters at −12°C and 1.72 atm.

13.73) The STP volume of a sample of nitrogen is 1.46 liters. If it is cooled to −44°C and compressed to 2.06 atm, what will be its new volume?

Section 13.10

13.26) What volume will be occupied by 0.16 mole of hydrogen at 751 torr and 22°C?

13.27) How many moles of methane, CH_4, are in a 2.55-liter cylinder if the temperature is 20°C and the pressure is 9.40 atmospheres?

13.28) 2.68 liters of an unknown gas, measured at 22°C and 745 torr, weigh 5.89 grams. Calculate the molar weight.

13.29) How many atmospheres of pressure will be exerted by 25.0 grams of sulfur dioxide when confined in a 2.15-liter cylinder at 20°C?

13.30) A 21.2-liter oxygen tank is installed in an unventilated corner of an industrial plant. The tank is fitted with a relief valve that opens at 24.0 atmospheres of pressure. On a very hot day, when

13.74) 1.26 moles of oxygen are compressed to 4.36 atmospheres at a temperature of 36°C. Calculate the volume.

13.75) Calculate the number of moles of ammonia in a 24.0-liter tank when the pressure is 894 torr and the temperature is 29°C.

13.76) 1.06 grams of an unknown compound are placed in a 0.500-liter testing vessel and vaporized by partial evacuation and heating. When the temperature is 86°C, the pressure is 0.390 atmosphere. Calculate the molar weight.

13.77) At what Celsius temperature will 0.200 mole of nitrogen exert a pressure of 545 torr in a 4.12-liter cylinder?

13.78) How many grams of chlorine are in a 0.716-liter cylinder if the pressure is 10.9 atm at 30°C?

the tank happened to contain 624 grams of oxygen, the valve opened. What was the Celsius temperature of the oxygen in the tank?

13.31) Calculate the number of kilograms of helium in a 1.75-cubic meter balloon at 9°C and 798 torr.

13.32)* The density of an unknown hydrocarbon gas at 25°C and 750 torr is 1.69 grams/liter. It consists of 85.6% carbon and 14.4% hydrogen. (a) What is the empirical formula of the gas? (b) Calculate the molar weight of the gas. (c) Write the molecular formula of the gas.

13.79) How much pressure (torr) will be exerted by 190 kilograms of methane, CH_4, when stored in a 269-cubic meter tank at 22°C?

13.80)* Analysis of 1.19 grams of hydrazine, a rocket fuel, shows that it contains 1.04 grams of nitrogen, and the balance is hydrogen. If the same quantity of hydrazine is vaporized at 96°C in a 1.47-liter chamber, it exerts a pressure of 0.766 atmosphere. Determine the molar weight, empirical formula, and molecular formula of hydrazine.

Section 13.11

13.33) What is the meaning of "molar volume?"

13.34) How many moles of ethane, C_2H_6, are in 16.9 liters at STP?

13.35) Estimate the molar weight of a gas having a density of 1.83 g/liter at STP.

13.36) 0.937 gram of an unknown gas occupies 0.744 liter at STP. Find the molar weight.

13.37)* 6.49 g NH_3 occupy 9.44 L at 22°C and 745 torr. Calculate its molar volume.

13.81) Explain the restrictions placed on the statement that 22.4 liters per mole is the molar volume of any gas.

13.82) Find the STP volume of 4.62 grams of ammonia.

13.83) What is the mass of 3.25 liters of neon (Z = 10) at STP?

13.84) Find the density of sulfur dioxide at STP.

13.85)* Find the molar volume of CH_4 at 22°C and 745 torr.

Section 13.12

13.38) How many liters of O_2, measured at STP, will be released by the decomposition of 2.65 grams of mercury(II) oxide: $2 HgO(s) \rightarrow 2 Hg(\ell) + O_2(g)$.

13.39) What quantity of manganese must a student react with excess hydrochloric acid to produce 85.0 ml hydrogen, measured at STP? $Mn(s) + 2 HCl(aq) \rightarrow H_2(g) + MnCl_2(aq)$.

13.40) What volume of oxygen, measured at 25°C and 752 torr, is required to "burn" 3.26 grams of magnesium by the reaction $2 Mg(s) + O_2(g) \rightarrow 2 MgO(s)$?

13.41) Copper(II) oxide may be "reduced" to copper by heating in a stream of hydrogen: $CuO(s) + H_2(g) \rightarrow Cu(s) + H_2O(g)$. How many grams of CuO will be reduced by 4.16 liters of hydrogen, measured at 1.65 atm and 243°C?

13.86) What mass of potassium chlorate must be decomposed to produce 1.50 liters of oxygen at STP? $2 KClO_3(s) \rightarrow 2 KCl(s) + 3 O_2(g)$.

13.87) When sodium hydrogen carbonate is treated with sulfuric acid, carbon dioxide bubbles off: $H_2SO_4(aq) + 2 NaHCO_3(s) \rightarrow Na_2SO_4(aq) + 2 H_2O(\ell) + 2 CO_2(g)$. What volume of CO_2, measured at STP, is available from 8.58 g $NaHCO_3$?

13.88) Chlorine combines directly with sodium to form sodium chloride: $2 Na(s) + Cl_2(g) \rightarrow 2 NaCl(s)$. How many grams of NaCl will result from the reaction of 0.745 liter Cl_2, measured at 1.21 atm and 22°C?

13.89) Calculate the volume of ammonia, measured at 746 torr and 26°C, released by the complete reaction of 15.8 grams of lime, CaO, with excess NH_4Cl solution: $2 NH_4Cl(aq) + CaO(s) \rightarrow 2 NH_3(g) + CaCl_2(aq) + H_2O(\ell)$.

Section 13.13

13.42) What is Avogadro's Hypothesis? Explain why the volume of gaseous reactants and products in a reaction, provided they are measured at the same temperature and pressure, are in the same ratio as the coefficients of the equation for the reaction.

13.43) How many liters of hydrogen will combine directly with 1.28 liters of oxygen to produce steam if both volumes are measured at 200°C and 1.00 atm? $2 H_2(g) + O_2(g) \rightarrow 2 H_2O(g)$.

13.44) In an all-gas phase reaction, 28.3 liters of steam at 540°C and 1.46 atm react with carbon monoxide to produce hydrogen and carbon dioxide: $CO(g) + H_2O(g) \rightarrow CO_2(g) + H_2(g)$. How many liters of CO, measured at 460°C and 2.19 atm, will be used?

Section 13.14

13.45) State Dalton's Law of Partial Pressures, either in words or as an equation. Explain how this law "fits" the ideal gas model.

13.46) Helium, neon, and argon are mixed in such a fashion that their respective partial pressures are 0.364 atm, 0.108 atm, and 0.529 atm. What is the total pressure of the system?

13.47) A laboratory hydrogen generator collects the gas produced by bubbling it through water. The total pressure of the gas collected is 751 torr. The temperature is 34°C, at which water vapor pressure is 40.0 torr. Calculate the partial pressure of hydrogen.

13.48)* Dry, clean air at sea level contains, by mole percent, 20.9% O_2, and 0.93% Ar. The rest is mostly nitrogen. If the total air pressure is 755 torr, what is the partial pressure of argon?

13.90)* Suggest a reason why Avogadro's Hypothesis is acceptable for gases, but not for solids or liquids.

13.91) How many liters of gaseous hydrogen chloride, measured at 1.10 atm and 31°C, will result from the combination of 0.734 liter of chlorine, also measured at 1.10 atm and 31°C, with excess hydrogen? $H_2(g) + Cl_2(g) \rightarrow 2 HCl(g)$.

13.92) Determine the volume of hydrogen, measured at 768 torr and 35°C, that is required to produce 4.80 liters of methane, CH_4, measured at 220°C and 1800 torr, by the reaction $4 H_2(g) + CS_2(g) \rightarrow CH_4(g) + 2 H_2S(g)$.

13.93)* A gaseous mixture contains only nitrogen and hydrogen at a total pressure of 1.00 atm. If the partial pressure of H_2 is 0.50 atm, find p_{N_2}. Which gas, if either, is present in the greatest mass? Explain your answer.

13.94) Find the partial pressure of oxygen in clean air if the total pressure is 757.0 torr, and partial pressures of other gases are (1) nitrogen, 590 torr; (2) argon, 7 torr; (3) all others, 0.2 torr.

13.95) If the total volume of gas collected in Problem 13.47 were 50.3 ml, what volume would be occupied by the dry hydrogen at STP?

13.96)* 0.350 liter of oxygen is collected over water at 750 torr and 20°C, using apparatus similar to that in Figure 13.15. Calculate the mass of the oxygen collected.

14

liquids and solids

The general differences between the three states of matter—gases, liquids, and solids—were first mentioned in Section 2.2. These were also related to the kinetic theory of matter. This theory pictures all matter as made up of particles in constant motion. Particle movement is restricted in a solid, partially limited in a liquid, but completely free in a gas, as shown in Figure 14.1. In Chapter 13 we examined the gaseous state in some detail. In this chapter we will turn our attention to the so-called condensed states of matter, liquids and solids.

SOLID   LIQUID   GAS

Figure 14.1
Characteristics of solids, liquids and gases. A solid has definite volume and definite shape. A liquid has definite volume, but assumes the shape of the bottom of its container, up to that total volume. A gas fills its container, acquiring both its shape and volume. Particles are in constant motion in all states: vibrating in fixed positions in a solid, moving randomly within the volume occupied by a liquid, and moving randomly within the entire volume of the container as a gas. Particles are very close to each other in a solid and liquid, but are widely separated in the gaseous state.

14.1 THE NATURE OF THE LIQUID STATE

PG 14 A   Explain the differences between the physical behavior of liquids and gases in terms of the relative distances between molecules and the effect of those distances on intermolecular forces.

The behavior of gases is neatly summarized in the ideal gas equation (p. 280) which may be applied to nearly all gases. No similar relationship is available for liquids. Two things about liquids are quite different from gases. First, liquid molecules are very close to each other, instead of being widely separated as in a gas. Second, because of this closeness, attractions between molecules are fairly strong and play a major role in determining the physical properties of a liquid. The properties of a gas can be explained only if we assume that intermolecular attractions are negligible.

The fact that gases can be compressed and liquids cannot is explained by these two differences. To compress a gas you need only to push the molecules closer to each other. In a liquid the molecules are already "touchingly close," so to compress a liquid you would have to crush or distort the actual molecules. This does not appear to occur. If you heat a gas at constant pressure or reduce the pressure on a gas at constant temperature, the gas expands readily. Neither an increase in temperature nor a reduction in pressure causes a major increase in the volume of a liquid. Evidently intermolecular attractions in a liquid hold the molecules together in an almost fixed total volume.

A gas may be condensed to a liquid by reducing its temperature, by compressing it, or by a combination of these actions.* Lowering the temperature slows the molecules in their random movement so that the intermolecular collisions are no longer elastic. Instead, the attractions between molecules cause them to stay together. This leads eventually to the formation of a liquid drop, and ultimately to nearly complete condensation. When a gas is compressed, the molecules are pushed closer to each other. At the smaller intermolecular distances the attractive forces are strong enough to cause the molecules to stick together, leading to condensation.

14.2 PHYSICAL PROPERTIES OF LIQUIDS

PG 14 B   For two liquids, given comparative values of physical properties that depend upon intermolecular attractions, predict the relative strengths of those attractions; or, given a comparison of the strengths of intermolecular attractions, predict the relative values of physical properties that depend upon them.

*There is a temperature, called critical temperature, above which a gas cannot be liquefied by compression alone.

Many physical properties of liquids are directly related to the strength of intermolecular attractions. Among them are the following:

Vapor pressure. Because of evaporation, the open space above any liquid contains some molecules in the gaseous, or vapor, state. The pressure exerted by these gaseous molecules is called **vapor pressure.** If the gas space above the liquid is closed, the vapor pressure will increase to a definite value, referred to as the **equilibrium vapor pressure** (see Section 14.4). Equilibrium vapor pressure is related to the strength of intermolecular forces. At a given temperature *liquids with relatively weak intermolecular attractions evaporate more readily, yielding higher vapor concentrations, and therefore higher vapor pressures than liquids with strong intermolecular forces.*

Boiling point. Liquids may be changed to gases by boiling. A liquid must be heated to make it boil. When the temperature reaches the **boiling point,** the average kinetic energy of the liquid particles is sufficient to overcome the forces of attraction that hold molecules in the liquid state, and it becomes a gas. *Liquids with strong intermolecular forces require higher temperatures for boiling than liquids with weak intermolecular attractions.*

Molar heat of vaporization. Even after the temperature of a liquid has been raised to its boiling point, additional energy is required to separate the liquid molecules from each other and keep them apart. As you continue to heat a boiling liquid, the temperatures of both the liquid and the vapor remain at the boiling point until all of the liquid has been changed to the gaseous state. The energy required to vaporize one mole of liquid is called the **molar heat of vaporization.** *A mole of a liquid with strong intermolecular attractions requires more energy to vaporize it than a mole of a liquid with weak intermolecular attractions.*

The trends in vapor pressure, boiling point, and molar heat of vaporization are correlated with strength of intermolecular attractions in Table 14.1.

Viscosity. One of the characteristics of liquids, according to the kinetic molecular theory, is that the liquid molecules are free to move about relative to each other within the body of the liquid. That freedom to move is not the same in all liquids. You are aware, for example, that water may be poured much

Table 14.1
Physical Properties of Liquids

Substance	Vapor Pressure at 20°C	Normal Boiling Point	Heat of Vaporization	Intermolecular Forces
Mercury	0.0012 torr	357°C	59 kJ/mol*	Strongest
Water	17.5	100	41	↑
Benzene	75	80	31	
Ether	442	35	26	
Ethane	27 000	−89	15	Weakest

* kJ is the abbreviation for kilojoule, a unit of energy (p. 445).

more freely than syrup, and syrup more readily than honey. The unique pouring characteristic of each liquid is the result of its **viscosity.** Viscosity may be thought of as an internal resistance to flow. *Liquids with strong intermolecular forces are generally more viscous than liquids with weak intermolecular attractions.*

Surface tension. When a liquid is broken into "small pieces" it forms drops. Ideally, isolated drops are perfect spheres. This is not readily apparent when we look at water drops because we usually see them in the act of falling, at which time they take on a "teardrop" form. A sphere has the smallest surface area possible for a drop of any given volume. This tendency toward a minimum surface is the result of **surface tension.** Within a liquid each molecule is attracted in all directions by the molecules that surround it. At the surface, however, the attraction is nearly all inward, pulling the surface molecules into a sort of tight skin over a standing liquid or around a drop. This is surface tension. The effect of surface tension in water may be seen when a needle floats if placed gently on a still surface, or when small bugs run across the surface of a quiet pond (Fig. 14.2). *Liquids with strong intermolecular attractions have higher surface tension than liquids with weak intermolecular forces.*

We now summarize these relationships between intermolecular attractions and physical properties. If you think in terms of the "stick togetherness" of molecules with strong attractive forces, you can usually reason to correct conclusions. With all other influences being equal, strong intermolecular attractions generally lead to:

1. Low vapor pressure. This is an "inverse" relationship because the vapor pressure is a result of the *gas* released by evaporation when the intermolecular

Figure 14.2
Surface tension. Unbalanced downward attractive forces at the surface of a liquid pull molecules into a difficult-to-penetrate skin capable of supporting small bugs or thin pieces of steel, such as a needle or razor blade. A bug literally runs on the water; it does not float in it. Molecules within the water are attracted in all directions, as shown.

attractions between liquid molecules are overcome. If the *stick togetherness* is high between liquid molecules, not much gas escapes, so the vapor pressure is low.

2. High boiling point. When *stick togetherness* is high, it takes a lot of agitation (high temperature) before the molecules can even *begin* to tear loose from each other within the liquid, where boiling occurs.

3. High heat of vaporization. Again, if *stick togetherness* is high, it takes still more energy to separate the molecules, even after they are shaking vigorously enough to boil.

4. High viscosity. High *stick togetherness* means more "gooey-ness."

5. High surface tension. High *stick togetherness* at the surface means more resistance to anything that would break through or stretch that surface (high surface tension).

14.3 TYPES OF INTERMOLECULAR FORCES

PG 14 C Identify and describe or explain dipole forces, dispersion forces, and hydrogen bonds.

14 D Given the structure of a molecule, or information from which it may be determined, identify the significant intermolecular forces present.

14 E Given the molecular structures of two substances, or information from which they may be obtained, compare or predict relative values of physical properties that are related to them.

As a group, the intermolecular forces that are responsible for so many physical properties of liquids are called **van der Waals forces.** These are the same forces that account for the nonideal behavior of many real gases, particularly when temperatures and/or pressures are near the values at which the gas would condense to a liquid. It is logical to ask, "Where do these forces come from? What causes them to exist?" It has already been stated that they are electrostatic in character; the attractions are between negative and positive charges. But molecules, you may protest, are electrically neutral. How can there be electrostatic attractions between them? To find an answer to that question you must recall that, even though the molecule may be neutral as a whole, the *distribution* of electrical charge within the molecule may be either symmetrical or nonsymmetrical—that is, balanced or unbalanced. Molecules with a symmetrical distribution of charge are nonpolar; if the charge distribution is unbalanced, the molecule is polar. This concept was introduced in Chapter 11, p. 238.

There are three kinds of intermolecular forces that can be traced to electrostatic attractions: dipole forces, dispersion forces, and hydrogen bonds.

1. Dipole forces A polar molecule is sometimes described as a **dipole.** This means that the molecule has two "poles," positive in one region and negative in another. Depending upon their size and shape, dipoles can interact with each other in a number of ways (Fig. 14.3).

Iodine chloride is a substance whose molecules are dipoles. In the solid state they are held as in Figure 14.3A. The forces between molecules are overcome by molecular motion at 27°C, and the substance melts. As a liquid the molecules still attract each other, but they "flow" freely. These attractions are overcome at 97°C, the boiling point of iodine chloride.

The effect of dipole forces is seen when the melting and boiling points of iodine chloride are compared with those of bromine. The molecules are similar in size, shape, and molecular weight. The main difference between them is that ICl molecules are polar and Br_2 molecules are not. Bromine melts at −7°C and boils at 59°C, both temperatures lower than the corresponding temperatures for iodine chloride. This shows that intermolecular forces are weaker in nonpolar bromine than in polar iodine chloride.

Table 14.2 shows additional polar-nonpolar boiling point comparisons between compounds having otherwise similar molecules. In each case the compound with polar molecules has the higher boiling point, supporting the explanation of stronger intermolecular forces between dipoles.

2. Dispersion (London) forces Returning to bromine, Br_2, what type of intermolecular attractions hold it in the liquid state at temperatures up to 59°C?

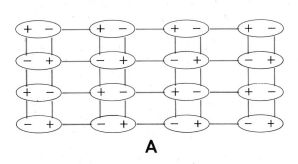

A

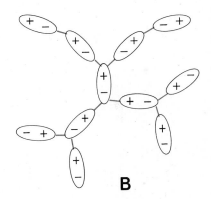

B

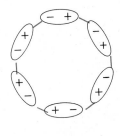

C

Figure 14.3
Dipole forces. A, The polar molecules are arranged in a regular geometric pattern, as they might be found in a molecular solid. B, Similar polar molecules are arranged in a random fashion, such as they might be found in a liquid. C, The molecules have grouped themselves into a larger unit, evidence of which is known to all three states of matter. In all cases the positive region of one polar molecule is attracted to the negative region of the second polar molecule.

Table 14.2
Boiling Points of Polar vs. Nonpolar Substances

Formulas	Polar or Nonpolar	Molecular Weight	Boiling Point (°C)	Formulas	Polar or Nonpolar	Molecular Weight	Boiling Point (°C)
N_2	Nonpolar	28	−196	GeH_4	Nonpolar	77	−90
CO	Polar	28	−192	AsH_3	Polar	78	−55
SiH_4	Nonpolar	32	−112	Br_2	Nonpolar	160	59
PH_3	Polar	34	−85	ICl	Polar	162	97

Why is it not, like chlorine, a gas at room temperature? The forces responsible for bromine being a liquid are called **dispersion forces,** or **London forces.**

Dispersion forces are believed to be the result of "temporary dipoles" that are formed by the shifting electron clouds within molecules that are, on the average, nonpolar (Fig. 14.4). These dipoles attract or repel the electron clouds of nearby nonpolar molecules, thereby inducing them to become dipoles temporarily. As long as these dipoles exist—a very small fraction of a second in each individual case—there is an attraction between them.

The strength of dispersion forces depends upon the ease with which electron distributions can be distorted, or "polarized." Large molecules, with many electrons, or with electrons far removed from the nucleus, are more easily polarized than small, compact molecules in which the nuclei hold electrons in position more firmly. Larger molecules are generally heavier. As a consequence,

Molecules

I   II

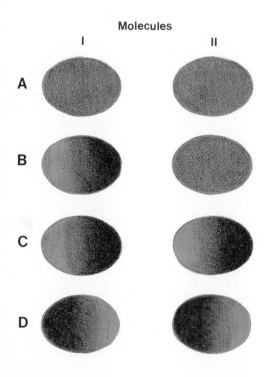

A

B

C

D

Figure 14.4
The origin of dispersion forces.
A, Molecules I and II are temporarily close to each other. Uniform shading indicates overall uniform distribution of electron cloud in molecule.

B, Electron cloud in Molecule I has shifted to right, forming a temporary dipole with concentration of negative charge to right, positive charge to left.

C, Negative charge concentration at the right of Molecule I repels electron cloud in Molecule II to right side of molecule, causing it to become a temporary dipole similar to Molecule I. At this time there is a weak dipole-dipole attraction between the two "instantaneous dipoles."

D, A small fraction of a second later the electron clouds may shift to opposite sides of the molecules. The instantaneous dipoles still attract each other weakly.

Again a small fraction of a second later the molecules interact similarly with outer nearby molecules, forming new weak intermolecular attractions.

intermolecular forces tend to increase with increasing molecular weight among otherwise similar substances. Figure 14.5 shows this for the halogens and for a series of organic compounds known as normal alkanes.

Dispersion forces exist between *all* kinds of molecules, polar or nonpolar. Among small, low molecular weight molecules they are weak, and important only when there are no dipole forces. But among compounds with large molecules and higher molecular weight, these forces can become quite strong— strong enough to be responsible for the existence at room-temperature of nonpolar liquids and solids, as bromine illustrates.

3. Hydrogen bonds Four hydrides of the Group 4A elements, in order of decreasing molecular weight, are SnH_4, GeH_4, SiH_4, and CH_4. These molecules are all tetrahedral in shape; they are nonpolar. The only intermolecular forces

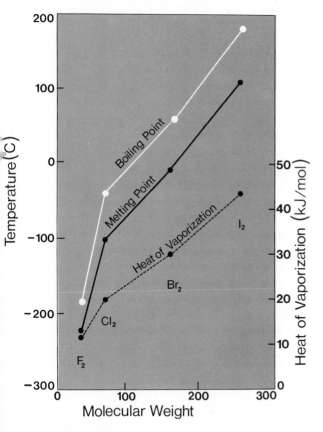

Boiling Points of Normal Alkanes

FORMULA	BOILING POINT (°C)
CH_4	−162
C_2H_6	−88
C_3H_8	−42
C_4H_{10}	0
C_5H_{12}	36
C_6H_{14}	69
C_7H_{16}	98
C_8H_{18}	126
C_9H_{20}	151
$C_{10}H_{22}$	174

Figure 14.5
Physical properties as a function of molecular size. The graph at the left shows the boiling points, melting points, and heats of vaporization of the halogens. The table at the right lists the boiling points of normal alkane hydrocarbons, which have the general formula C_nH_{2n+2}. When the molecular sizes increase, as they do with increasing molecular weight, intermolecular attractions due to dispersion forces increase. This causes an increase in the magnitude of physical properties related to these attractions.

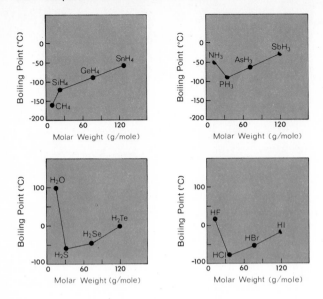

Figure 14.6
Boiling points of different hydrides. In the graph at the upper left, there is no hydrogen bonding in the compound with the smallest molecule (lowest molecular weight), so it has the lowest boiling point in its family. In the other three graphs, the compound with the smallest molecule exhibits hydrogen bonding between molecules. These strong intermolecular forces cause the boiling points to be abnormally high compared to other members of their respective families.

are dispersion forces. We would expect their boiling points to decrease in the order shown, matching the decreasing molecular sizes. They conform to this expectation, as shown in the first graph in Figure 14.6.

With some modification, similar reasoning would lead us to expect that ammonia, NH_3, should have the lowest boiling point of the hydrides of the Group 5A elements. Also, water should have the lowest boiling point of the Group 6A hydrides, and hydrogen fluoride, HF, should be the lowest boiling of the Group 7A hydrides. Obviously, from Figure 14.6, this is not the case. Why not? What is responsible for the unexpectedly strong intermolecular attractions in ammonia, water, and hydrogen fluoride?

The name **hydrogen bond** is given to the abnormally strong intermolecular forces in NH_3, H_2O, and HF. The attraction is between the hydrogen atom of one molecule and the nitrogen, oxygen, or fluorine atom of another. Figure 14.7 shows the hydrogen bonding effect in water. The oxygen atom in a water molecule is considerably more electronegative (3.5) than the hydrogen atom (2.1). The electron pairs that form bonds between the oxygen atom and each hydrogen atom are drawn closely to the oxygen atom, giving that region of the molecule a negative charge. This leaves each hydrogen nucleus—nothing more than a proton—as a small, highly concentrated region of positive charge. The negatively charged oxygen region of one water molecule can get quite close to the small, positively charged region of a neighboring molecule. The result is a hydrogen bond, a dipole-dipole attraction that is much stronger than ordinary dipole forces. Hydrogen bonds between molecules are roughly one tenth as strong as covalent bonds between atoms within a molecule.

To recognize the possibility of hydrogen bonding in a substance, examine its structure. Figure 14.8 shows the structures present in most hydrogen bonding situations. Normally a hydrogen atom is covalently bonded to an atom of a

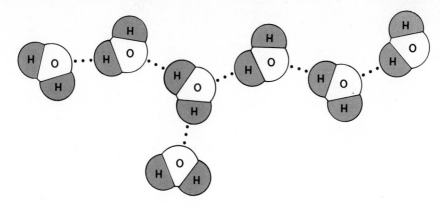

Figure 14.7
Hydrogen bonding in water. Intermolecular hydrogen bonds are present between the electronegative oxygen region of one molecule and the electropositive hydrogen region of the second molecule.

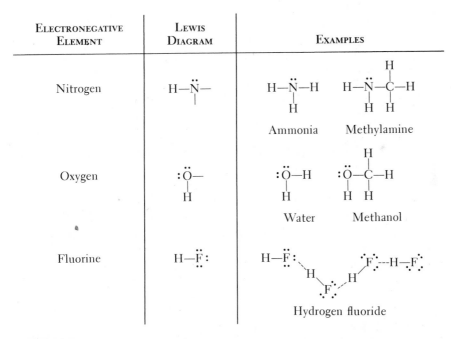

Figure 14.8
Recognizing hydrogen bonding. Hydrogen bonding most frequently occurs when hydrogen is covalently bonded to nitrogen, oxygen, or fluorine in a molecule. The temporary intermolecular bonds are established between one of these strongly electronegative atoms in one molecule and a hydrogen atom of a different molecule. Ammonia and methylamine are examples of compounds of nitrogen that exhibit hydrogen bonding; water and methanol are oxygen compounds that are hydrogen-bonded. In hydrogen fluoride there is evidence that molecules form chains of variable length, as shown.

highly electronegative element that has one or more unshared electron pairs; fluorine, oxygen, and nitrogen are the most common. The hydrogen bond forms between that electronegative atom and the hydrogen atom in a nearby molecule. If the hydrogen atom is covalently bonded to a nitrogen atom, the nitrogen may be bonded to two other species, as shown in Figure 14.8. If both of these are hydrogen atoms, the compound is, of course, ammonia, NH_3. If one is a hydrogen atom and the other is a CH_3 group, methylamine is formed. If the hydrogen atom is bonded to oxygen, the oxygen will be bonded to one other species, which may be another hydrogen atom, producing water, H_2O. Another example is methanol, as shown.

Hydrogen bonding involving fluorine is unique because both hydrogen and fluorine form only one covalent bond. A linkage of indefinite length is possible between several HF molecules, as indicated in Figure 14.8. Such chains are also known to form closed rings.

The abnormal boiling point of water—abnormal in the sense that it departs from what would be expected when compared to other Group 6A hydrides—is but one example of a long list of "abnormal" physical properties of water. Water is so common a substance we take it much for granted. But, in relative trends of physical properties predicted from the periodic table, water is one of the most "uncommon" substances known. Nearly all of its unique properties are at least partly the result of hydrogen bonding.

14.4 LIQUID-VAPOR EQUILIBRIUM

PG 14 F Describe or explain the equilibrium between a liquid and its own vapor, and the process by which the equilibrium is reached.

In Section 14.2, vapor pressure and boiling point were identified as physical properties that are related to intermolecular attractions. In this section and the next we will examine these properties in greater detail, and thereby learn a bit more about the liquid state.

In Chapter 13, temperature was described as a measure of the average kinetic energy of the particles in a sample of matter. The word *average* suggests that at a given temperature the particles in the sample do not all have the same kinetic energy, but rather a range of energies. Kinetic energy is expressed mathematically as $\frac{1}{2}mv^2$, where m is the mass of the particle and v is its velocity. If the sample is a pure substance, all particles have the same mass, except for minor variations due to isotopes. Their different kinetic energies therefore indicate that they have different velocities.

We will now look at what happens at the molecular level when a liquid evaporates. Intermolecular forces tend to keep the substance in the liquid state. But a few of the faster moving molecules near the surface have enough kinetic energy to overcome these attractions and escape (evaporate) from the liquid. At a given temperature the fraction, or percentage, of the total sample that is able to evaporate is constant. As a consequence the rate of evaporation per unit of surface area is also constant at that temperature. If the temperature rises, a larger portion of the sample has enough kinetic energy to evaporate, and the evaporation rate is greater.

If a liquid with weak intermolecular forces, such as benzene, is placed in an Erlenmeyer flask, which is then stoppered as in Figure 14.9, the benzene will begin to evaporate. If temperature is held constant, the *rate* of evaporation will remain constant throughout the experiment. The constant evaporation rate is represented by the fixed length of black arrows pointing upward from the liquid in each view of the flask, and also as the horizontal black line in a graph of evaporation rate versus time in Figure 14.9.

At first (Time 0) the movement of molecules is entirely in one direction, from the liquid to the vapor. However, as the concentration of molecules in the vapor builds up, an occasional molecule hits the surface and reenters the liquid. The change of state from a gas to a liquid is called **condensation**. The *rate* of condensation depends upon the concentration of molecules in the vapor state. At Time 1 there will be a small number of molecules in the vapor state, so the condensation rate will be more than zero, but much less than the evaporation rate. This is shown by the arrow lengths in the Time 1 flask of Figure 14.9.

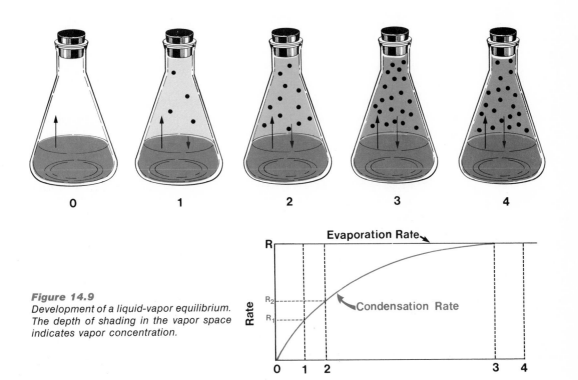

Figure 14.9
Development of a liquid-vapor equilibrium. The depth of shading in the vapor space indicates vapor concentration.

As long as the rate of evaporation is greater than the rate of condensation, the vapor concentration will rise over the next interval of time. Therefore the rate of return from vapor to liquid rises with time (Time 2). Eventually the rates of vaporization and condensation become equal (Time 3). The number of molecules moving from vapor to liquid in unit time just balances the number moving in the opposite direction. We describe this situation, **when opposing rates of change are equal**, as a condition of **dynamic equilibrium** between liquid and vapor. Once equilibrium is reached, the concentration of molecules in the vapor has a certain fixed value which does not change. The rates therefore remain equal (Time 4).

Changes that occur in either direction, such as the change from a liquid to a vapor and the opposite change from a vapor to a liquid, are called **reversible changes**; if the change is chemical it is a **reversible reaction**. Chemists write equations describing reversible changes with a double arrow, one pointing in each direction. For example, the reversible change between liquid benzene, $C_6H_6(\ell)$, and benzene vapor, $C_6H_6(g)$, is represented by the equation

$$C_6H_6(\ell) \rightleftarrows C_6H_6(g) \tag{14.1}$$

The ideal gas law predicts that the partial pressure exerted by the vapor at any time depends upon the concentration of molecules in the vapor state, expressed as the number of moles of gas per unit volume. Solving $pV = nRT$ for partial pressure,

$$p = \frac{nRT}{V} = RT \times \frac{n}{V}$$

With both R and T constant, it follows that the partial pressure of the vapor is proportional to n/V, which is moles per unit volume, or concentration. When the vapor concentration becomes constant at equilibrium, the vapor pressure also becomes constant. **The partial pressure exerted by a vapor in equilibrium with its liquid phase at a given temperature is the equilibrium vapor pressure of the substance at that temperature.**

THE EFFECT OF TEMPERATURE

PG 14 G   Describe the relationship between vapor pressure and temperature for a liquid-vapor system in equilibrium; explain this relationship in terms of the kinetic molecular theory.

The vapor pressure of a liquid always increases as temperature rises. This is understood in terms of how equilibrium is reached in a liquid-vapor system. It was noted earlier that only a fraction of the molecules present have enough kinetic energy to evaporate at any given temperature, and that this fraction is larger at higher temperatures. Evaporation rate therefore increases with temperature. If the system reaches equilibrium, the condensation rate must also be greater at higher temperatures. This takes a higher equilibrium vapor concentration, which in turn exerts a larger vapor pressure.

The equilibrium vapor pressures of many liquids have been measured at different temperatures. Some of these are plotted in Figure 14.10.

The extent to which vapor pressure increases with temperature may be surprising. The vapor pressure of water at 25°C, for example, is 24 torr, but at 60°C it is 149 torr, about 500% greater. The absolute temperature increase from 298 K to 333 K is only 35°, about 12%. Clearly the vapor pressure change is not a gas law pressure increase, in which pressure is proportional to absolute temperature. It is a fact, however, that the percentage of water molecules with enough kinetic energy to evaporate increases by about 600% over the same temperature range, matching much more closely the percentage increase in vapor pressure. It is the *number of particles with enough energy to evaporate* that is responsible for the vapor pressure dependence upon temperature.

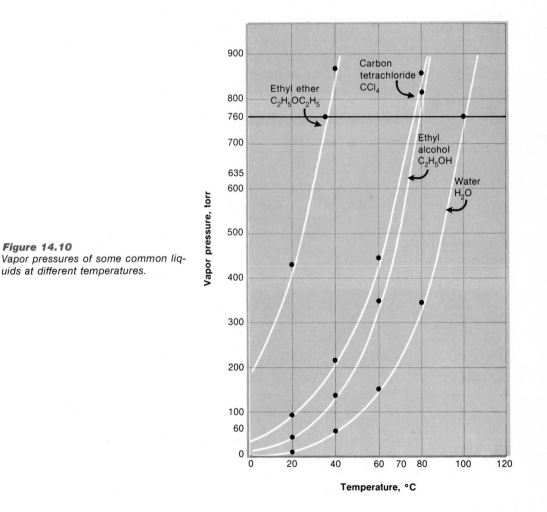

Figure 14.10
Vapor pressures of some common liquids at different temperatures.

Perhaps this idea of percentage increases may be more easily understood by comparing it to the heights of one hundred 14-year-old students. Suppose their average height is 5'4", or 64 inches, and only five students (5%) are six feet tall. Four years later they are measured again. The average height has increased to 5'8", or 68 inches, and fifteen (15%) have passed the six-foot mark. At both times the large majority are shorter than six feet, and the average height has increased only 4" out of 64", or about 6%, over the four-year period; but the *number* of students having reached six feet has *tripled*, or increased by 200%, over the same period.

14.5 THE PHENOMENON OF BOILING

PG 14 H Describe the process of boiling and the relationships among boiling point, vapor pressure, and surrounding pressure.

When a liquid is heated in an open container, bubbles form, usually at the base of the container where heat is being applied. The first bubbles that we see are often air, driven out of solution by an increase in temperature. Eventually, when a certain temperature is reached, vapor bubbles form throughout the liquid, rise to the surface, and break. When this happens we say the liquid is boiling.

In order for a stable bubble to form in a boiling liquid, the vapor pressure within the bubble must be high enough to push back the surrounding liquid and the atmosphere above the liquid. The minimum temperature at which this can occur is called the **boiling point: the boiling point is that temperature at which the vapor pressure of the liquid is equal to the pressure above its surface.** Actually the vapor pressure within a bubble must be a tiny bit greater than the surrounding pressure, which suggests that bubbles probably form in local "hot spots" within the boiling liquid. The boiling temperature at one atmosphere— the temperature at which the vapor pressure is equal to one atmosphere—is called the **normal boiling point.** From Figure 14.10 we see that the normal boiling point of water is 100°C; of ethyl alcohol, 78°C; of carbon tetrachloride, 77°C; and of ethyl ether, 35°C.

According to the definition, the boiling point of a liquid depends upon the pressure above it. If that pressure is reduced, the temperature at which the vapor pressure equals the lower surrounding pressure comes down also, and the liquid will boil at that lower temperature. This is why liquids at higher altitudes boil at reduced temperatures. In mile-high Denver, where atmospheric pressure is typically about 630 torr, water boils at 95°C. It is possible to boil water at room temperature by creating a vacuum in the space above it. When pressure is reduced to 20 torr, water boils at 22°C, about 72°F. A method for purifying a compound that might decompose or oxidize at its normal boiling point is to boil it at reduced temperature in a vacuum, and then condense the vapor.

It is also possible to *raise* the boiling point of a liquid by *increasing* the pressure above it. The pressure cooker used in the kitchen takes advantage of this effect. By allowing the pressure to build up within the cooker, it is possible to reach temperatures as high as 110°C without boiling off the water. At this temperature, foods cook in about half the time required at 100°C.

14.6 THE NATURE OF THE SOLID STATE

PG 14 I Distinguish between crystalline and amorphous solids.

A solid whose particles are arranged in a geometric pattern that repeats itself over and over in three dimensions is a **crystalline solid.** Each particle occupies a fixed position in the crystal. It can vibrate about that site but cannot move past its neighbors. The high degree of order often leads to large crystals which have a precise geometric shape. In ordinary table salt we can distinguish small cubic crystals of sodium chloride. Large, beautifully formed crystals of such minerals as quartz (SiO_2) and fluorite (CaF_2) are found in nature. Figure 14.11 has photographs of three crystalline solids.

In an **amorphous solid** such as glass, rubber, or plastic, there is no long-range order. Even though the arrangement around a particular site may resemble that in a crystal, the pattern does not repeat itself throughout the solid (Fig. 14.12). From a structural standpoint, we may regard an amorphous solid as intermediate between the crystalline and the liquid states. In many amorphous solids, the particles have some freedom to move with respect to one another. The elasticity of rubber and the tendency of glass to flow when subjected to stress over a long period of time suggest that the particles in these materials are not rigidly fixed in position.

Figure 14.11
Crystals of NaCl, $NH_4H_2PO_4$, and $CuSO_4 \cdot 5\,H_2O$.

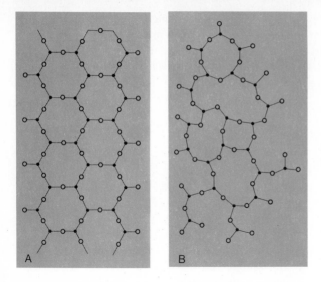

Figure 14.12
Crystalline (A) and amorphous (B) solids.

Crystalline solids have characteristic physical properties which can serve to identify them. Sodium chloride, for example, melts sharply at 800°C. This is in striking contrast to glass, which first softens and then slowly liquifies over a wide range of temperatures. The physical properties of a few crystalline solids are listed in Table 14.3.

14.7 TYPES OF CRYSTALLINE SOLIDS

PG 14 J Distinguish among the following types of crystalline solids: ionic, molecular, macromolecular, and metallic.

Solids such as those listed in Table 14.3 can be divided into four classes on the basis of their particle structure and the type of forces that hold these particles together in the crystal lattice.

Ionic crystals. Examples are NaF, $CaCO_3$, AgCl, and NH_4Br. Oppositely charged ions are held together by strong electrostatic forces (recall Fig. 10.2). As pointed out earlier, ionic crystals are typically high melting, frequently water-soluble, and have very low electrical conductivities. Their melts and water solutions, in which the ions can move around, conduct electricity readily.

Molecular crystals. Examples are I_2 and ICl. Small, discrete molecules are held together by relatively weak intermolecular forces of the types discussed in Section 14.3. Molecular crystals are typically soft, low melting, and generally (but not always) insoluble in water. They usually dissolve in nonpolar or slightly

Table 14.3
Physical Properties of Crystalline Solids

Substance	Formula	Melting Point (°C)	Heat of Fusion*	Water Solubility**	Electrical Conductivity
Naphthalene	$C_{10}H_8$	80	19	0.03	nonconductor
Iodine	I_2	114	17	0.3	nonconductor
Potassium nitrate	KNO_3	333	11	316	nonconductor
Sodium chloride	NaCl	800	29	360	nonconductor†
Copper	Cu	1083	13	0	conductor
Iron	Fe	1535	11	0	conductor
Silicon dioxide	SiO_2	1710	—	0	nonconductor
Magnesium oxide	MgO	2800	—	0.006	nonconductor†
Diamond	C	3500	—	0	nonconductor

* kJ/mol.
** g/1000 g water at room temperature.
† Conducts when melted.

polar organic solvents such as carbon tetrachloride or chloroform. Molecular substances, with rare exceptions, are nonconductors when pure, even in the liquid state.

Macromolecular crystals. Examples are diamond, C, and quartz, SiO_2. Atoms are covalently bonded to each other to form one "huge molecule" making up the entire crystal. The prefix *macro-* means large.

There are no small, discrete molecules in macromolecular crystals. In diamonds each carbon atom is covalently bonded to four other carbon atoms to give a structure that repeats throughout the entire crystal (Fig. 14.13A). The

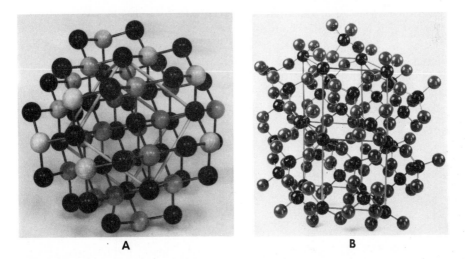

A   **B**

Figure 14.13
A, Model of diamond (carbon) crystal. B, Model of quartz (SiO_2) crystal. (Diamond photograph by Judy Serface; quartz photograph, courtesy of Klinger Scientific Apparatus Corporation, Jamaica N.Y.)

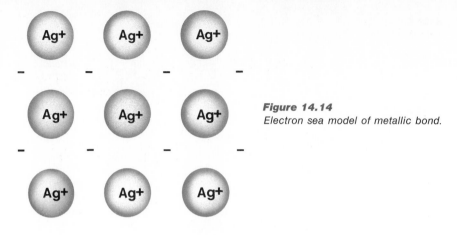

Figure 14.14
Electron sea model of metallic bond.

structure of silicon dioxide resembles that of diamond in that the atoms are held together by a continuous series of covalent bonds. Each silicon atom is bonded to four oxygen atoms, each oxygen to two silicons, as shown in Figure 14.13B.

Macromolecular solids, like ionic crystals, are high-melting. A very high temperature is needed to break the covalent bonds in crystals of quartz (1710°C) or diamond (3500°C). Macromolecular crystals are almost always insoluble in water or any common solvent. They are generally poor conductors of electricity in either the solid or liquid state.

Metallic crystals. A simple model of bonding in a metal consists of a crystal of positive ions through which valence electrons move freely. This so-called "electron sea" model of a metallic crystal is illustrated in Figure 14.14 for silver. The Ag^+ ions form the backbone of the crystal; the electrons surrounding these ions are not tied down to any particular ion and hence are not restricted to a particular location. It is because of these freely moving electrons that metals are excellent conductors of electricity.

The general properties of the four kinds of crystalline solids are summarized in Table 14.4.

CHAPTER 14 IN REVIEW

14.1 THE NATURE OF THE LIQUID STATE

14 A Explain the differences between the physical behavior of liquids and gases in terms of the relative distances between molecules and the effect of those distances on intermolecular forces. (307)

Table 14.4
Types of Crystals

Type	Examples	Properties
Ionic	KNO_3, NaCl, MgO	High melting; generally water-soluble; brittle; conduct only when melted or dissolved in water.
Molecular	$C_{10}H_8$, I_2	Low melting; usually more soluble in organic solvents than in water; nonconductors in pure state.
Macromolecular	SiO_2, C	Very high melting; insoluble in all common solvents; brittle; non- or semiconductors.
Metallic	Cu, Fe	Wide range of melting points; insoluble in all common solvents; malleable; ductile; good electrical conductors.

14.2 PHYSICAL PROPERTIES OF LIQUIDS

14 B For two liquids, given comparative values of physical properties that depend on intermolecular attractions, predict the relative strengths of those attractions; or, given a comparison of the strengths of intermolecular attractions, predict the relative values of physical properties that depend upon them. (307)

14.3 TYPES OF INTERMOLECULAR FORCES

14 C Identify and describe or explain dipole forces, dispersion forces, and hydrogen bonds. (310)

14 D Given the structure of a molecule, or information from which it may be determined, identify the significant intermolecular forces present. (310)

14 E Given the molecular structures of two substances, or information from which they may be obtained, compare or predict relative values of physical properties that are related to them. (310)

14.4 LIQUID-VAPOR EQUILIBRIUM

14 F Describe or explain the equilibrium between a liquid and its own vapor, and the process by which the equilibrium is reached. (316)

14 G Describe the relationship between vapor pressure and temperature for a liquid-vapor system in equilibrium; explain this relationship in terms of the kinetic molecular theory. (318)

14.5 THE PHENOMENON OF BOILING

14 H Describe the process of boiling and the relationships among boiling point, vapor pressure, and surrounding pressure. (320)

14.6 THE NATURE OF THE SOLID STATE

14 I Distinguish between crystalline and amorphous solids. (321)

14.7 TYPES OF CRYSTALLINE SOLIDS

14 J Distinguish among the following types of crystalline solids: ionic, molecular, macromolecular, metallic. (322)

TERMS AND CONCEPTS

Evaporate; evaporation (308)
Vapor pressure (308)
Equilibrium vapor pressure (308)
Boiling point (308)
Molar heat of vaporization (308)
Viscosity (308)
Surface tension (309)
Van der Waals forces (310)
Dipole; dipole forces (311)
Dispersion (London) forces (311)
Hydrogen bond (314)

Condense; condensation (317)
Dynamic equilibrium (318)
Reversible change; reversible reaction (318)
Boiling (320)
Normal boiling point (320)
Crystalline solid (321)
Amorphous solid (321)
Ionic crystal (322)
Molecular crystal (322)
Macromolecular crystal (323)
Metallic crystal (323)

Most of these terms and many others appear in the Glossary.

QUESTIONS AND PROBLEMS

An asterisk (*) identifies a question that is relatively difficult, or that extends beyond the performance goals of the chapter.

Section 14.1

14.1) Explain why gases are less dense than liquids.

14.2) Explain why water is less compressible than air.

Section 14.2

14.3) Identify and explain the relationship between intermolecular attractions and equilibrium vapor pressure.

14.4) What is meant by molar heat of vaporization?

14.5) Which liquid is more viscous, water or motor oil? In which liquid do you suppose the intermolecular attractions are stronger? Explain.

14.32) Explain why two gases will mix with each other more rapidly than two liquids.

14.33) Explain why intermolecular attractions are stronger in the liquid state than in the gas state.

14.34) How do intermolecular attractions influence the boiling point of a pure substance?

14.35) Why does molar heat of vaporization depend upon the strength of intermolecular attractions?

14.36) A tall glass cylinder is filled to a depth of 1 meter with water. Another tall glass cylinder is filled to a depth of 1 meter with syrup. Identical ball bearings are dropped into each tube at the same instant. In which tube will the ball bearing reach the bottom first? Explain your prediction in terms of viscosity and intermolecular attractions.

14.6) A falling drop of any liquid tends to be spherical in shape. If the liquid is water there is a visible elongation of the drop into a "teardrop" form. A drop of mercury, however, remains more spherical. Compare the surface tension of water with that of mercury. What does this suggest about the strength of intermolecular attractions in water compared to the interatomic attractions in mercury?

14.7) One of the functions of soap is to change the surface tension of water. Considering the purpose of laundry soap, do you think soap increases or decreases intermolecular attractions in water? Explain.

The normal boiling and melting points for three nitrogen oxides are given at the right. Refer to this table in answering the next two questions in each column.

14.8)* In which physical state, gas, liquid, or solid, would you find each nitrogen oxide at $-90.0°C$?

14.9)* Which of the three oxides would you expect to have the highest viscosity at $-90°C$? Explain how you reached your conclusion.

Section 14.3

14.10) Identify the three major types of intermolecular forces and explain why they exist.

14.11) Identify the principal intermolecular forces in each of the following compounds: HBr; C_2H_2; NF_3; C_2H_5OH.

14.12) Given that ionic compounds generally have higher melting points than molecular compounds of similar molar weight, compare dipole forces and forces between ions. How are they alike and how are they different?

14.37) If water is spilled on a laboratory desk top, it usually spreads over the surface, wetting any papers or books that may be in its path. If mercury is spilled, it neither spreads nor makes paper it contacts wet, but rather forms little drops that are easily combined into pools by pushing them together. Suggest an explanation for these facts in terms of the apparent surface tension and intermolecular attraction in mercury and in water.

14.38) The level at which a duck floats on water is determined more by the thin oil film that covers its feathers than by a lower density of its body compared to water. The water does not "mix" with the oil, and therefore does not penetrate the feathers. If, however, a few drops of "wetting agent" are placed in the water near the duck, the poor bird will sink. State the effect of wetting agent on surface tension and intermolecular attractions of water.

	NO	N_2O	NO_2
Boiling point	$-152°C$	$-88.5°C$	$+21.2°C$
Melting point	$-164°C$	$-90.8°C$	$-11.2°C$

14.39)* Which of the three oxides would you expect to have the highest molar heat of vaporization? Explain how you reached your conclusion.

14.40)* Which of the three oxides would you expect to have a measurable vapor pressure at $-90.0°C$? Explain your answer.

14.41) Under what circumstances are dispersion forces likely to produce stronger intermolecular attractions than dipole forces, and when are dispersion forces likely to be weaker?

14.42) Identify the principal intermolecular forces in each of the following compounds: $NH(CH_3)_2$; CH_2F_2; C_3H_8.

14.43) Compare dipole forces and hydrogen bonds. How are they different, and how are they similar?

For the next two questions in each column predict, on the basis of molecular size and polarity, and hydrogen bonding, which member of each of the following pairs has the higher boiling point and state the reason for your choice. Assume that molecular size is related to molar weight.

14.13) CH_4 and CCl_4.

14.14) H_2S and PH_3.

14.15) Hydrogen is usually covalently bonded to one of what three elements in order for hydrogen

14.44) CH_4 and NH_3.

14.45) Ar and Ne.

14.46) What physical feature of the hydrogen atom, when covalently bonded to an appropriate

bonding to be present? What unique feature do these elements share that sets them apart from other elements?

14.16) Of the three types of intermolecular forces, which one(s) (a) operate in all molecular substances, and (b) operate between all polar molecules?

14.17) In which of the following substances would you expect dipole forces to operate?

(a) H—C≡N   (b) O=C=O

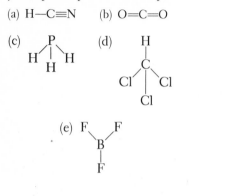

(c) P / H H, H

(d) H, C, Cl Cl, Cl

(e) F, F, B, F

14.18) Predict which compound, C_3H_8 or C_6H_{14}, has the higher melting and boiling points. Explain your prediction.

14.19) Predict which compound, SO_2 or CO_2, has the higher vapor pressure as a liquid at a given temperature. Explain your prediction.

Section 14.4

14.20) What essential condition exists when a system is in a state of *dynamic* equilibrium?

second element, is largely responsible for the strength of hydrogen bonding between molecules?

14.47) Of the three types of intermolecular forces, which one(s) (a) account for the high melting point, boiling point, and other abnormal properties of water, and (b) increase with molecular size?

14.48) In which of the following substances would you expect hydrogen bonds to form?

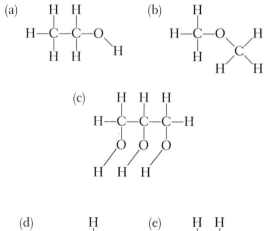

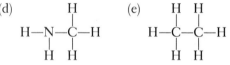

14.49) Predict which compound, CO_2 or CS_2, has the higher melting and boiling points. Explain your prediction.

14.50) Predict which compound, CH_4 or CH_3F, has the higher vapor pressure as a liquid at a given temperature. Explain your prediction.

14.51) Explain why the rate of evaporation from a liquid depends upon temperature. Explain why the rate of condensation depends upon concentration in the vapor state.

The next two questions in each column are based on the apparatus shown in Figure 14.15. Study the caption that describes how vapor pressure is measured, and then answer the questions.

14.21)* A student uses the apparatus in Figure 14.15 to determine the equilibrium vapor pressure of a volatile liquid. He observes that the pressure increases rapidly at first, but more slowly as equilibrium is approached. Suggest a reason for this.

14.52)* Why would the apparatus in Figure 14.15 be of little or no value in determining the equilibrium vapor pressure of water if used as described? Under what conditions might the apparatus give acceptable values for water vapor pressure?

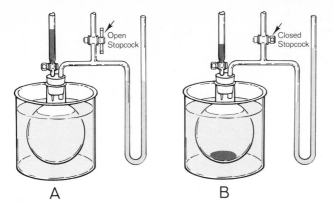

A   B

Figure 14.15
Measurement of vapor pressure. A, The buret contains the liquid whose vapor pressure is to be measured. The flask, tubes, and manometer above the mercury in the left leg are all at atmospheric pressure through the open stopcock. The mercury in the open right leg of the manometer is also at atmospheric pressure, so the mercury levels are the same in the two legs. B, To measure vapor pressure, stopcock is closed, trapping air in flask in space above the left mercury level. Liquid is introduced to flask from buret. Evaporation occurs until equilibrium is reached. Vapor causes increase in pressure, which is measured directly by the difference in mercury levels.

14.22)* Suppose in making the vapor pressure measurement described in Figure 14.15 that all of the liquid introduced into the flask evaporates. Explain what this means in terms of evaporation and condensation rates. How does the vapor pressure in the flask compare with the equilibrium vapor pressure at the existing temperature?

14.23) Using the ideal gas equation, show why the partial pressure of a gaseous component depends upon its vapor concentration at a given temperature.

14.24)* Three closed boxes have identical volumes. A beaker containing a small quantity of acetone, an easily vaporized liquid, is placed in one. Over a period of time it evaporates completely. A medium quantity of acetone is placed in the second. Eventually it evaporates until only a small portion of liquid remains. A larger quantity is placed in the third, and about half of it eventually evaporates. (a) Which box or boxes develop the greatest acetone vapor pressure? (b) Which probably has the least? (c) Explain both answers.

14.25)* The equilibrium vapor pressure of water at 50°C is 93 torr. A sealed flask under vacuum—assume the pressure inside is zero—contains a vial of liquid water. The vial is broken, and some of the water vaporizes. What is the maximum pressure that could be reached in this system?

14.53)* After the system has come to equilibrium, as in Figure 14.15B an additional volume of liquid is introduced into the flask. Describe and explain what will happen to the pressure indicated by the manometer. Disregard any Boyle's Law effect; assume that the change in gas volume resulting from increased liquid volume is negligible.

14.54) Using the ideal gas equation, show why equilibrium vapor pressure is dependent upon temperature.

14.55)* Three closed boxes have different volumes: one is small, one medium-sized, and one large. Beakers containing equal quantities of acetone are placed in the boxes. Eventually all the acetone evaporates in one box, but equilibrium is reached in the other two. (a) In which box does complete evaporation occur? (b) Compare the eventual vapor pressures in the three boxes. (c) Explain both answers.

14.56)* Suppose, in Problem 14.25, the sealed flask contained air at 760 torr instead of a vacuum, and the same vial of liquid water. What then would be the maximum that could be reached when some of the liquid vaporized?

Section 14.5

14.26) Define boiling point. Draw a vapor pressure-temperature curve and locate the boiling point on it.

14.27) The vapor pressure of a certain compound at 20°C is 906 torr. Is the substance a gas or a liquid at 760 torr? Explain.

14.28) Explain why high boiling liquids usually have low vapor pressures.

14.29) The molar heat of vaporization of substance X is 34 kJ/mol; of substance Y, 27 kJ/mol. Which substance would be expected to have the higher normal boiling point? the higher vapor pressure at 25°C?

14.57) An industrial process requires boiling a liquid whose boiling point is so high that maintenance costs on associated pumping equipment are prohibitive. Suggest a way this problem might be solved.

14.58) Normally a gas may be condensed by cooling it. Suggest a second method, and explain why it will work.

14.59) Explain why low-boiling liquids usually have low molar heats of vaporization.

14.60) At 20°C the vapor pressure of substance M is 520 torr; of substance N, 634 torr. Which substance will have the lower boiling point? the lower molar heat of vaporization?

Section 14.6

14.30) Compare amorphous and crystalline solids in terms of structure. How do crystalline and amorphous solids differ in physical properties? Explain the difference.

14.61) Is ice a crystalline solid or an amorphous solid? On what properties do you base your conclusion?

Problems 14.31 and 14.62: For each solid whose physical properties are tabulated below, state whether it is most likely to be ionic, molecular, macromolecular, or metallic:

	SOLID	MELTING POINT	WATER SOLUBILITY	CONDUCTIVITY (PURE)	TYPE OF SOLID
14.31)	A	150°C	Insoluble	Nonconductor	_____
	B	1450°C	Insoluble	Excellent	_____
14.62)	C	2000°C	Insoluble	Nonconductor	_____
	D	1050°C	Soluble	Nonconductor	_____

solutions

15.1 THE CHARACTERISTICS OF A SOLUTION

Solutions abound in nature. We are surrounded by the gaseous solution known as air. The oceans are a water solution of sodium chloride and other substances. Some of these are present in sufficient concentration to make it commercially profitable to extract them. Magnesium is a notable example. Even what we call "fresh" water is a solution, although the concentrations are so low we tend to think of the water we drink as "pure." "Hard" water may be sufficiently pure for human consumption, but there are enough calcium and magnesium salts present to form solid deposits in hot water pipes and boilers. Even rain water is a solution, containing dissolved gases. Oxygen is not very soluble in water but what little there is in solution is mighty important to fish, who cannot survive without it.

A **solution is a homogeneous mixture.** This implies uniform distribution of solution components, so that a sample taken from any part of the solution will have the same composition. Two solutions made up of the same substances, however, may have different compositions. Variable composition is an important feature that distinguishes solutions from compounds. A solution of ammonia in water, for example, may contain 1% ammonia by weight, or 2%, 5%, 20.3% . . . up to the 29% solution we call "concentrated ammonia." In contrast, the composition of pure compounds is always the same (the Law of Constant Composition, p. 18); for ammonia the composition is always 82.4% nitrogen and 17.6% hydrogen.

A solution may exist in any of the three states, gas, liquid, or solid. Air is a gaseous solution, made up of nitrogen, oxygen, carbon dioxide, and other gases in small amounts. Odors are usually the result of liquids (perfume, benzene) or solids (moth balls) that vaporize and become part of the air solution. In addition to oxygen in water, dissolved carbon dioxide in carbonated beverages is a familiar liquid solution of a gas. Alcohol in water is an example of the solution of two liquids, and the oceans of the world are natural liquid solutions of solids. Solid state solutions are common in the form of metal alloys.

Particle size distinguishes solutions from other mixtures. Dispersed particles in solutions, which may be atoms, ions, or molecules, are very small—generally less than 5×10^{-7} cm in diameter. Particles of this size do not settle on standing, and they are too small to be seen.

In contrast, a **suspension** is a heterogeneous distribution of larger particles, usually 10^{-4} cm or larger in diameter, in a second medium. In the liquid or gaseous state, suspended particles will settle over a period of time. Dispersions of particles between solution and suspension size are called **colloids.** Milk, fog, and jelly are common examples of colloids. Colloidal particles, which do not settle on standing, may consist of a single huge molecule, or more commonly an aggregate of smaller molecules. Colloidal particles, like dissolved particles, are too small to be seen, but they are large enough to reflect light and produce a visible beam (Fig. 15.1).

15.2 SOLUTION TERMINOLOGY

PG 15 B   Distinguish among terms in the following groups:

Solute and solvent
Concentrated and dilute
Solubility, saturated, unsaturated, and supersaturated
Miscible and immiscible

In discussing solutions a language of closely related and sometimes overlapping terms is used. We will now identify and define these terms.

Solute and solvent. When solids or gases are dissolved in liquids, the solid or gas is said to be the **solute** and the liquid the **solvent.** More generally, the solute is taken to be the substance present in a relatively small amount. The medium

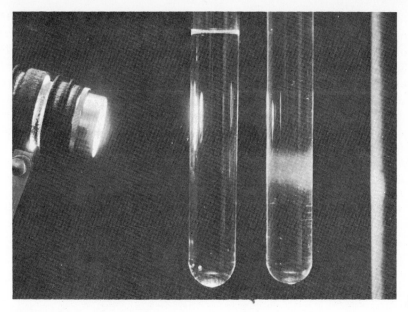

Figure 15.1
Reflection of light by a colloidal dispersion. A beam of light is invisible as it passes through a solution (left tube) because the dissolved particles are too small to reflect light. Colloidal particles (right tube) are too small to be seen with the naked eye, but they are large enough to reflect light and produce a visible beam. The effect is similar to the beam of a searchlight, which is visible only because the light is reflected by the dust particles in the air.

in which the solute is dissolved is usually larger in quantity, and is called the solvent. The distinction is not precise, however. Water is capable of dissolving more than its own weight of some solids, but the water continues to be called the solvent. In alcohol-water solutions, either liquid may be the more abundant and, in a given context, either might be called the solute or solvent.

Concentrated and dilute. A **concentrated** solution has a *relatively* large quantity of a specific solute per unit amount of solution, and a **dilute** solution has a *relatively* small quantity of the same solute per unit amount of solution. The terms compare concentrations of two solutions of the *same solute and solvent*. They carry no other quantitative meaning.

Solubility, Saturated, and Unsaturated. **Solubility** is a measure of how much solute will dissolve in a given amount of solvent at a given temperature. It is sometimes expressed by giving the number of grams of solute that will dissolve in 100 grams of solvent. A solution which can exist in equilibrium with undissolved solute is a **saturated** solution. A solution whose concentration corresponds to the solubility limit is therefore saturated. If the concentration of a solute is less than the solubility limit it is **unsaturated**.

Supersaturated solutions. Under carefully controlled conditions, a solution can be produced in which the concentration of solute is greater than the normal solubility limit. Such a solution is said to be **supersaturated.** A supersaturated solution of sodium acetate, for example, may be prepared by dissolving 80 grams of the salt in 100 grams of water at about 50°C. If the solution is then cooled to 20°C without stirring, shaking, or other disturbance, all 80 grams of solute will remain in solution even though the solubility at 20°C is only 46.5 grams/100 grams of water. The supersaturated solution can be maintained indefinitely so long as there are no tiny particles upon which crystallization can start. If a small seed crystal of sodium acetate is added, crystallization takes place until equilibrium is attained by the formation of a saturated solution.

Miscible and Immiscible. Miscible and immiscible are terms customarily limited to solutions of liquids in liquids. If two liquids dissolve in each other in all proportions they are said to be **miscible** in each other. Alcohol and water, for example, are miscible liquids. Liquids that are insoluble in each other, as oil and water, are **immiscible.** Some liquid pairs will mix appreciably with each other, but in limited proportions; these are said to be *partially miscible.*

15.3 THE FORMATION OF A SOLUTION

PG 15 C   Describe the formation of a saturated solution from the time excess solid solute is first placed into a liquid solvent.

   15 D   Identify and explain the factors that determine the time required to dissolve a given amount of solute, or to reach equilibrium.

We will now examine how a solution is formed. When a soluble ionic crystal is placed in water, the negatively charged ions at the surface are attracted by the positive region of the polar water molecules (Fig. 15.2). A "tug of war" for the negative ions begins; water molecules tend to pull them from the crystal, while neighboring positive ions tend to hold them in the crystal. In a similar way, positive ions at the surface are attacked by the negative portion of the water molecules and are torn from the crystal. Once released, the ions are surrounded by the polar water molecules. Such ions are said to be **hydrated.**

The dissolving process is reversible. As the dissolved solute particles move randomly through the solution, they come into contact with the undissolved solute and crystallize—return to the solid state. The rate at which crystallization occurs depends upon the concentration of solute at the surface of the undissolved solid. If, before all the solute is dissolved, the concentration increases to the point that the crystallization rate is equal to the rate at which the solid is dissolving, an equilibrium is established. For NaCl, this equilibrium may be represented by the "reversible reaction" equation (p. 318),

$$NaCl(s) \rightleftharpoons Na^+(aq) + Cl^-(aq) \tag{15.1}$$

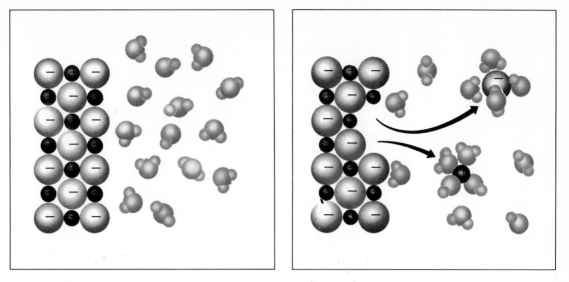

Figure 15.2
Dissolving of an ionic solute in water.

The concentration at equilibrium identifies a saturated solution, as noted in the previous section, and represents the solubility at the existing temperature (Fig. 15.3).

    The time required to dissolve a given amount of solute—or to reach equilibrium concentration, if excess solute is present—depends upon several factors:

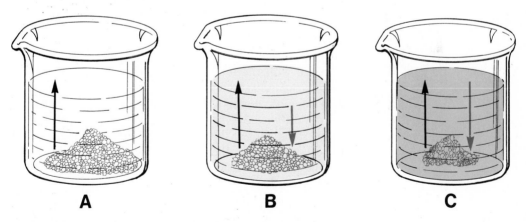

A   **B**   **C**

Figure 15.3
Development of equilibrium in producing a saturated solution. The solution dissolves at a constant rate, shown by the black arrows. In A, when dissolving has just begun, the solute concentration in the solvent is zero, so no crystallization can occur. In B the solute concentration has risen to yield a crystallization rate indicated by the colored arrow—still less than the dissolving rate. Eventually, in C, the concentration has increased to the point that the dissolving and crystallization rates are equal. Equilibrium has been reached, and the solution is saturated.

1. The dissolving process depends upon surface area. Therefore a finely divided solid, which offers more surface per unit of mass than a coarsely divided solid, will dissolve more rapidly.

2. Stirring or agitating the solution prevents a localized build-up of concentration to near saturation level at the solute surface, thereby minimizing the rate of crystallization. The net dissolving rate is therefore maximized.

3. At higher temperatures, particle movement is more rapid, thereby speeding up all physical processes.

15.4 FACTORS THAT DETERMINE SOLUBILITY

The extent to which a particular solute will dissolve in a given solvent is related to three factors: the strength of intermolecular forces between solvent molecules and solute molecules, the temperature and, in cases of gases dissolved in liquids, the partial pressure of the solute gas.

INTERMOLECULAR FORCES

> **PG 15 E** Given the structural formulas of two molecular substances, or other information from which the strength of their intermolecular forces may be estimated, predict if they will dissolve appreciably in each other, and state the criteria on which your prediction is based.

You saw in Chapter 14 that physical properties of a molecular substance usually may be associated with the intermolecular forces resulting from the geometry of its molecules. Solubility is among these properties. Generally speaking, *if the forces between molecules of substance A are roughly equal to the intermolecular forces of substance B, substances A and B will probably dissolve in each other.* From the standpoint of these forces, the molecules appear to be able to replace each other. On the other hand, if the intermolecular forces between solute molecules are quite different from the forces between solvent molecules, it is unlikely that the substances will dissolve in each other.

If we consider, for example, intermolecular attractions between such substances as hexane, C_6H_{14}, and decane, $C_{10}H_{22}$, we find that each substance has only relatively weak dispersion forces. These forces are roughly equal for the two substances, which are soluble in each other. Neither, however, is soluble in water or methanol, CH_3OH, two liquids that exhibit strong hydrogen bonding. But water and methanol are soluble in each other, again supporting the correlation between solubility and similarity of intermolecular forces.

TEMPERATURE

Temperature exerts a major influence on most chemical equilibria, including solution equilibria. Consequently solubility is temperature-dependent. Figure 15.4 indicates that solubility of most solids increases with rising temperature, but there are notable exceptions. The solubilities of gases in liquids, on the other hand, are generally lower at higher temperatures. The explanation of the interrelationship between temperature and solubility involves energy changes in the solution process, as well as other factors. We will look into this matter qualitatively in Chapter 16.

PRESSURE

PG 15 F Predict how the solubility of a gas in a liquid will be affected by a change in the partial pressure of that gas over the liquid.

Changes in partial pressure (p. 296) of a solute gas over a liquid solution have a pronounced effect on the solubility of the gas (Fig. 15.5). This is some-

Figure 15.4
Temperature-solubility curves for various salts in water. Note that the solubility of most salts increases with temperature, but there are exceptions.

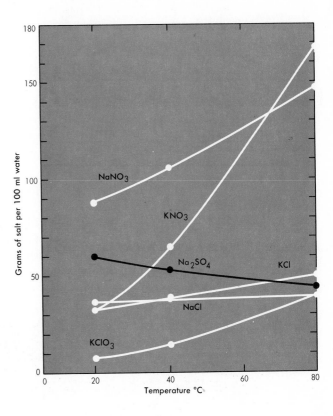

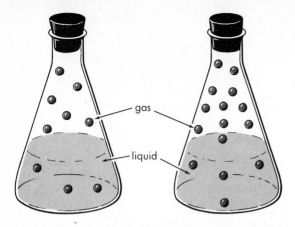

Figure 15.5
Effect of partial pressure of a gas on its solubility in a liquid. Solute gas concentration, and therefore its partial pressure, is lower in the left flask than in the right flask. Consequently the solute concentration in solution is lower in the left flask.

times startlingly apparent on opening a bottle or can of a carbonated beverage. Such beverages are bottled or canned under carbon dioxide partial pressure slightly greater than one atmosphere, which increases the solubility of the gas. This is what is meant by "carbonated." As the pressure is released on opening, solubility decreases, resulting in bubbles of carbon dioxide escaping from the solution.

In an "ideal" solution, the solubility of a gaseous solute in a liquid is directly proportional to the partial pressure of the gas over the surface of the liquid. A state of equilibrium is reached that is very similar to the vapor pressure equilibrium described by Figure 14.9 (p. 317) and the solid-in-liquid saturated solution shown in Figure 15.3. Neither the partial pressure nor the total pressure caused by other gases has a significant effect on the solubility of the solute gas. This is what would be expected for an ideal gas, where all molecules are widely separated and completely independent.

Pressure has little or no effect on the solubility of solids or liquids in a liquid solvent. None of the events described in Figure 15.3 are influenced by gas pressure above the liquid surface.

15.5 SOLUTION CONCENTRATION: PERCENTAGE BY WEIGHT

The concentration of a solution is ordinarily expressed in terms of amount of solute present in a given quantity of solvent or of total solution. Amount of solute may be given in grams or moles; quantity of solvent or solution may be stated in mass or volume units. In this and the next three sections we will examine four ways to express solution concentration. We begin with percentage by weight.

> **PG 15 G** Given grams of solute and grams of solvent or solution, calculate percentage concentration.
>
> **15 H** Given grams of solution and percentage concentration, calculate grams of solute and grams of solvent.

If the concentration of a solute in a solution is given in percent, you may assume it to be percent by weight unless specifically stated otherwise. As always, percent is $\dfrac{\text{part quantity}}{\text{total quantity}} \times 100$. Accordingly, the percentage by weight of solute is defined by the following equations:

$$\% \text{ by weight of solute} = \frac{\text{grams solute}}{\text{grams solution}} \times 100 \qquad (15.2)$$

$$= \frac{\text{grams solute}}{\text{grams solute} + \text{grams solvent}} \times 100 \qquad (15.3)$$

EXAMPLE 15.1   125 grams of solution, when evaporated to dryness, was found to contain 42.3 grams of solute. What was the percentage by weight of the solute?

This involves only a direct substitution into one of the defining equations.

$$\frac{\text{g solute}}{\text{g solution}} \times 100 = \frac{42.3}{125} \times 100 = 33.8\%$$

There are 42.3 grams of solute; the total quantity is 125 grams of solution.

EXAMPLE 15.2   3.50 grams of potassium nitrate are dissolved in 25.0 grams of water. Calculate the percentage concentration of potassium nitrate.

Again, one of the defining equations may be used.

$$\frac{\text{g solute}}{\text{g solute} + \text{g solvent}} \times 100 = \frac{3.50}{3.50 + 25.0} \times 100 = 12.3\%$$

EXAMPLE 15.3   You are to prepare 250 grams of 7.00% sodium carbonate solution. How many grams of sodium carbonate and how many milliliters of water do you use? (Recall that the density of water is 1.00 g/mL.)

You may approach this as a typical percentage problem, or from a dimensional analysis viewpoint, where percentage is grams of solute per 100 grams of solution. On this basis, a 7.00% solution means 7.00 grams of solute ≏ 100 grams of solution (p. 583).

$$250 \text{ g solution} \times \frac{7.00 \text{ g Na}_2\text{CO}_3}{100 \text{ g solution}} = 17.5 \text{ g Na}_2\text{CO}_3; \text{ or}$$

$$7.00\% \text{ of } 250 = 0.0700 \times 250 = 17.5 \text{ g Na}_2\text{CO}_3$$

grams water = grams solution − grams solute = 250 − 17.5 = 232 g water

At 1.00 g/mL, 232 grams of water = 232 milliliters.

15.6 SOLUTION CONCENTRATION: MOLALITY (OPTIONAL)

Many physical properties of solutions are related to the solution concentration expressed as **molality. Molality is the number of moles of solute dissolved in one kilogram of solvent.** Mathematically,

$$m = \frac{\text{moles of solute}}{\text{kilograms of solvent}} \tag{15.4}$$

where m is the symbol of molality. In essence, the molality of a solution is *the number of moles of solute that are equivalent to one kilogram of solvent.* In a 1.5 molal solution, for example, 1.5 moles of solute ≏ 1 kilogram of solvent.

The quantity of solvent in molality problems is usually given in *grams*, or in volume, which may be converted to grams by multiplying by density. Molality, however, is based on *kilograms* of solvent. To convert grams to kilograms you divide by 1000—or more simply, move the decimal three places to the left. It is convenient to do this mentally in the initial setup of the problem.

EXAMPLE 15.4   Calculate the molality of a solution prepared by dissolving 15.0 grams of sugar, $C_{12}H_{22}O_{11}$ (MW = 342 g/mole), in 350 milliliters of water.

Solution: The data of the problem may be interpreted as stating the concentration of the solution in grams of solute per milliliter of solvent: 15.0 grams solute per 350 mL H_2O. Because the density of water is 1.00 g/mL, the concentration is 15.0 grams solute per 350 g H_2O, or

$$\frac{15.0 \text{ g } C_{12}H_{22}O_{11}}{0.350 \text{ kg } H_2O}$$

Molality is moles of solute per kilogram of solvent. To convert the above expression to molality it is necessary only to change 15.0 grams of $C_{12}H_{22}O_{11}$ to moles:

$$\frac{15.0\ g\ C_{12}H_{22}O_{11}}{0.350\ kg\ H_2O} \times \frac{1\ mol\ C_{12}H_{22}O_{11}}{342\ g\ C_{12}H_{22}O_{11}} = 0.125\ m$$

EXAMPLE 15.5   How many grams of KCl (MW = 74.6 g/mole) must be dissolved in 250 grams of water to make a 0.400 m solution?

Solution: Molality provides a unit conversion between moles of solute and kilograms of solvent. Beginning with the given quantity, 250 grams of water, or 0.250 kilogram, the unit path becomes kilograms water → moles solute → grams solute, the second conversion being via molar weight. The entire setup is

$$0.250\ kg\ H_2O \times \frac{0.400\ mol\ KCl}{1\ kg\ H_2O} \times \frac{74.6\ g\ KCl}{1\ mol\ KCl} = 7.46\ g\ KCl$$

15.7 SOLUTION CONCENTRATION: MOLARITY

PG 15 I   Given two of the following, calculate the third: volume of solution, molarity, and moles (or grams with known or calculable molar weight) of solute.

Percentage and molality concentrations are both based on mass of solute and mass of solvent, each considered separately. In using liquids, volume is much more easily measured than mass. Therefore solution concentrations are frequently expressed in terms of their **molarity,** which relates to a certain volume of solution. Abbreviated M, molarity is defined as **moles of solute per liter of solution.** In the form of an equation,

$$M = \frac{moles\ solute}{liter\ solution} = \frac{mol}{L} \tag{15.5}$$

In essence, the molarity of a solution is *the number of moles of solute that are equivalent to one liter of solution.* In an X molar solution, X moles of solute are equivalent to one liter of solution:

$$X\ mol\ solute \simeq 1\ L\ solution \tag{15.6}$$

The preparation of a solution of known molarity is a frequent procedure in a laboratory. The volume of solution to be prepared is often expressed in milliliters. Because molarity is defined in liters, volume must be changed to match the larger unit. Conversion is simply division by 1000, which requires moving the decimal three places to the left. Thus, 750 milliliters is 0.750 liter:

$$750 \text{ mL} \times \frac{1 \text{ L}}{1000 \text{ mL}} = 0.750 \text{ L}$$

This milliliters-to-liters conversion will be performed without comment in the examples that follow.

EXAMPLE 15.6   How many grams of silver nitrate, $AgNO_3$ (MW = 170 g/mol), must be dissolved to prepare 500 milliliters of 0.150 M $AgNO_3$?

The given quantity is 500 milliliters. If this is expressed in liters, the unit path becomes liters → moles $AgNO_3$ → grams $AgNO_3$. Complete the problem.

$$0.500 \text{ L} \times \frac{0.150 \text{ mol } AgNO_3}{1 \text{ L}} \times \frac{170 \text{ g } AgNO_3}{1 \text{ mol } AgNO_3} = 12.8 \text{ g } AgNO_3$$

A clearer idea of the meaning of molarity may be gained by imagining the preparation of a solution of known molarity. Building on Example 15.6, the first step in the process is to weigh out 12.8 grams of silver nitrate (Fig. 15.6). This is transferred to a 500-mL volumetric flask containing *less than* 500 mL water. After dissolving the solute, water is added to the 500-milliliter mark on the neck of the flask. Notice the definition of molarity is based on the volume of *solu-*

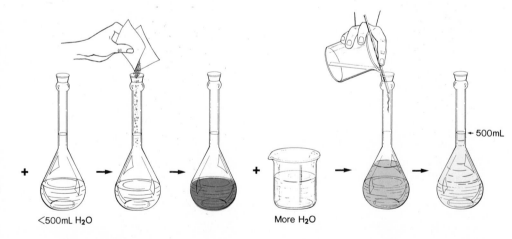

12.8 g AgNO$_3$

<500mL H$_2$O

More H$_2$O

500mL

Figure 15.6
Preparation of 500 mL 0.150 M AgNO$_3$.

tion, not the volume of *solvent*. This is why the solute is dissolved in less than 500 milliliters of water and then diluted to that volume. If it were dissolved *in* 500 milliliters of water, the final volume would be slightly more than 500 milliliters of solution.

EXAMPLE 15.7 15.8 grams of sodium hydroxide (MW = 40.0 g/mol) are dissolved in water and diluted to 100 milliliters. Calculate the molarity.

From the data presented, the concentration may be expressed in grams of solute per liter of solution. Write the concentration in these units.

$$\frac{15.8 \text{ g NaOH}}{100 \text{ mL solution}} = \frac{15.8 \text{ g NaOH}}{0.100 \text{ L solution}} \times \underline{\hspace{2cm}}$$

The only requirement for converting the above expression to molarity, or moles per liter, is to convert grams of solute to moles. Complete the problem.

$$\frac{15.8 \text{ g NaOH}}{0.100 \text{ L}} \times \frac{1 \text{ mol NaOH}}{40.0 \text{ g NaOH}} = 3.95 \text{ mol NaOH/L} = 3.95 \text{ M}$$

One of the more important functions of molarity is the conversion between volume of solution and moles of solute.

EXAMPLE 15.8 How many moles of solute are there in 45.3 mL 0.550 M solution?

From the given quantity it is a one-step conversion to moles. Set up and solve.

$$0.0453 \text{ L} \times \frac{0.550 \text{ mol}}{1 \text{ L}} = 0.0249 \text{ mol}$$

The reverse problem is also significant.

EXAMPLE 15.9   Find the number of milliliters of 1.40 M solution that contains 0.287 mole of solute.

Converting the given quantity to volume is easy enough, but be sure you answer the question in the units required.

- - - - - -   - - - - - -

$$0.287 \; \cancel{mol} \times \frac{1 \cancel{L}}{1.40 \; \cancel{mol}} \times \frac{1000 \; mL}{1 \cancel{L}} = 205 \; mL$$

This setup may be shortened by one step by recognizing that moles/liter is the same as moles/1000 milliliters:

$$0.287 \; \cancel{mol} \times \frac{1000 \; mL}{1.40 \; \cancel{mol}} = 205 \; mL$$

15.8 SOLUTION CONCENTRATION: NORMALITY (OPTIONAL)

PG 15 J   Define an equivalent of an acid or base.

15 K   Given an equation for an acid-base reaction, state the number of equivalents of acid or base per mole and calculate the equivalent weight of the acid or base.

15 L   Given an equation for an acid-base reaction and two of the following, calculate the third: volume of solution, normality, and equivalents of acid or base (or grams with known or calculable molar weight).

Normality is a particularly convenient way to express solution concentration in analytical work. Designated N, it is **the number of equivalents of solute per liter of solution.** Mathematically,

$$N = \frac{\text{equivalents solute}}{\text{liter solution}} = \frac{eq}{L} \tag{15.7}$$

In essence, the normality of a solution is the number of equivalents of solute that are the same as one liter of solution. In an X normal solution,

$$X \text{ equivalents of solute} \simeq 1 \text{ liter of solution} \tag{15.8}$$

The concept of normality leads immediately to the meaning of **equivalent. One equivalent of an acid is that quantity that yields one mole of hydrogen ions in a chemical reaction; one equivalent of a base is that quantity that reacts with one mole**

of hydrogen ions. Because hydrogen and hydroxide ions combine on a one-to-one basis, one mole of hydroxide ions is one equivalent. According to these definitions, both one mole of HCl and one mole of NaOH are each one equivalent because they yield, respectively, one mole of hydrogen ions and one mole of hydroxide ions in a reaction. H_2SO_4, on the other hand, has two equivalents per mole because each mole of the compound can release two moles of hydrogen ions. Similarly, one mole of $Al(OH)_3$ may represent three equivalents because three moles of hydroxide ion may react.

Notice that the number of equivalents in a mole of an acid or base depends upon a specific reaction, not simply upon the number of moles of hydrogen or hydroxide ions present in one mole of solute. The number of moles of these ions that actually react may or may not be the same as those present. For example, we might expect that phosphoric acid, H_3PO_4, has three equivalents per mole—and indeed it does in the reaction

$$H_3PO_4(aq) + 3\,NaOH(aq) \rightarrow Na_3PO_4(aq) + 3\,HOH(\ell) \qquad (15.9)$$

For reasons beyond the scope of this discussion, this reaction is difficult to perform quantitatively in the laboratory. It is possible, however, to control the reaction in a titration experiment (p. 352) so that 1.00 mole of H_3PO_4 reacts with 1.00 mole of NaOH in

$$H_3PO_4(aq) + NaOH(aq) \rightarrow NaH_2PO_4(aq) + HOH(\ell) \qquad (15.10)$$

In Equation 15.10, one mole of phosphoric acid yields one mole of hydrogen ions to react with one mole of sodium hydroxide. This is one equivalent of sodium hydroxide. Phosphoric acid has one equivalent per mole in Equation 15.10.

Under other conditions the phosphoric acid-sodium hydroxide reaction yields a one-to-two mole ratio between the acid and the base:

$$H_3PO_4(aq) + 2\,NaOH(aq) \rightarrow Na_2HPO_4(aq) + 2\,HOH(\ell) \qquad (15.11)$$

In this reaction one mole of phosphoric acid releases two moles of hydrogen ions, so there are two equivalents per mole.

EXAMPLE 15.10 For each of the following equations state the number of equivalents of acid and base per mole:

Eq. acid/mol   Eq. base/mol

$$2\,HBr + Ba(OH)_2 \rightarrow BaBr_2 + 2\,HOH$$

$$H_3C_6H_5O_7 + 2\,KOH \rightarrow K_2HC_6H_5O_7 + 2\,HOH$$

Remember, you are interested only in the number of moles of H^+ or OH^- that *react*, not the number present, in one mole of acid or base. The formula of citric acid, $H_3C_6H_5O_7$, is written as the inorganic chemist is most apt to write it, with three ionizable hydrogens first.

	Eq. acid/mol	Eq. base/mol
$2\,HBr + Ba(OH)_2 \rightarrow BaBr_2 + 2\,HOH$	1	2
$H_3C_6H_5O_7 + 2\,KOH \rightarrow K_2HC_6H_5O_7 + 2\,HOH$	2	1

In the first equation each mole of $Ba(OH)_2$ yields two OH^- ions, so there are two equivalents per mole. Each mole of HBr produces one H^+, so there is one equivalent per mole. In the second equation there could be one, two, or three equivalents per mole of $H_3C_6H_5O_7$, depending upon how many ionizable hydrogens are released in the reaction. That number is two: $H_3C_6H_5O_7 \rightarrow 2\,H^+ + HC_6H_5O_7{}^{2-}$. The two H^+ ions released combine with the two OH^- ions from two moles of KOH to form two HOH molecules. There are therefore two equivalents of acid per mole. In potassium hydroxide there is only one OH^- in a formula unit, so there can be only one equivalent per mole.

Example 15.10 illustrates an important fact we will identify here but not use until Section 15.11. In both reactions, notice that *the total number of equivalents of each reacting species is the same*. In the first reaction there is one equivalent for each of two moles of acid, or two equivalents of acid. There are also two equivalents of base, coming from one mole of $Ba(OH)_2$. In the second equation you again have two equivalents of acid, coming this time from one mole of acid, and two equivalents of base from two moles of base. The "same number of equivalents of all reactants" idea extends to the product species, too. Once you find the number of equivalents of one species in a reaction, you have the number of equivalents of *all* species. It is this fact that makes normality such a useful tool in quantitative work.

It is sometimes convenient to find the **equivalent weight** of a substance, **the number of grams per equivalent.** Equivalent weight, g/eq, is similar to molar weight, g/mol. Equivalent weight (EW) is readily calculated by dividing molar weight by equivalents per mole:

$$\frac{g/mol}{eq/mol} = \frac{g}{mol} \times \frac{mol}{eq} = g/eq \tag{15.12}$$

In ordinary acid-base reactions there are one, two, or three equivalents per mole. It follows that the equivalent weight of an acid or base is the same as, one half of, or one third of the molar weight. The molar weight of phosphoric acid is 98.0 g/mol. For the three reactions of phosphoric acid (Equations 15.9, 15.10, and 15.11) the equivalent weights are:

Equation 15.10: $\dfrac{98.0\ g\ H_3PO_4/mol}{1\ eq\ H_3PO_4/mol} = \dfrac{98.0\ g\ H_3PO_4}{1\ eq\ H_3PO_4} = 98.0\ g\ H_3PO_4/eq\ H_3PO_4$

Equation 15.11: $\dfrac{98.0\ g\ H_3PO_4/mol}{2\ eq\ H_3PO_4/mol} = \dfrac{98.0\ g\ H_3PO_4}{2\ eq\ H_3PO_4} = 49.0\ g\ H_3PO_4/eq\ H_3PO_4$

Equation 15.9: $\dfrac{98.0\ g\ H_3PO_4/mol}{3\ eq\ H_3PO_4/mol} = \dfrac{98.0\ g\ H_3PO_4}{3\ eq\ H_3PO_4} = 32.7\ g\ H_3PO_4/eq\ H_3PO_4$

EXAMPLE 15.11   Calculate the equivalent weight of KOH (56.1 g/mol), $Ba(OH)_2$ (171 g/mol), and $H_3C_6H_5O_7$ (192 g/mol) for the reactions in Example 15.10.

KOH $Ba(OH)_2$ $H_3C_6H_5O_7$

$$\text{KOH: } \frac{56.1 \text{ g KOH}}{1 \text{ eq KOH}} = 56.1 \text{ g KOH/eq; Ba(OH)}_2\text{: } \frac{171 \text{ g Ba(OH)}_2}{2 \text{ eq Ba(OH)}_2} = 85.5 \text{ g Ba(OH)}_2\text{/eq}$$

$$\text{H}_3\text{C}_6\text{H}_5\text{O}_7\text{: } \frac{192 \text{ g H}_3\text{C}_6\text{H}_5\text{O}_7}{2 \text{ eq H}_3\text{C}_6\text{H}_5\text{O}_7} = 96.0 \text{ g H}_3\text{C}_6\text{H}_5\text{O}_7\text{/eq}$$

Just as molar weight makes it possible to convert in either direction between grams and moles, equivalent weight sets the path between grams and equivalents. Equation 7.4 for molar weight (p. 138) has its counterpart for equivalent weight:

$$\text{EW grams} \simeq 1 \text{ equivalent} \tag{15.13}$$

In practice it is often more convenient to use the fractional form for equivalent weight—the molar weight over the number of equivalents per mole, which are boxed in above. We will use both setups in the next example, but only the fractional setup thereafter. If your instructor emphasizes equivalent weight as a quantity, you should, of course, follow those instructions.

EXAMPLE 15.12 Calculate the number of equivalents in 68.5 grams of Ba(OH)_2.

The numbers you need are in Example 15.11. Complete the problem.

_ _ _ _ _ _   _ _ _ _ _ _

Using equivalent weight, $68.5 \text{ g Ba(OH)}_2 \times \dfrac{1 \text{ eq Ba(OH)}_2}{85.5 \text{ g Ba(OH)}_2} = 0.801 \text{ eq Ba(OH)}_2$

Using the fractional setup, $68.5 \text{ g Ba(OH)}_2 \times \dfrac{2 \text{ eq Ba(OH)}_2}{171 \text{ g Ba(OH)}_2} = 0.801 \text{ eq Ba(OH)}_2$

You are now ready to use the equivalent concept in normality problems.

EXAMPLE 15.13 Calculate the normality of a solution prepared by dissolving 2.50 g NaOH in 500 mL of solution.

Solution: The data of the problem give the concentration in grams of solute per liter, g/L, after converting milliliters to liters. It must be changed to equivalents per liter, eq/L.

In other words, grams must be changed to equivalents. This is done by equivalent weight, g/eq:

$$\frac{2.50 \text{ g NaOH}}{0.500 \text{ L}} \times \frac{1 \text{ eq NaOH}}{40.0 \text{ g NaOH}} = 0.125 \text{ eq NaOH/L} = 0.125 \text{ N NaOH}$$

In essence, grams per liter has been divided by equivalent weight, grams per equivalent.

EXAMPLE 15.14 250 mL of a sulfuric acid solution contains 10.5 g H_2SO_4. Calculate its normality for the reaction $H_2SO_4 + 2\,NaOH \rightarrow Na_2SO_4 + 2\,HOH$.

The procedure is the same. Express the concentration in grams per liter and convert to equivalents per liter. In doing so you must determine the number of equivalents in one mole of sulfuric acid. Complete the problem.

$$\frac{10.5 \text{ g } H_2SO_4}{0.250 \text{ L}} \times \frac{2 \text{ eq } H_2SO_4}{98.1 \text{ g } H_2SO_4} = 0.856 \text{ eq } H_2SO_4/L = 0.856 \text{ N } H_2SO_4$$

The setup is the same as dividing the concentration in grams per liter by the equivalent weight of the acid:

$$\frac{10.5 \text{ g } H_2SO_4}{0.250 \text{ L}} \times \frac{1 \text{ eq } H_2SO_4}{49.1 \text{ g } H_2SO_4} = 0.856 \text{ eq } H_2SO_4/L = 0.856 \text{ N } H_2SO_4$$

Just as molarity provides a way to convert in either direction between moles of solute and volume of solution, normality offers a unit path between equivalents of solute and volume of solution. Equation 15.8 is the link.

EXAMPLE 15.15 How many equivalents are in 18.6 mL 0.856 N H_2SO_4?

The unit path is L → eq H_2SO_4. Set up and solve.

$$0.0186 \text{ L} \times \frac{0.856 \text{ eq } H_2SO_4}{1 \text{ L}} = 0.0159 \text{ eq } H_2SO_4$$

Example 15.14 presents an important relationship involving normality— the product of volume (L) times normality (eq/L) is equivalents of solute:

$$V \times N = \text{equivalents} \qquad (15.14)$$

We will use this relationship in the next section.

Naturally, it would be nice to know how to prepare a solution of specified normality.

EXAMPLE 15.16   How many grams of phosphoric acid must be used to prepare 100 mL 0.350 N H_3PO_4 to be used in the reaction $H_3PO_4 + NaOH \rightarrow NaH_2PO_4 + HOH$?

This problem may be solved by the conventional dimensional analysis approach. Your given quantity is volume of solution, which must be converted to grams of H_3PO_4. There is no direct conversion, but Equation 15.8 and Example 15.14 showed you how to get from liters to equivalents. Equivalent weight, grams per equivalent, works very well for the second step. The unit path is L $\rightarrow$ eq $H_3PO_4 \rightarrow$ g H_3PO_4.

$$0.100 \, \cancel{L} \times \frac{0.350 \, \cancel{\text{eq } H_3PO_4}}{1 \, \cancel{L}} \times \frac{98.0 \text{ g } H_3PO_4}{1 \, \cancel{\text{eq } H_3PO_4}} = 3.43 \text{ g } H_3PO_4$$

Only one of the three available hydrogen ions in phosphoric acid reacts, so there is only one equivalent per mole.

EXAMPLE 15.17   How many grams of oxalic acid, $H_2C_2O_4$, must you use to prepare 2.50 liters of 0.440 N solution for the reaction $H_2C_2O_4 + 2 \, KOH \rightarrow K_2C_2O_4 + 2 \, HOH$?

This problem is similar to Example 15.16. Set up and solve.

$$2.50 \, \cancel{L} \times \frac{0.440 \, \cancel{\text{eq } H_2C_2O_4}}{1 \, \cancel{L}} \times \frac{90.0 \text{ g } H_2C_2O_4}{2 \, \cancel{\text{eq } H_2C_2O_4}} = 49.5 \text{ g } H_2C_2O_4$$

The equation shows that both hydrogens in $H_2C_2O_4$ react, so there are two equivalents per mole.

15.9 SOLUTION STOICHIOMETRY

PG 15 M   Given the quantity of any species participating in a chemical reaction for which the equation may be written, find the quantity of any other species, either quantity being measured in (a) grams, (b) volume of gas at specified temperature and pressure, or (c) volume of solution at specified molarity.

The pattern for solving stoichiometry problems was introduced in Section 9.2. It is repeated here for ready reference:

1. Convert the quantity of given species to moles.
2. Convert the moles of given species to moles of wanted species.
3. Convert the moles of wanted species to the quantity of units required.

Equation 9.2 (p. 183) established a unit path matching the stoichiometric pattern for quantities measured in grams. It was expanded to include volumes of gases at specified temperature and pressure in Equation 13.24 (p. 288). In Examples 15.8 and 15.9 you learned how to convert in either direction between volume of solution of known molarity and moles of solute. You now have three ways to perform Steps 1 and 3 of the stoichiometric pattern. They are summarized in this unit path equation:

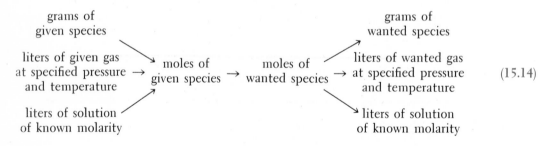

$$(15.14)$$

By the methods used earlier you should be able to calculate from any quantity on the left to any quantity on the right.

EXAMPLE 15.18   How many grams of lead(II) iodide will be precipitated from the addition of excess potassium iodide solution to 50.0 mL 1.22 M $Pb(NO_3)_2$? The equation is:

$$Pb(NO_3)_2(aq) + 2\ KI(aq) \rightarrow PbI_2(s) + 2\ KNO_3(aq)$$

Begin with the given quantity from the statement of the problem. The unit path is liters given → moles given (Ex. 15.8) → moles wanted → grams wanted. Set up the first step to moles of given species, but do not solve.

$$0.0500\,\cancel{L} \times \frac{1.22 \text{ mol Pb(NO}_3)_2}{1\,\cancel{L}} \times \underline{}$$

The remainder of the problem is like the first stoichiometry problems: convert moles of given to moles of wanted, and then to grams. Complete the setup and the solution.

$$0.0500\,\cancel{L} \times \frac{1.22 \text{ } \cancel{\text{mol Pb(NO}_3)_2}}{1\,\cancel{L}} \times \frac{1 \text{ } \cancel{\text{mol PbI}_2}}{1 \text{ } \cancel{\text{mol Pb(NO}_3)_2}} \times \frac{461 \text{ g PbI}_2}{1 \text{ } \cancel{\text{mol PbI}_2}} = 28.1 \text{ g PbI}_2$$

EXAMPLE 15.19 Calculate the number of milliliters of 0.842 M NaOH that will be required to precipitate as $Cu(OH)_2$ all the copper in 30.0 mL 0.635 M $CuSO_4$. The equation is

$$2\,NaOH(aq) + CuSO_4(aq) \rightarrow Cu(OH)_2(s) + Na_2SO_4(aq)$$

The first two steps in the unit path this time are as in the last example: L given → mol given → mol wanted. Set up that far, but do not solve.

$$0.0300\,\cancel{L} \times \frac{0.635 \text{ } \cancel{\text{mol CuSO}_4}}{1\,\cancel{L}} \times \frac{2 \text{ mol NaOH}}{1 \text{ } \cancel{\text{mol CuSO}_4}} \times \underline{}$$

At this point you have the number of moles of sodium hydroxide required. It is "packaged" at 0.842 mol/L, or 0.842 mol/1000 mL. How many such "packages" are required to obtain the number of moles indicated by the above setup? Recall Example 15.9, in which you performed the identical operation. Complete the setup and solve the problem—remembering, of course, that the answer is required in milliliters.

$$0.0300\,\cancel{L} \times \frac{0.635 \text{ } \cancel{\text{mol CuSO}_4}}{1\,\cancel{L}} \times \frac{2 \text{ } \cancel{\text{mol NaOH}}}{1 \text{ } \cancel{\text{mol CuSO}_4}} \times \frac{1\,\cancel{L}}{0.842 \text{ } \cancel{\text{mol NaOH}}} \times \frac{1000 \text{ mL}}{1\,\cancel{L}} = 45.2 \text{ mL NaOH}$$

Alternatively, using the direct 0.842 mol NaOH/1000 mL conversion, as in Example 15.9:

$$0.0300 \cancel{L} \times \frac{0.635 \cancel{\text{mol CuSO}_4}}{1 \cancel{L}} \times \frac{2 \cancel{\text{mol NaOH}}}{1 \cancel{\text{mol CuSO}_4}} \times \frac{1000 \text{ mL}}{0.842 \cancel{\text{mol NaOH}}} = 45.2 \text{ mL NaOH}$$

EXAMPLE 15.20   How many liters of dry hydrogen, measured at STP, will be released by the complete reaction of 45.0 mL 0.486 M H_2SO_4 with excess granular zinc? The equation is

$$Zn(s) + H_2SO_4(aq) \rightarrow ZnSO_4(aq) + H_2(g)$$

Recalling that the molar volume of any ideal gas at STP is 22.4 liters per mole, set up and solve this problem completely.

$$0.0450 \cancel{L} \times \frac{0.486 \cancel{\text{mol H}_2\text{SO}_4}}{1 \cancel{L}} \times \frac{1 \cancel{\text{mol H}_2}}{1 \cancel{\text{mol H}_2\text{SO}_4}} \times \frac{22.4 \text{ L H}_2}{1 \cancel{\text{mol H}_2}} = 0.490 \text{ L H}_2$$

15.10 TITRATION USING MOLARITY

PG 15 N   Given the volume of a solution that reacts with a known mass of a primary standard and the equation for the reaction, calculate the molarity of the solution.

15 O   Given the volumes of two solutions that react with each other in a titration, the molarity of one solution, and the equation for the reaction, calculate the molarity of the second solution.

One of the more important laboratory operations in analytical chemistry is called **titration**. Titration is the very careful addition of one solution into another by means of a buret (Fig. 15.7). The buret accurately measures the volume of a solution required to react with a certain quantity of another dissolved substance. When precisely that volume has been reached, a substance known as an **indicator** changes color, and the operator stops the flow from the buret. Phenolphthalein is a typical indicator for acid-base titrations. It is colorless in an acid solution and pink in a basic solution.

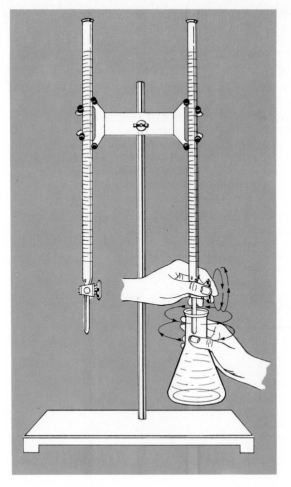

Figure 15.7
Titration from a buret into a flask.

Titration can be used to **standardize** a solution, which means that its precise concentration for use in later titrations is found. Sodium hydroxide cannot be weighed accurately because it absorbs moisture from the air and increases in weight during the weighing process. Therefore, it is not possible to prepare a sodium hydroxide solution whose molarity is known precisely. Instead the solution is standardized against a weighed quantity of something that can be weighed accurately. Such a substance is called a **primary standard.** Oxalic acid, $H_2C_2O_4$, is commonly used.* When used to standardize sodium hydroxide the equation is

$$H_2C_2O_4(s) + 2\,NaOH(aq) \rightarrow Na_2C_2O_4(aq) + 2\,HOH(\ell)$$

*Actually oxalic acid dihydrate. $H_2C_2O_4 \cdot 2\,H_2O$, is used, but to avoid confusion we will assume that the anhydrous compound is suitable.

The following example illustrates the standardization process.

EXAMPLE 15.21   0.839 g $H_2C_2O_4$ is dissolved in water, and the solution is titrated with a solution of NaOH of unknown concentration. 28.3 mL NaOH(aq) are required to neutralize the acid. Calculate the molarity of the NaOH.

Solution: Your goal in this problem is to find the moles of NaOH per liter of solution. The reaction involves 28.3 mL, or 0.0283 L. If you can find the number of moles in 0.0283 L, all you have to do is to divide that number of moles by 0.0283 to get molarity.

How do you find the number of moles of NaOH in 0.0283 L? Whatever that number is, you know it is the number that reacts with 0.839 g $H_2C_2O_4$. We need a unit path from grams of $H_2C_2O_4$ to moles of NaOH. This path is the first two steps of the stoichiometric pattern: g $H_2C_2O_4 \rightarrow$ mol $H_2C_2O_4 \rightarrow$ mol NaOH. The setup is

$$0.839 \text{ g } H_2C_2O_4 \times \frac{1 \text{ mol } H_2C_2O_4}{90.0 \text{ g } H_2C_2O_4} \times \frac{2 \text{ mol NaOH}}{1 \text{ mol } H_2C_2O_4} \times \underline{\hspace{2cm}}$$

Now that you have moles of NaOH, all that remains is to divide by volume, 0.0283 L. To divide by 0.0283 L you multiply by its inverse:

$$0.839 \text{ g } H_2C_2O_4 \times \frac{1 \text{ mol } H_2C_2O_4}{90.0 \text{ g } H_2C_2O_4} \times \frac{2 \text{ mol NaOH}}{1 \text{ mol } H_2C_2O_4} \times \frac{1}{0.0283 \text{ L}} = 0.659 \text{ M NaOH}$$

Now that you have seen the procedure, try a similar problem.

EXAMPLE 15.22   A potassium hydroxide solution is standardized by titrating against sulfamic acid, HSO_3NH_2 (97.1 g/mol). The equation is

$$HSO_3NH_2 + KOH \rightarrow KSO_3NH_2 + HOH$$

34.2 mL of solution are required to neutralize 0.395 g HSO_3NH_2. Find the molarity of the KOH.

$$0.395 \text{ g } HSO_3NH_2 \times \frac{1 \text{ mol } HSO_3NH_2}{97.1 \text{ g } HSO_3NH_2} \times \frac{1 \text{ mol KOH}}{1 \text{ mol } HSO_3NH_2} \times \frac{1}{0.0342 \text{ L}} = 0.119 \text{ M KOH}$$

Once a solution is standardized it may be used to find the concentration of other solutions. This is a widely used procedure in industrial laboratories.

EXAMPLE 15.23 25.0 milliliters of an electroplating solution are analyzed for the sulfuric acid concentration. 46.8 mL 0.659 M NaOH are required to neutralize the sample. Calculate the molarity of H_2SO_4 in the bath. The equation is

$$2\,NaOH + H_2SO_4 \rightarrow Na_2SO_4 + 2\,HOH.$$

Again you are seeking a molarity in moles of sulfuric acid per liter. If you can find the number of moles of acid in the 25.0-mL (0.0250 L) sample you can divide moles by liters to find concentration. The problem appears to have two given quantities, both volumes of solution. Only one, 46.8 mL NaOH, is accompanied by a molarity that permits you to convert to moles. This is your starting point. Think in terms of the unit path from L NaOH to mol H_2SO_4, the first two steps of the stoichiometric pattern. Set up the problem that far, but do not calculate the answer.

$$0.0468\,\cancel{L} \times \frac{0.659\,\cancel{\text{mol NaOH}}}{1\,\cancel{L}} \times \frac{1\,\text{mol}\,H_2SO_4}{2\,\cancel{\text{mol NaOH}}} \times \underline{\hspace{3cm}}$$

The unit path is L NaOH → mol NaOH → mol H_2SO_4.

The final step is to divide by the volume of solution that has the above number of moles of sulfuric acid, as in the last two examples. Complete the problem.

$$0.0468\,\cancel{L} \times \frac{0.659\,\cancel{\text{mol NaOH}}}{1\,\cancel{L}} \times \frac{1\,\text{mol}\,H_2SO_4}{2\,\cancel{\text{mol NaOH}}} \times \frac{1}{0.0250\,\text{L}} = 0.617\;M\;H_2SO_4$$

15.11 TITRATION USING NORMALITY (OPTIONAL)

PG 15 P Given the volume of a solution that reacts with a known mass of a primary standard and the equation for the reaction, calculate the normality of the solution.

15 Q Given the volumes of two solutions that react with each other in a titration, the normality of one solution, and the equation for the reaction, calculate the normality of the second solution.

In Section 15.8 it was noted that normality is a convenient concentration unit in analytical work. This is particularly true in commercial laboratories where the same titration is performed again and again. The laboratory operations are the same, but the calculations are somewhat different. To illustrate this we will repeat Examples 15.21 and 15.23, using the same titration data with equivalents and normality. In solving these problems, recall the important fact that was pointed out on page 346: the number of equivalents of all species in a reaction is the same. If you find the number of equivalents of one substance, you have found the number of equivalents of all species.

EXAMPLE 15.24   0.839 g $H_2C_2O_4$ is dissolved in water and the solution is titrated with a solution of NaOH of unknown concentration. 28.3 mL NaOH(aq) are required to neutralize the acid. Calculate the normality of the NaOH for the reaction.

$$H_2C_2O_4 + 2\,NaOH \rightarrow Na_2C_2O_4 + 2\,HOH$$

Your goal this time is to find the number of equivalents of NaOH per liter. Find first the number of equivalents of NaOH in 0.0283 liters. Then divide the equivalents by the volume and you have normality.

The quantity you know about is 0.839 g $H_2C_2O_4$. Look at the equation and determine the number of equivalents of acid per mole. Put these together and calculate the number of equivalents of $H_2C_2O_4$ in 0.839 grams. Try to figure it out without looking back, but if you have trouble check Example 15.12, p. 347.

- - - - - -   - - - - - -

$$0.839 \text{ g } H_2C_2O_4 \times \frac{2 \text{ eq } H_2C_2O_4}{90.0 \text{ g } H_2C_2O_4} = 0.0186 \text{ eq } H_2C_2O_4$$

The equation shows that each mole of $H_2C_2O_4$ releases two moles of H^+, so oxalic acid has two equivalents per mole.

How many equivalents of sodium hydroxide react with 0.0186 eq $H_2C_2O_4$?

- - - - - -   - - - - - -

0.0186 eq NaOH

Once you have found the equivalents of one species in a reaction you have found the equivalents of all species.

Dividing equivalents by liters gives normality. Complete the problem.

- - - - - -   - - - - - -

$$\frac{0.0186 \text{ eq NaOH}}{0.0283 \text{ L}} = 0.657 \text{ eq NaOH/L} = 0.657 \text{ N NaOH}$$

The answer to Example 15.21 was 0.659 M NaOH. The molarity and normality should be the same when there is one equivalent per mole. The difference is caused in the rounding off that occurred in calculating the number of equivalents. The same problem in a single calculation setup is

$$0.839 \text{ g } H_2C_2O_4 \times \frac{2 \text{ eq } H_2C_2O_4}{90.0 \text{ g } H_2C_2O_4} \times \frac{1 \text{ eq NaOH}}{1 \text{ eq } H_2C_2O_4} \times \frac{1}{0.0283 \text{ L}} = 0.659 \text{ N NaOH}$$

Compare this setup with the setup for Example 15.21. The numbers are the same, but the arrangement and units differ slightly.

Once you have the normality of a solution, you can use it to find the normality of other solutions. Equation 15.13 (p. 347) indicates that the number of equivalents of a species in a reaction is the product of the solution volume times normality. If the number of equivalents of all species in the reaction is the same, then

$$V_1N_1 = V_2N_2 \tag{15.15}$$

where subscripts 1 and 2 identify the reacting solutions. Solving for the second normality,

$$N_2 = \frac{V_1N_1}{V_2} \tag{15.16}$$

That's all it takes to calculate normality in this follow-up to Example 15.23.

EXAMPLE 15.25 25.0 milliliters of an electroplating solution are analyzed for its sulfuric acid concentration. 46.8 mL 0.659 N NaOH are required to neutralize the sample. Calculate the normality of H_2SO_4 in the bath.

Substitute directly into Equation 15.16 and solve. Omit units. They can be included, but they do not make the problem easier and the setup becomes needlessly cluttered.

$$N_2 = \frac{46.8 \times 0.659}{25.0} = 1.23 \text{ N } H_2SO_4$$

Notice that volume in milliliters was used in the setup. There is a volume factor in both the numerator and denominator of Equation 15.16, so the conversion to liters cancels.

The normality of sulfuric acid in Example 15.25 is twice the molarity calculated in Example 15.23. This is as it should be, since

$$\frac{0.617 \text{ mol } H_2SO_4}{1 \text{ L}} \times \frac{2 \text{ eq } H_2SO_4}{1 \text{ mol } H_2SO_4} = 1.23 \text{ eq } H_2SO_4/L = 1.23 \text{ N } H_2SO_4$$

15.12 COLLIGATIVE PROPERTIES OF SOLUTIONS (OPTIONAL)

A pure solvent has certain distinct, definite physical properties, as does any pure substance. The introduction of a solute into the solvent affects these properties. The properties of the solution, a mixture, depend upon the relative quantities of solvent and solute. It has been found experimentally that, in *dilute* solutions of certain solutes, the *change* in some of these properties is proportional to the molal concentration of the solute particles. The proportionality constant is independent of the solute; it is a property of the solvent. Solution properties that are determined solely by the *number* of solute particles dissolved in a fixed quantity of solvent are called **colligative properties.**

Freezing and boiling points of solutions are colligative properties. Perhaps the most common example is the antifreeze used in the cooling systems of automobiles. The solute dissolved in the radiator water reduces the freezing temperature to a level well below the normal freezing point of pure water, and also raises the boiling point above the normal boiling point.

The mathematical relationship for the change in freezing point between a solution and a pure solvent is

$$\Delta T_f = K_f m \qquad (15.17)$$

where ΔT_f is the **freezing point depression,** as it is called, K_f is a proportionality constant known as the **molal freezing point constant,** and m is the molality of the solution. Similarly,

$$\Delta T_b = K_b m \qquad (15.18)$$

in which ΔT_b is the **boiling point elevation** and K_b is the **molal boiling point constant.** For water $K_f = -1.86$, and $K_b = 0.52$. Boiling and freezing point problems may be solved algebraically by direct substitution into one or the other of the above equations.

EXAMPLE 15.26   Determine the freezing point of a solution of 12.0 grams of urea, $CO(NH_2)_2$, in 250 grams of water.

Solution: To use Equation 15.17 it is necessary to express the solution concentration in molality:

$$\frac{12.0 \text{ g } CO(NH_2)_2}{0.250 \text{ kg } H_2O} \times \frac{1 \text{ mol } CO(NH_2)_2}{60.0 \text{ g } CO(NH_2)_2} = 0.800 \text{ m}$$

If $K_f = -1.86$ for water,

$$\Delta T_f = K_f m = -1.86 \times 0.800 = -1.49°C$$

This indicates the freezing point of the solution is 1.49°C below the normal freezing point of water, 0°C. The solution freezes at $-1.49°C$.

Notice that Equations 15.17 and 15.18 do not yield freezing or boiling points, but indicate *changes* in these properties. ΔT_f and the freezing point of the solution are the same in Example 15.26 only because the solvent happens to freeze at 0°C, a convenience that will not be true for any solvent other than water. Notice also that we have not used units in the usual way. They can be included, but they are awkward and tend to confuse the calculation rather than aid it.

CHAPTER 15 IN REVIEW

15.1 THE CHARACTERISTICS OF A SOLUTION

15 A Distinguish among a pure substance, a solution, a suspension, and a colloid. (331)

15.2 SOLUTION TERMINOLOGY

15 B Distinguish between terms in the following groups: solute and solvent; concentrated and dilute; solubility, saturated, unsaturated, and supersaturated; miscible and immiscible. (332)

15.3 THE FORMATION OF A SOLUTION

15 C Describe the formation of a saturated solution from the time excess solid solute is first placed into a liquid solvent. (334)

15 D Identify and explain the factors that determine the time required to dissolve a given amount of solute, or to reach equilibrium. (334)

15.4 FACTORS THAT DETERMINE SOLUBILITY

15 E Given the structural formulas of two molecular substances, or other information from which the strength of their intermolecular forces may be estimated, predict if they will dissolve appreciably in each other and state the criteria on which your prediction is based. (336)

15 F Predict how the solubility of a gas in a liquid will be affected by a change in the partial pressure of that gas over the liquid. (337)

15.5 SOLUTION CONCENTRATION: PERCENTAGE BY WEIGHT

15 G Given grams of solute and grams of solvent or solution, calculate percentage concentration. (338)

15 H Given grams of solution and percentage concentration, calculate grams of solute and grams of solvent. (338)

15.6 SOLUTION CONCENTRATION: MOLALITY (OPTIONAL) (340)

15.7 SOLUTION CONCENTRATION: MOLARITY

15 I Given two of the following, calculate the third: volume of solution, molarity, and moles (or grams with known or calculable molar weight) of solute. (341)

15.8 SOLUTION CONCENTRATION: NORMALITY (OPTIONAL)

15 J Define an equivalent of an acid or a base. (344)

15 K Given an equation for an acid-base reaction, state the number of equivalents of acid or base per mole and calculate the equivalent weight of the acid or the base. (344)

15 L Given an equation for an acid-base reaction and two of the following, calculate the third: volume of solution, normality, and equivalents of acid or base (or grams with known or calculable molar weight). (344)

15.9 SOLUTION STOICHIOMETRY

15 M Given the quantity of any species participating in a chemical reaction for which the equation may be written, find the quantity of any other species, either quantity being measured in (a) grams, (b) volume of gas at specified temperature and pressure, or (c) volume of solution at specified molarity. (350)

15.10 TITRATION USING MOLARITY

15 N Given the volume of a solution that reacts with a known mass of a primary standard and the equation for the reaction, calculate the molarity of the solution. (352)

15 O Given the volumes of two solutions that react with each other in a titration, the molarity of one solution, and the equation for the reaction, calculate the molarity of the second solution. (352)

15.11 TITRATION USING NORMALITY (OPTIONAL)

15 P Given the volume of a solution that reacts with a known mass of a primary standard and the equation for the reaction, calculate the normality of the solution. (355)

15 Q Given the volumes of two solutions that react with each other in a titration, the normality of one solution, and the equation for the reaction, calculate the normality of the second solution. (355)

15.12 COLLIGATIVE PROPERTIES OF SOLUTIONS (OPTIONAL) (358)

TERMS AND CONCEPTS

Note: Terms or concepts in italics are from optional sections in the text.

Solution (331)
Suspension (332)
Colloid (332)
Solute (332)
Solvent (332)
Concentrated (333)
Dilute (333)
Solubility (333)
Saturated (333)
Unsaturated (333)
Supersaturated (334)
Miscible, immiscible (334)
Hydrated (334)
Percent concentration (338)

Molality (340)
Molarity (341)
Normality (344)
Equivalent (344)
Equivalent weight (346)
Titration (352)
Indicator (352)
Standardize (353)
Primary standard (353)
Colligative property (358)
Freezing point depression (358)
Molal freezing point constant (358)
Boiling point elevation (358)
Molal boiling point constant (358)

Most of these terms and many others appear in the Glossary.

QUESTIONS AND PROBLEMS

An asterisk (*) identifies a question that is relatively difficult, or that extends beyond the performance goals of the chapter.

Section 15.1

15.1) As you dissolve sugar in clear water, you can obtain a suspension, a colloid, or a solution, depending upon when you make your observation. Explain.

15.56) "Mixtures of gases are always true solutions." True or false? Explain why.

Section 15.2

15.2) Explain why the distinction between solute and solvent is not clearly defined in many solutions.

15.3) Solution A contains 10 grams of solute dissolved in 100 grams of solvent, while solution B has only 5 grams of a different solute per 100 grams of solvent. Under what circumstances can solution A be classified as dilute and solution B as concentrated?

15.4) Suggest simple laboratory tests by which you could determine if a solution is unsaturated, saturated, or supersaturated. Explain why your suggestions would distinguish between the different classifications.

15.57) Identify the *solute* and the *solvent* in each of the following solutions: (a) salt water [NaCl(aq)]; (b) sterling silver (92.5% Ag, 7.5% Cu); (c) air (about 80% N_2, 20% O_2).

15.58) Would it be proper to say that a saturated solution is a concentrated solution? or that a concentrated solution is a saturated solution? Point out the distinctions between these sometimes confused terms.

15.59) What happens if you add a very small amount of solid salt (NaCl) to each of the beakers described below? Include a statement about the *amount* of solid eventually found in the beaker, compared with the amount you added: (a) a beaker containing *saturated* NaCl solution; (b) a beaker with *unsaturated* NaCl solution; (c) a beaker containing supersaturated NaCl solution.

15.5) Suggest units in which solubility might be expressed other than grams per 100 grams of solvent.

15.6) Contrast the terms miscibility, miscible, and immiscible with their counterparts, solubility, soluble, and insoluble.

Section 15.3

15.7) Describe the forces that promote the dissolving of a solid solute in a liquid solvent.

15.8) "A dynamic equilibrium exists when a saturated solution is in contact with excess solute." Explain the meaning of that statement. Is it possible to have a saturated solution *without* excess solute? Explain.

15.9) Why is it customary to stir coffee or tea after putting sugar into it?

Section 15.4

15.10) Would water or carbon tetrachloride be a better solvent for benzene, C_6H_6? Why? The structural formula of benzene may be represented as

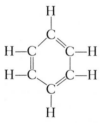

15.11) Suppose you have a spot on some clothing, and water will not take it out. If you have available cyclopentane and methanol (see structures below), which would you choose as the more promising solvent to try? Why?

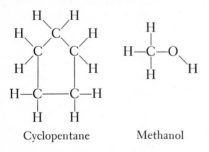

Cyclopentane   Methanol

15.60) In stating solubility, an important variable must be specified. What is that variable, and how does solubility of a solid solute *usually* depend upon it?

15.61) Given an example of two immiscible substances other than oil and water.

15.62) Describe the forces that oppose the dissolving of a solute in a liquid solvent.

15.63) Bakers use confectioner's sugar because it is more finely powdered than the crystals of table (granulated) sugar. Do you think confectioner's sugar would dissolve more or less quickly than table sugar? Why?

15.64) Explain the effect of heating on the rate at which a solid dissolves in a liquid.

15.65) Suggest why water and liquid HF are good solvents for many ionic salts, but not for waxes and oils having structures such as

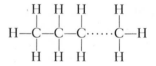

15.66) Glycerin and normal hexane (see structures below) are organic compounds of approximately the same molecular weight. Which of these is more apt to be miscible with carbon tetrabromide? Why?

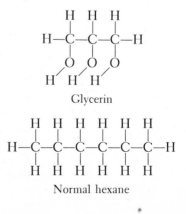

Glycerin

Normal hexane

15.12) "The solubility of carbon dioxide in a carbonated beverage may be increased by raising the air pressure under which it is bottled." Criticize the foregoing statement.

15.67) On opening a bottle of carbonated beverage, many bubbles are released, suggesting that the beverage is bottled under high pressure. Yet for safety reasons the pressure cannot be more than slightly greater than one atmosphere. How, then, do you account for the substantial reduction in CO_2 solubility in an opened bottle?

Section 15.5

15.13) A sodium chloride solution weighing 32.4 grams was carefully evaporated to dryness. 2.78 grams of NaCl were recovered. Calculate the percentage NaCl in the original solution.

15.68) Calculate the concentration in percent by weight if a solution is prepared by dissolving 4.23 grams of silver nitrate in 78.4 grams of water.

15.14) How many grams of boric acid are required to prepare 75.0 grams of 4.00% solution?

15.69) Calculate the mass of magnesium sulfate and the milliliters of water required to prepare 245 grams of 10.6% solution.

Section 15.6

15.15)* Calculate the molality of the solution prepared by dissolving 16.9 grams of glucose, $C_6H_{12}O_6$, a type of sugar, in 87.5 grams of water.

15.70) 24.6 grams of ethanol, C_2H_5OH, the alcohol of distilled spirits fame, are dissolved in 55.5 grams of water. Find the molality of the solution.

15.16)* Find the molality of a 10% NaCl solution. (Hint: Determine first the grams of salt and grams of water in a definite quantity—say 100 grams—of solution.)

15.71) Calculate the molality of the solution described in Problem 15.69.

15.17)* How many grams of formic acid, HCOOH—a compound first formed by distilling, of all things, red ants, and a source of irritation from their bites—must be dissolved in 50.0 milliliters of water to produce a 1.50 m solution?

15.72) Calculate the weight of malonic acid, $C_3H_4O_4$, that must be dissolved in 0.600 liter of water to produce a 0.75 m solution.

Section 15.7

15.18) How many grams of silver nitrate, widely used in the manufacture of photographic chemicals, must be dissolved in the preparation of 150 mL of 0.125 M solution?

15.73) Calculate the mass of potassium hydroxide required for the preparation of 250 mL of a 2.50 M solution.

15.19) Oxalic acid, a compound used to remove iron stain from fabrics and porcelain, and as a bleach for leather goods, is a dihydrate with the formula $H_2C_2O_4 \cdot 2\,H_2O$. Determine the weight of oxalic acid required for 7.50×10^2 mL of a 0.480 M solution.

15.74) How many grams of $NiCl_2 \cdot 6\,H_2O$, widely used in nickel plating automotive hardware, are dissolved in 0.250 liter of a 1.12 M solution?

15.20) 16.2 grams of ammonium sulfate, a compound sold at local garden centers for lawn fertilizer, are dissolved in water and diluted to 300.0 mL. Find the molarity of the solution.

15.75) 25.0 mL of a solution of potassium bromide, used in manufacturing photographic paper and film, were carefully evaporated to dryness. 2.35 grams of solute were recovered. What was the molarity of the initial solution?

15.21) Calculate the molarity of the solution prepared by dissolving 56.2 g $CuSO_4 \cdot 5\,H_2O$, a compound used in electroforming printing plates, in 500.0 mL of solution.

15.76) A solution is prepared by dissolving 5.11 grams of potassium chromate, K_2CrO_4, in 150 mL of water and then diluting to 250.0 mL. Find the molarity.

15.22)* The density of 22.0% $NH_4Cl(aq)$ is 1.06 g/mL. Calculate the molarity.

15.23) How many moles of potassium nitrate, a fertilizer and food preservative, are in 45.1 mL of 0.378 M KNO_3?

15.24) Concentrated sulfuric acid is 18 molar. What volume of concentrated acid is required to obtain 1.24 moles of H_2SO_4?

15.25)* 150 mL 0.633 M $(NH_4)_2CO_3$ are diluted to 450 mL. Calculate the molarity of the diluted solution.

15.26)* How many milliliters of concentrated sulfuric acid (18.0 M) must be used to prepare 500 mL 0.450 M H_2SO_4?

15.77)* The density of 2.70 M NH_4NO_3 is 1.09 g/mL. Find its percentage concentration.

15.78) Calculate the moles of ammonia present in 8.00×10^2 mL of 2.25 M NH_3.

15.79) How many milliliters of 0.642 M NaOH are necessary to obtain 0.0395 mole of sodium hydroxide?

15.80)* What is the molarity of the solution produced by diluting 88 mL 0.756 M KOH to 650 mL?

15.81)* What volume of commercial HCl, which is 12 M, must be diluted to 500 mL to produce a 2.5 molar solution?

Section 15.8

15.27) Explain why the number of equivalents in a mole of acid or base is not always the same.

15.28) 65.6 grams of KOH are dissolved in water and diluted to 1.50 liters. Find the normality of the solution.

15.29) What is the number of equivalents per mole of citric acid, $H_3C_6H_5O_7$ (192 g/mol) in

$$H_3C_6H_5O_7(s) + 3\,NaOH(\ell) \rightarrow$$
$$Na_3C_6H_5O_7(aq) + 3\,HOH(\ell)?$$

What is the equivalent weight of the acid?

15.30) What is the equivalent weight of the base in Problem 15.29?

15.31) 11.9 g H_3PO_4 are dissolved in 100 mL solution and used in the reaction

$$H_3PO_4(aq) + 2\,NaOH(aq) \rightarrow$$
$$2\,HOH(\ell) + NaH_2PO_4(aq)$$

What is the normality of the phosphoric acid?

15.32) How many equivalents per mole of sulfuric acid are used in the reaction

$$H_2SO_4(aq) + Na_3PO_4(aq) \rightarrow$$
$$Na_2SO_4(aq) + NaH_2PO_4(aq)$$

Find the equivalent weight of the acid.

15.33) Find the equivalent weight of the base, Na_3PO_4, in the reaction of Problem 15.32.

15.34) Find the normality of 0.284 M H_2SO_4 in the reaction of Problem 15.32.

15.35) Calculate the number of equivalents in 50.0 mL 0.114 N HCl.

15.82) What is equivalent weight? Why can you state positively the equivalent weight of LiOH, but not H_2SO_4?

15.83) What is the normality of a solution containing 10.2 grams of HCl per liter?

15.84) What is the number of equivalents per mole of oxalic acid, $H_2C_2O_4$ (90 g/mol), in

$$NaOH(aq) + H_2C_2O_4(aq) \rightarrow$$
$$NaHC_2O_4(aq) + HOH(\ell)$$

Calculate the equivalent weight of the acid.

15.85) What is the equivalent weight of the base in Problem 15.84?

15.86) What is the normality of 0.350 M $H_2C_2O_4$ if used in the reaction of Problem 15.84?

15.87) Calculate the number of equivalents per mole of HCl in the reaction

$$2\,HCl(aq) + Na_2CO_3(aq) \rightarrow$$
$$CO_2(g) + H_2O(\ell) + 2\,NaCl(aq)$$

Also calculate the equivalent weight of the acid.

15.88) Na_2CO_3 is the base in Problem 15.87. Calculate its equivalent weight.

15.89) Calculate the normality of a solution of 18.2 g Na_2CO_3 dissolved in water, diluted to 250 mL, and then used in the reaction of Problem 15.87.

15.90) How many equivalents are in 50.0 mL of 0.114 N H_2SO_4?

15.36) How many grams of HCl are required to prepare the solution in Problem 15.35?

15.37) How would you prepare 750 mL 0.250 N $H_2C_2O_4$ for the reaction

$$H_2C_2O_4(aq) + 2\,NaOH(aq) \rightarrow$$
$$Na_2C_2O_4(aq) + 2\,HOH(\ell)$$

15.38) What is the equivalent weight of $H_2C_2O_4$ in Problem 15.37?

15.39) What volume of 0.200 N H_3PO_4 represents 0.005 00 equivalents in the reaction

$$2\,NaOH(aq) + H_3PO_4(aq) \rightarrow$$
$$2\,HOH(\ell) + Na_2HPO_4(aq)$$

Section 15.9

15.40) Calculate the number of grams of barium chromate that can be precipitated by adding excess potassium chromate, K_2CrO_4, to 50.0 mL 0.424 M $BaCl_2$.

15.41) Find the number of milliliters of 0.246 M $AgNO_3$ required to precipitate as silver phosphate all the phosphate ion in a solution containing 2.10 grams of sodium phosphate.

15.42) How many milliliters of 6.2 M NaOH must react with aluminum to liberate 2.4 liters of hydrogen, measured at STP? The equation is

$$2\,Al(s) + 6\,NaOH(aq) \rightarrow$$
$$2\,Na_3AlO_3(aq) + 3\,H_2(g)$$

15.43)* How many grams of "milk of magnesia," $Mg(OH)_2$, will precipitate when 50.0 mL 0.240 M $MgCl_2$ are added to 50.0 mL 0.420 M NaOH?

15.44)* An industrial waste solution is essentially 1.13 M in HNO_3. 61.2 kilograms of $NaHCO_3$ are used to neutralize 140 gallons of the solution. How many liters of carbon dioxide, measured at

15.91) How many grams of H_2SO_4 are needed to prepare the solution in Problem 15.90 if it is used in the reaction of Problem 15.32?

15.92) How many grams of NaOH are needed to prepare 400 mL 0.550 N NaOH?

15.93) Calculate the equivalent weight of Na_2CO_3 in

$$Na_2CO_3(aq) + HC_2H_3O_2(aq) \rightarrow$$
$$NaHCO_3(aq) + NaC_2H_3O_2(aq)$$

15.94) What volume of 0.200 N H_3PO_4 represents 0.005 00 equivalents in the reaction

$$NaOH(aq) + H_3PO_4(aq) \rightarrow$$
$$HOH(\ell) + NaH_2PO_4(aq)$$

15.95) Excess sodium hydroxide solution is added to 20.0 mL 0.184 M $ZnCl_2$. Calculate the number of grams of zinc hydroxide that will precipitate.

15.96) In a reaction that produces hydrogen gas, how many milliliters of 0.569 M HCl are required to react with 0.104 gram of magnesium? How many milliliters of hydrogen, corrected to STP, will be released?

15.97) If you warm the vessel in which solutions of sodium sulfite and hydrochloric acid react in a controlled oxygen-free atmosphere, sulfur dioxide is driven off as a gas:

$$Na_2SO_3(aq) + 2\,HCl(aq) \rightarrow$$
$$2\,NaCl(aq) + H_2O(\ell) + SO_2(g)$$

What volume of SO_2, measured after the gas has been adjusted to STP, will be produced by the complete reaction of 35.0 mL 0.924 M Na_2SO_3?

15.98)* In an effort to recover dissolved silver salts—you may consider them to be $AgNO_3(aq)$ for the purpose of the problem—by the "salting out" process, a workman tosses a cupful of salt into a 25.0-liter recovery tank. If the cupful contained 425 grams of NaCl, and the solution was 0.12 molar in silver ion (or silver nitrate, if you wish), calculate the grams of silver chloride that will precipitate. Was one cupful of salt enough to precipitate all the silver ion, or should he have used more?

15.99)* The phosphate ion is to be precipitated from a 280-liter solution containing sodium and potassium phosphate and ammonium chloride. The phosphate ion concentration, expressed as

STP, will be released, assuming the process proceeds to the complete consumption of the limiting reagent in

$$NaHCO_3(s) + HNO_3(aq) \rightarrow$$
$$H_2O(\ell) + CO_2(g) + NaNO_3(aq)$$

Section 15.10

15.45) Potassium hydrogen phthalate, $KHC_8H_4O_4$ (204 g/mol), is used as a primary standard for bases. In one titration 5.34 grams of dissolved $KHC_8H_4O_4$ required 32.5 milliliters of a sodium carbonate solution:

$$2\, KHC_8H_4O_4(aq) + Na_2CO_3(aq) \rightarrow$$
$$2\, NaKC_8H_4O_4(aq) + H_2O(\ell) + CO_2(g)$$

Calculate the molarity of the sodium carbonate solution.

15.46) Find the molarity of a solution of hydrochloric acid if 16.8 mL are required to react with 10.0 mL 0.862 M NaOH in a titration experiment.

15.47) In a chemical analysis, 14.9 mL 0.518 M $AgNO_3$ are required to react with all of the nickel chloride in a 10.0-mL sample of a plating solution:

$$2\, AgNO_3(aq) + NiCl_2(aq) \rightarrow$$
$$2\, AgCl(s) + Ni(NO_3)_2(aq)$$

Find the molarity of the nickel chloride solution.

15.48)* Iron is present as Fe_2O_3 in a 0.504-g sample of iron ore. The iron is reduced to Fe^{2+} in what may be considered as a solution of $FeCl_2$. This solution is titrated with 34.2 mL 0.0271 M $KMnO_4$ in an acid solution:

$$KMnO_4(aq) + 8\, HCl(aq) + 5\, FeCl_2(aq) \rightarrow$$
$$MnCl_2(aq) + 4\, H_2O(\ell) + 5\, FeCl_3(aq) + KCl(aq)$$

Calculate the percentage of iron in the sample.

Section 15.11

15.49) 0.512 gram of oxalic acid, $H_2C_2O_4$, is dissolved in water. The solution is titrated with NaOH(aq). 16.2 milliliters of the base are used in the reaction

$$H_2C_2O_4(s) + 2\, NaOH(aq) \rightarrow$$
$$Na_2C_2O_4(aq) + 2\, HOH(\ell)$$

What is the normality of the base?

molarity of Na_3PO_4, is 0.704 M. If 55.2 liters of 3.50 M $Ca(NO_3)_2$ are added to the solution, how many kilograms of calcium phosphate will precipitate? Will all of the phosphate ion be precipitated? If not, how many moles of $PO_4{}^{3-}$ will remain in solution?

15.100) Sodium carbonate is used as a primary standard in determining the concentration of hydrochloric acid. If 0.254 g Na_2CO_3 require 32.4 mL HCl in standardization, find the molarity of the acid.

15.101) 21.6 mL 0.655 M NaOH are required to titrate a 25.0-mL sample of oxalic acid solution by the following reaction:

$$H_2C_2O_4(aq) + 2\, NaOH(aq) \rightarrow$$
$$Na_2C_2O_4(aq) + 2\, HOH(\ell)$$

Calculate the molarity of the oxalic acid.

15.102) 50.0 milliliters of a solution of silver nitrate require 28.6 mL 0.134 M KSCN in the titration

$$AgNO_3(aq) + KSCN(aq) \rightarrow$$
$$AgCl(s) + KNO_3(aq)$$

What is the molarity of the $AgNO_3$?

15.103)* 0.479 gram of impure $CaCO_3$ is dissolved in 50.0 mL 0.131 M HCl, which is in excess:

$$CaCO_3(s) + 2\, HCl \rightarrow$$
$$CaCl_2(aq) + CO_2(g) + H_2O(\ell)$$

The excess HCl is titrated with 6.8 mL 0.110 M NaOH. Calculate the percentage of $CaCO_3$ in the original sample. Assume that the sample contains nothing that reacts with HCl or NaOH. (Hint: Find the moles of HCl added, the moles of HCl in excess, and the moles of HCl that reacted with the $CaCO_3$ in the sample.)

15.104) $NaHCO_3$ is used as a primary standard in finding the normality of H_2SO_4 in the reaction

$$2\, NaHCO_3(s) + H_2SO_4(aq) \rightarrow$$
$$Na_2SO_4(aq) + 2\, CO_2(g) + 2\, HOH(\ell)$$

What is the normality if 0.618 g $NaHCO_3$ requires 20.6 mL H_2SO_4?

15.50) Calculate the normality of a NaOH solution if 21.9 mL are required to titrate 25.0 mL 0.324 N H_2SO_4.

15.105) 25.0 milliliters of an HCl solution of unknown concentration require 31.4 mL 0.372 N Na_2CO_3 in the titration

2 HCl(aq) + Na_2CO_3(aq) →
2 NaCl(aq) + CO_2(g) + HOH(ℓ)

Find the normality of the acid.

15.51) Find the equivalent weight of an unknown base if 0.452 gram of it requires 31.8 mL 0.169 N HCl for neutralization.

15.106) A solution known to contain 0.268 gram of H_3PO_4 requires 18.9 mL 0.289 N NaOH in a titration reaction. (a) Calculate the equivalent weight of the H_3PO_4. (b) Write the equation for the reaction.

Section 15.12

15.52) The specific gravity of any solution of NaCl is greater than 1.00. The specific gravity of any solution of NH_3 is less than 1.00. Is specific gravity a colligative property? Why or why not?

15.107) Is the partial pressure exerted by one component of a gaseous mixture at a given temperature and volume a colligative property? Justify your answer, pointing out in the process what classifies a property as "colligative."

15.53) What will be the boiling temperature of a solution of 96.1 grams of ethylene glycol, $C_2H_6O_2$ (permanent antifreeze), in 100 grams of water? The molal boiling point constant for water is 0.52.

15.108) Calculate the freezing point of a solution of 2.16 grams of naphthalene (moth balls), $C_{10}H_8$, in 31.0 grams of benzene. Pure benzene freezes at 5.50°C, and its molal freezing point constant is −5.10.

15.54) The normal freezing point of naphthalene is 80.2°C, and $K_f = -6.9$. Calculate the freezing point of the solution formed when 4.34 grams of paradichlorobenzene, $C_6H_4Cl_2$, are dissolved in 70.0 grams of naphthalene.

15.109) Determine the freezing point of 0.65 m $C_3H_8O_3$, an aqueous solution of glycerol, a substance used in making cosmetics and nitroglycerin. $K_f = -1.86$ for water.

15.55) Calculate the molal concentration of an aqueous solution that boils at 100.89°C. $K_b = 0.52$ for water.

15.110) What is the molality of a solution of an unknown solute in acetic acid if the solution freezes at 13.4°C? The normal freezing point of acetic acid is 16.6°C, and $K_f = -3.90$.

16

reactions that occur in water solutions; net ionic equations

In Chapter 8 you wrote chemical equations for some reactions that occur in water solutions. At that time you were promised a description of these reactions in the form of net ionic equations. That promise will be fulfilled in this chapter.

Net ionic equations give us an understanding of what really happens when solutions react. To write a net ionic equation we must identify precisely what is in a solution, the exact nature of the tiniest solute particle. Our study therefore begins by examining the electrical properties of solutions and asking what makes them what they are.

16.1 ELECTROLYTES AND SOLUTION CONDUCTIVITY

PG 16 A Distinguish among strong electrolytes, weak electrolytes, and nonelectrolytes.

PG 16 B Describe or explain electrical conductivity through a solution.

Suppose two metal strips, called **electrodes**, are placed in a liquid and wired to a battery and an electric light bulb (Fig. 16.1). To light the bulb a

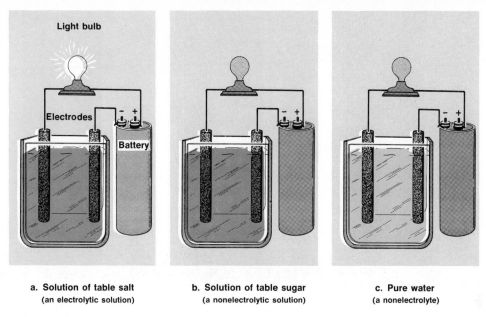

a. **Solution of table salt**
(an electrolytic solution)

b. **Solution of table sugar**
(a nonelectrolytic solution)

c. **Pure water**
(a nonelectrolyte)

Figure 16.1
Electrolytes and nonelectrolytes. A, NaCl solution conducts electricity, as shown by the glowing light bulb. NaCl is an electrolyte. B, Sugar solution does not conduct electricity, so the bulb does not glow. Sugar is a nonelectrolyte. C, Pure water does not conduct electricity; it is a nonelectrolyte. Therefore the light bulb does not glow.

current must pass through the liquid from one metal strip to the other. If the liquid is pure water, no appreciable current passes through; the bulb does not light up. If the electrodes are in a solution of sodium chloride, the bulb glows brightly—almost as much as if the metal strips were "shorted" with a screwdriver. A salt solution is an excellent **conductor** of electricity. A solution of sugar produces no such effect; like pure water, a sugar solution is a **nonconductor**.

Passage of electricity through a liquid is called **electrolysis**. A solute that yields a solution that is a good conductor is called a **strong electrolyte**. Sugar, whose solution is a nonconductor, is a **nonelectrolyte**. Some solutes, such as acetic acid, are **weak electrolytes**. Their solutions conduct electricity, but poorly, permitting only a dim glow of the lamp in Figure 16.1. The term *electrolyte* is also used to refer to the *solution* through which the current passes. The acid solution in an automobile storage battery is an electrolyte in this sense. In commercial electroplating, *electrolyte* is almost always used in this way.

What is called an electric *current* is a movement of electric charge. In a metal wire it is a movement of electrons that makes up a current. In a solution the charge is carried by ions rather than electrons. When an ionic compound dissolves, its positive ions (cations) and negative ions (anions) are free to move. When a pair of electrodes is connected to a source of direct current, such as a storage battery, one electrode, called an **anode**, has a positive charge, and the

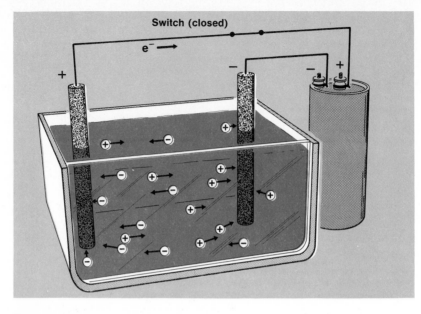

Figure 16.2
Conductivity in an ionic solution. The conductivity of a solution is positive evidence that mobile ions are present.

other, the **cathode,** gains a negative charge. The *a*node attracts the *a*nions in the solution, while the *ca*tions move toward the *ca*thode (Fig. 16.2). This movement of ions is the electric current that flows through the solution. *Solution conductivity is positive evidence of the presence of ions.*

When sugar, a nonelectrolyte, dissolves, the crystal breaks into individual sugar molecules. These molecules are electrically neutral and do not move toward either electrode. Even if there were movement there would be no current, because the molecules have no charge. If the solute is a weak electrolyte, such as acetic acid, only a small fraction of the molecules separate into ions. This allows a weak current flow, which accounts for the dim glow of the bulb in Figure 16.1. The large percentage of solute particles in a weak electrolyte are unionized (un-ionized, not union-ized) molecules.

The equations that describe solution reactions include the ions and/or molecules that are the actual solute particles in the solution. These particles must be identified by their chemical formulas. A list of the particles in any solution is its **solution inventory.** You will learn how to write solution inventories for different kinds of solutes in the next two sections.

16.2 SOLUTION INVENTORIES OF IONIC COMPOUNDS

PG 16 C Given the formula of an ionic compound, write the solution inventory when it is dissolved in water.

If an ionic compound dissolves, its solution inventory always consists of ions. These ions are identified simply by separating the compound into its ions. When sodium chloride dissolves, the solution inventory is sodium ions and chloride ions:

$$NaCl(s) \xrightarrow{H_2O} Na^+(aq) + Cl^-(aq) \qquad (16.1)$$

If the solute is barium chloride, the inventory is barium ions and chloride ions:

$$BaCl_2(s) \xrightarrow{H_2O} Ba^{2+}(aq) + 2\ Cl^-(aq) \qquad (16.2)$$

Notice that no matter where it comes from, the formula of the chloride ion is *always* Cl^-, never Cl_2^- or Cl_2^{2-}. The subscript after an ion in a formula, as the 2 in $BaCl_2$, tells us how many ions are present in the formula unit. The subscript is not part of the ion formula.

In this chapter you should include state symbols for all species in equations. It helps in writing net ionic equations. Also remember to include the charge every time you write the formula of an ion.

EXAMPLE 16.1 Write solution inventories for the following ionic compounds by writing their dissolving equations.

$$NaOH(s) \xrightarrow{H_2O} \quad Na^+ + OH^-$$

$$K_2SO_4(s) \xrightarrow{H_2O} \quad 2K^+ + SO_4^{2-}$$

$$(NH_4)_2CO_3(s) \xrightarrow{H_2O} \quad 2NH_4^+ + CO_3^{2-}$$

- - - - - -   - - - - - -

$$NaOH(s) \xrightarrow{H_2O} Na^+(aq) + OH^-(aq)$$

$$K_2SO_4(s) \xrightarrow{H_2O} 2\ K^+(aq) + SO_4^{2-}(aq)$$

$$(NH_4)_2CO_3(s) \xrightarrow{H_2O} 2\ NH_4^+(aq) + CO_3^{2-}(aq)$$

Polyatomic and monatomic ions are handled in exactly the same way in writing solution inventories. In $(NH_4)_2CO_3$, notice that the subscript outside the parentheses tells us how many ammonium ions are present, while the subscript 4 inside the parentheses is part of the polyatomic ion formula.

EXAMPLE 16.2 Write the solution inventories of the following ionic solutes *without* writing the dissolving equations.

$$MgSO_4 \qquad Mg^{2+} + SO_4^{2-} \qquad\qquad AlBr_3 \qquad Al^{3+} + 3Br^-$$

$$Ca(NO_3)_2 \qquad Ca^{2+} + 2NO_3^- \qquad\qquad Fe_2(SO_4)_3 \quad 2\ Fe^{3+} + 3\ SO_4^{2-}$$

This question is the same as Example 16.1. It asks simply for the products of the reaction without writing an equation. This is how you will write these inventories later in the chapter. Be sure to show the number of each kind of ion released by a formula unit—the coefficient if the equation were written. Also remember state symbols.

$MgSO_4$: $Mg^{2+}(aq) + SO_4^{2-}(aq)$ $AlBr_3$: $Al^{3+}(aq) + 3\ Br^-(aq)$

$Ca(NO_3)_2$: $Ca^{2+}(aq) + 2\ NO_3^-(aq)$ $Fe_2(SO_4)_3$: $2\ Fe^{3+}(aq) + 3\ SO_4^{2-}(aq)$

16.3 STRONG ACIDS AND WEAK ACIDS

PG 16 D Explain why the solution of an acid may be a good conductor or a poor conductor of electricity.

16 E Given the formula of a soluble acid, write the solution inventory when it is dissolved in water.

To write the solution inventory of an acid, you must be able to do two things: first, recognize the compound as an acid, and second, classify it as a strong acid or weak acid. An acid, as we have used the term so far, is a hydrogen-bearing compound that releases hydrogen ions in water solution. Its formula has the form HX, H_2X, or H_3X, where X represents anything that becomes an anion when the acid ionizes. In HCl, X is Cl^-; in H_2SO_4, X is SO_4^{2-}. Notice that the anion may or may not contain oxygen. Even hydrogen may be present; acetic acid, $HC_2H_3O_2$, for example, yields $C_2H_3O_2^-$, the acetate ion. Do not be concerned if the anion is unfamiliar. It will behave in exactly the same way as the ion from any other acid, and should be treated accordingly.

Hydrochloric acid is the water solution of hydrogen chloride, a gaseous molecular compound. Hydrochloric acid is a **strong acid.** This means it is almost completely ionized in water, and is an excellent conductor. The ionization equation (p. 111) is

$$HCl(g) \xrightarrow{\ H_2O\ } HCl(aq) \rightarrow H^+(aq) + Cl^-(aq) \tag{16.3}$$

Hydrofluoric acid is the water solution of hydrogen fluoride, another gaseous molecular compound. Hydrofluoric acid is a **weak acid.** It is only slightly ionized in water, and is a poor conductor. The ionization process may be described by an equation similar to that for hydrochloric acid, but with an important difference:

$$HF(g) \xrightarrow{\ H_2O\ } HF(aq) \xrightleftharpoons{\longrightarrow} H^+(aq) + F^-(aq) \tag{16.4}$$

The double arrow shows that the ionization process is reversible (p. 318). Its greater length from right to left says it is much more likely for hydrogen and fluoride ions to form dissolved hydrogen fluoride molecules than for the molecules to form ions. In other words, the dissolved particles in hydrofluoric acid are mostly unionized HF molecules, and few ions. This is why the solution is a poor conductor.

In describing acids, the words *strong* and *weak* are used exactly the same way they are used in describing electrolytes. A strong acid is almost completely ionized in water, and a weak acid is ionized only slightly. *The solution inventory of a strong acid is the ions it forms; the solution inventory of a weak acid is the unionized molecule.*

Now, how do you determine if an acid is strong or weak? The answer is, by knowing the acids that are strong. We will consider only six:

HCl	hydrochloric acid	HNO_3	nitric acid
HBr	hydrobromic acid	H_2SO_4	sulfuric acid
HI	hydroiodic acid	$HClO_4$	perchloric acid

These acids are easily memorized. HCl, HNO_3 and H_2SO_4 are the three best known acids. HBr and HI are, like HCl, made up of hydrogen and a halogen. $HClO_4$ is not widely discussed in an introductory course; it simply must be remembered as a strong acid. In classifying an acid, you first check to see if it is one of the strong acids. If not, it must be weak.*

EXAMPLE 16.3 Write the solution inventories of the following acids:

HNO_2 $\xrightarrow{H_2O}$ $H^+ + NO_2^-$ HI

$H_2SO_4 \longrightarrow$ $HC_3H_5O_2$

You could write these inventories with equations, as in Example 16.1. It is better to just identify the ions—or molecules—as in Example 16.2. If a molecule produces more than one ion of a kind, show it. Use state symbols.

HNO_2: $HNO_2(aq)$ HI: $H^+(aq) + I^-(aq)$
H_2SO_4: $2\,H^+(aq) + SO_4^{2-}(aq)$ $HC_3H_5O_2$: $HC_3H_5O_2(aq)$

*The dividing line between strong and weak acids is arbitrary, and at least three acids are marginal in their classifications. Sulfuric acid is definitely strong in its first ionization step, but questionable in the second. In all reactions in this book the second ionization does occur, so we regard it as a strong acid that releases both hydrogen ions. Both oxalic acid, $H_2C_2O_4$, and phosphoric acid, H_3PO_4, are marginal in their first ionizations, but definitely weak in the second and, in the case of phosphoric acid, third. We regard them as weak, but we will avoid questions in which they might have to be classified.

HNO$_2$ and HC$_3$H$_5$O$_2$ are not among the six strong acids, so their solution inventories are unionized molecules. HI and H$_2$SO$_4$ are strong acids; they break up into ions. A common mistake with H$_2$SO$_4$ is to write H$_2$$^+$(aq) or something like that for the hydrogen ion, H$^+$(aq). Like the chloride ion in Equations 16.1 and 16.2, the hydrogen ion has the same formula no matter where it comes from.

16.4 NET IONIC EQUATIONS: WHAT THEY ARE AND HOW TO WRITE THEM

If a solution of lead nitrate is added to a solution of sodium chloride, lead chloride precipitates. In Chapter 8 you learned how to write the equation for a precipitation reaction. In this case it is

$$Pb(NO_3)_2(aq) + 2\,NaCl(aq) \rightarrow PbCl_2(s) + 2\,NaNO_3(aq) \qquad (16.5)$$

In this chapter we will identify this kind of equation as a *conventional* equation.

A conventional equation serves many useful purposes, including its essential role in solving stoichiometry problems. For a reaction that occurs in water solutions, however, it is not always satisfactory. Usually it does not identify the reactants and/or products correctly, and it generally does not tell exactly what chemical changes take place. A conventional equation has no value when it comes to solving chemical equilibrium problems, such as those in Chapter 20 of this book.

To illustrate the shortcomings of a conventional equation, the lead nitrate solution in the above reaction contains no substance whose formula is Pb(NO$_3$)$_2$. Actually present in that solution are the solution inventory species, lead ion, Pb^{2+}(aq), and nitrate ion, NO$_3$$^-$(aq). Similarly, the sodium chloride solution really contains Na$^+$(aq) and Cl$^-$(aq). The reaction does not produce anything whose formula is NaNO$_3$. The ions, Na$^+$(aq) and NO$_3$$^-$(aq), are present in the solution inventory of the combined solutions. The only substance in Equation 16.5 that is *really there* is solid lead chloride, PbCl$_2$(s). If you perform the reaction you can see the precipitate.

To write an equation that describes the above reaction more accurately, the formulas of the dissolved substances are replaced with their solution inventories. This produces the **ionic equation:**

$$Pb^{2+}(aq) + 2\,NO_3^-(aq) + 2\,Na^+(aq) + 2\,Cl^-(aq) \rightarrow$$
$$PbCl_2(s) + 2\,Na^+(aq) + 2\,NO_3^-(aq) \qquad (16.6)$$

Notice that, in order to keep the ionic equation balanced, the coefficients from the conventional equation are repeated. One formula unit of Pb(NO$_3$)$_2$ gives one Pb^{2+} ion and two NO$_3$$^-$ ions, two formula units of NaCl give two Na$^+$ ions and two Cl$^-$ ions, and two formula units of NaNO$_3$ give two Na$^+$ ions and two NO$_3$$^-$ ions.

An ionic equation tells more than just what happens in a chemical change. It includes **spectator ions,** or simply **spectators.** A spectator is an ion that is present at the scene of a reaction, but experiences no chemical change. It appears on both sides of the ionic equation. $Na^+(aq)$ and $NO_3^-(aq)$ are spectators in Equation 16.6. To change an ionic equation into a **net ionic equation** you simply remove the spectators:

$$Pb^{2+}(aq) + 2\,Cl^-(aq) \rightarrow PbCl_2(s) \qquad\qquad (16.7)$$

A net ionic equation tells exactly what chemical change took place, and nothing else.

In Equations 16.5 to 16.7 you have the three steps to be followed in writing a net ionic equation. They are:

1. Write the conventional equation, including designations of state—(g), (ℓ), (s), or (aq). Balance the equation.

2. Write the ionic equation by replacing each *dissolved* substance, designated (aq) in the conventional equation, with its solution inventory species. Never change solids (s), liquids (ℓ), or gases (g) in this step. Be sure the equation is balanced in both atoms and charge. (Charge balance is discussed in the next section.)

3. Write the net ionic equation by removing the spectators from the ionic equations. Reduce coefficients to lowest terms, if necessary. Be sure the equation is balanced in both atoms and charge.

16.5 REDOX REACTIONS THAT ARE DESCRIBED BY "SINGLE REPLACEMENT" EQUATIONS

PG 16 F   Given two substances that may engage in a redox reaction and an activity series by which the reaction may be predicted, write the conventional, ionic, and net ionic equations for the reaction that will occur, if any.

A piece of zinc is dropped into sulfuric acid. Hydrogen gas bubbles out vigorously. When the reaction ends, the vessel contains a solution of zinc sulfate. The question is, what happened?

To "answer" the question by the conventional equation (Step 1):

$$Zn(s) + H_2SO_4(aq) \rightarrow H_2(g) + ZnSO_4(aq)$$

But this doesn't really answer the question. There were no H_2SO_4 molecules in the sulfuric acid. Instead, each H_2SO_4 molecule released two $H^+(aq)$ ions and one $SO_4^{2-}(aq)$ ion. At the end of the reaction there were no white $ZnSO_4$ crystals present. There was only a clear colorless solution of $Zn^{2+}(aq)$ and $SO_4^{2-}(aq)$ ions. So the ionic equation (Step 2), in which the dissolved substances in the conventional equation are replaced by their solution inventories, provides a better explanation of what happened:

$$Zn(s) + 2\,H^+(aq) + SO_4{}^{2-} \rightarrow H_2(g) + Zn^{2+} + SO_4{}^{2-} \qquad (16.8)$$

The ionic equation remains balanced in both atoms and charge. The net charge is zero on both sides.

At this point the answer to what happened is cluttered with something that was present, but was no part of the chemical change. The sulfate ion, $SO_4{}^{2-}$, appears on both sides of the ionic equation. It is a spectator. It may be taken away (Step 3):

$$Zn(s) + 2\,H^+(aq) \rightarrow H_2(g) + Zn^{2+}(aq) \qquad (16.9)$$

This is the net ionic equation. This is what happened—and no more.

Notice that all equations are balanced.* You have been balancing atoms since Chapter 8, but charge is something new. Neither protons nor electrons are created or destroyed in a chemical change, so the total charge among the reactants must be equal to the total charge among the products. In Equation 16.8, two plus charges in $2\,H^+(aq)$ added to two negative charges in $SO_4{}^{2-}(aq)$ give a net zero charge on the left. Two plus charges in $Zn^{2+}(aq)$ and two negative charges in $SO_4{}^{2-}(aq)$ on the right also total zero. The equation is balanced in charge. Equation 16.9 is also balanced, this time with a net charge of $+2$ on each side. Notice that the net charge does not have to be zero for the equation to be balanced.

EXAMPLE 16.4 A reaction occurs when a piece of zinc is dipped into $Cu(NO_3)_2(aq)$. Write the conventional, ionic, and net ionic equations.

Begin by writing the conventional equation. Do you know what the products are? What might they be? You know that the equation for this kind of redox reaction is a single replacement type (p. 166). Something must replace something else in a compound. One other hint—back on page 169 is the statement, "In solution reactions, any ions not otherwise accounted for may be considered as a compound in solution." Now, the conventional equation, please. Don't forget the state designations.

----   ----

$$Zn(s) + Cu(NO_3)_2(aq) \rightarrow Cu(s) + Zn(NO_3)_2(aq)$$

In the single replacement equation, zinc appears to replace copper in $Cu(NO_3)_2$.

Now write the ionic equation, replacing formulas of dissolved substances with their solution inventories.

---

*If the purpose is simply to write the net ionic equation, it is not necessary to balance the conventional and ionic equations. Most instructors, however, recommend balancing all three equations, both in atoms and charge.

$$Zn(s) + Cu^{2+}(aq) + 2 NO_3^-(aq) \rightarrow Cu(s) + Zn^{2+}(aq) + 2 NO_3^-(aq)$$

The ionic equation will always be balanced if the solution inventories are written with coefficients to show how many ions come from all formula units. You should always check, though. The total charge is zero on both sides.

Examine the ionic equation. Are there any spectators? If so, remove them and write the net ionic equation.

$$Zn(s) + Cu^{2+}(aq) \rightarrow Cu(s) + Zn^{2+}(aq)$$

If you were to place a strip of copper into a solution of zinc nitrate and go through the identical thought process, the equations would be exactly the reverse of those in Example 16.4. Does the reaction occur in both directions? If not, in which way does it occur? And how can you tell?

Technically both reactions occur, or can be made to occur. Practically, only the reaction in Example 16.4 takes place. The best way to find out which of two reversible reactions occurs is to try them and see. These experiments have been done, and the results are summarized in the **activity series** in Table 16.1. In solution, under normal conditions, any element in the table will replace the ions of any element beneath it. Zinc is above copper in the table, so zinc will replace $Cu^{2+}(aq)$ ions in solution. Copper, being below zinc, will not replace $Zn^{2+}(aq)$ in solution. You may use Table 16.1 to predict whether or not a redox reaction will take place. If asked to write the equation for a reaction that does not occur, write NR for "no reaction" on the product side: $Cu(s) + Zn^{2+}(aq) \rightarrow NR$.

Table 16.1

Activity Series

Li
K
Ba
Ca
Na
Mg
Al
Zn
Ni
Pb
H_2
Cu
Ag

EXAMPLE 16.5   Write the conventional, ionic, and net ionic equations for the reaction that will occur, if any, between calcium and hydrochloric acid.

First, will a reaction occur? Check the activity series; then answer.

Yes. Calcium is above hydrogen in the series, so calcium will replace hydrogen ions from solution.

Write the conventional equation. Remember the state designation.

-------   -------

$$Ca(s) + 2\,HCl(aq) \rightarrow H_2(g) + CaCl_2(aq)$$

Now write the ionic equation.

-------   -------

$$Ca(s) + 2\,H^+(aq) + 2\,Cl^-(aq) \rightarrow H_2(g) + Ca^{2+} + 2\,Cl^-(aq)$$

Each $HCl(aq)$ yields one $H^+(aq)$ and one $Cl^-(aq)$. The conventional equation has two $HCl(aq)$, so there will be $2\,H^+(aq)$ and $2\,Cl^-(aq)$ in the ionic equation.

Now eliminate the spectators and write the net ionic equation.

-------   -------

$$Ca(s) + 2\,H^+(aq) \rightarrow H_2(g) + Ca^{2+}(aq)$$

Chloride ion, $Cl^-(aq)$, is the only spectator.

EXAMPLE 16.6 Write the three equations for the reaction between nickel and a solution of $MgCl_2$.

See what you can do this time without hints.

$$Ni(s) + MgCl_2(aq) \rightarrow NR$$

Nickel is beneath magnesium in the activity series, so no redox reaction occurs.

EXAMPLE 16.7 Copper is placed into a solution of silver nitrate. Write the three equations.

-------   -------

$$Cu(s) + 2\,AgNO_3(aq) \rightarrow 2\,Ag(s) + Cu(NO_3)_2(aq)$$

$$Cu(s) + 2\,Ag^+(aq) + 2\,NO_3^-(aq) \rightarrow 2\,Ag(s) + Cu^{2+}(aq) + 2\,NO_3^-(aq)$$

$$Cu(s) + 2\,Ag^+(aq) \rightarrow 2\,Ag(s) + Cu^{2+}(aq)$$

EXAMPLE 16.8 If potassium is placed into water, hydrogen gas bubbles out. Write all three equations.

This reaction is more clearly seen if you write the formula of water as HOH. Treat it as a weak acid, with only the first hydrogen ionizable. Go for all three equations, but watch those state symbols.

$$2\,K(s) + 2\,HOH(\ell) \rightarrow H_2(g) + 2\,KOH(aq)$$

$$2\,K(s) + 2\,HOH(\ell) \rightarrow H_2(g) + 2\,K^+(aq) + 2\,OH^-(aq)$$

Water is a liquid molecular compound, $HOH(\ell)$. It does not break into ions. Never separate into ions a gas (g), liquid (ℓ), or solid (s). This particular ionic equation has no spectators, so it is also the net ionic equation.

16.6 ION COMBINATIONS THAT FORM PRECIPITATES

PG 16 G Predict whether or not a precipitate will form when known solutions are combined; if a precipitate forms, write the net ionic equation. (Reference to a solubility table may or may not be allowed.)

PG 16 H Given the product of a precipitation reaction, write the net ionic equation.

An **ion combination reaction** occurs when the cation from one reactant combines with the anion from another to form a particular kind of product compound. The conventional equation is a double displacement type in which the ions appear to "change partners" (p. 169): $MY + NX \rightarrow MX + NY$. In this section the product is an insoluble ionic compound that settles to the bottom of the mixed solutions. A solid formed this way is called a **precipitate;** the reaction is a **precipitation reaction.**

EXAMPLE 16.9 When hydrochloric acid and silver nitrate solutions are mixed, a white precipitate of silver chloride is produced. Develop the net ionic equation for the reaction.

The same three steps you used on redox reactions are applied to precipitation reactions. Write all three equations.

––––––   ––––––

$HCl(aq) + AgNO_3(aq) \rightarrow HNO_3(aq) + AgCl(s)$

$H^+(aq) + Cl^-(aq) + Ag^+(aq) + NO_3^-(aq) \rightarrow H^+(aq) + NO_3^-(aq) + AgCl(s)$

$Ag^+(aq) + Cl^-(aq) \rightarrow AgCl(s)$

$H^+(aq)$ and $NO_3^-(aq)$ are spectators in the ionic equation.

Let's try another that has two interesting features at the end.

EXAMPLE 16.10 When solutions of silver chlorate, $AgClO_3(aq)$, and aluminum chloride, $AlCl_3(aq)$, are combined, silver chloride precipitates. Write the three equations.

Proceed as usual. When you get to the net ionic equation, which will have something you haven't seen so far, see if you can figure out what to do about it. If necessary, read Step 3 of the procedure on page 375.

––––––   ––––––

$3\,AgClO_3(aq) + AlCl_3(aq) \rightarrow 3\,AgCl(s) + Al(ClO_3)_3(aq)$

$3\,Ag^+(aq) + 3\,ClO_3^-(aq) + Al^{3+}(aq) + 3\,Cl^-(aq) \rightarrow 3\,AgCl(s) + Al^{3+}(aq) + 3\,ClO_3^-(aq)$

$3\,Ag^+(aq) + 3\,Cl^-(aq) \rightarrow 3\,AgCl(s);$ or $Ag^+(aq) + Cl^-(aq) \rightarrow AgCl(s)$

The net ionic equation this time has 3 for the coefficient of all species. The third step of the procedure says to reduce coefficients to lowest terms. The equation may be divided by 3, as shown. That leads to the second interesting feature.

Examples 16.9 and 16.10 produced the same net ionic equation even though the reacting solutions were completely different. Actually only the spectators were different, but they are not part of the chemical change. Getting them out of the way shows that both reactions are exactly the same. It will be this way whenever a solution containing $Ag^+(aq)$ ion is added to a solution containing $Cl^-(aq)$ ion. Therefore, if asked to write the net ionic equation for the reaction between such solutions, you can go directly to $Ag^+(aq) + Cl^-(aq) \rightarrow AgCl(s)$. (It is possible that a second reaction may occur between the other pair of ions, yielding a second net ionic equation.)

This simple and direct procedure may be used to write the net ionic equation for the precipitation of any insoluble ionic compound. The compound is the product, and the reactants are the ions in the compound. It is like writing a solution inventory equation (Equations 16.1 and 16.2, and Example 16.1) in reverse. Try it on the following:

EXAMPLE 16.11 Write the net ionic equations for the precipitation of the following from aqueous solutions:

CuS:

$Mg(OH)_2$:

Li_3PO_4:

_____

$Cu^{2+}(aq) + S^{2-}(aq) \rightarrow CuS(s)$
$Mg^{2+}(aq) + 2\,OH^-(aq) \rightarrow Mg(OH)_2(s)$
$3\,Li^+(aq) + PO_4{}^{3-}(aq) \rightarrow Li_3PO_4(s)$

If we knew in advance those combinations of ions that yield insoluble compounds, we could predict precipitation reactions. These compounds have been identified in the laboratory. Table 16.2 shows the result of such experiments for a large number of ionic compounds. These solubilities have also been summarized in a set of "solubility rules" that your instructor may ask you to memorize. These rules are in Table 16.3.

In the next four examples use either table to predict precipitation reactions. We recommend that you use each table at least once.

Table 16.2
Solubilities of Ionic Compounds*

Ions	Acetate	Bromide	Carbonate	Chlorate	Chloride	Fluoride	Hydrogen Carbonate	Hydroxide	Iodide	Nitrate	Nitrite	Phosphate	Sulfate	Sulfide	Sulfite
Aluminum	I	S		S	S	I		I	−	S		I	S	−	
Ammonium	S	S	S	S	S	S	S	−	S	S	S	S	S	S	S
Barium	−	S	I	S	S	I		S	S	S	S	I	I	−	I
Calcium	S	S	I	S	S	I		I	S	S	S	I	I	−	I
Cobalt(II)	S	S	I	S	S	−		I	S	S		I	S	I	I
Copper(II)	S	S			S	S		I		S		I	S	I	
Iron(II)	S	S	I		S	I		I	S	S		I	S	I	I
Iron(III)	−	S			S	I		I	S	S		I	S	−	
Lead(II)	S	I	I	S	I	I		I	I	S	S	I	I	I	I
Lithium	S	S	S	S	S	S	S	S	S	S	S	I	S	S	
Magnesium	S	S	I	S	S	I		I	S	S	S	I	S	−	S
Nickel		S	I	S	S	S		I	S	S		I	S	I	I
Potassium	S	S	S	S	S	S	S	S	S	S	S	S	S	S	S
Silver	I	I	I	S	I	S		−	I	S	I	I	I	I	I
Sodium	S	S	S	S	S	S	S	S	S	S	S	S	S	S	S
Zinc	S	S	I	S	S	S		I	S	S		I	S	I	I

* Compounds having solubilities of 0.1 M or more at 20°C are listed as soluble (S); if the solubility is less than 0.1 M, the compound is listed as insoluble (I). A dash (−) identifies an unstable species in aqueous solution, and a blank space indicates lack of data. In writing equations for reactions that occur in water solution, insoluble substances, shown by I in this table, have the state symbol for a solid, (s). Dissolved substances, S in the table, are designated by (aq) in an equation.

Table 16.3
Solubility Rules for Ioniz Compounds*

The following compounds are Soluble	Except	The following compounds are Insoluble	Except
Ammonium salts		Carbonates	NH_4^+ and alkali metals
Alkali metal salts (Column IA)	Li^+		
		Phosphates	Na^+, K^+, and NH_4^+
Nitrates		Hydroxides	Ba^{2+} and alkali metals
Halides (Column 7A)	Ag^+, Hg_2^{2+}, Pb^{2+}		
Acetates	Ag^+, Al^{3+}	Sulfides	NH_4^+ and salts of alkali metals
Chlorates			
Sulfates	Ba^{2+}, Sr^{2+}, Ca^{2+}, Pb^{2+}, Ag^+, Hg^{2+}, Hg_2^{2+}		

* For purposes of these rules, a compound is considered to be soluble if it dissolves to a concentration of 0.1 M or more at 20°C.

EXAMPLE 16.12   Solutions of lead(II) nitrate and sodium fluoride are combined. Write the net ionic equation for any precipitation reaction that may occur.

Start with the conventional equation.

$$Pb(NO_3)_2(aq) + 2\,NaF(aq) \rightarrow PbF_2(\quad) + 2\,NaNO_3(\quad)$$

The spaces between parentheses after the products have been left blank. You must determine if the compound is soluble or insoluble. If soluble, the state symbol should be (aq); if insoluble, (s). You will have to use Table 16.2 to find out about the solubility of lead(II) fluoride; fluoride solubilities are not easily included in solubility rules. The solubility of sodium nitrate is readily found from either table. Fill in the state symbols.

$$Pb(NO_3)_2(aq) + 2\,NaF(aq) \rightarrow PbF_2(s) + 2\,NaNO_3(aq)$$

In Table 16.2 the intersection of the lead(II) ion line and the fluoride column shows that lead(II) fluoride is insoluble, so the designation (s) follows PbF_2. The intersection of the sodium ion line and nitrate ion column shows that sodium nitrate is soluble in water. This is confirmed by the solubility rules, which indicate that all sodium compounds and all nitrates are soluble. The designation (aq) therefore follows $NaNO_3$.

Complete the example by writing the ionic and net ionic equations.

$$Pb^{2+}(aq) + 2\,NO_3^-(aq) + 2\,Na^+(aq) + 2\,F^-(aq) \rightarrow PbF_2(s) + 2\,Na^+(aq) + 2\,NO_3^-(aq)$$

$$Pb^{2+}(aq) + 2\,F^-(aq) \rightarrow PbF_2(s)$$

EXAMPLE 16.13   Write the net ionic equation for any reaction that will occur when solutions of nickel chloride and ammonium carbonate are combined.

This example does not ask for all three equations, but only the net ionic equation. You might wish to try for the net ionic equation without the other two. Can you, simply by looking at the names of the reactants, identify the names of the products as the ions rearrange themselves? If so, determine from Table 16.2 or 16.3 which, if either, is insoluble. For any insoluble compound you can write the net ionic equation directly, as in Example 16.11. If you are not ready for this step, write all three equations, as before. The three steps are shown in the answer.

$NiCl_2(aq) + (NH_4)_2CO_3(aq) \rightarrow NiCO_3(s) + 2\,NH_4Cl(aq)$

$Ni^{2+}(aq) + 2\,Cl^-(aq) + 2\,NH_4^+(aq) + CO_3^{2-}(aq) \rightarrow NiCO_3(s) + 2\,NH_4^+(aq) + 2\,Cl^-(aq)$

$Ni^{2+}(aq) + CO_3^{2-}(aq) \rightarrow NiCO_3(s)$

EXAMPLE 16.14   Write the net ionic equation for any reaction that occurs between solutions of aluminum sulfate and calcium acetate, $Al_2(SO_4)_3(aq)$ and $Ca(C_2H_3O_2)_2(aq)$.

Be careful here!

———   ———

$Al_2(SO_4)_3(aq) + 3\,Ca(C_2H_3O_2)_2(aq) \rightarrow 3\,CaSO_4(s) + 2\,Al(C_2H_3O_2)_3(s)$

$2\,Al^{3+}(aq) + 3\,SO_4^{2-}(aq) + 3\,Ca^{2+}(aq) + 6\,C_2H_3O_2^-(aq) \rightarrow 3\,CaSO_4(s) + 2\,Al(C_2H_3O_2)_3(s)$

This time *both* new combinations of ions precipitate. There are no spectators, so the ionic equation is the net ionic equation. Actually there are two separate reactions taking place at the same time:

$$Al^{3+}(aq) + 3\,C_2H_3O_2^-(aq) \rightarrow Al(C_2H_3O_2)_3(s)$$
$$Ca^{2+}(aq) + SO_4^{2-}(aq) \rightarrow CaSO_4(s)$$

EXAMPLE 16.15   Write the net ionic equation for any reaction that occurs when solutions of ammonium sulfate and potassium nitrate are combined.

Again, be careful.

———   ———

$(NH_4)_2SO_4(aq) + KNO_3(aq) \rightarrow NR$

This time both new combinations of ions are soluble, so there is no precipitation reaction. If you complete the conventional equation, as you would have done in Chapter 8, and from it write the ionic equation, you will find that *all* ions are spectators.

16.7 ION COMBINATIONS THAT FORM MOLECULES

PG 16 I   Given reactants that yield a molecular product, write the net ionic equation.

The reaction of an acid often leads to an ion combination that yields a molecular product instead of an insoluble ionic compound. Except for the difference in the product, the equations are written in exactly the same way. Just as you had to recognize an insoluble product and not break it up in the ionic equations, you must now recognize a molecular product and not break it into ions. Water or a weak acid are the two kinds of molecular products you will find.

The neutralization reaction between an acid and a hydroxide to produce water and a salt (p. 171) is the most common molecular product reaction.

EXAMPLE 16.16   Write the conventional, ionic, and net ionic equations for the reaction between hydrochloric acid and a solution of sodium hydroxide.

Proceed just as you did for precipitation reactions. Watch your state designations.

$$HCl(aq) + NaOH(aq) \rightarrow HOH(\ell) + NaCl(aq)$$

$$H^+(aq) + Cl^-(aq) + Na^+(aq) + OH^-(aq) \rightarrow HOH(\ell) + Na^+(aq) + Cl^-(aq)$$

$$H^+(aq) + OH^-(aq) \rightarrow HOH(\ell)$$

Water is the molecular product. It is unionized and is in the liquid state.

The acid in a neutralization may be a weak acid. You must then recall that the solution inventory of a weak acid is the acid molecule; it is not broken into ions.

EXAMPLE 16.17   Write the three equations leading to the net ionic equation for the reaction between acetic acid, $HC_2H_3O_2(aq)$, and a solution of sodium hydroxide.

$$HC_2H_3O_2(aq) + NaOH(aq) \rightarrow HOH(l) + NaC_2H_3O_2(aq)$$
$$HC_2H_3O_2(aq) + Na^+(aq) + OH^-(aq) \rightarrow HOH(l) + Na^+(aq) + C_2H_3O_2^-(aq)$$
$$HC_2H_3O_2(aq) + OH^-(aq) \rightarrow HOH(l) + C_2H_3O_2^-(aq)$$

The weak acid molecule appears in its molecular form in the net ionic equation.

When the reactants are a strong acid and the salt of a weak acid, that weak acid is formed as the molecular product. You must recognize it as a weak acid and leave it in molecular form in the solution inventory.

EXAMPLE 16.18   Develop the net ionic equation for the reaction between hydrochloric acid and a solution of sodium acetate, $NaC_2H_3O_2(aq)$.

$$HCl(aq) + NaC_2H_3O_2(aq) \rightarrow HC_2H_3O_2(aq) + NaCl(aq)$$
$$H^+(aq) + Cl^-(aq) + Na^+(aq) + C_2H_3O_2^-(aq) \rightarrow HC_2H_3O_2(aq) + Na^+(aq) + Cl^-(aq)$$
$$H^+(aq) + C_2H_3O_2^-(aq) \rightarrow HC_2H_3O_2(aq)$$

Compare the reactions in Examples 16.16 and 16.18. The only difference between them is that Example 16.16 has the hydroxide ion as a reactant and Example 16.18 has the acetate ion as a reactant. In the first case the molecular product is water, formed when the hydrogen ion bonds to the hydroxide ion. In the second case the molecular product is acetic acid, a weak acid, formed when the hydrogen ion bonds to the acetate ion. Actually water is a weak acid too, weaker than acetic acid or any other acid that dissolves in water. "Weak," you recall, means that the acid does not ionize to a large extent. It is because water ionizes less than acetic acid that weak acids can neutralize hydroxides, as in Example 16.17.

Just as you can write the net ionic equation for a precipitation reaction without the conventional and ionic equations, so you can write the net ionic equation for a molecule formation reaction. Again you must recognize the product from the formulas of the reactants. The acid will contribute a hydrogen ion to the molecular product. It will form a molecule with the anion from the other reactant. If the molecule is water or a weak acid, you have the reactants and products of the net ionic equation.

EXAMPLE 16.19   Without writing the conventional and ionic equations, write the net ionic equations for each of the following pairs of reactants:

H_2SO_4 and LiOH:

KNO_2 and HBr:

- - - - - -   - - - - - -

$$H^+(aq) + OH^-(aq) \rightarrow HOH(\ell) \qquad\qquad H^+(aq) + NO_2^-(aq) \rightarrow HNO_2(aq)$$

$Li^+(aq)$ and $SO_4{}^{2-}(aq)$ are spectators in the first equation for a neutralization reaction. HNO_2 is recognized as a weak acid in the second reaction because it is not one of the six strong acids. $K^+(aq)$ and $Br^-(aq)$ are spectators.

There are two points by which you identify a molecular product reaction: (1) one reactant is an acid, usually strong; (2) one product is water or a weak acid. One of the most common mistakes in writing net ionic equations is the failure to recognize a weak acid as a molecular product. If one reactant is a strong acid, you can be sure there will be a molecular product. If it isn't water, look for a weak acid.

16.8 ION COMBINATIONS THAT FORM UNSTABLE PRODUCTS

PG 16 J   Given reactants that form H_2CO_3, H_2SO_3, or "NH_4OH" by ion combination, write the net ionic equation for the reaction.

Three ion combinations yield molecular products that are not the products you would expect. Two of the expected products are carbonic and sulfurous acids. If hydrogen ions from one reactant reach carbonate ions from another, H_2CO_3, carbonic acid, should form:

$$2\,H^+(aq) + CO_3{}^{2-}(aq) \rightarrow H_2CO_3(aq)$$

But carbonic acid is unstable and decomposes to carbon dioxide gas and water. The correct net ionic equation is therefore

$$2\,H^+(aq) + CO_3{}^{2-}(aq) \rightarrow CO_2(g) + H_2O(\ell)$$

Sulfurous acid, H_2SO_3, decomposes in the same way to sulfur dioxide and water, but the sulfur dioxide remains in solution:

$$2\,H^+(aq) + SO_3{}^{2-}(aq) \rightarrow SO_2(aq) + H_2O(\ell)$$

The third ion combination that yields unexpected molecular products occurs when ammonium and hydroxide ions meet:

$$NH_4^+(aq) + OH^-(aq) \rightarrow \text{"}NH_4OH\text{"}$$

In spite of printed labels and laboratory bottles with NH_4OH etched on them, and wide use of the name "ammonium hydroxide," no substance having the formula NH_4OH exists at ordinary temperatures. The actual product is a solution of ammonia molecules, $NH_3(aq)$. The proper net ionic equation is therefore

$$NH_4^+(aq) + OH^-(aq) \rightarrow NH_3(aq) + H_2O(\ell)$$

The reaction is reversible, and actually reaches an equilibrium in which the solution inventory is overwhelmingly dissolved ammonia molecules.

There is no system by which these three "different" molecular product reactions can be recognized. You simply must be alert to them and catch them when they appear. Once again, the predicted but unstable formulas are H_2CO_3, H_2SO_3, and NH_4OH.

EXAMPLE 16.20   Write the conventional, ionic, and net ionic equations for the reaction between solutions of sodium carbonate and hydrochloric acid.

— — — — — —   — — — — — —

$2\,HCl(aq) + Na_2CO_3(aq) \rightarrow 2\,NaCl(aq) + CO_2(g) + H_2O(\ell)$

$2\,H^+(aq) + 2\,Cl^-(aq) + 2\,Na^+(aq) + CO_3^{2-}(aq) \rightarrow 2\,Na^+(aq) + 2\,Cl^-(aq) + CO_2(g) + H_2O(\ell)$

$2\,H^+(aq) + CO_3^{2-}(aq) \rightarrow CO_2(g) + H_2O(\ell)$

16.9 ION COMBINATION REACTIONS WITH UNDISSOLVED SOLUTES

In every ion combination reaction considered so far, it has been assumed that both reactants are in solution. This is not always the case. Sometimes the description of the reaction will indicate that a reactant is a solid, liquid, or gas, even though it may be soluble in water. In such a case write the correct state symbol after the formula in the conventional equation and carry the formula through all three equations unchanged.

A common example of this kind of reaction occurs with a compound which the handbooks say is insoluble in water but soluble in acids. The net ionic equation shows why.

EXAMPLE 16.21   Write the net ionic equation to describe the reaction when solid aluminum hydroxide dissolves in hydrochloric acid, nitric acid, and sulfuric acid.

This is a neutralization reaction between a strong acid and a solid hydroxide. Write the conventional equation for hydrochloric acid only.

‒ ‒ ‒ ‒ ‒ ‒   ‒ ‒ ‒ ‒ ‒ ‒

$3\ HCl(aq) + Al(OH)_3(s) \rightarrow 3\ HOH(\ell) + AlCl_3(aq)$

Now write the ionic and net ionic equations. Remember that you replace only dissolved substances, designated (aq), with their solution inventories.

‒ ‒ ‒ ‒ ‒ ‒   ‒ ‒ ‒ ‒ ‒ ‒

$3\ H^+(aq) + 3\ Cl^-(aq) + Al(OH)_3(s) \rightarrow 3\ HOH(\ell) + Al^{3+}(aq) + 3\ Cl^-(aq)$
$3\ H^+(aq) + Al(OH)_3(s) \rightarrow 3\ HOH(\ell) + Al^{3+}(aq)$

Now think a bit before writing the net ionic equation for the reaction between solid aluminum hydroxide and nitric acid. The question asks only for the net ionic equation. Can you write it directly, without the conventional and ionic equations? If not, write all three equations.

‒ ‒ ‒ ‒ ‒ ‒   ‒ ‒ ‒ ‒ ‒ ‒

$3\ H^+(aq) + Al(OH)_3(s) \rightarrow 3\ HOH(\ell) + Al^{3+}(aq)$

This is the same as the equation for hydrochloric acid. Sulfuric acid will also produce the same equation. The chloride, nitrate, and sulfate ions are spectators in the three reactions.

16.10 SUMMARY

In this chapter you have learned how to write net ionic equations for three kinds of reactions. The process is simpler if you recognize the kind of reaction and know what to expect. This summary should help:

If the reactants are an *element* and a *compound*, write the single replacement equation for what must be a redox reaction. Proceed with solution inventories, the ionic equation, and the net ionic equation as you learned in Section 16.5.

If the reactants are *two ionic compounds*, write the double displacement equation for what must be an ion combination-precipitation reaction. Identify the insoluble product and proceed as in Section 16.6.

If the reactants are an *acid* and an *ionic compound*, write the double displacement equation for what must be an ion combination-molecule formation reaction. The hydrogen-bearing product is the unionized molecule; it will be water or a weak acid. Proceed as in Section 16.7.

There are four pitfalls to avoid. They are:

1. Be sure your reactant states are correct. If the question says a reactant is a solid, liquid, or gas, it must appear as such in the net ionic equation, even if it is soluble in water.
2. Be sure there is a reaction. Check the activity series to be sure a possible redox reaction will occur. Check ionic products to be sure one will precipitate. Check molecular products to be sure one is water or a *weak* acid.
3. Watch for double reactions. A double replacement equation can have two insoluble products, or an insoluble product and an unionized molecule.
4. Watch for the unstable ion combination products H_2CO_3, H_2SO_3, or "NH_4OH." If you have one, proceed as in Section 16.8.

CHAPTER 16 IN REVIEW

16.1   ELECTROLYTES AND SOLUTION CONDUCTIVITY

     16 A   Distinguish between strong electrolytes, weak electrolytes, and nonelectrolytes. (368)
     16 B   Describe or explain electrical conductivity through a solution. (368)

16.2   SOLUTION INVENTORIES OF IONIC COMPOUNDS

     16 C   Given the formula of an ionic compound, write the solution inventory when it is dissolved in water. (370)

16.3   STRONG ACIDS AND WEAK ACIDS

     16 D   Explain why the solution of an acid may be a good conductor or a poor conductor of electricity. (372)
     16 E   Given the formula of a soluble acid, write the solution inventory when it is dissolved in water. (372)

16.4   NET IONIC EQUATIONS: WHAT THEY ARE AND HOW TO WRITE THEM (374)

16.5   REDOX REACTIONS THAT ARE DESCRIBED BY "SINGLE REPLACEMENT" EQUATIONS

     16 F   Given two substances that may engage in a redox reaction and an activity series by which the reaction may be predicted, write the conventional, ionic, and net ionic equations for the reaction that will occur, if any. (375)

16.6   ION COMBINATIONS THAT FORM PRECIPITATES

16 G Predict whether or not a precipitate will form when known solutions are combined; if a precipitate forms, write the net ionic equation. (Reference to a solubility table may or may not be allowed.) (379)

16 H Given the product of a precipitation reaction, write the net ionic equation. (379)

16.7   ION COMBINATIONS THAT FORM MOLECULES

16 I Given reactants that yield a molecular product, write the net ionic equation. (385)

16.8   ION COMBINATIONS THAT FORM UNSTABLE PRODUCTS

16 J Given reactants that form H_2CO_3, H_2SO_3, or "NH_4OH" by ion combination, write the net ionic equation for the reaction. (387)

16.9   ION COMBINATION REACTIONS WITH UNDISSOLVED SOLUTES (388)

16.10   SUMMARY (389)

TERMS AND CONCEPTS

Electrode (368)
Conductor, nonconductor (369)
Electrolysis (369)
Electrolyte: strong, weak, non- (369)
Anode, cathode (369, 370)
Solution inventory (370)
Strong acid, weak acid (372)

Conventional equation (374)
Ionic equation (374)
Spectator ion; spectator (375)
Net ionic equation (375)
Activity series (377)
Ion combination reaction (379)
Precipitate, precipitation reaction (379)

Most of these terms and many others appear in the Glossary.

QUESTIONS AND PROBLEMS

Section 16.1

16.1) If solid solute A is a strong electrolyte and solute B is a nonelectrolyte, state how these solutes differ.

16.2) How do you explain the passage of "electricity" through a solution? The ability of a solution to conduct is considered evidence of what?

16.35) How does a weak electrolyte differ from a strong electrolyte?

16.36) All soluble ionic compounds are electrolytes. A molecular solute may or may not be an electrolyte. Show how both of these statements are true.

For the next four questions in each column, write the solution inventory for the water solution of each substance given.

Section 16.2

16.3) LiF; $Mg(NO_3)_2$; $FeCl_3$.

16.4) NH_4NO_3; $Ca(OH)_2$; Li_2SO_3.

16.37) Na_2S; $(NH_4)_2SO_4$; $MnCl_2$.

16.38) $NiSO_4$; K_3PO_4; KNO_2.

Section 16.3

16.5) $HCHO_2$; HCl; $HC_4H_4O_6$.

16.6) HI; H_2SO_4; $H_3C_2H_5O_7$.

16.39) HNO_3; $HC_2H_3O_2$; HBr.

16.40) $HClO_4$; $H_2C_4H_4O_4$; HF.

Section 16.5

For each pair of reactants given in the next three questions in each column, write the net ionic equation for any redox reaction that may be predicted by Table 16.1, p. 377. If no redox reaction occurs, write NR.

16.7) Ba(s), $Zn(NO_3)_2$(aq).

16.8) Pb(s), $Ca(NO_3)_2$(aq).

16.9) Mg(s), $Al_2(SO_4)_3$(aq).

16.41) Cu(s), Li_2SO_4(aq).

16.42) Ba(s), HCl(aq).

16.43) Ni(s), $AgNO_3$(aq).

Section 16.6

For each pair of reactants given in the next four questions in each column, write the net ionic equation for any precipitation reaction that may be predicted by Tables 16.2 and 16.3, p. 382. If no precipitation occurs, write NR.

16.10) $BaCl_2$(aq), Na_2CO_3(aq).

16.11) $CoSO_4$(aq), NaOH(aq).

16.12) $FeCl_2$(aq), $(NH_4)_2S$(aq).

16.13) $NiCl_2$(aq), $CuSO_4$(aq).

16.14) Write the net ionic equations for the precipitation of each of the following insoluble ionic compounds: MgF_2; $Zn_3(PO_4)_2$.

16.44) $Pb(NO_3)_2$(aq), KI(aq).

16.45) $KClO_3$(aq), $Mg(NO_2)_2$(aq).

16.46) $AgNO_3$(aq), LiBr(aq).

16.47) $ZnCl_2$(aq), Na_2SO_3(aq).

16.48) Write the net ionic equations for the precipitation of each of the following insoluble ionic compounds: $PbCO_3$; $Ca(OH)_2$.

Section 16.7

For each pair of reactants given in the next three questions in each column, write the net ionic equation for the molecule formation reaction that will occur.

16.15) $NaC_6H_5O_7$(aq), HNO_3(aq).

16.16) $Ca(OH)_2$(aq), HBr(aq).

16.17) $KC_4H_4O_6$(aq), HCl(aq).

16.49) $NaNO_2$(aq), HI(aq).

16.50) $KC_3H_5O_3$(aq), $HClO_4$(aq).

16.51) RbOH(aq), H_2SO_4(aq).

Section 16.8

For each pair of reactants given in the next two questions in each column, write the net ionic equation for the reaction that will occur.

16.18) K_2CO_3(aq), HNO_3(aq).

16.19) LiOH(aq), $(NH_4)_2SO_4$(aq).

16.52) NH_4SO_3(aq), HBr(aq).

16.53)* $MgSO_3$(aq), H_3PO_4(aq).

The remaining questions include all types of reactions covered in this chapter. Use the suggestions in Section 16.10 to identify the reaction type, and then develop the net ionic equation. Use the activity series and solubility tables to predict whether or not redox or precipitation reactions will take place. If you find a "no reaction" combination, mark it NR.

16.20) Silver nitrate solution is added to a solution of ammonium carbonate.

16.21) Aluminum nitrate solution is added to potassium phosphate solution.

16.22) Metallic calcium is dropped into a solution of silver nitrate.

16.23) When combined in an oxygen-free atmosphere, sodium sulfite solution and hydrochloric acid react.

16.24) Sulfuric acid neutralizes potassium hydroxide.

16.25) A solution of potassium iodide is able to dissolve solid silver chloride and form a new precipitate.

16.26) Lithium metal is immersed in a solution of copper(II) sulfate.

16.27) Sodium benzoate solution, $NaC_7H_5O_2(aq)$, is treated with hydrochloric acid.

16.28) Ammonium chloride solution is added to a solution of potassium hydroxide.

16.29) A solution of sodium fluoride is poured into a solution of calcium nitrate.

16.30) A piece of potassium is dropped into a solution of zinc nitrate.

16.31) Dilute nitric acid is poured over solid barium hydroxide.

16.32) Sulfuric acid is able to dissolve copper(II) hydroxide.

16.33) Aluminum reacts with a solution of sodium hydroxide to produce hydrogen gas and a solution of sodium aluminate, Na_3AlO_3.

16.34) Consider H_3PO_4 to be a weak acid and write the net ionic equations for Equations 15.10 and 15.11, page 345.

16.54) Barium chloride and sodium sulfite solutions are combined in an oxygen-free atmosphere.

16.55) Copper(II) sulfate and sodium hydroxide solutions are combined.

16.56) Carbon dioxide bubbles appear as hydrochloric acid is poured onto solid magnesium carbonate.

16.57) Nitric acid is able to dissolve solid lead(II) hydroxide.

16.58) Oxalic acid, $H_2C_2O_4(s)$, is neutralized by sodium hydroxide solution.

16.59) What happens if barium is placed into hydrochloric acid?

16.60) Hydrochloric acid is poured into a solution of sodium hydrogen sulfite.

16.61) Solutions of magnesium sulfate and ammonium bromide are combined.

16.62) Magnesium ribbon is placed in hydrochloric acid.

16.63) Solid nickel hydroxide is readily dissolved by hydrobromic acid.

16.64) Sodium fluoride solution is poured into nitric acid.

16.65) Silver wire is dropped into hydrochloric acid.

16.66) When metallic lithium is added to water, hydrogen is released, leaving behind an aqueous solution of lithium hydroxide.

16.67) Aluminum shavings are dropped into a solution of copper(II) nitrate.

16.68) (a) Hydrochloric acid reacts with a solution of sodium hydrogen carbonate. (b) Hydrochloric acid reacts with a solution of sodium carbonate. (c) A limited amount of hydrochloric acid reacts with a solution of sodium carbonate, yielding $NaHCO_3(aq)$ as one product.

17

acid-base (proton transfer) reactions

We have used names and formulas of acids since Chapter 6. On page 111 you learned that one kind of acid releases hydrogen ions when dissolved in water. A base is a substance that reacts with the hydrogen ion, and compounds that contain the hydroxide ion were given as the best examples. In this chapter we will look more deeply into acids and bases, including reactions that go well beyond those between hydrogen and hydroxide ions.

The study of acid-base chemistry requires the use of net ionic equations, which were introduced in Chapter 16. If these equations were not included in your earlier assignments, you should become familiar with them now. The following should help.

When sodium hydroxide solution reacts with hydrochloric acid, the products are a solution of sodium chloride and water:

$$NaOH(aq) + HCl(aq) \rightarrow NaCl(aq) + HOH(\ell)$$

It is a fact that the dissolved particles in $NaOH(aq)$ are not particles of NaOH. Instead they are individual $Na^+(aq)$ ions and $OH^-(aq)$ ions, which separate as the solid dissolves. The ions that make up a solution are called its solution inventory (p. 370). Similarly, the solution inventory of $NaCl(aq)$ is $Na^+(aq) + Cl^-(aq)$. Hydrochloric acid is a *strong acid* (p. 372), which means that its molecules separate into ions: $H^+(aq) + Cl^-(aq)$. Water is a molecular substance; it does not separate into ions to an appreciable extent. If solution inventories are substituted for ionic and molecular formulas in the above equation, we get the *ionic equation* (p. 374):

$$Na^+(aq) + OH^-(aq) + H^+(aq) + Cl^-(aq) \rightarrow Na^+(aq) + Cl^-(aq) + HOH(\ell)$$

The sodium and chloride ions appear on both sides of the equation. They are present, but are not changed. Such ions are called *spectator ions* (p. 375). To write a net ionic equation the spectator ions are removed:

$$H^+(aq) + OH^-(aq) \rightarrow HOH(\ell)$$

17.1 THE ARRHENIUS THEORY OF ACIDS AND BASES

PG 17 A   Distinguish between an acid and a base in terms of their general properties and the ions associated with those properties.

A water solution is called an acid if it has certain properties. These include:

1. A sour taste, as shown by vinegar (acetic acid) and lemon juice (citric acid). It is reliably reported that water solutions of HCl, HNO_3, and H_2SO_4 (hydrochloric, nitric, and sulfuric acids) also taste sour.
2. The ability to give a certain color to substances known as acid-base indicators. Perhaps the most familiar of these is litmus, an organic dye that is red in acid solutions. Needless to say, this property affords a safer and a more pleasant test for acidity than taste.
3. The ability to release gaseous carbon dioxide when carbonate ions are added to the solution.
4. The ability to react with and neutralize a base.
5. The ability to react with certain metals, releasing hydrogen gas as one of the products.

A water solution is called a base if it has the following properties:

1. A bitter taste. Sodium bicarbonate, sometimes used to relieve stomach acidity, is mildly basic, and better tasting than most other bases.
2. A "soapy" or slippery feeling. This property is shown by such familiar household products as liquid detergents, ammonia, and lye. Since feeling a base actually involves the dissolving of a surface layer of skin, it is not recommended as a test for basicity.
3. The ability to give a certain color to an acid-base indicator. A piece of red litmus paper turns blue when touched by a basic solution.
4. The ability to react with and neutralize an acid.
5. The ability to form a precipitate when added to solutions of certain cations.

Long ago, in 1884, it was proposed by Svante Arrhenius that solutions having the properties of acids have one thing in common: they all release hydrogen ions, H^+, in water solutions. Furthermore, all solutions containing hydrogen ions behave like acids. Specifically, in regard to the reaction with carbonate ion, it is the hydrogen ion that combines with the carbonate ion to yield carbonic acid, which is unstable and decomposes to water and carbon dioxide:

$$2\,H^+(aq) + CO_3{}^{2-}(aq) \rightarrow H_2CO_3(aq) \rightarrow H_2O(\ell) + CO_2(g)$$

On evidence such as this, **the Arrhenius concept of an acid is a solution containing hydrogen ions.**

Basic solutions also have their ion that is always present: **the Arrhenius concept of a base is a solution containing hydroxide ion, OH^-.** It is the hydroxide ion that gives a base its properties. The fourth property listed for acids and bases, their ability to neutralize each other, is associated with the net ionic equation for a strong acid-hydroxide neutralization developed in Example 16.16, p. 385

$$H^+(aq) + OH^-(aq) \rightarrow HOH(\ell)$$

In this reaction the H^+ and OH^- ions are literally cancelling each other and all the acid and base properties associated with each other. Hence the name, *neutralization.*

The fifth properties in the above lists are indeed properties of the hydrogen and hydroxide ions, and therefore properties of acids and bases. But the reactions are not acid-base reactions. When zinc is placed in hydrochloric acid, for example, hydrogen is released:

$$Zn(s) + 2\,HCl(aq) \rightarrow H_2(g) + ZnCl_2(aq)$$

This is actually a redox reaction between an active metal, one above hydrogen in the activity series (p. 377), and the hydrogen ion of an acid. The precipitation of a metallic hydroxide when a base is added to a solution of a metal cation shows only that most hydroxides are insoluble (p. 382). They are precipitations that happen to involve a base, rather than acid-base reactions.

17.2 THE BRÖNSTED-LOWRY THEORY OF ACIDS AND BASES

PG 17 B   State the Brönsted-Lowry concept of acids and bases.

The simple hydrogen-hydroxide ion concept of acids and bases is not adequate to describe all of the acid-base reactions we know about, particularly those that occur without water. In 1923 Johannes N. Brönsted and Thomas M. Lowry independently proposed a broader acid-base theory that is known by both their names, the Brönsted-Lowry theory. By this concept an **acid-base reaction is one in which there is a transfer of a proton from one species to another. The acid is**

the proton donor; the base is the proton acceptor. In this sense an acid-base reaction may be thought of as a **proton transfer reaction.**

The proton is the true identity of what we have been calling a hydrogen ion. A hydrogen atom consists of one proton and one electron. Remove the electron and all that is left is the proton, the nucleus of the atom. Lone protons do not actually exist in solutions; rather, they become hydrated. This means each proton is bonded to one or more water molecules to form an ion whose formula might be H_3O^+, $H_5O_2^+$, $H_7O_3^+$, or $H \cdot (H_2O)_n^+$. Many chemists emphasize the hydrated character of the hydrogen ion by writing it as H_3O^+ and calling it the **hydronium ion.** Because H^+ is easier to write and use, and because the actual formula of the hydrated proton is not known, we will continue to use H^+ with the understanding that it is hydrated. Exceptions will be made when the role played by water is important—as it is now.

We saw earlier that acid solutions are formed when hydrogen-bearing compounds react with water. Hydrogen chloride is typical:

$$HCl(g) + H_2O(\ell) \rightarrow H_3O^+(aq) + Cl^-(aq) \qquad (17.1)$$

Equation 17.1 for the ionization of HCl is an acid-base or proton transfer reaction in the Brönsted-Lowry sense. HCl gives up its hydrogen ion, a proton, and therefore qualifies as the acid; the water molecule receives the proton, thereby satisfying the definition of a base. Lewis diagrams show how this is done:

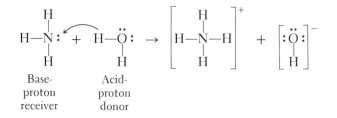

When ammonia dissolves in water another proton transfer reaction occurs:

$$NH_3(aq) + HOH(\ell) \rightleftharpoons NH_4^+(aq) + OH^-(aq) \qquad (17.2)$$

Water is the acid. It donates a proton to ammonia, the base. The transfer is seen clearly with Lewis diagrams:

The double arrow in Equation 17.2 indicates that the reaction is reversible (p. 318), and suggests correctly that it reaches equilibrium. When a reversible reaction equation is read from left to right you describe the **forward reac-**

tion or **direction**; from right to left it is the **reverse reaction** or **direction**. In 1 M NH_3(aq) less than 1% of the ammonia is changed to ammonium ions. The solution inventory is mainly ammonia molecules. The reverse reaction is therefore said to be **favored**. The favored direction of an equilibrium points to the more abundant substances.

In Equation 17.1 water is a proton receiver, a base. In the forward direction of Equation 17.2 water is a proton donor, an acid. Substances that can behave as an acid in one proton transfer reaction and a base in another are said to be **amphoteric**.

In Equation 17.1 we saw HCl as an acid, a proton donor. Equation 17.2 had NH_3 as a base, a proton acceptor. Do you suppose that if we brought HCl and NH_3 together, the HCl would give its proton to NH_3? The answer, yes, is readily evident if bottles of concentrated ammonia and hydrochloric acid are opened next to each other. The solid product is often found on the outside of reagent bottles in the laboratory. Conventional and Lewis diagram equations describe the reaction:

$$NH_3(g) + HCl(g) \rightarrow NH_4Cl(s) \tag{17.3}$$

$$
\begin{array}{ccc}
\overset{\displaystyle H}{\underset{\displaystyle H}{H-N:}} + H-\overset{..}{\underset{..}{Cl}}: & \rightarrow & \left[\overset{\displaystyle H}{\underset{\displaystyle H}{H-N-H}} \right]^{+} + \left[:\overset{..}{\underset{..}{Cl}}: \right]^{-}
\end{array}
$$

Base-
proton
receiver

Acid-
proton
donor

NH_4Cl(s) is an assembly of NH_4^+ and Cl^- ions. Equation 17.3 is an acid-base reaction in the absence of both water and the hydroxide ion.

Reviewing Equations 17.1 to 17.3, we find that all fit into the general equation for a Brönsted-Lowry proton transfer reaction:

$$
\begin{array}{ccccccc}
B & + & HA & \rightarrow & HB^+ & + & A^- \tag{17.4}
\end{array}
$$

Base-
proton
receiver

Acid-
proton
donor

In this equation, note that the charges are not "absolute" charges. They indicate, rather, that the acid species, in losing a proton, leaves a species having a charge one less than the acid, and that the base, in gaining a proton, increases by 1 in charge.

CONJUGATE ACID-BASE PAIRS

PG 17 C Define or identify conjugate acid-base pairs.

It was noted that Equation 17.2 reaches equilibrium. The fact is that most acid-base reactions reach a state of equilibrium. Let's look at the reverse reaction in Equation 17.4:

$$\text{B} \quad + \quad \text{HA} \quad \rightleftharpoons \quad \text{HB}^+ \quad + \quad \text{A}^- \qquad (17.5)$$

Acid-   Base-
proton   proton
donor   receiver

Is not HB^+ donating a proton to A^- in the reverse reaction? In other words, HB^+ is an acid in the reverse reaction, and A^- is a base. From this we see that the products of any proton transfer acid-base reaction are *another* acid and base for the reverse reaction.

Combinations such as acid HA and base A^- that result from an acid losing a proton or a base gaining one are called **conjugate acid-base pairs.** In the forward reaction of Equation 17.2, $HOH(l)$ releases a hydrogen ion, leaving the hydroxide ion:

$$HOH(l) \rightarrow H^+(aq) + OH^-(aq)$$

Water is an acid, a proton donor. What is left after the proton is gone, $OH^-(aq)$, is the **conjugate base** of water. In the reverse direction $OH^-(aq)$ is a base because it gains a proton. The **conjugate acid** of $OH^-(aq)$ is $HOH(l)$, the species formed when the base gains a proton. In the other part of the reaction, $NH_3(aq) + H^+(aq) \rightleftharpoons NH_4^+(aq)$, the base, $NH_3(aq)$, in the forward direction gains a proton to form its conjugate acid, $NH_4^+(aq)$; and $NH_3(aq)$ is the conjugate base of acid $NH_4^+(aq)$ in the reverse direction. For the whole reaction, two conjugate acid-base pairs can be identified:

┌────────conjugate acid-base pair────────┐

$$NH_3(aq) \quad + \quad HOH(l) \quad \rightleftharpoons \quad NH_4^+(aq) \quad + \quad OH^-(aq) \qquad (17.2)$$

└────────conjugate acid-base pair────────┘

You can write the formula of the conjugate base of any acid simply by removing a proton. If HCO_3^- acts as an acid, its conjugate base is CO_3^{2-}. You can also write the formula of the conjugate acid of any base by adding a proton. If HCO_3^- is a base, its conjugate acid is H_2CO_3.

EXAMPLE 17.1 (a) Write the formula of the conjugate base of H_3PO_4.

(b) Write the formula of the conjugate acid of $C_7H_5O_2^-$.

In (b) don't let $C_7H_5O_2^-$ confuse you just because it is unfamiliar. Just do what must be done to find the formula of a conjugate acid.

(a) $H_3PO_4 \rightarrow H^+ + H_2PO_4^-$, the conjugate base of H_3PO_4

(b) $C_7H_5O_2^- + H^+ \rightarrow HC_7H_5O_2$, the conjugate acid of $C_7H_5O_2^-$

Remove a proton to get a conjugate base, and add one to get a conjugate acid.

EXAMPLE 17.2 Nitrous acid engages in a proton transfer reaction with sulfite ion: $HNO_2(aq) + SO_3^{2-}(aq) \rightarrow NO_2^-(aq) + HSO_3^-(aq)$. Answer the questions about this reaction in the steps that follow:

For the forward reaction, identify the acid and the base.

HNO_2 is the acid; it donates a proton to SO_3^{2-}, the base, or proton receiver.

Identify the acid and base for the reverse reaction.

HSO_3^- is the acid; it donates a proton to NO_2^-, the base, or proton receiver.

Identify the conjugate of HNO_2. Is it a conjugate acid or a conjugate base?

NO_2^- is the conjugate *base* of HNO_2. It is the base that remains after the proton has been donated by the acid.

Identify the other conjugate acid-base pair, and classify each species as the acid or the base.

SO_3^{2-} and HSO_3^- are the other conjugate acid-base pair. SO_3^{2-} is the base, and HSO_3^- is the conjugate acid—the species produced when the base accepts a proton.

EXAMPLE 17.3 Identify the conjugate acid-base pairs in

$$HCHO_2 + PO_4^{3-} \rightleftharpoons HPO_4^{2-} + CHO_2^-$$

$HCHO_2$ and CHO_2^- are one conjugate acid-base pair; PO_4^{3-} and HPO_4^{2-} are the second pair.

17.3 LEWIS THEORY OF ACIDS AND BASES

PG 17 D   Describe the Lewis theory of acids and bases and the structural features associated with it; identify potential Lewis acids and Lewis bases.

The Arrhenius theory of acids and bases identifies acids as substances yielding a hydrogen ion, and bases as sources of hydroxide ions. The Brönsted-Lowry theory expands the idea of bases to include anything that can receive protons. A third concept, the Lewis theory of acids and bases, expands the idea of acids to include anything that can accept an electron pair to form a covalent bond with a Brönsted-Lowry base.

The identifying characteristic of a Lewis acid or a Lewis base rests in the structure of the particle. Lewis diagrams for OH^-, H_2O, and NH_3, all of which have been identified as bases in this chapter, are

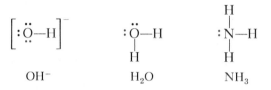

$$OH^- \qquad\qquad H_2O \qquad\qquad NH_3$$

They have in common an unshared electron pair that can form a covalent bond with anything that can accept an electron pair. The hydrogen ion, H^+, with its empty 1s orbital, qualifies:

$$H_2O \qquad\qquad H_3O^+ \qquad\qquad NH_4^+$$

In each case the base has contributed the electron pair to the bond. **A Lewis base is an electron pair donor.** The hydrogen ion has accepted the electron pair from the base in forming the bond. **A Lewis acid is an electron pair acceptor.**

Boron trifluoride is an example of a Lewis acid. The boron atom is surrounded by three electron pairs, one pair short of the noble gas octet of electrons. It may acquire these by sharing with a Lewis base, such as ammonia:

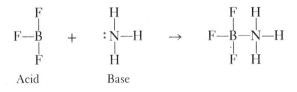

$$\text{Acid} \qquad\qquad \text{Base}$$

Other examples of Lewis acid-base reactions appear in the formation of some complex ions and in organic reactions. You will study them in more advanced courses.

17.4 RELATIVE STRENGTHS OF ACIDS AND BASES

PG 17 E   Distinguish between a strong acid and a weak acid; between a strong base and a weak base.

17 F   Given a table of relative strengths of acids and bases, identify the stronger and weaker of two acids or two bases.

In Section 16.3, the distinction was made between the relatively few *strong* acids and the many *weak* acids. Strong acids are those that ionize almost completely, whereas weak acids ionize but slightly. Hydrochloric acid is a strong acid; 0.10 M HCl is almost 100% ionized. Acetic acid is a weak acid; only 1.3% of the molecules ionize in 0.10 M $HC_2H_3O_2$. In a Brönsted-Lowry sense, acid strength is a measure of the tendency of an acid to lose protons. *A strong acid loses protons readily—ionizes more completely; a weak acid clings to its protons—does not ionize significantly.*

Table 17.1 is a list of acids arranged in order of decreasing strength. The strongest acids are at the top and the weakest are at the bottom. Each ionization is shown as an equilibrium equation of the general form $HA(aq) \rightleftharpoons H^+(aq) + A^-(aq)$. Notice the ionization of **polyprotic acids,** those able to release more than one proton. Sulfuric acid loses its first proton almost completely, according to the sixth equation in Table 17.1. This places H_2SO_4 among the strong acids. The HSO_4^- ion, four lines lower in the table, releases the second proton less readily. The second ionization step occurs to about 29% in 0.10 M HSO_4^-. **Diprotic** (two-proton) and **triprotic** (three-proton) acids always ionize in a stepwise fashion like this, and each succeeding acid is weaker than the one before.

The species on the right side of each equation in Table 17.1 is the conjugate base of the acid on the left. Being a base, it is able to accept a proton from a potential donor. *A strong base is one that has a strong attraction for protons; a weak base has a weak attraction for protons.* It follows that the conjugate base of a strong acid is a weak base, and that a weak acid has a strong conjugate base. HCl, for example, a strong acid, loses its proton readily to yield the weak base, Cl^-. The chloride ion has no tendency to gain a proton to form HCl by the reverse reaction. Conversely the carbonate ion, CO_3^{2-}, has a strong attraction for protons to form weak acid HCO_3^-. Thus CO_3^{2-} is a strong base. By this reasoning we conclude that the bases in Table 17.1 are listed in order of *increasing* strength; the weaker bases are at the top, and the stronger bases are at the bottom.

The strongest base that is stable in water solution is the hydroxide ion. Accordingly, solutions of soluble metal hydroxides are the most common bases used in the laboratory. As shown on page 382, there aren't many soluble hydroxides; only the hydroxides of the alkali metals, plus barium, strontium, and calcium, can qualify—and the last three are questionable, having solubility limits of about 0.17 M, 0.04 M, and 0.03 M, respectively. But these are considerably

Table 17.1
Relative Strengths of Acids and Bases

Acid Name	Acid Formula	Base Formula
Perchloric	$HClO_4$	$\rightleftarrows H^+ + ClO_4^-$
Hydroiodic	HI	$\rightleftarrows H^+ + I^-$
Hydrobromic	HBr	$\rightleftarrows H^+ + Br^-$
Hydrochloric	HCl	$\rightleftarrows H^+ + Cl^-$
Nitric	HNO_3	$\rightleftarrows H^+ + NO_3^-$
Sulfuric	H_2SO_4	$\rightleftarrows H^+ + HSO_4^-$
Hydronium ion	H_3O^+	$\rightleftarrows H^+ + H_2O$
Oxalic	$H_2C_2O_4$	$\rightleftarrows H^+ + HC_2O_4^-$
Sulfurous	H_2SO_3	$\rightleftarrows H^+ + HSO_3^-$
Hydrogen sulfate ion	HSO_4^-	$\rightleftarrows H^+ + SO_4^{2-}$
Phosphoric	H_3PO_4	$\rightleftarrows H^+ + H_2PO_4^-$
Hydrofluoric	HF	$\rightleftarrows H^+ + F^-$
Nitrous	HNO_2	$\rightleftarrows H^+ + NO_2^-$
Formic (methanoic)	$HCHO_2$	$\rightleftarrows H^+ + CHO_2^-$
Benzoic	$HC_7H_5O_2$	$\rightleftarrows H^+ + C_7H_5O_2^-$
Hydrogen oxalate ion	$HC_2O_4^-$	$\rightleftarrows H^+ + C_2O_4^{2-}$
Acetic (ethanoic)	$HC_2H_3O_2$	$\rightleftarrows H^+ + C_2H_3O_2^-$
Propionic (propanoic)	$HC_3H_5O_2$	$\rightleftarrows H^+ + C_3H_5O_2^-$
Carbonic	H_2CO_3	$\rightleftarrows H^+ + HCO_3^-$
Hydrosulfuric	H_2S	$\rightleftarrows H^+ + HS^-$
Dihydrogen phosphate ion	$H_2PO_4^-$	$\rightleftarrows H^+ + HPO_4^{2-}$
Hydrogen sulfite ion	HSO_3^-	$\rightleftarrows H^+ + SO_3^{2-}$
Hypochlorous	$HClO$	$\rightleftarrows H^+ + ClO^-$
Boric	H_3BO_3	$\rightleftarrows H^+ + H_2BO_3^-$
Ammonium ion	NH_4^+	$\rightleftarrows H^+ + NH_3$
Hydrocyanic	HCN	$\rightleftarrows H^+ + CN^-$
Hydrogen carbonate ion	HCO_3^-	$\rightleftarrows H^+ + CO_3^{2-}$
Monohydrogen phosphate ion	HPO_4^{2-}	$\rightleftarrows H^+ + PO_4^{3-}$
Hydrogen sulfide ion	HS^-	$\rightleftarrows H^+ + S^{2-}$
Water	HOH	$\rightleftarrows H^+ + OH^-$
Hydroxide ion	OH^-	$\rightleftarrows H^+ + O^{2-}$

(Left margin: STRENGTH — Increasing ↑ / Decreasing ↓. Right margin: STRENGTH — Decreasing ↑ / Increasing ↓.)

more soluble than the next in the list; the solubility of $Mg(OH)_2$ is about 0.000 003 M—and the solubilities of other hydroxides go down from there.

EXAMPLE 17.4 Using Table 17.1, list the following acids in order of decreasing strength (strongest first): $HC_2O_4^-$; NH_4^+; H_3PO_4.

Find the three acids among those listed in the table, and list them from strongest (first) to weakest (last).

_ _ _ _ _ _   _ _ _ _ _ _

H_3PO_4; $HC_2O_4^-$; NH_4^+

EXAMPLE 17.5 Using Table 17.1, list the following bases in order of decreasing strength (strongest first): $HC_2O_4^-$; SO_3^{2-}; F^-.

SO_3^{2-}; F^-; $HC_2O_4^-$

The ion $HC_2O_4^-$ appears in both Example 17.4 and 17.5, first as an acid and second as a base. It is the intermediate ion in the two-step ionization of oxalic acid, $H_2C_2O_4$. The $HC_2O_4^-$ ion is amphoteric (p. 398), as are most intermediate ions in the stepwise ionization of polyprotic acids.

17.5 PREDICTING ACID-BASE REACTIONS

PG 17 G Given a table of the relative strengths of acids and bases and the identity of an acid and a base from the table, (a) write the equation for the single-proton transfer reaction between them, and (b) predict which direction, forward or reverse, will be favored at equilibrium.

The chemist likes to know if an acid-base reaction will occur if certain reactants are brought together. Obviously there must be a potential proton donor and acceptor—there can be no proton transfer reaction without both. From there the decision is based on the relative strengths of the conjugate acid and base pairs. The stronger acid and base are the most reactive. They *do* what they must do to behave as an acid and a base. The weaker acid and base are more stable—less reactive. It follows that *the stronger acid will always transfer a proton to the stronger base, yielding the weaker acid and base as favored species at equilibrium.* Figure 17.1 summarizes the proton transfer from the stronger acid to the stronger base from the standpoint of positions in Table 17.1.

If a hydrogen sulfate ion, HSO_4^-, a relatively strong acid trying to lose a proton, finds a hydroxide ion, OH^-, a strong base seeking a proton, both tendencies can be satisfied by the transfer of the proton from HSO_4^- to OH^-:

$$HSO_4^-(aq) + OH^-(aq) \rightleftharpoons SO_4^{2-}(aq) + HOH(\ell)$$

Now identify the conjugate acid-base pairs. In the forward direction the acid is HSO_4^-. Its conjugate base for the reverse reaction is SO_4^{2-}. Similarly OH^- is the base in the forward reaction, and HOH is the conjugate acid for the reverse

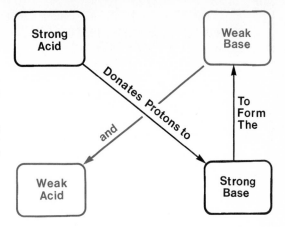

Figure 17.1
Correlation between prediction of an acid-base reaction and position in Table 17.1. Of the two conjugate acid-base pairs in the proton transfer equation, the stronger acid (upper left) will do-nate protons to the stronger base (lower right), yielding the weaker base (upper right) and weaker acid (lower left). That direction, forward or reverse, will be favored that has the weaker acid and weaker base as products.

reaction. These acid-base roles for the different directions are shown with the letters A for acid and B for base:

$$HSO_4^-(aq) + OH^-(aq) \rightleftharpoons SO_4^{2-}(aq) + HOH(\ell)$$

$$\text{A} \qquad\quad \text{B} \qquad\qquad \text{B} \qquad\qquad \text{A}$$

Now compare the two acids in strength. HSO_4^- is near the top of the list, a much stronger acid than water. We therefore label HSO_4^- with SA for strong acid, and water with WA for weak acid. Similarly, compare the bases: OH^- is a stronger base (SB) than SO_4^{2-} (WB).

$$HSO_4^-(aq) + OH^-(aq) \rightleftharpoons SO_4^{2-}(aq) + HOH(\ell)$$

$$\text{SA} \qquad\quad \text{SB} \qquad\qquad \text{WB} \qquad\qquad \text{WA}$$

As you see, the weaker combination is on the right in this equation. This indi-cates that the reaction is favored in the forward direction: the proton transfers spontaneously from the strong proton donor to the strong proton receiver, yield-ing as products in greater abundance the weaker conjugate base and conjugate acid.

The following procedure is recommended in the prediction of acid-base reactions:

1. For a given pair of reactants, write the equation for the transfer of *one* proton from one species to the other. (Do not transfer two protons.)

2. Identify the acid and base on each side of the equation.

3. Determine which side of the equation has *both* the weaker acid and the weaker base (they must both be on the same side). That side identifies the products in the favored direction.

EXAMPLE 17.6 Write the net ionic equation for the reaction between hydrofluoric acid, HF, and the sulfite ion, SO_3^{2-}, and predict which side will be favored at equilibrium.

The first step is to write the equation for the single-proton transfer reaction between HF and SO_3^{2-}. Complete this step.

$$HF(aq) + SO_3^{2-}(aq) \rightleftharpoons F^-(aq) + HSO_3^-(aq)$$

Next we identify the acid and base on each side of the equation. Do so with letters A and B, as in the preceding discussion.

$$\underset{A}{HF(aq)} + \underset{B}{SO_3^{2-}(aq)} \rightleftharpoons \underset{B}{F^-(aq)} + \underset{A}{HSO_3^-(aq)}$$

In each case the acid is the species with a proton to donate. It is transferred from acid HF to base SO_3^{2-} in the forward reaction, and from acid HSO_3^- to base F^- in the reverse reaction.

Finally determine which reaction, forward or reverse, is favored at equilibrium. It is the side with the weaker acid and base. Refer to Table 17.1.

The forward reaction is favored at equilibrium.

HSO_3^- is a weaker acid than HF, and F^- is a weaker base than SO_3^{2-}. These species are the products in the favored direction.

EXAMPLE 17.7   Write the net ionic equation for the acid-base reaction between HCO_3^- and ClO^- and predict which side will be favored at equilibrium.

Complete the example as before.

$$HCO_3^-(aq) + ClO^-(aq) \rightleftharpoons CO_3^{2-}(aq) + HClO(aq)$$

The reverse reaction will be favored. This conclusion is based on HCO_3^- being a weaker acid than HClO, and ClO^- being a weaker base than CO_3^{2-}.

Up to this point most attention has been given to the direction in which an equilibrium is favored. This does not mean that we can ignore the unfavored direction. Consider, for example, the reaction described by Equation 17.2: $NH_3(aq) + HOH(\ell) \rightleftharpoons OH^-(aq) + NH_4^+(aq)$. Although this reaction proceeds only slightly in the forward direction, many of the properties of household ammonia, in particular its cleaning power, depend upon the presence of OH^- ions, few as they are.

17.6 THE WATER EQUILIBRIUM: pH

PG 17 H   Write the equation for the ionization of water, the equation for K_w, and state the value of K_w at 25°C.

17 I   Distinguish among an acidic solution, a basic solution, and a neutral solution in terms of pH and $[H^+]$.

17 J   Given one of the following, calculate any or all of the others: pH, $[H^+]$, $[OH^-]$, pOH.

17 K   Given the pH, $[H^+]$, $[OH^-]$, or pOH of two or more solutions, rank them in order of acidity, basicity, pH, $[H^+]$, $[OH^-]$, or pOH.

In this section and the next you will be multiplying and dividing exponentials, and taking the square root of an exponential. You may wish to review these procedures in Appendix I, page 580.

One of the most critical equilibria in all of chemistry is represented by the next to last line in Table 17.1, the ionization of water. Careful control of tiny traces of hydrogen and hydroxide ions marks the difference between success and failure in an untold number of industrial chemical processes; and in biochemical systems, these concentrations are vital to survival itself.

Though pure water is generally regarded as a nonconductor, a sufficiently sensitive detector shows that even water contains a tiny concentration of ions. These ions come from the ionization of the water molecule:

$$HOH(\ell) \rightleftarrows H^+(aq) + OH^-(aq) \qquad (17.6)$$

Each chemical equilibrium has an **equilibrium constant** that is calculated from the concentrations of one or more species in the equation.* The equilibrium constant for water, K_w, at 25°C is

$$K_w = [H^+][OH^-] = 1.0 \times 10^{-14}, \qquad (17.7)$$

where $[H^+]$ and $[OH^-]$ are the concentrations of the hydrogen and hydroxide ions in moles per liter. At first we will consider $K_w = 10^{-14}$ and work only with concentrations that can be expressed with whole-number exponents. In the next section, after you have become familiar with the mathematical procedures, concentrations will be written in the usual exponential notation form, including coefficients.

---

*Equilibrium constants are covered in more detail in Chapter 20.

8

8

The stoichiometry of Equation 17.6 indicates that the theoretical concentrations of hydrogen and hydroxide ions in pure water must be equal. If $x = [H^+] = [OH^-]$, then substituting into Equation 17.7 gives

$$x^2 = 10^{-14}$$

$$x = \sqrt{10^{-14}} = 10^{-7} \text{ moles/liter}$$

Water or water solutions in which $[H^+] = [OH^-] = 10^{-7}$ are said to be neutral solutions, neither acidic nor basic. A solution in which $[H^+]$ is greater than $[OH^-]$ is acidic; a solution in which $[OH^-]$ is greater than $[H^+]$ is basic.

Equation 17.7 indicates an inverse relationship between $[H^+]$ and $[OH^-]$; if one concentration goes up, the other must go down. In fact, if we know either the hydrogen or hydroxide ion concentration, we can, by Equation 17.7, find the other.

EXAMPLE 17.8 Find the hydroxide ion concentration in a solution in which $[H^+] = 10^{-5}$.

This problem requires little explanation. Solve Equation 17.7 for $[OH^-]$, substitute, and calculate.

From Equation 17.7, $[OH^-] = \dfrac{10^{-14}}{[H^+]} = \dfrac{10^{-14}}{10^{-5}} = 10^{-14-(-5)} = 10^{-9}$

Chemists use base 10 logarithms* in the form of "p" values to express the very small values of $[H^+]$ and $[OH^-]$. By this system, if Q is a number, then

$$pQ = -\log Q \tag{17.8}$$

Applied to $[H^+]$ and $[OH^-]$, it follows that

$$pH = -\log [H^+] \qquad\qquad pOH = -\log [OH^-] \tag{17.9}$$

$[H^+] = \text{antilog} - pH = 10^{-pH}$ $[OH^-] = \text{antilog} - pOH = 10^{-pOH}$ (17.10)

The following procedures result:

1. If hydrogen ion concentration in moles per liter is expressed as 10 raised to some negative exponent, pH is the exponent with its sign changed to positive. Thus, if $[H^+] = 10^{-x}$, then $pH = x$.

*A base 10 logarithm is an exponent of 10. Any number, N, can be expressed as an exponential: $N = 10^x$. The logarithm of N is x: $\log N = x$. In reverse, if $\log N = x$, then $N = 10^x$. N is called the antilogarithm—or antilog—of x.

2. Given the pOH of a solution, the hydroxide ion concentration is 10 raised to a power that is the negative of the pOH. Thus, if pOH $= y$, then $[OH^-] = 10^{-y}$. Also, $[OH^-] = $ antilog $-$ pOH.*

EXAMPLE 17.9   (a) The hydrogen ion concentration of a solution is 10^{-9}. What is its pH? (b) The pOH of a certain electroplating bath is 4. What is the hydroxide ion concentration?

- - - - - -   - - - - - -

(a) $[H^+] = 10^{-pH} = 10^{-9}$. pH $= 9$, the negative of the exponent of 10.

(b) $[OH^-] = 10^{-pOH} = 10^{-4}$. The exponent of 10 is the negative of the pOH.

The nature of the equilibrium constant and the logarithmic relationship between pH and pOH yield a simple equation that ties the two together:

$$pH + pOH = 14.0 \qquad (17.11)$$

Between Equations 17.7, 17.9, 17.10, and 17.11, if you know any one of the group consisting of pH, pOH, $[OH^-]$, or $[H^+]$, you can calculate the others. Figure 17.2 is a "pH loop" that summarizes these calculations.

EXAMPLE 17.10   Assuming complete ionization of 0.01 M NaOH, find its pH, pOH, $[OH^-]$, and $[H^+]$.

Solution: (a) Starting with $[OH^-]$, if 0.01 mole of NaOH is dissolved in one liter of solution, the concentration of the hydroxide ion is 0.01 molar: $[OH^-] = 0.01 = 10^{-2}$.

(b) If $[OH^-] = 10^{-2}$, pOH $= -\log 10^{-2} = 2$.

(c) pH $= 14 - $ pOH $= 14 - 2 = 12$.

(d) If pH $= 12$, $[H^+] = 10^{-pH} = 10^{-12}$.

Two things are worth noting about Example 17.10. First, if we extend the problem by one more step we complete the full pH loop. We began with

---

*Mathematically, pH and pOH are negative numbers when the corresponding molarities are 1 or more. The "p" value is not ordinarily used for concentrations greater than 1 molar, so we need not be concerned with negative values of pH or pOH.

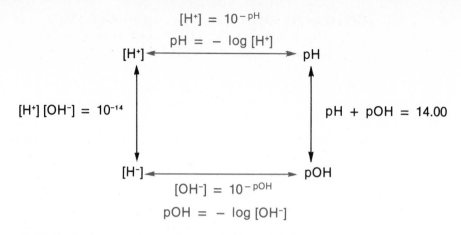

Figure 17.2
The "pH loop." Given the value for any corner of the pH loop, all other values may be calculated by progressing around the loop in either direction. Conversion equations are shown for each step around the loop.

[OH⁻] and went counterclockwise through pOH, pH, and [H⁺]. The [H⁺] of 10^{-12} can be converted to [OH⁻] by Equation 17.7:

$$[OH^-] = \frac{K_w}{[H^+]} = \frac{10^{-14}}{10^{-12}} = 10^{-2}$$

This is the same as the [OH⁻] from which we began. The completion of the loop may therefore be used to check the correctness of the other steps in the process.

The second observation from Example 17.10 is that the loop may be circled in either direction. Starting with [OH⁻] $= 10^{-2}$ and moving clockwise,

$$[H^+] = \frac{K_w}{[OH^-]} = \frac{10^{-14}}{10^{-2}} = 10^{-12}$$

It follows that pH = 12 and pOH = 14 − 12 = 2, the same results reached by circling the loop in the opposite direction.

You should now be able to make a complete trip around the pH loop.

EXAMPLE 17.11   The pH of a solution is 3. Calculate the pOH, [H⁺], and [OH⁻] in any order. Confirm your result by calculating the starting pH—by completing the loop.

You may go either way around the loop, but complete it whichever way you choose, making sure you return to the starting point.

COUNTERCLOCKWISE	CLOCKWISE
From pH $= 3$, $[H^+] = 10^{-3}$	$pOH = 14 - 3 = 11$
$[OH^-] = \dfrac{10^{-14}}{10^{-3}} = 10^{-11}$	From $pOH = 11$, $[OH^-] = 10^{-11}$
From $[OH^-] = 10^{-11}$, $pOH = 11$	$[H^+] = \dfrac{10^{-14}}{10^{-11}} = 10^{-3}$
$pH = 14 - 11 = 3$	From $[H^+] = 10^{-3}$, $pH = 3$

Most of the solutions we work with in the laboratory and all of those involved in biochemical systems have pH values between 1 and 14. This corresponds to H^+ concentrations between 10^{-1} and 10^{-14}, as shown in Table 17.2.

Let's pause for a moment to develop a "feeling" for pH, what it means. pH is a measure of acidity, at least in the sense that hydrogen ion concentration expresses acidity. It is an inverse sort of measurement; the higher the pH, the lower the acidity, and vice versa. Table 17.3 brings out this relationship.

On examining Table 17.3, we see that each pH unit represents a factor of 10. Thus a solution of pH 2 is 10 times as acidic as a solution with pH $= 3$, and 100 times as acidic as the solution of pH 4. In general, the relative acidity in terms of $[H^+]$ is 10^x, where x is the difference between the two pH measurements. From this we conclude that a 0.1 M solution of a strong acid, with pH $= 1$, is one million times as acidic as a neutral solution, with pH $= 7$. (One million is based on the pH difference, $7 - 1 = 6$. As an exponential, $10^6 = 1\ 000\ 000$.)

If you understand the idea behind pH, you should be able to make some comparisons.

EXAMPLE 17.12   Arrange the following solutions in order of decreasing acidity (i.e., highest $[H^+]$ first, lowest last): Solution A, pH $= 8$; Solution B, $pOH = 4$; Solution C, $[H^+] = 10^{-6}$; Solution D, $[OH^-] = 10^{-5}$.

To make comparisons, all values should be converted to the same basis, pH, pOH, $[H^+]$, or $[OH^-]$. Because the question asks for a list based on acidity, find the $[H^+]$ for each solution.

$[H^+] = 10^{-8}$ for A; 10^{-10} for B; 10^{-6} for C; 10^{-9} for D.

Now arrange these $[H^+]$ values in decreasing order. Remember that the exponents are negative.

C, A, D, B $10^{-6} > 10^{-8} > 10^{-9} > 10^{-10}$

Table 17.2
pH Values of Common Liquids

Liquid	pH	Liquid	pH
Human gastric juices	1.0–3.0	Cow's milk	6.3– 6.6
Lemon juice	2.2–2.4	Human blood	7.3– 7.5
Vinegar	2.4–3.4	Sea water	7.8– 8.3
Carbonated drinks	2.0–4.0	Saturated $Mg(OH)_2$	10.5
Orange juice	3.0–4.0	Household ammonia (1–5%)	10.5–11.5
Black coffee	3.7–4.1	0.1 M Na_2CO_3	11.7
Tomato juice	4.0–4.4	1 M NaOH	14.0

Table 17.3
pH and Hydrogen Ion Concentration

$[H^+]$	$[H^+]$	pH	Acidity or Basicity*
1.0	10^0	0	
0.1	10^{-1}	1	
0.01	10^{-2}	2	Strongly Acid
0.001	10^{-3}	3	pH $<$ 4
0.000 1	10^{-4}	4	
0.000 1	10^{-4}	4	
0.000 01	10^{-5}	5	Weakly acid
0.000 001	10^{-6}	6	4 $\leq$ pH $<$ 6
0.000 001	10^{-6}	6	Neutral
0.000 000 1	10^{-7}	7	(or near neutral)
0.000 000 01	10^{-8}	8	6 $\leq$ pH $<$ 8
0.000 000 01	10^{-8}	8	
0.000 000 001	10^{-9}	9	Weakly Basic
0.000 000 000 1	10^{-10}	10	8 $\leq$ pH $<$ 11
0.000 000 000 1	10^{-10}	10	
0.000 000 000 01	10^{-11}	11	
0.000 000 000 001	10^{-12}	12	Strongly basic
0.000 000 000 000 1	10^{-13}	13	11 $\leq$ pH
0.000 000 000 000 01	10^{-14}	14	

*Ranges of acidity and basicity are arbitrary.

Various methods are used to measure pH in the laboratory. Acid-base indicators have been mentioned already. Each indicator has a specific pH range over which it changes color. pH papers (Fig. 17.3), strips impregnated with an indicator dye that functions over a narrow or broad range, selected for the pH being measured, are widely used for rough pH readings. More accurate measurements are made with pH meters. Figure 17.4 shows a student model in use.

Figure 17.3
pH papers are impregnated with dyes that, when dipped into solutions, turn certain colors that depend upon the pH of the solution. By comparing the color with a standard, the pH of the solution may be estimated. (Photograph by Gregory A. Peters.)

17.7 NONINTEGER pH-[H⁺] AND pOH-[OH⁻] CONVERSIONS (OPTIONAL)

It is sad but true that in the real world solutions do not come neatly packaged in concentrations that can be expressed as whole-number powers of ten. $[H^+]$ is more apt to have a value such as 2.7×10^{-4}, or the pH of a solution is more likely to be 6.24. The chemist must be able to convert each form to the other. Most late-model hand calculators are able to make these conversions. If you do not have such a calculator you will have to use a table of logarithms. A "two-place log table" is suitable for two significant figure concentrations, the usual precision of hydrogen ion concentration measurements and their pH values. Table 17.4 is a two-place log table.

A word about significant figures in logarithms: The logarithm of a number, N, has two parts, the *characteristic* and *mantissa*. The characteristic appears to the left of

Figure 17.4
A pH meter for student use. (Photograph by Gregory A. Peters.)

Table 17.4
Two-Place Logarithms

	0.0	0.1	0.2	0.3	0.4	0.5	0.6	0.7	0.8	0.9
1	.00	.04	.08	.11	.15	.18	.20	.23	.26	.28
2	.30	.32	.34	.36	.38	.40	.41	.43	.45	.46
3	.48	.49	.51	.52	.53	.54	.56	.57	.58	.59
4	.60	.61	.62	.63	.64	.65	.66	.67	.68	.69
5	.70	.71	.72	.72	.73	.74	.75	.76	.76	.77
6	.78	.79	.79	.80	.81	.81	.82	.83	.83	.84
7	.85	.85	.86	.86	.87	.88	.88	.89	.89	.90
8	.90	.91	.91	.92	.92	.93	.93	.94	.94	.95
9	.95	.96	.96	.97	.97	.98	.98	.99	.99	1.00

the decimal point. It is related to the size of N. Its only purpose is to locate the decimal point if N is expressed in ordinary decimal form, or to establish the exponent if N is shown in exponential notation. The characteristic has nothing to do with significant figures. The mantissa appears to the right of the decimal point. It expresses the precision of N. If N is a two-significant figure number, its logarithm should have two digits after the decimal. 1.23 and 11.23 are both two-significant figure logarithms.

Both calculator and log table conversions are shown in detail in the next example. Sequences vary on different calculators. Below are the procedures for the Texas Instrument TI-30 II and Hewlett-Packard HP-32E calculators described in Appendix I, page 572. If the procedure for your calculator differs from these, check your instruction book.

EXAMPLE 17.13   Calculate the pH of a solution if $[H^+] = 2.7 \times 10^{-4}$.
Solution by Calculator: $pH = -\log (2.7 \times 10^{-4})$

TI-30 II			HP-32E	
Press	Display		Press	Display
2.7	*2.7*		2.7	*2.7*
EE	*2.7 00*		EEX	*2.7 00*
4	*2.7 04*		4	*2.7 04*
+/−	*2.7 -04*		CHS	*2.7 -04*
log	*−3.5686 00*		gold f	*2.7 -04*
+/−	*3.5686 00*		log	*−3.5686*
			CHS	*3.5686*

The answer should be rounded off to two significant figures, 3.57.
Solution by Logarithm Table

$$pH = -\log (2.7 \times 10^{-4}) = -(\log 2.7 + \log 10^{-4}) = -[0.43 + (-4)] = 3.57$$

The logarithm of 2.7 is found in Table 17.3 at the intersection of the line that begins with 2 and the column headed by 0.7. The logarithm of 10^{-4} is −4.

Be sure you understand the sequence you are using, and then try this example:

EXAMPLE 17.14   Find the pOH of a solution if its hydroxide ion concentration is 7.9×10^{-5} moles per liter.

$$pOH = -\log (7.9 \times 10^{-5}) = -(\log 7.9 + \log 10^{-5}) = -[0.90 + (-5)] = 4.10$$

The calculator display shows 4.1024, which rounds off to 4.10.

The next example shows how to convert a number in "p" notation to exponential notation. The TI-30 II has two sequences by which a pOH can be changed to $[OH^-]$. One calculates the antilogarithm of $-pOH$; the other raises 10 to the $-pOH$ power: $[OH^-] = 10^{-pOH}$. The HP-32E uses only the second method.

EXAMPLE 17.15   The pOH of a solution is 6.24. Find $[OH^-]$.

Solution by Calculator: $[OH^-]$ = antilogarithm of $-6.24 = 10^{-6.24}$

TI-30 II ANTILOG SEQUENCE		TI-30 II 10^{-pOH} SEQUENCE		HP-32E SEQUENCE	
Press	Display	Press	Display	Press	Display
6.24	*6.24*	10	*10*	6.24	*6.24*
+/−	*−6.24*	y^x	*10*	CHS	*−6.24*
EE	*−6.24 00*	6.24	*6.24*	Blue g	*−6.24*
INV	*−6.24 00*	+/−	*−6.24*	10^x	*5.7544-07*
log	*5.7544-07*	EE	*−6.24 00*		
		=	*5.7544-07*		

Solution by Logarithm Table.   To solve $[OH^-] = 10^{-pOH} = 10^{-6.24}$ by a logarithm you must evaluate $10^{-6.24}$. The procedure and mathematical explanation for each step follow:

$10^{-6.24} = 10^{0.76-7}$ $= 10^{0.76+(-7)}$	The exponent is rewritten as the sum of a positive decimal fraction and the closest integer more negative than the exponent.
$= 10^{0.76} \times 10^{-7}$	From algebra, $a^{m+m} = a^m \times a^n$ (Appendix I, p. 580).
$= 5.7 \times 10^{-7}$ or 5.8×10^{-7}	The antilogarithm of 0.76—the number that is equal to $10^{0.76}$—is found by locating 0.76 in Table 17.4. It appears twice in the 5 line of the table, under 0.7 and 0.8. The result may therefore be expressed as 5.7×10^{-7} or 5.8×10^{-7}.

It is always wise to check the final answer to be sure it is in the right range. pOH 6.24 is between 6 and 7. This means that $[OH^-]$ is between 10^{-7} and 10^{-6}, more than 10^{-7} but less than 10^{-6}. 5.8×10^{-7} falls in this range.

EXAMPLE 17.16 Find the hydrogen ion concentration of a solution if its pH is 11.62.

$[H^+] = 10^{-11.62} = 10^{0.38-12}$.   Antilog $0.38 = 2.4$. $[H^+] = 2.4 \times 10^{-12}$.

EXAMPLE 17.17 $[OH^-] = 5.2 \times 10^{-9}$ for a certain solution. Calculate in order pOH, pH, and $[H^+]$, and then complete the pH loop by recalculating $[OH^-]$ from $[H^+]$.

$\text{pOH} = -\log (5.2 \times 10^{-9}) = 8.28$

$\text{pH} = 14.00 - 8.28 = 5.72$

$[H^+] = \text{antilog} -5.72 = 10^{-5.72} = 1.9 \times 10^{-6}$

$[OH^-] = \dfrac{1.0 \times 10^{-14}}{1.9 \times 10^{-6}} = 5.3 \times 10^{-9}$

The variation between 5.2×10^{-9} and 5.3×10^{-9} comes from rounding off in expressing intermediate answers. If the calculator sequence is completed without rounding off, the loop returns to 5.2×10^{-9} for $[OH^-]$.

CHAPTER 17 IN REVIEW

17.1   THE ARRHENIUS THEORY OF ACIDS AND BASES

17 A Distinguish between an acid and a base in terms of their general properties and the ions associated with those properties. (395)

17.2   BRÖNSTED-LOWRY THEORY OF ACIDS AND BASES

17 B State the Brönsted-Lowry concept of acids and bases. (396)
17 C Define or identify conjugate acid-base pairs. (398)

17.3 LEWIS THEORY OF ACIDS AND BASES

17 D Describe the Lewis theory of acids and bases and the structural features associated with it; identify potential Lewis acids and Lewis bases. (401)

17.4 RELATIVE STRENGTHS OF ACIDS AND BASES

17 E Distinguish between a strong acid and a weak acid; between a strong base and a weak base. (402)

17 F Given a table of relative strengths of acids and bases, identify the stronger and weaker of two acids or two bases. (402)

17.5 PREDICTING ACID-BASE REACTIONS

17 G Given a table of the relative strengths of acids and bases and the identity of an acid and a base from the table, (a) write the equation for the single-proton transfer reaction between them, and (b) predict which direction, forward or reverse, will be favored at equilibrium. (404)

17.6 THE WATER EQUILIBRIUM: pH

17 H Write the equation for the ionization of water, the equation for K_w, and state the value of K_w at 25°C. (407)

17 I Distinguish between an acidic solution, a basic solution, and a neutral solution in terms of pH and $[H^+]$. (407)

17 J Given one of the following, calculate any or all of the others: pH; $[H^+]$; $[OH^-]$; pOH. (407)

17 K Given the pH, $[H^+]$, $[OH^-]$, or pOH of two or more solutions, rank them in order of acidity, basicity, pH, $[H^+]$, $[OH^-]$, or pOH.

17.7 NONINTEGER pH-$[H^+]$ AND pOH-$[OH^-]$ CONVERSIONS (OPTIONAL) (413)

TERMS AND CONCEPTS

Arrhenius acid-base theory (396)
Brönsted-Lowry acid-base theory (396)
Proton donor; proton receiver (397)
Proton transfer reaction (397)
Hydronium ion, H_3O^+ (397)
Forward or reverse reaction or direction (397)
Amphoteric (398)
Conjugate acid-base pairs (399)
Conjugate acid; conjugate base (399)

Lewis acid-base theory (401)
Lewis acid; Lewis base (401)
Electron pair donor; electron pair acceptor (401)
Polyprotic; diprotic; triprotic (402)
Equilibrium constant (402)
Water constant (407)
Acidic, basic, and neutral solutions (408)
pH; pOH (408)

Most of these terms and many others appear in the Glossary.

QUESTIONS AND PROBLEMS

An asterisk (*) identifies a question that is relatively difficult, or that extends beyond the performance goals of the chapter.

Section 17.1

17.1) Identify the ions traditionally present in solutions called acids and bases. List three compounds that are commonly regarded as acids and three that are regarded as bases which contain these ions in their water solutions.

17.30) Identify at least two of the classical properties of acids and two of bases.

Section 17.2

17.2) Distinguish between an Arrhenius acid and a Brönsted-Lowry acid. Are the two concepts in agreement? Justify your answer.

17.31) Distinguish between an Arrhenius base and a Brönsted-Lowry base. Are the two concepts in agreement? Justify your answer.

17.3) What are the conjugate acids of OH^- and HCO_3^-? Write the formulas of the conjugate bases of H_3O^+ and HCO_3^-.

17.32) Give the formula of the conjugate base of HF; of $H_2PO_4^-$. Give the formula of the conjugate acid of NO_2^-; of $H_2PO_4^-$.

17.4) For the reaction

$$HNO_2(aq) + CN^-(aq) \rightleftharpoons NO_2^-(aq) + HCN(aq)$$

identify the acid and base on each side of the equation—i.e., the acid and base for the forward reaction and the acid and base for the reverse reaction.

17.33) For the reaction

$$HSO_4^-(aq) + C_2O_4^{2-}(aq) \rightleftharpoons SO_4^{2-}(aq) + HC_2O_4^-(aq)$$

identify the acid and the base on each side of the equation—i.e., the acid and base for the forward reaction and the acid and base for the reverse reaction.

17.5) Identify the conjugate acid-base pairs in Question 17.4.

17.34) Identify the conjugate acid-base pairs in Question 17.32.

17.6) Identify both conjugate acid-base pairs in the reaction

$$HSO_4^-(aq) + C_2O_4^{2-}(aq) \rightleftharpoons SO_4^{2-}(aq) + HC_2O_4^-(aq)$$

17.35) For the reaction

$$HNO_2(aq) + CN^-(aq) \rightleftharpoons NO_2^-(aq) + HCN(aq)$$

identify both conjugate acid-base pairs.

17.7) Identify the conjugate acid-base pairs in

$$H_2PO_4^-(aq) + HCO_3^-(aq) \rightleftharpoons HPO_4^{2-}(aq) + H_2CO_3(aq)$$

17.36) Identify the conjugate acid-base pairs in

$$NH_4^+(aq) + HPO_4^{2-}(aq) \rightleftharpoons NH_3(aq) + H_2PO_4^-(aq)$$

Section 17.3

17.8) What structural feature must be present in a compound for it to qualify as a Lewis acid? Lewis base?

17.37) Explain and/or illustrate by an example what is meant by identifying a Lewis acid as an electron pair acceptor, and a Lewis base as an electron pair donor.

17.9) Is the Arrhenius theory of acids and bases consistent with the Lewis acid-base theory? Explain.

17.38) Is the Brönsted-Lowry acid-base theory consistent with the Lewis acid-base theory? Explain.

17.10)* Aluminum chloride, $AlCl_3$, behaves more as a molecular compound than an ionic one. This

17.39) Silver ions form covalent bonds with two ammonia molecules to form a strong complex

is illustrated in its ability to form a fourth covalent bond with a chloride ion:

$$AlCl_3 + Cl^- \rightarrow AlCl_4^-.$$

From the Lewis diagram of the aluminum chloride molecule and the electron configuration of the chloride ion, show that this is an acid-base reaction in the Lewis sense, and identify the Lewis acid and the Lewis base.

17.11)* The reaction between calcium oxide, an ionic compound, and covalent sulfur trioxide can be described by the following equation, which shows to some extent the bonding structure. Explain how this reaction is an acid-base reaction in the Lewis sense.

ion, $Ag(NH_3)_2^+$, in a Lewis acid-base reaction. Identify the acid and base, and explain why this qualifies as an acid-base reaction in the Lewis sense.

17.40)* Diethyl ether reacts with boron trifluoride by forming a covalent bond between the molecules. Describe the reaction from the standpoint of the Lewis acid-base theory, based on the following "structural" equation:

$$Ca^{2+} \left[:\overset{..}{\underset{..}{O}}: \right]^{2-} + \underset{:\overset{..}{O}:}{\overset{:\overset{..}{O}:}{S{=}O:}} \rightarrow Ca^{2+} \left[:\overset{..}{\underset{..}{O}}{-}S{-}\overset{..}{\underset{..}{O}}: \right]^{2-}$$

$$:\overset{..}{\underset{..}{F}}{-}B \;+\; :\overset{..}{\underset{..}{O}}{-}C_2H_5 \rightarrow :\overset{..}{\underset{..}{F}}{-}B{-}\overset{..}{\underset{..}{O}}{-}C_2H_5$$

Section 17.4

Refer to Table 17.1 when answering questions in this section.

17.12) What is the difference between a strong acid and a weak acid, according to the Brönsted-Lowry concept? Identify two examples of strong acids and two examples of weak acids.

17.13) List the following bases in order of their decreasing strength (strongest base first): SO_4^{2-}; Br^-; $H_2PO_4^-$; CO_3^{2-}.

17.14) List the following acids in order of their decreasing strength (strongest acid first): $H_2C_2O_4$; HSO_3^-; H_2O; HI; NH_4^+.

17.41) What is the difference between a strong base and a weak base, according to the Brönsted-Lowry concept? Identify two examples of strong bases and two examples of weak bases.

17.42) List the following acids in order of their increasing strength (weakest acid first): $HC_2O_4^-$; H_2SO_3; HOH; $HClO$.

17.43) List the following bases in order of their decreasing strength (strongest base first): CN^-; H_2O; HSO_3^-; ClO^-; Cl^-.

Section 17.5

For each acid and base given in this section, complete a proton transfer equation for the transfer of one proton. Using Table 17.1 predict the direction in which the resulting equilibrium will be favored.

17.15) $HC_7H_5O_2(aq) + SO_4^{2-}(aq) \rightleftharpoons$

17.16) $H_2C_2O_4(aq) + NH_3(aq) \rightleftharpoons$

17.17) $H_3PO_4(aq) + CN^-(aq) \rightleftharpoons$

17.18) $H_2BO_3^-(aq) + NH_4^+(aq) \rightleftharpoons$

17.19) $HPO_4^{2-}(aq) + HC_2H_3O_2(aq) \rightleftharpoons$

17.44) $HC_3H_5O_2(aq) + PO_4^{3-}(aq) \rightleftharpoons$

17.45) $HSO_4^-(aq) + CO_3^{2-}(aq) \rightleftharpoons$

17.46) $H_2CO_3(aq) + NO_3^-(aq) \rightleftharpoons$

17.47) $NO_2^-(aq) + H_3O^+(aq) \rightleftharpoons$

17.48) $HSO_4^-(aq) + HC_2O_4^-(aq) \rightleftharpoons$

Section 17.6

17.20) How is it that we can classify water as a nonconductor of electricity and yet talk about the ionization of water? If it ionizes, why does it not conduct?

17.49) Of what significance is the very small value of 10^{-14} for K_w, the ionization equilibrium constant for water?

17.21) Identify the ranges of the pH scale that we classify as strongly acidic, weakly acidic, strongly basic, weakly basic, and neutral, or close to neutral. Select any integer from 1 to 14 and explain what is meant by saying that this number is the pH of a certain solution.

17.50) What do we mean when we say that one solution is acidic, another is neutral, and another is basic?

In the next four questions the pH, pOH, [OH⁻], or [H⁺] of a solution is given. Find each of the other values. Also classify each solution as strongly acidic, weakly acidic, neutral (or close to neutral), weakly basic, or strongly basic, as these terms are used in Table 17.3.

17.22) $[H^+] = 10^{-6}$

17.23) $pOH = 3$

17.24) $pH = 1$

17.25) $[OH^-] = 10^{-9}$

17.51) $[OH^-] = 10^{-5}$

17.52) $pH = 2$

17.53) $pOH = 4$

17.54) $[H^+] = 10^{-11}$

Section 17.7

In the questions in this section the pH, pOH, [OH⁻], or [H⁺] of a solution is given. Find each of the other values.

17.26) $[H^+] = 3.4 \times 10^{-9}$

17.27) $pOH = 11.82$

17.28) $[OH^-] = 0.26$

17.29) $pH = 12.05$

17.55) $[OH^-] = 5.9 \times 10^{-4}$

17.56) $pH = 8.28$

17.57) $[H^+] = 7.2 \times 10^{-13}$

17.58) $pOH = 7.35$

oxidation-reduction (electron transfer) reactions

In Chapter 17 you found that acid-base reactions may be regarded as proton transfer reactions, in the Brönsted-Lowry sense. Now we will study oxidation-reduction reactions. You will see that they are electron transfer reactions, with many parallels to the proton transfers of the previous chapter.

18.1 ELECTRON TRANSFER REACTIONS

If we place a strip of zinc into a solution of copper sulfate, the zinc becomes coated with a reddish deposit of copper metal. The blue color of the solution, which is characteristic of the hydrated Cu^{2+} ion, fades as the reaction proceeds (Fig. 18.1). These two observations suggest that Cu^{2+} ions in the solu-

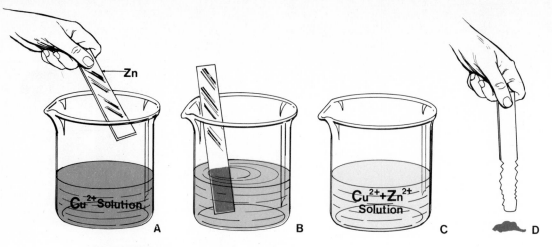

Figure 18.1
An electron transfer reaction.
A, A strip of clean zinc is placed into a solution of Cu²⁺ ion. This ion makes the solution blue.
B, The zinc immediately is coated with finely divided copper metal produced when the copper ion gains two electrons to become a copper atom. The blue color becomes lighter as the Cu²⁺ concentration becomes lower.
C, After the metal is removed, the solution contains a reduced concentration of blue Cu²⁺ ions plus some colorless Zn²⁺ ions formed as zinc atoms gave up electrons to the copper ions.
D, After removing the copper from the zinc, the corroded zinc surface appears.

tion have been changed to copper atoms. Such a change is possible only if each Cu^{2+} ion gains two electrons:

$$Cu^{2+}(aq) + 2e^- \rightarrow Cu(s)$$

The source of these electrons appears when we examine what has happened to the zinc strip. If we carefully scrape off the copper deposit, we find that the zinc weighs less than it did originally. Moreover, we can show by chemical tests that Zn^{2+} ions have entered the solution. In short, zinc atoms have given up electrons and become Zn^{2+} ions:

$$Zn(s) \rightarrow Zn^{2+}(aq) + 2e^-$$

The net ionic equation for the total reaction that occurs when Cu^{2+} ions come into contact with Zn atoms may be derived by adding the two **half-reaction equations** written above:

$$Cu^{2+}(aq) + 2e^- \rightarrow Cu(s)$$
$$\underline{Zn(s) \rightarrow Zn^{2+}(aq) + 2e^-}$$
$$Cu^{2+}(aq) + Zn(s) \rightarrow Cu(s) + Zn^{2+}(aq) \qquad (18.1)$$

This reaction is an **electron transfer reaction**. Electrons have been transferred from zinc atoms to copper(II) ions. Notice that, while no electrons appear in the final equation, the electron transfer character of the reaction is quite clear in the half-reactions. Notice also that the number of electrons lost by one species is exactly equal to the number of electrons gained by the other.

There is another way in which this reaction can be carried out. In Figure 18.2 the electron transfer from zinc atoms to copper(II) ions takes place through a wire rather than on the surface of the zinc. The voltmeter in the circuit not only detects the passage of electrons, but also confirms the direction in which they move and measures the "strength" of the electron transfer tendency. In this so-called *voltaic cell* the "chemical energy" stored in the two reactants is converted to useful electrical energy. This cell was once used as a source of direct current to operate such devices as telegraph relays and doorbells.

Reactions in which electrons are transferred from one species to another are **oxidation-reduction** reactions, often called "**redox**" reactions. Any such reaction can always be considered to be the sum of two **half-reactions: a reduction half-reaction in which a species gains electrons** ($Cu^{2+} + 2e^- \rightarrow Cu$), and an **oxidation half-reaction, in which another species loses electrons** ($Zn - 2e^- \rightarrow Zn^{2+}$ or, as it is usually written, $Zn \rightarrow Zn^{2+} + 2e^-$). We can apply this analysis to several reactions that take place in water solutions:

1. The evolution of hydrogen gas on adding zinc to sulfuric acid (the "acid property" that is a redox reaction, mentioned on page 166):

Reduction: $2\,H^+(aq) + 2e^- \rightarrow H_2(g)$

Oxidation: $Zn(s) \rightarrow Zn^{2+}(aq) + 2e^-$

Redox: $2\,H^+(aq) + Zn(s) \rightarrow H_2(g) + Zn^{2+}(aq)$ (18.2)

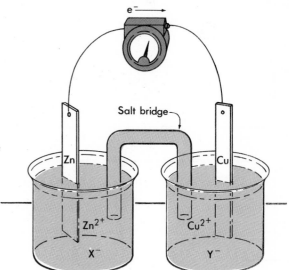

Salt bridge

Figure 18.2
The Zn-Cu^{2+} reaction as part of an electrolytic circuit.

2. The preparation of bromine by bubbling chlorine gas through a solution of NaBr:

Reduction: $Cl_2(g) + 2e^- \rightarrow 2\,Cl^-(aq)$

Oxidation: $\underline{2\,Br^-(aq) \rightarrow Br_2(\ell) + 2e^-}$

Redox: $Cl_2(g) + 2\,Br^-(aq) \rightarrow 2\,Cl^-(aq) + Br_2(\ell)$ (18.3)

3. The formation of a "chemical pine tree" (needles of silver—Fig. 8.6) by placing a copper wire into silver nitrate solution:

Reduction: $2\,Ag^+(aq) + 2e^- \rightarrow 2\,Ag(s)$

Oxidation: $\underline{Cu(s) \rightarrow Cu^{2+}(aq) + 2e^-}$

Redox: $2\,Ag^+(aq) + Cu(s) \rightarrow 2\,Ag(s) + Cu^{2+}(aq)$ (18.4)

The development of Equation 18.4 needs special comment. The usual reduction equation for silver ion is $Ag^+(aq) + e^- \rightarrow Ag(s)$. Because two moles of electrons are lost in the oxidation reaction, *two moles of electrons must be gained in the reduction reaction.* As has already been noted, the number of electrons lost by one species must equal the number gained by the other species. It is therefore necessary to multiply the usual Ag^+ reduction equation by 2 to bring about this equality in electrons gained and lost. They then cancel when the half-reaction equations are added.

EXAMPLE 18.1   Combine the following half-reactions to produce a balanced redox reaction equation. Indicate which half-reaction is an oxidation reaction, and which is a reduction.

$Co^{2+}(aq) + 2e^- \rightarrow Co(s)$

$\underline{Sn(s) \rightarrow Sn^{2+}(aq) + 2e^-}$

- - - - - -   - - - - - -

Reduction: $Co^{2+}(aq) + 2e^- \rightarrow Co(s)$

Oxidation: $\underline{Sn(s) \rightarrow Sn^{2+}(aq) + 2e^-}$

Redox: $Co^{2+}(aq) + Sn(s) \rightarrow Co(s) + Sn^{2+}(aq)$

EXAMPLE 18.2   Combine the following half-reactions to produce a balanced redox equation. Identify the oxidation half-reaction and reduction half-reaction.

$Fe^{2+}(aq) \rightarrow Fe^{3+}(aq) + e^-$

$\underline{Al^{3+}(aq) + 3e^- \rightarrow Al(s)}$

- - - - - -   - - - - - -

Oxidation:	$3\ Fe^{2+}(aq) \rightarrow 3\ Fe^{3+}(aq) + 3e^-$
Reduction:	$Al^{3+}(aq) + 3e^- \rightarrow Al(s)$
Redox:	$Al^{3+}(aq) + 3\ Fe^{2+}(aq) \rightarrow Al(s) + 3\ Fe^{3+}(aq)$

In this example it is necessary to multiply the oxidation half-reaction equation by 3 in order to balance the electrons gained and lost.

Another reaction involving iron and aluminum introduces an additional technique.

EXAMPLE 18.3   Arrange and modify the following half-reactions as necessary, so they add up to produce a balanced redox equation. Identify the oxidation half-reaction and the reduction half-reaction.

$$Fe^{2+}(aq) + 2e^- \rightarrow Fe(s); \quad Al(s) \rightarrow Al^{3+}(aq) + 3e^-$$

This will extend you a bit when it comes to balancing electrons. Two electrons are transferred for each atom of iron and three per atom of aluminum. In what ratio must the atoms be used to equate the electrons gained and lost? Multiply and add the half-reaction equations accordingly.

Reduction:	$3\ Fe^{2+}(aq) + 6e^- \rightarrow 3\ Fe(s)$
Oxidation:	$2\ Al(s) \rightarrow 2\ Al^{3+}(aq) + 6e^-$
Redox:	$3\ Fe^{2+}(aq) + 2\ Al(s) \rightarrow 3\ Fe(s) + 2\ Al^{3+}(aq)$

In this example electrons are transferred two at a time in the iron half-reaction and three at a time in the aluminum half-reaction. The simplest way to equate these is to take the iron half-reaction three times and the aluminum half-reaction twice. This gives six electrons for both half-reactions—just as two Al^{3+} and three O^{2-} balance the positive and negative charges in the ions making up the formula of Al_2O_3.

18.2 OXIDATION NUMBERS AND REDOX REACTIONS

PG 18 C   Given the formula of a chemical species, determine the oxidation number of each element it contains.

18 D   Distinguish between oxidation and reduction in terms of oxidation number change.

The redox reactions that we have discussed up to this point have been rather simple ones involving only two reactants. With Equations 18.1 to 18.4 we can see at a glance which species has gained and which has lost electrons. Some oxidation-reduction reactions are not so readily analyzed. Consider, for example, a reaction that is sometimes used in the general chemistry laboratory to prepare chlorine gas from hydrochloric acid:

$$MnO_2(s) + 4\,H^+(aq) + 2\,Cl^-(aq) \rightarrow Mn^{2+}(aq) + Cl_2(g) + 2\,H_2O(\ell); \quad (18.5)$$

or the reaction, taking place in a lead storage battery, that produces the electrical spark to start an automobile:

$$Pb(s) + PbO_2(s) + 4\,H^+(aq) + 2\,SO_4{}^{2-}(aq) \rightarrow 2\,PbSO_4(s) + 2\,H_2O(\ell) \quad (18.6)$$

Looking at these equations, it is by no means obvious which species are gaining and which are losing electrons.

"Electron bookkeeping" in redox reactions like Equations 18.5 and 18.6 is accomplished by using oxidation numbers, which were introduced in Section 12.2. By following a set of rules, oxidation numbers may be assigned to each element in a molecule or ion. The rules are repeated here for your convenience:

1. **The oxidation number of any elemental substance is 0.** For example, the oxidation number of sodium in Na, of chlorine in Cl_2, or of phosphorus in P_4 is 0.

2. **The oxidation number of an element as a monatomic ion is equal to the charge on that ion.** The oxidation number of chlorine in Cl^- is -1, that of manganese in the Mn^{2+} ion is $+2$.

3. **The oxidation number of combined oxygen is -2** except in peroxides (-1) and superoxides $(-\frac{1}{2})$. In water, H_2O, sulfuric acid, H_2SO_4, sodium hydroxide, NaOH, nitrate ion, $NO_3{}^-$, hypochlorite ion, ClO^-, and nearly all other species, the oxidation state of oxygen is -2. We will not encounter peroxides or superoxides in this text.

4. **The oxidation number of combined hydrogen is $+1$,** except in hydrides (-1). In water, sulfuric acid, sodium hydroxide, ammonium ion, $NH_4{}^+$, and nearly all other species, the oxidation state of hydrogen is $+1$. We will have no further encounter with hydrides in this text.

5. **In any molecular or ionic species, the sum of the oxidation numbers of all elements in the species is equal to the charge on the species.** With this rule we are able to figure out the oxidation numbers of elements not fixed by the first four rules, as the following examples illustrate.

EXAMPLE 18.4 What is the oxidation number of manganese in MnO_2?

Oxidation rules 1, 2, and 4 do not apply. Rule 5 does not give us a direct answer, but it tells us the *sum* of all oxidation numbers in the species. It is:

0. MnO_2 is a compound and therefore has 0 charge.

We now see that the oxidation number contributed by oxygen plus that from manganese is 0. If we could find the amount from oxygen . . . . Rule 3 helps. Oxygen contributes how much to the *total*? (Careful!)

-4.   Each oxygen atom contributes -2, so the total from oxygen is $2(-2) = -4$.

The oxidation state of manganese should now be apparent.

The total is 0, and oxygen is -4. Therefore $-4 +$ oxidation state of $Mn = 0$. Consequently the oxidation state of Mn in MnO_2 is $+4$.

EXAMPLE 18.5   Find the oxidation state of:

   (a) S in $SO_4{}^{2-}$ (b) Cr in $HCr_2O_7{}^-$ (c) Fe in $Fe_2(SO_4)_3$

   (a) $+6$; (b) $+6$; (c) $+3$

(a) $SO_4{}^{2-}$. Oxygen contributes $4(-2) = -8$. The ion has a charge of -2. Therefore $-8 +$ ox. no. of $S = -2$. Ox. no. of $S = +6$.

(b) $HCr_2O_7{}^-$. Hydrogen contributes $+1$ to the total, by rule 4. Oxygen contributes $7(-2) = -14$. The total is the charge on the ion, -1. Therefore $+1 + 2$ (ox. no. of Cr) $- 14 = -1$. Solving, the oxidation number of Cr is $+6$.

(c) $Fe_2(SO_4)_3$. The charge on the sulfate ion is -2, and there are three of them. The sulfate ion contributes $3(-2) = -6$ to the total oxidation number of the compound, which, by Rule 5, is 0. Two iron ions must contribute a total of $+6$ to that 0 total: 2 (ox. no. of Fe) $- 6 = 0$, or $+3$ for each iron(III) ion.

Let's glance back to Equations 18.2 and 18.3 and to the redox reaction in Example 18.2 to see what happens to oxidation numbers during a redox reaction. Looking first at the oxidation half-reaction from Equation 18.2, $Zn \rightarrow Zn^{2+} + 2e^-$, the oxidation number of Zn is 0, by Rule 1; the oxidation number of Zn^{2+} is $+2$, by Rule 2. The oxidation state of zinc *increased* from 0 to $+2$ as it was *oxidized*. Similarly, from

1. Equation 18.3, bromine was oxidized from −1 to 0: $2\,Br^- \rightarrow Br_2$. The oxidation number increased.
2. Example 18.2, Fe^{2+} was oxidized from +2 to +3: $Fe^{2+} \rightarrow Fe^{3+}$. The oxidation number increased.

If we examine an endless number of *oxidation* reactions we find that in each case there is an *increase* in the oxidation number of the element oxidized.

At this point you probably suspect that reduction is accompanied by a reduction in oxidation number. So it is. For the same reactions:

Equation 18.2: $H^+ \rightarrow H_2$, a reduction from +1 to 0

Equation 18.3: $Cl_2 \rightarrow 2\,Cl^-$, a reduction from 0 to −1

Example 18.2: $Al^{3+} \rightarrow Al$, a reduction from +3 to 0

We can now state a second and broader definition of oxidation and reduction. **Oxidation is an increase in oxidation number; reduction is a reduction in oxidation number.** These definitions are more useful in identifying the elements oxidized and reduced in a redox reaction when electron transfer is not apparent. All you must do is find the elements that change oxidation number and determine the direction of the change.

There are some techniques that enable you to spot quickly an element that changes oxidation number, or to dismiss quickly some elements that do not change. These are:

1. An element that is in its elemental state must change. As an element on one side of the equation, its oxidation number is 0; as anything other than an element on the other side, it is *not* 0.
2. In other than elemental form, hydrogen is +1 and oxygen is −2. Unless they are elements on one side, they do not change. In more advanced courses you will have to be alert to the hydride, peroxide, and superoxide exceptions noted in the oxidation number rules.
3. A Group 1A or 2A element has only one oxidation state other than 0. If it does not appear as an element, it does not change. This observation is helpful when you must find the element oxidized or reduced in a conventional equation.

We will now use these ideas to find the elements oxidized and reduced in Equation 18.5,

$$MnO_2(s) + 4\,H^+(aq) + 2\,Cl^-(aq) \rightarrow Mn^{2+}(aq) + Cl_2(g) + 2\,H_2O(\ell)$$

Chlorine is an element on the right, so it must be something else on the left. It is—the chloride ion, Cl^-. The oxidation number change is −1 to 0, an *increase*. Chlorine is *oxidized*.

Neither hydrogen nor oxygen appear as elements, so we conclude that they do not change oxidation state. That leaves manganese. Its oxidation state is +4 in MnO_2 on the left, and +2 as Mn^{2+} on the right. This is a *decrease*, from +4 to +2; manganese is *reduced*.

EXAMPLE 18.6 Determine the element oxidized and the element reduced in a lead storage battery, Equation 18.6:

$$Pb(s) + PbO_2(s) + 4 H^+(aq) + 2 SO_4{}^{2-}(aq) \rightarrow 2 PbSO_4(s) + 2 H_2O(\ell)$$

Assign oxidation numbers to as many elements as necessary until you come up with the pair that changed. Then identify the oxidation and reduction changes. (Careful. This one is a bit tricky.)

------   ------

Lead is both oxidized (0 in Pb to $+2$ in $PbSO_4$) and reduced ($+4$ in PbO_2 to $+2$ in $PbSO_4$).

The oxidation of lead can be spotted quickly, as it is an element on the left. You might have thought sulfur to be the element reduced, but its oxidation state is $+6$ in the sulfate ion whether the ion is by itself on the left, or part of a solid ionic compound on the right.

While oxidation number is a very useful device to keep track of what the electrons are up to in a redox reaction, we should emphasize that it is a man-made concept that has no real physical basis. Unlike the charge of a monatomic ion, the oxidation number of an atom in a molecule or polyatomic ion cannot be measured in the laboratory. It is all very well to talk about "$+4$ manganese" in MnO_2 or "$+6$ sulfur" in the $SO_4{}^{2-}$ ion, but we must not fall into the trap of thinking that the elements in these species actually carry positive charges equal to their oxidation numbers.

18.3 OXIDIZING AGENTS (OXIDIZERS); REDUCING AGENTS (REDUCERS)

PG 18 E Distinguish between an oxidizing agent and a reducing agent.

The two essential reactants in a redox reaction are given special names to indicate the role they play. The species that accepts electrons is referred to as an **oxidizing agent,** or **oxidizer;** the species that donates the electrons so reduction can occur is called a **reducing agent,** or **reducer.** For example, in Equation 18.2,

$$2 H^+(aq) + Zn(s) \rightarrow H_2(g) + Zn^{2+}(aq)$$

H^+ has accepted electrons from Zn—it has *oxidized* Zn to Zn^{2+}—and is therefore the oxidizing agent. Conversely, Zn has donated electrons to H^+—it has reduced H^+ to H_2—and is therefore the reducing agent. In Equation 18.5,

$$MnO_2(s) + 4 H^+(aq) + 2 Cl^-(aq) \rightarrow Mn^{2+}(aq) + Cl_2(g) + 2 H_2O(\ell)$$

Cl⁻ is the reducer, reducing manganese from +4 to +2. The oxidizer is MnO_2—the whole compound, not just the Mn; it oxidizes chlorine from −1 to 0.

The following example summarizes the redox concepts:

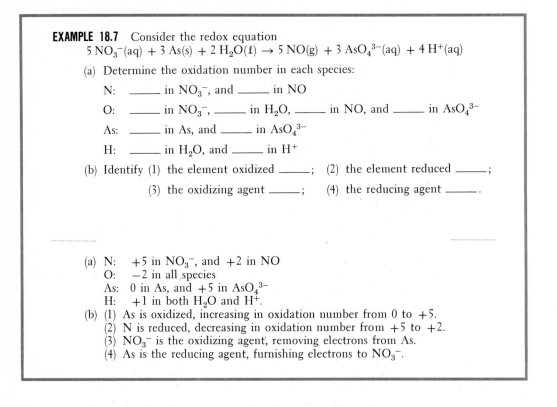

EXAMPLE 18.7   Consider the redox equation
$$5\,NO_3^-(aq) + 3\,As(s) + 2\,H_2O(\ell) \rightarrow 5\,NO(g) + 3\,AsO_4^{3-}(aq) + 4\,H^+(aq)$$

(a) Determine the oxidation number in each species:

N: _____ in NO_3^-, and _____ in NO

O: _____ in NO_3^-, _____ in H_2O, _____ in NO, and _____ in AsO_4^{3-}

As: _____ in As, and _____ in AsO_4^{3-}

H: _____ in H_2O, and _____ in H^+

(b) Identify (1) the element oxidized _____;   (2) the element reduced _____;

(3) the oxidizing agent _____;   (4) the reducing agent _____.

- - - - -   - - - - -

(a) N: +5 in NO_3^-, and +2 in NO
    O: −2 in all species
    As: 0 in As, and +5 in AsO_4^{3-}
    H: +1 in both H_2O and H^+.
(b) (1) As is oxidized, increasing in oxidation number from 0 to +5.
    (2) N is reduced, decreasing in oxidation number from +5 to +2.
    (3) NO_3^- is the oxidizing agent, removing electrons from As.
    (4) As is the reducing agent, furnishing electrons to NO_3^-.

18.4 REDOX REACTIONS AND ACID-BASE REACTIONS COMPARED

PG 18 F   Identify the essential difference between a redox reaction and an acid-base reaction.

18 G   Distinguish between a strong oxidizing agent and a weak oxidizing agent; between a strong reducing agent and a weak reducing agent.

At this stage in our study of oxidation-reduction reactions, it may be useful to pause briefly and point out how redox reactions resemble acid-base reactions.

1. Acid-base reactions involve a transfer of protons; redox reactions, a transfer of electrons.

2. In both cases, the reactants are given special names to indicate their role in the transfer process. An acid is a proton donor; a base is a proton acceptor. A reducing agent is an electron donor; an oxidizing agent is an electron acceptor.

3. Just as certain species (e.g., HCO_3^-, H_2O) can either donate or accept protons and thereby behave as an acid in one reaction and a base in another, so we observe that certain species can either accept or donate electrons, acting as an oxidizing agent in one reaction and a reducing agent in another. An example is the Fe^{2+} ion, which can oxidize Zn atoms to Zn^{2+} in the reaction

$$Fe^{2+}(aq) + Zn(s) \rightarrow Fe(s) + Zn^{2+}(aq)$$

Fe^{2+} can also reduce Cl_2 molecules to Cl^- ions in another reaction:

$$Cl_2(g) + 2\,Fe^{2+}(aq) \rightarrow 2\,Cl^-(aq) + 2\,Fe^{3+}(aq)$$

4. Just as we classify acids and bases as "strong" or "weak" depending upon how readily they donate or accept protons, we use the same adjectives to compare the relative strengths of different oxidizing and reducing agents. A "strong" oxidizing agent is one that has a strong attraction for electrons; a "strong" reducing agent gives up electrons readily.

5. Just as most acid-base reactions in solution reach a state of equilibrium, so most aqueous redox reactions reach equilibrium. Just as the favored side of an acid-base equilibrium can be predicted from acid-base strength, so the favored side of a redox equilibrium can be predicted from oxidizer-reducer strength.

Let's look more closely into the matter of the strength of oxidizing and reducing agents.

18.5 STRENGTHS OF OXIDIZING AGENTS AND REDUCING AGENTS

PG 18 H Given a table of relative strengths of oxidizing and reducing agents, identify the stronger and weaker of two oxidizing or two reducing agents.

With a proper selection of electrodes, a pH meter (p. 412) can be used to measure electrical potential, or voltage, in a redox reaction. The instrument is, in fact, the voltmeter that appears in Figure 18.2. Voltages from such cells enable us to determine the relative strengths of oxidizing and reducing agents.

Table 18.1 is a list of oxidizing agents in order of decreasing strength on the left side of the equation, and of reducing agents in order of increasing

Table 18.1
Relative Strengths of Oxidizing and Reducing Agents

Oxidizing Agent		Reducing Agent
$F_2(g) + 2e^-$	$\rightarrow$	$2\,F^-$
$Cl_2(g) + 2e^-$	$\rightarrow$	$2\,Cl^-$
$\frac{1}{2}\,O_2(g) + 2\,H^+ + 2e^-$	$\rightarrow$	H_2O
$Br_2(\ell) + 2e^-$	$\rightarrow$	$2\,Br^-$
$NO_3 + 4\,H^+ + 3e^-$	$\rightarrow$	$NO(g) + 2\,H_2O$
$Ag^+ + e^-$	$\rightarrow$	$Ag(s)$
$Fe^{3+} + e^-$	$\rightarrow$	Fe^{2+}
$I_2(s) + 2e^-$	$\rightarrow$	$2\,I^-$
$Cu^{2+} + 2e^-$	$\rightarrow$	$Cu(s)$
$2\,H^+ + 2e^-$	$\rightarrow$	$H_2(g)$
$Ni^{2+} + 2e^-$	$\rightarrow$	$Ni(s)$
$Co^{2+} + 2e^-$	$\rightarrow$	$Co(s)$
$Cd^{2+} + 2e^-$	$\rightarrow$	$Cd(s)$
$Fe^{2+} + 2e^-$	$\rightarrow$	$Fe(s)$
$Zn^{2+} + 2e^-$	$\rightarrow$	$Zn(s)$
$Al^{3+} + 3e^-$	$\rightarrow$	$Al(s)$
$Na^+ + e^-$	$\rightarrow$	$Na(s)$
$Ca^{2+} + 2e^-$	$\rightarrow$	$Ca(s)$
$Li^+ + e^-$	$\rightarrow$	$Li(s)$

(Left margin, vertical: STRENGTH — Increasing (up) / Decreasing (down). Right margin, vertical: STRENGTH — Decreasing (up) / Increasing (down).)

strength on the right. (Compare to Table 17.1, p. 403), which lists acids in order of decreasing strength on the left and bases in order of increasing strength on the right.) The strongest oxidizing agent in the table is the F_2 molecule, located at the upper left. Indeed, fluorine is such a powerful oxidizing agent that it is unstable in water solution; it oxidizes water molecules to liberate oxygen. Chlorine, Cl_2, listed just below F_2, is used as a disinfectant in water supplies because of its ability to oxidize harmful organic matter.

The strongest reducing agents are found at the bottom of the right column. Their reducing strength (ability to donate electrons) decreases steadily as we move up the column. The three metals listed at the bottom (and, indeed, all the Group 1A and 2A elements except magnesium and beryllium) are such powerful reducing agents that they react with water. In reaction they reduce H_2O molecules to form elementary hydrogen.

The metals between Na and H_2 in the right column of Table 18.1 are stable in pure water; they are, however, strong enough reducing agents to liberate hydrogen gas from solutions of strong acids. The reducing agents above H_2 in Table 18.1 are relatively weak. In particular, the fluoride ion, F^-, at the top of the right column holds on to its electrons so tightly that it cannot be oxidized in water solution.

Perhaps you recognize Table 18.1 as the source of the activity series (Table 16.1, p. 377) that you used to predict simple redox reactions in writing net ionic equations. Such predictions are based on the relative strengths of the reducing agents. The activity series corresponds with the right side of Table 18.1, from the bottom to the top.

18.6 PREDICTING REDOX REACTIONS

PG 18 I   Given a table of relative strengths of oxidizing and reducing agents and the identity of an oxidizer and a reducer from the table, (a) write the net ionic equation for the redox reaction between them, and (b) predict whether the forward or reverse reaction will be favored at equilibrium.

As Table 17.1 enables us to write acid-base reaction equations and predict the direction that will be favored at equilibrium, Table 18.1 enables us to do the same things for redox reactions. The redox table has a limitation, however. Acid-base reactions are all *single*-proton transfer reactions and are automatically balanced if taken directly from the table. Redox half-reactions, on the other hand, frequently involve unequal numbers of electrons. They must be balanced as in Equation 18.4 and Example 18.2. Examples 18.8 and 18.9 illustrate the process.

EXAMPLE 18.8   Write the net ionic equation for the redox reaction between the cobalt(II) ion, Co^{2+}, and metallic silver, Ag.

Solution: First, as in a Brönsted-Lowry acid-base reaction there must be a proton giver and a proton taker, so in a redox reaction there must be an electron giver (reducer) and an electron taker (oxidizer). Consulting Table 18.1, we find Co^{2+} among the oxidizers and Ag among the reducers. The reduction half-reaction is

Reduction: $Co^{2+}(aq) + 2e^- \rightarrow Co(s)$

To obtain the oxidation half-reaction for silver it is necessary to *reverse* the reduction half-reaction found in the table:

Oxidation: $Ag(s) \rightarrow Ag^+(aq) + e^-$

Multiplying the oxidation equation by 2 to equalize electrons gained and lost, and adding to the reduction equation yields

Reduction: $Co^{2+}(aq) + 2e^- \rightarrow Co(s)$
2 × Oxidation: $\underline{2\,Ag(s) \rightarrow 2\,Ag^+(aq) + 2e^-}$
Redox: $Co^{2+}(aq) + 2\,Ag(s) \rightarrow Co(s) + 2\,Ag^+(aq)$

The principle underlying the prediction of the favored direction of a redox reaction is the same as for an acid-base reaction. The stronger oxidizing agent—strong in its attraction for electrons—will take the electrons from a strong reducing agent—strong in its tendency to donate electrons— to produce the weaker oxidizer and reducer. This is shown in Figure 18.3, which is strikingly similar to Figure 17.1, page 405. In the reaction

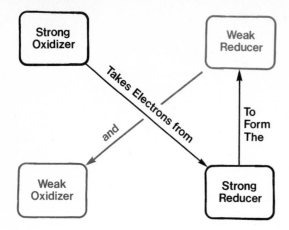

Figure 18.3
Correlation between prediction of a redox reaction and position in Table 18.1. The stronger oxidizing agent (upper left) takes electrons from the stronger reducing agent (lower right), yielding the weaker reducing agent (upper right) and weaker oxidizing agent (lower left). *That direction, forward or reverse, will be favored that has* the weaker reducer and oxidizer as products. *The similarity between acid-base and redox reaction predictions may be seen by comparing this illustration with Figure 17.3 (page 405).*

$2\,H^+(aq) + Zn(s) \rightleftharpoons H_2(g) + Zn^{2+}(aq)$ (Equation 18.2), the positions in the table establish H^+ and Zn as the stronger oxidizer and reducer. The reaction is favored in the forward direction, yielding the weaker oxidizer and reducer, Zn^{2+} and H_2. (Note: We will use double arrows when necessary to indicate the equilibrium character of redox reactions.)

EXAMPLE 18.9   In which direction, forward or reverse, will the redox reaction in Example 18.8 be favored?

———   ———

   Reverse

Ag^+ is a stronger oxidizer than Co^{2+}, and is therefore able to take electrons from cobalt atoms. Also, cobalt atoms are a stronger reducer than silver atoms, and therefore readily release electrons to Ag^+. The weaker reducer and oxidizer, Ag and Co^{2+}, are favored.

EXAMPLE 18.10   Write the redox reaction equation between metallic copper and a strong acid, H^+, and indicate the direction that is favored.

———   ———

Reduction:	$2 H^+(aq) + 2e^- \rightleftharpoons H_2(g)$
Oxidation:	$Cu(s) \rightleftharpoons Cu^{2+}(aq) + 2e^-$
Redox:	$2 H^+(aq) + Cu(s) \rightleftharpoons H_2(g) + Cu^{2+}(aq)$

The reverse reaction is favored.

EXAMPLE 18.11   Write the net ionic equation for the redox reaction between Al(s) and $Ni^{2+}(aq)$, and predict the favored direction, forward or reverse.

Reduction half-reaction:	$Ni^{2+}(aq) + 2e^- \rightleftharpoons Ni(s)$
Oxidation half-reaction:	$Al(s) \rightleftharpoons Al^{3+}(aq) + 3e^-$
$3 \times$ Reduction:	$3 Ni^{2+}(aq) + 6e^- \rightleftharpoons 3 Ni(s)$
$2 \times$ Oxidation:	$2 Al(s) \rightleftharpoons 2 Al^{3+}(aq) + 6e^-$
	$3 Ni^{2+}(aq) + 2 Al(s) \rightleftharpoons 3 Ni(s) + 2 Al^{3+}(aq)$

The forward reaction is favored.

One of the properties of acids listed in Section 17.1 is their ability to release hydrogen gas on reaction with "active" metals. Judging from the result of Example 18.10, copper does not qualify as an active metal. Those metals loosely classified as *active* are the reducers located below hydrogen in Table 18.1. But there is more to the reactions between metals and acids.

EXAMPLE 18.12   Write the equation for the reaction between copper and nitric acid, and predict which direction is favored.
   Copper is in our table of oxidizing and reducing agents, but you will search in vain for HNO_3. The solution inventory species of nitric acid (NO_3^- and H^+) are present, though. We will comment on the imbalance between hydrogen ions and nitrate ions shortly. This reaction summarizes our equation writing methods to this point. Take it all the way.

2 × Reduction: $2 NO_3^-(aq) + 8 H^+(aq) + 6e^- \rightleftharpoons 2 NO(g) + 4 H_2O(\ell)$

3 × Oxidation: $3 Cu(s) \rightleftharpoons 3 Cu^{2+}(aq) + 6e^-$

Redox: $2 NO_3^-(aq) + 8 H^+(aq) + 3 Cu(s) \rightleftharpoons 2 NO(g) + 4 H_2O(\ell) + 3 Cu^{2+}(aq)$

The forward reaction is favored.

Don't worry about those missing nitrate ions, the six unaccounted for from the 8 moles of HNO_3 that furnished the 8 H^+. They're in there as spectators, just enough to balance the 3 Cu^{2+}.

Before leaving this series of examples, let's note that any strong acid is theoretically able to oxidize an active metal, one below hydrogen in Table 18.1, while it won't affect copper. Copper is attacked by nitric acid, however—not by the hydrogen ion alone, but by the combination of the nitrate and hydrogen ions in the acid. The combination is an even stronger oxidizer. In actual practice we find that some predictions based on Table 18.1 must be modified because of other redox possibilities not included in the table, the concentrations of reaction species, and reaction conditions.

18.7 WRITING REDOX EQUATIONS

PG 18 J Given the identity of an oxidizer and reducer, and the identity of their reduced and oxidized products in an acidic solution, write the net ionic equation for the reaction.

Thus far we have considered only redox reactions for which we know the oxidation and reduction half-reactions. We are not always this fortunate. Sometimes we know only the reactants and products. Considering nitric acid, for example, suppose we knew only that the product of the reduction of nitric acid is NO(g). How do we get from this information to the reduction half-reaction given in Table 18.1 so that we can write an equation such as the one in the last example?

Let's develop our method using the reduction of NO_3^- to NO in an acidic aqueous solution. We begin the reduction equation by writing

$$NO_3^-(aq) \rightarrow NO(g)$$

and then go about balancing the equation. Nitrogen is in balance, but there is a shortage of two oxygen atoms on the right. The reaction takes place in an acidic water solution, so there is a good supply of hydrogen ions and water molecules that may be either reactants or products in the half-reaction. The H^+ ions can form water with the oxygen released by the NO_3^- ion. We therefore **balance oxygen by adding one water molecule for each oxygen atom required,** two in this case.

$$NO_3^-(aq) \rightarrow NO(g) + 2 H_2O(\ell)$$

We turn next to hydrogen: **hydrogen is balanced with H^+ ions.** In this case four H^+ ions are required to make two water molecules.

$$4\,H^+(aq) + NO_3^-(aq) \rightarrow NO(g) + 2\,H_2O(\ell)$$

Hydrogen, nitrogen, and oxygen are now all in balance—but the equation is not. We've encountered quite a few net ionic equations by now, and without any special effort or attention on our part, the total charge of all species on one side of the equation has usually been equal to the total charge of all species on the other side. The charges have not always been zero, but they have always been equal, and therefore balanced. But this is not true in the equation above. Checking charge, we find a net charge of $+3$ on the left ($+4 - 1$), and zero on the right. **Charge is balanced by adding the required number of electrons to the side that has the more positive charge.** In this case it is the left side:

$$4\,H^+(aq) + NO_3^-(aq) + 3e^- \rightarrow NO(g) + 2\,H_2O(\ell)$$

We have now developed the reduction equation in Table 18.1. Oxidation equations are written in exactly the same way.

When the reactants and products of a redox reaction are known we may write a balanced redox equation by following these steps:

1. Identify the element reduced and the element oxidized. For each, write a partial half-reaction equation, using the element in its original form (element, monatomic ion, or part of a polyatomic ion or compound) on the left, and in its final form on the right.
2. Balance each half-reaction equation separately.
   a. First balance the element oxidized or reduced.
   b. Balance elements other than hydrogen and oxygen, if any.
   c. Balance oxygen by adding water molecules where necessary.
   d. Balance hydrogen by adding hydrogen ions, H^+, where necessary.
   e. Balance charges by adding electrons where necessary.
3. Equalize the electrons gained or lost by multiplication of the half-reaction equations.
4. Add the half-reaction equations to get the net ionic equation.

Steps 3 and 4 are what you did with the half-reaction equations in earlier sections.

Notice that the above instructions are for redox reactions in *acid* solutions. The procedure is somewhat different with basic solutions, but we will omit that procedure in this introductory text.

EXAMPLE 18.13   Write the net ionic equation for the redox reaction between iodide and sulfate ions in an acidic solution. The products are iodine and sulfur.

$$I^-(aq) + SO_4^{2-}(aq) \rightarrow I_2(s) + S(s).$$

First, identify the element reduced and the element oxidized.

- - - - - -   - - - - - -

Sulfur is reduced (ox. no. change +6 to 0) and iodine is oxidized (-1 to 0). Balance atoms first, and then charges, in the oxidation half-reaction:

$$I^-(aq) \rightarrow I_2(s)$$

- - - - - -   - - - - - -

$2\,I^-(aq) \rightarrow I_2(s) + 2e^-$. (This one happens to be in Table 18.1.) Now for the reduction half-reaction:

$$SO_4^{2-}(aq) \rightarrow S(s)$$

Sulfur is already in balance. The only other element is oxygen. According to Step 2c, oxygen is balanced by adding the necessary water molecules. Complete that step.

- - - - - -   - - - - - -

$$SO_4^{2-}(aq) \rightarrow S(s) + 4\,H_2O(\ell)$$

Four oxygen atoms in a sulfate ion require four water molecules.
Next comes the hydrogen balancing, using H^+ ions.

- - - - - -   - - - - - -

$$8\,H^+(aq) + SO_4^{2-}(aq) \rightarrow S(s) + 4\,H_2O(\ell)$$

Finally, add to the positive side the electrons that will bring the charges into balance.

- - - - - -   - - - - - -

$$8\,H^+(aq) + SO_4^{2-}(aq) + 6e^- \rightarrow S(s) + 4\,H_2O(\ell)$$

On the left there are 8 + charges from hydrogen ion and 2 − charges from sulfate ion, a net of 6 +. On the right the net charge is zero. Charge is balanced by adding 6 electrons to the left (positive) side.
Now that you have the two half-reaction equations, finish writing the net ionic equation (Steps 3 and 4) as you did before.

$$2\,I^-(aq) \rightarrow I_2(s) + 2e^-$$
$$8\,H^+ + SO_4^{2-}(aq) + 6e^- \rightarrow S(s) + 4\,H_2O(\ell)$$

- - - - - -   - - - - - -

Reduction: $8\,H^+(aq) + SO_4^{2-}(aq) + 6e^- \rightarrow S(s) + 4\,H_2O(\ell)$

$3 \times$ Oxidation: $6\,I^-(aq) \rightarrow 3\,I_2(s) + 6e^-$

Redox: $8\,H^+(aq) + SO_4^{2-}(aq) + 6\,I^-(aq) \rightarrow 4\,H_2O(\ell) + 3\,I_2(s)$

Let's check to make sure the equation is, indeed, balanced:
—the atoms balance (six I, one S, four O, and eight H atoms on each side);
—the charge balances $(+8 - 2 - 6 = 0 + 0 + 0)$.

EXAMPLE 18.14 The permanganate ion, MnO_4^-, is a strong oxidizing agent, and will oxidize chloride ion to chlorine in an acidic solution. Manganese ends up as a monatomic manganese(II) ion. Write the net ionic equation for the redox reaction.

This is a challenging example, but watch how it falls into place following the procedure we have outlined. Begin by separating the formulas of the starting and ending species in the oxidation half-reaction from the verbal description of the reaction. Write them on opposite sides of an arrow. Complete the oxidation half-reaction equation.

$$2\,Cl^-(aq) \rightarrow Cl_2(g) + 2e^-$$

With a switch from iodine to chlorine, this is the same oxidation half-reaction as in the last example.

Now write the formulas of the starting and ending species for the reduction reaction on opposite sides of the arrow.

$$MnO_4^-(aq) \rightarrow Mn^{2+}(aq)$$

Can you take the reduction to a complete half-reaction equation? First do oxygen, then hydrogen, and finally charge.

$$8\,H^+(aq) + MnO_4^-(aq) + 5e^- \rightarrow Mn^{2+}(aq) + 4\,H_2O(\ell)$$

By steps: Oxygen: $MnO_4^-(aq) \rightarrow Mn^{2+}(aq) + 4\,H_2O(\ell)$ (four waters for four oxygen atoms).

Hydrogen: $8\,H^+(aq) + MnO_4^-(aq) \rightarrow Mn^{2+}(aq) + 4\,H_2O(\ell)$ (eight H^+ for four waters).

Charge: $+8 - 1$ on left is $+7$; $+2 + 0$ on right is $+2$. Charge is balanced by adding 5 negatives on the left, or five electrons, as in the final answer.

Now combine the half-reaction equations for the net ionic equation.

2 × Reduction: $16 H^+(aq) + 2 MnO_4^-(aq) + 10e^- \rightarrow 2 Mn^{2+}(aq) + 8 H_2O(l)$

5 × Oxidation: $\underline{\hspace{5cm} 10 Cl^-(aq) \rightarrow 5 Cl_2(g) + 10e^-}$

Redox: $16 H^+(aq) + 2 MnO_4^-(aq) + 10 Cl^-(aq) \rightarrow 2 Mn^{2+}(aq) + 8 H_2O(l) + 5 Cl_2(g)$

Checking:

—atoms balance (sixteen H, two Mn, eight O, and ten Cl);
—charges balance ($+16 - 2 - 10 = +4$).

Can you imagine a trial and error approach to an equation such as this?

CHAPTER 18 IN REVIEW

18.1 ELECTRON TRANSFER REACTIONS

18 A Distinguish between oxidation and reduction in terms of electrons gained or lost. (421)

18 B Given an oxidation half-reaction equation and a reduction half-reaction equation, add them to obtain a balanced redox reaction equation. (421)

18.2 OXIDATION NUMBERS AND REDOX REACTIONS

18 C Given the formula of a chemical species, determine the oxidation number of each element it contains. (425)

18 D Distinguish between oxidation and reduction in terms of oxidation number change. (425)

18.3 OXIDIZING AGENTS (OXIDIZERS); REDUCING AGENTS (REDUCERS)

18 E Distinguish between an oxidizing agent and a reducing agent. (429)

18.4 REDOX REACTIONS AND ACID-BASE REACTIONS COMPARED

18 F Identify the essential difference between a redox reaction and an acid-base reaction. (430)

18 G Distinguish between a strong oxidizing agent and a weak oxidizing agent; between a strong reducing agent and a weak reducing agent. (430)

18.5 STRENGTHS OF OXIDIZING AGENTS AND REDUCING AGENTS

18 H Given a table of relative strengths of oxidizing and reducing agents, identify the stronger and weaker of two oxidizing or two reducing agents. (431)

18.6   PREDICTING REDOX REACTIONS

18 I   Given a table of relative strengths of oxidizing and reducing agents and the identity of an oxidizer and a reducer from the table, (a) write the net ionic equation for the redox reaction between them, and (b) predict whether the forward or reverse reaction will be favored at equilibrium. (433)

18.7   WRITING REDOX EQUATIONS

18 J   Given the identity of an oxidizer and reducer, and the identity of their reduced and oxidized products in an acidic solution, write the net ionic equation for the reaction. (436)

TERMS AND CONCEPTS

Half-reaction equation (422)
Electron transfer reaction (422)
Oxidation-reduction (redox) reaction (423)
Half-reaction (423)
Reduction (423, 428)

Oxidation (423, 428)
Oxidation number (426)
Oxidizing agent; oxidizer (429)
Reducing agent; reducer (429)

Most of these terms and many others appear in the Glossary.

QUESTIONS AND PROBLEMS

To save space, the designations (s), (l), (g), and (aq) are omitted. All ions below are in aqueous solution.

Section 18.1

18.1) Define oxidation; define reduction. It is sometimes said that oxidation and reduction are simultaneous processes—that you cannot have one without the other. From the standpoint of your definitions, explain why.

18.2) Classify each of the following half-reaction equations as oxidation or reduction half-reactions:
(a) $2\,Cl^- \rightarrow Cl_2 + 2e^-$
(b) $Na \rightarrow Na^+ + e^-$
(c) $Sn^{2+} \rightarrow Sn^{4+} + 2e^-$
(d) $O_2 + 4\,H^+ + 4e^- \rightarrow 2\,H_2O$

18.25) Using any example of a redox reaction, explain why such reactions are described as electron transfer reactions.

18.26) Classify each of the following half-reaction equations as oxidation or reduction half-reactions:
(a) $Zn \rightarrow Zn^{2+} + 2e^-$
(b) $2\,H^+ + 2e^- \rightarrow H_2$
(c) $Fe^{2+} \rightarrow Fe^{3+} + e^-$
(d) $NO + 2\,H_2O \rightarrow NO_3^- + 4\,H^+ + 3e^-$

For the next two questions in each column, classify the equation given as an oxidation or a reduction half-reaction equation.

18.3) Tarnishing of silver:
$$2\,Ag + S^{2-} \rightarrow Ag_2S + 2e^-$$

18.4) One side of an automobile battery:
$$PbO_2 + SO_4^{2-} + 4\,H^+ + 2e^- \rightarrow PbSO_4 + 2\,H_2O$$

18.27) Dissolving ozone, O_3, in water:
$$O_3 + H_2O + 2e^- \rightarrow O_2 + 2\,OH^-$$

18.28) Dissolving gold (Z = 79):
$$Au + 4\,Cl^- \rightarrow AuCl_4^- + 3e^-$$

18.5) Combine the following half-reaction equations to produce a balanced redox equation: $Ni^{2+} + 2e^- \rightarrow Ni$; $Mg \rightarrow Mg^{2+} + 2e^-$.

18.6) The half-reaction at one electrode of a lead storage battery, used in automobiles and boats, is given in Question 18.4. The reaction at the other electrode is

$$Pb + SO_4^{2-} \rightarrow PbSO_4 + 2e^-$$

Write the equation for the overall battery reaction.

18.29) Combine the following half-reaction equations to produce a balanced redox equation: $Cr \rightarrow Cr^{3+} + 3e^-$; $Cl_2 + 2e^- \rightarrow 2\,Cl^-$.

18.30) The half-reactions that take place at the electrodes of an alkaline cell, widely used in flashlights, calculators, etc., are

$$Ni_2O_3 + 3\,H_2O + 2e^- \rightarrow 2\,Ni(OH)_2 + 2\,OH^-$$
$$Cd + 2\,OH^- \quad \rightarrow \quad Cd(OH)_2 + 2e^-$$

Which equation is for the oxidation half-reaction? Write the overall equation for the cell.

Section 18.2

In the next two questions in each column, give the oxidation state of the element whose symbol is underlined in each formula.

18.7) $\underline{Mg}^{2+}$; $\underline{Cl}^-$; $\underline{Cl}O^-$; $K\underline{Cl}O_3$

18.8) $\underline{N}_2O_5$; $\underline{N}H_4^+$; $\underline{Mn}O_4^-$; $Na_2H\underline{P}O_3$

18.31) $\underline{Al}^{3+}$; $\underline{S}^{2-}$; $\underline{S}O_3^{2-}$; $Na_2\underline{S}O_4$

18.32) $\underline{N}_2O_3$; $\underline{N}O_3^-$; $\underline{Cr}O_4^{2-}$; $NaH_2\underline{P}O_4$

In the next three questions in each column, (1) identify the element experiencing oxidation or reduction; (2) state "oxidized" or "reduced"; and (3) show the change in oxidation number. Example: $2\,Cl^- \rightarrow Cl_2 + 2e^-$. Chlorine oxidized from -1 to 0.

18.9) a) $Cu^{2+} + 2e^- \rightarrow Cu$
b) $Co^{3+} + e^- \rightarrow Co^{2+}$

18.10) a) $H_2O + SO_3^{2-} \rightarrow SO_4^{2-} + 2\,H^+ + 2e^-$
b) $PH_3 \rightarrow P + 3\,H^+ + 3e^-$

18.11) a) $2\,HF \rightarrow F_2 + 2\,H^+ + 2e^-$
b) $MnO_4^{2-} + 2\,H_2O + 2e^- \rightarrow$
$MnO_2 + 4\,OH^-$

18.33) a) $Br_2 + 2e^- \rightarrow 2\,Br^-$
b) $Pb^{2+} + 2\,H_2O \rightarrow PbO_2 + 4\,H^+ + 2e^-$

18.34) a) $8\,H^+ + IO_4^- + 8e^- \rightarrow I^- + 4\,H_2O$
b) $4\,H^+ + O_2 + 4e^- \rightarrow 2\,H_2O$

18.35) a) $NO_2 + H_2O \rightarrow NO_3^- + 2\,H^+ + e^-$
b) $2\,Cr^{3+} + 7\,H_2O \rightarrow$
$Cr_2O_7^{2-} + 14\,H^+ + 6e^-$

Section 18.3

18.12) In the reaction between copper(II) oxide and hydrogen, identify the oxidizing agent and the reducing agent:

$$CuO + H_2 \rightarrow Cu + H_2O$$

18.13) Name the oxidizing and reducing agents in the reaction:

$$BrO_3^- + 3\,HNO_2 \rightarrow Br^- + 3\,NO_3^- + 3\,H^+$$

18.36) Identify the oxidizing and reducing agents in

$$Cl_2 + 2\,Br^- \rightarrow 2\,Cl^- + Br_2$$

18.37) What is the oxidizing agent in the equation for the storage battery, $Pb + PbO_2 + 4\,H^+ + 2\,SO_4^{2-} \rightarrow 2\,PbSO_4 + 2\,H_2O$? What does it oxidize? Also name the reducing agent and the species it reduces.

Section 18.4

18.14) Show how redox and acid-base reactions parallel each other—how they are similar, but also what makes them different.

18.38) Explain how a strong acid is similar to a strong reducer. Also, explain how a strong base compares to a strong oxidizer.

Section 18.5

18.15) Which is the stronger oxidizing agent, Ag^+ or H^+? What is the basis of your selection? What is the meaning of the statement that one substance is a stronger oxidizer than another?

18.39) Identify the stronger reducer between Zn and Fe^{2+}. On what basis do you make your decision? What is the significance of one reducer being stronger than another?

18.16) Arrange the following reducing agents in order of *decreasing* strength—i.e., strongest reducer first: H_2; Al; Cl^-; Fe^{2+}.

18.40) Arrange the following oxidizers in order of *increasing* strength—i.e., weakest oxidizing agent first: Na^+; Br_2; Fe^{2+}; Cu^{2+}.

Section 18.6

In this section, write the redox equation for the two redox reactants given, using Table 18.1 as a source of the required half-reactions. Then predict the direction in which the reaction will be favored at equilibrium.

18.17) $Ni + Zn^{2+} \rightleftharpoons$

18.18) $Fe^{3+} + Co \rightleftharpoons$

18.19) $O_2 + H^+ + Ca \rightleftharpoons$

18.41) $Br_2 + I^- \rightleftharpoons$

18.42) $H^+ + Br^- \rightleftharpoons$

18.43) $NO + H_2O + Fe^{2+} \rightleftharpoons$

Section 18.7

In this section, each "equation" identifies an oxidizer and a reducer, as well as the oxidized and reduced products of the redox reaction. Write the separate oxidation and reduction half-reaction equations, assuming that the reaction takes place in an acidic solution, and add them to produce a balanced redox equation.

18.20) $Ag + SO_4^{2-} \rightarrow Ag^+ + SO_2$

18.21) $NO_3^- + Zn \rightarrow NH_4^+ + Zn^{2+}$

18.22) $Cr_2O_7^{2-} + Fe^{2+} \rightarrow Cr^{3+} + Fe^{3+}$

18.23) $I^- + MnO_4^- \rightarrow I_2 + MnO_2$

18.24) $BrO_3^- + Br^- \rightarrow Br_2$

18.44) $S_2O_3^{2-} + Cl_2 \rightarrow SO_4^{2-} + Cl^-$

18.45) $Sn + NO_3^- \rightarrow H_2SnO_3 + NO_2$

18.46) $C_2O_4^{2-} + MnO_4^- \rightarrow CO_2 + Mn^{2+}$

18.47) $Cr_2O_7^{2-} + NH_4^+ \rightarrow Cr_2O_3 + N_2$

18.48) $As_2O_3 + NO_3^- \rightarrow AsO_4^{3-} + NO$

energy

part i: the nature of energy

19.1 PHYSICAL AND CHEMICAL ENERGY—A REVIEW

When the subject of *energy* was introduced qualitatively in Chapter 2, many of its basic ideas were described. Energy was defined as the ability to do *work*, or *to exert a force* over a distance. Two kinds of *mechanical energy* were identified. The energy possessed by an object in motion is *kinetic energy*, whereas energy possessed by an object because of its position in a gravitational, electrical, or magnetic force field is *potential energy*. *Temperature* is a measure of the average kinetic energy of the particles in a sample of matter (p. 277). *Chemical energy* turns out to be associated with potential energy changes resulting from the rearrangement of electrons, protons, atoms, ions, and molecules in electric fields.

A natural tendency for a system to reach its lowest energy state was identified as one of the major driving forces leading to the formation of chemical bonds (p. 211). The energy of a chemical change is the combination of energies absorbed and released as old bonds are broken and new bonds are formed. In Chapter 14 we considered intermolecular forces, which have their origin in electrostatic attractions and repulsions. Molar heat of vaporization is one physical property that is directly concerned with energy; two others will be introduced in this chapter.

The Law of Conservation of Energy (p. 23) reminds us that in ordinary (nonnuclear) physical and chemical changes, energy may be converted from one form to another, but it is neither created nor destroyed. This suggests that chemical energy may be changed to such things as electrical energy in car batteries, kinetic energy in a moving automobile, sound in a firecracker, light in flames, and heat in the warming of homes. A detailed study of energy and work related to chemical change takes nearly all of these into account. For our purposes, however, we will limit consideration to the most obvious of these forms of energy, heat. In fact, the terms *endothermic* and *exothermic* (p. 22) refer respectively to the absorption or release of heat energy.

19.2 UNITS OF ENERGY

PG 19 A State the relationship between the two principal scientific units for heat energy. Given one, calculate the other.

So far our descriptions of energy have been entirely qualitative. In this chapter we will approach the quantitative side of the subject. This means we will have to identify a unit of energy. The SI (p. 29) energy unit is the **joule**, which is assigned the symbol J. The joule is formally defined in terms of mechanical work, a force times a distance: **one joule is a force of one newton** (another SI unit) **acting over a distance of one meter.** The joule is gradually taking the place of the more familiar unit, the *calorie*. **One calorie is the quantity of heat required to raise the temperature of one gram of water one degree Celsius.** The relationship between a calorie and a joule is

$$1 \text{ calorie} = 4.184 \text{ joules} \tag{19.1}$$

Both the calorie and the joule are small quantities of energy, so energy is often expressed in terms of the *kilo-* unit, which is 1000 times larger—the kilocalorie (kcal) and kilojoule (kJ):

$$1 \text{ kcal} = 1000 \text{ cal}; \quad 1 \text{ kJ} = 1000 \text{ J} \tag{19.2}$$

It follows that

$$1 \text{ kilocalorie} = 4.184 \text{ kilojoules} \tag{19.3}$$

The term *Calorie* used in nutrition is actually the kilocalorie. Capitalization of the first letter is used to distinguish the larger unit from the smaller. Caloric food requirements vary considerably among individuals. A small adult doing average physical work needs as little as 2000 to 4000 Calories per day. This requirement may increase to 5000 to 6000 Calories per day for a large man engaged in strenuous physical labor.

In this text we will use the SI energy units, the joule and the kilojoule. Each example or problem answer, however, will be followed by its equivalent in calories or kilocalories, enclosed in square brackets.

EXAMPLE 19.1 Convert 859 calories to kilojoules.

Between Equations 19.1 to 19.3 you should be able to chart a unit path from the given quantity to the answer. Set up and solve.

$$859 \; \cancel{cal} \times \frac{4.184 \; \cancel{J}}{1 \; \cancel{cal}} \times \frac{1 \; kJ}{1000 \; \cancel{J}} = 3.59 \; kJ$$

A precautionary note: there is a strong tendency to think of the temperature degree as a unit of heat. It is not. The distinction between temperature and energy can probably be seen in a practical kitchen procedure. If you place a small quantity of water on a stove to bring it from room temperature to boiling for, say, two cups of tea, you must deliver a certain amount of energy from your electric or gas range. The final temperature is 100°C. If you perform the identical act that will provide 20 cups of tea for a party, you will heat to boiling ten times the quantity of water required for two cups. This will require ten times as much heat as is needed for two cups. Yet the 100°C temperatures reached are the same. Clearly temperature degrees are not suitable units for quantity of heat energy.

part ii: energy in physical change

19.3 ENERGY, TEMPERATURE, AND CHANGE OF STATE

When you take an ice cube from a refrigerator it has a dull surface and feels dry. This is because it *is* dry; it is all solid at a temperature below the melting point. If the cube is placed in a beaker and allowed to warm up, the first visible change is that the surface becomes shiny. The ice at the surface has melted and has changed from the solid to the liquid state. Eventually the entire cube melts. Experiments show that this occurs at the melting point; there is no increase in temperature during melting.

If the liquid water is allowed to sit, its temperature rises to room temperature. If heat is applied, the temperature continues to rise without a change in appearance until the boiling point is reached. At that time bubbles appear as the liquid is changed to the invisible gas we call steam. If you continue to add heat, the liquid eventually disappears when it is changed entirely to steam.

Experiments show that the change from liquid to gas also occurs at constant temperature, the boiling point. If heat is added to the steam, its temperature rises without further change in appearance.

There are not many pure substances that we are familiar with in all three states, as we are with water. But if we were to follow the same procedure with all elements and those compounds that do not decompose on heating, we would find that they all behave like water. Two kinds of change can be identified: (1) as a solid, liquid, or gas is heated, temperature increases without a change of state; and (2) as a substance is heated at the melting point or boiling point, there is a change of state without a change in temperature. We will now examine these two kinds of change separately.

19.4 ENERGY AND CHANGE OF TEMPERATURE: SPECIFIC HEAT

PG 19 B   Given (1) the mass of a pure substance, (2) its specific heat, and (3) its temperature change, or initial and final temperatures, calculate the heat flow.

19 C   Given the heat flow to or from a known mass of a substance, and its temperature change, or initial and final temperatures, calculate the specific heat of the substance.

In order to raise the temperature of a substance you must heat it; heat must flow into the substance. If the substance is to be cooled, there must be heat flow from the substance to its surroundings. Experiments such as those illustrated in Figures 19.1 and 19.2 indicate that heat flow is proportional to both the mass of the sample and its temperature change. Combining these proportionalities and introducing a proportionality constant yields the equation

$$Q = m \times c \times \Delta T \tag{19.4}$$

where Q is the heat flow in joules, m is the mass in grams, c is the proportionality constant, and ΔT is the temperature change, or final temperature minus initial temperature, $T_f - T_i$.

The proportionality constant, c, is a property of a pure substance called **specific heat. Specific heat is the heat flow required to change the temperature of one gram of a substance one degree Celsius.** The units of specific heat may be determined by solving Equation 19.4 for the proportionality constant:

$$\text{specific heat} = c = \frac{Q}{m \times \Delta T} = \frac{\text{joules}}{\text{grams } °C} = \frac{J}{g \; °C} \tag{19.5}$$

These units are read "joules per gram degree." Specific heats of selected substances are given in Table 19.1.

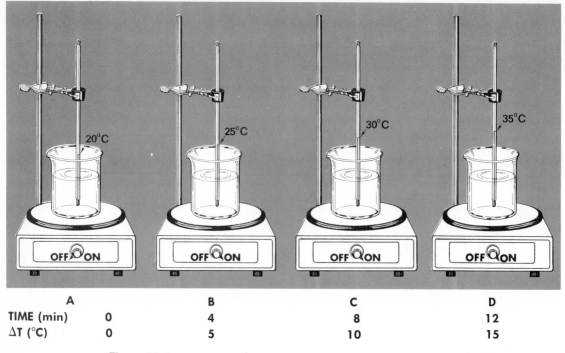

	A	B	C	D
TIME (min)	0	4	8	12
ΔT (°C)	0	5	10	15

Figure 19.1

Experiment demonstrating relationship between heat absorbed by a sample of matter and its temperature change. A beaker containing 100 grams of water at 20°C is placed on a hot plate (A). Assume that, when the hot plate is turned on, water absorbs heat at a constant rate. This makes time proportional to heat absorbed; the longer the water is heated, the more heat it absorbs. Therefore time is a "measure" of heat flow to the water. After 4 minutes (B), the temperature has risen to 25°C. At 8 minutes (C) the temperature is 30°C, and at 12 minutes (D), 35°C. These data and the total change in temperature, ΔT, are tabulated beneath the illustrations. The data show that ΔT is proportional to time. At 4 minutes, ΔT is 5°C. At 8 minutes (twice 4), ΔT is 10°C (twice 5°); and at 12 minutes (3 times 4), ΔT is 15°C (3 × 5). Because time is a measure of heat, we conclude that ΔT is proportional to heat absorbed for a fixed quantity of water.

Specific heat is a measure of the relative ease with which a given quantity of a substance may be heated or cooled. A substance with a low specific heat absorbs little energy in warming through a given temperature range, and loses little as it cools. This is why metals, which generally have specific heats less than 0.8 J/(g °C), are well suited for pots and pans used for cooking.

To store heat energy, a substance with a high specific heat is best. In solar heating systems, rocks are used to store heat from the sun in daytime for use at night. Their specific heats are relatively high. They are therefore able to absorb a large amount of heat per unit of mass, which is later given back when needed. At 4.18 J/g °C, water has one of the highest specific heats of all substances. This is why air temperatures near large bodies of water are usually warmer in winter and cooler in summer than nearby inland temperatures.

Specific heat problems may be solved algebraically by direct substitution into Equation 19.4 or 19.5.

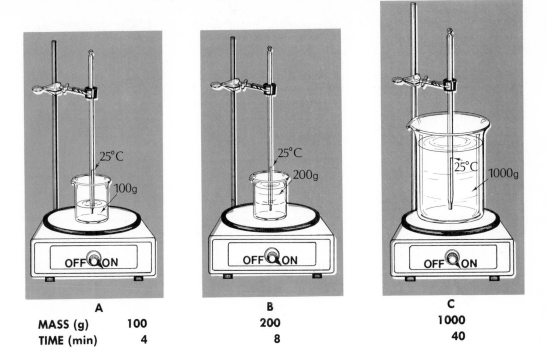

A   **B**   **C**

	A	B	C
MASS (g)	100	200	1000
TIME (min)	4	8	40

Figure 19.2
Experiment demonstrating relationship between mass and heat absorbed for a constant temperature change. All temperature changes are from 20°C to 25°C. Beaker A contains 100 grams of water, and requires 4 minutes. Beaker B contains 200 grams (twice as much as A) and requires 8 minutes (twice as long as A). Beaker C holds 1000 grams (10 times A) and requires 40 minutes (10 times A). Assuming that heat absorbed is proportional to time, as in Figure 19.1, we conclude that heat absorbed is proportional to mass for a constant temperature change.

Table 19.1
Selected Specific Heats

Substance	$\frac{J}{g\ °C}$	$\frac{cal}{g\ °C}$	Substance	$\frac{J}{g\ °C}$	$\frac{cal}{g\ °C}$
Elements			Sulfur	0.732	0.175
Aluminum	0.88	0.21	Zinc	0.38	0.092
Carbon			*Compounds*		
Diamond	0.50	0.12	Acetone	2.1	0.51
Graphite	0.71	0.17	Benzene	1.8	0.42
Cobalt	0.46	0.11	Carbon		
Copper	0.38	0.092	tetrachloride	0.84	0.20
Gold	0.13	0.031	Ethanol	2.5	0.59
Iron	0.444	0.106	Methanol	2.6	0.61
Lead	0.16	0.038	Ice [$H_2O(s)$]	2.1	0.49
Magnesium	1.0	0.24	Steam [$H_2O(g)$]	2.0	0.48
Silicon	0.71	0.17	Water [$H_2O(\ell)$]	4.18	1.00
Silver	0.24	0.057			

EXAMPLE 19.2   How much heat is required to raise the temperature of 500 grams of water for a pot of tea from 14°C to 90°C? Answer in both joules and kilojoules.

Direct substitution into Equation 19.4 yields the heat flow in joules, which may be expressed in kilojoules too. For ΔT, substitute $T_{final} - T_{initial}$.

-------   -------

$$Q = m \times c \times \Delta T = 500\,g \times \frac{4.18\ J}{g\,°C} \times (90 - 14)\,°C = 1.6 \times 10^5\ J = 1.6 \times 10^2\ kJ$$

$$[3.8 \times 10^4\ cal = 38\ kcal]$$

As a quantitative physical property, the specific heat of a substance must be found by laboratory measurements. This is done in a device known as a **calorimeter** (Fig. 19.3). If you find the joules of heat gained or lost by a known mass of substance as it passes through a measured temperature change, you can calculate the specific heat by direct substitution into Equation 19.5.

EXAMPLE 19.3   In a college laboratory experiment a student observes an increase from 25.0°C to 31.7°C when 141 grams of aluminum absorb 803 joules (192 cal) of heat. Calculate the specific heat of aluminum *from these data.*

-------   -------

$$c = \frac{803\ J}{(141\ g)(31.7 - 25.0)\,°C} = 0.85\ J/g\,°C \quad [0.20\ cal/g\,°C]$$

If any three of the four quantities—heat flow, mass, specific heat, and temperature change—are known, the fourth may be found algebraically from Equation 19.4. Notice that ΔT will be negative if the final temperature is lower than the initial temperature. This gives a negative heat flow, which indicates that the substance is cooling rather than heating.

19.5 ENERGY AND CHANGE OF STATE

Heat flows into or out of a substance when it changes state. It takes heat to melt a solid, and heat is released by a liquid when it freezes. This is also true for the

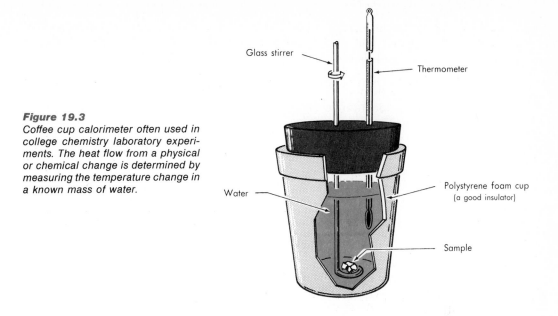

Figure 19.3
Coffee cup calorimeter often used in college chemistry laboratory experiments. The heat flow from a physical or chemical change is determined by measuring the temperature change in a known mass of water.

change between a liquid and a gas. Heat is absorbed as a liquid vaporizes, and released when a gas condenses. In this section you will learn how to calculate the heat flow that accompanies a change of state.

HEAT OF VAPORIZATION

PG 19 D   Given two of the following, calculate the third: (a) mass of a pure substance changing between the liquid and vapor states; (b) heat of vaporization or condensation; (c) heat flow.

Any time you lift an object from the floor and place it on a table, you do work (Fig. 2.8). There is a gravitational attraction between the object and the earth. You must expend energy—do work—against that attractive force when you increase the distance of separation between the objects. Similarly, there is an attractive force between molecules in a liquid. To separate those molecules and convert the liquid to a gas requires energy. To boil water, you heat it. When a liquid evaporates at room temperature, heat is absorbed from the surroundings. This is why you feel cold when you come out of a shower or swimming pool; the evaporation of moisture draws heat from the body surface, lowering its temperature.

In Section 14.2, page 308, **molar heat of vaporization** was defined as **the energy required to vaporize one mole of a liquid.** From the definition, molar heat of vaporization is measured in energy units per mole, usually kilojoules per mole. Alternately, we may speak of **heat of vaporization as the energy required to vaporize one gram of a liquid.** According to this definition, heat of vaporization units are kilojoules per gram. Heats of vaporization for several substances are given in Table 19.2.

When a vapor condenses to a liquid at the boiling point, the reverse energy change occurs. The heat flow is then referred to as **heat of condensation.** Values are identical to heats of vaporization, except that they are negative, indicating that heat is flowing *from* the substance (an exothermic change), rather than into it.

The energy required to vaporize a given quantity of a liquid may be calculated from the equation

$$Q = m \times \Delta H_{vap} \qquad (19.6)$$

where m is mass in grams and ΔH_{vap} is heat of vaporization in kilojoules per gram. Applications are straightforward "substitute and solve" problems.

EXAMPLE 19.4 Calculate the heat of vaporization of a substance if 241 kilojoules (57.6 kcal) are required to vaporize a 78.2-gram sample.

We recommend that you solve Equation 19.6 for ΔH_{vap} first, and then substitute numbers and units and calculate the answer. Equally correct, you may substitute numbers and units directly into the equation and solve for the missing item. Handling units is slightly more difficult by the second method. Take your choice and solve the problem completely.

$$\Delta H_{vap} = \frac{Q}{m} = \frac{241 \text{ kJ}}{78.2 \text{ g}} = 3.08 \text{ kJ/g} \quad [737 \text{ cal/g}]$$

EXAMPLE 19.5 Calculate the energy required to vaporize 250 grams of water at its boiling point.

$$Q = m \times \Delta H_{vap} = 250 \text{ g} \times \frac{2.26 \text{ kJ}}{\text{g}} = 565 \text{ kJ} \quad [135 \text{ kcal}]$$

Table 19.2
Latent Heats of Selected Substances

Substance	Melting Point (°C)	Boiling Point (°C)	Heat of Fusion		Heat of Vaporization	
			(J/g)	(cal/g)	(kJ/g)	(cal/g)
$H_2O(s)$	0		335	80		
$H_2O(\ell)$		100			2.26	540
Na	98	892	113	27	4.27	1020
NaCl	801	1413	519	124		
Cu(s)	1083		205	49		
Cu(ℓ)		2595			4.81	1150
Zn	419	907	100	24	1.76	420
Bi	271	1560	54	13		
Pb	327	1744	23	5.5		
Ni	1453	2732	310	74		

EXAMPLE 19.6   If the molar heat of vaporization of benzene is 31 kJ/mol (7.3 kcal/mol), how much energy is absorbed as 62.5 grams of benzene, C_6H_6, vaporize at the boiling point?

This problem is again a direct substitution into Equation 19.6, but be sure to watch your units carefully.

$$Q = m \times \Delta H_{vap} = 62.5 \; g \; \cancel{C_6H_6} \times \frac{1 \; mol \; \cancel{C_6H_6}}{78.0 \; g \; \cancel{C_6H_6}} \times \frac{31 \; kJ}{1 \; mol \; \cancel{C_6H_6}} = 25 \; kJ \quad [5.8 \; kcal]$$

HEAT OF FUSION

PG 19 E   Given two of the following, calculate the third: (a) mass of a pure substance changing between the solid and liquid states; (b) heat of fusion or solidification; (c) heat flow.

To melt a solid, its crystal form must be broken down. This is an endothermic process; energy must be supplied. This is evident when you consider that ice must be warmed to change it to liquid water, and candle wax melts from the heat of the burning candle. Conceptually similar to heat of vaporization, the **heat of fusion, ΔH_{fus}, of a substance is the energy required to melt one gram of** a solid at its melting point. Again the definition indicates units of joules per gram. Heat of fusion may also be given as molar heat of fusion in kilojoules per mole. Heats of fusion are generally much smaller than heats of vaporization, as may be seen in Table 19.2, where both values are given for several substances.

Just as condensation is the opposite of vaporization, freezing is the opposite of melting. The quantity of heat released in freezing a sample is identical to the heat required to melt that sample. Accordingly, **heat of solidification** is numerically equal to heat of fusion, but the sign is negative.

The heat flow equation for the change between solid and liquid is

$$Q = m \times \Delta H_{fus} \tag{19.7}$$

Calculation methods are identical to those in vaporization problems.

EXAMPLE 19.7   Calculate the energy required to melt 135 grams of sodium at its melting point. Express the answer in both joules and kilojoules.

The heat of fusion of sodium may be found in Table 19.2. That and equation 19.7 are all you need.

$$Q = m \times \Delta H_{fus} = 135\,\cancel{g} \times \frac{113\ J}{\cancel{g}} = 1.5 \times 10^4\ J = 15\ kJ \quad [3.6 \times 10^3\ cal = 3.6\ kcal]$$

19.6 CHANGE IN TEMPERATURE PLUS CHANGE OF STATE

PG 19 F   Sketch, interpret, and/or identify regions in a graph of temperature versus energy for a pure substance over a temperature range from below the melting point to above the boiling point.

19 G   Given (1) the mass of a pure substance, (2) ΔH_{vap} and/or ΔH_{fus} of the substance, and (3) the average specific heat of the substance in the solid, liquid, and/or vapor state, calculate the total heat flow in going from one state and temperature to another state and temperature.

The temperature changes and changes of state we have been considering occur independently of each other, and sometimes one after the other. It helps to follow them on a graph of temperature versus energy. Figure 19.4 is drawn for 1.00 gram of ice at $-10°C$ that is heated until it becomes steam at $120°C$.

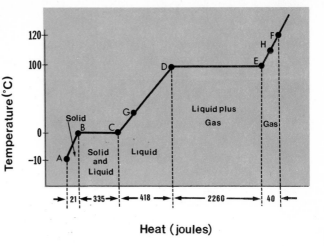

Figure 19.4
Temperature vs. energy absorbed as one gram of ice at −10°C is warmed, melted to the liquid state, heated to the boiling point, boiled, and then heated to 120°C. The curve is typical of most pure substances passing between the solid and vapor states. (Axes are not drawn to scale.)

The changes to be examined are the same as those described previously for ice taken from a refrigerator.

Point A to Point B, warming ice from −10°C to 0°C. Warming ice from −10°C to 0°C is a change in temperature without a change of state. The energy absorbed may be calculated from Equation 19.4, $Q = m \times c \times \Delta T$. From Table 19.1 the specific heat of ice, $H_2O(s)$, is 2.1 J/g °C. The heat absorbed from point A to point B, Q_{A-B}, is

$$Q_{A-B} = 1.00 \text{ g} \times \frac{2.1 \text{ J}}{\text{g } °C} \times [0 - (-10)]°C = 21 \text{ J}$$

Point B to Point C, melting ice at 0°C. This is a change of state at constant temperature. The energy absorbed may be calculated from Equation 19.7, $Q = m \times \Delta H_{fus}$. From Table 19.2, the heat of fusion of water is 335 J/g. The heat absorbed from Point B to Point C, Q_{B-C}, is

$$Q_{B-C} = 1.00 \text{ g} \times \frac{335 \text{ J}}{\text{g}} = 335 \text{ J}$$

Point C to Point D, warming water from 0°C to 100°C. Like A to B, this is a temperature change without a change of state. Using the specific heat of $H_2O(\ell)$,

$$Q_{C-D} = 1.00 \text{ g} \times \frac{4.18 \text{ J}}{\text{g } °C} \times (100 - 0)°C = 418 \text{ J}$$

Point D to Point E, boiling water at 100°C. Like B to C, this is a change of state at constant temperature. Using the heat of vaporization of water,

$$Q_{D-E} = 1.00 \text{ g} \times \frac{2.26 \text{ kJ}}{\text{g}} = 2.26 \text{ kJ} = 2260 \text{ J}$$

Point E to Point F, heating steam from 100°C to 120°C. Like A to B and D to E, this is a temperature change without a change of state. Using the specific heat of $H_2O(g)$,

$$Q_{E-F} = 1.00\,g \times \frac{2.0\ J}{g\,°C} \times (120 - 100)°C = 40\ J$$

To find the heat flow from any point on the graph to any other point you need only to calculate the heat flows of the individual steps in the process and add them. For example, the heat required to melt one gram of ice at 0°C (point B to C), heat it to boiling (C to D), and change it to steam at the boiling point (D to E) is

$$Q_{B-E} = Q_{B-C} + Q_{C-D} + Q_{D-E} = 335\ J + 418\ J + 2260\ J = 3013\ J = 3.01\ kJ$$

(rounded off because of 2.26 kJ for Q_{D-E}).

EXAMPLE 19.8   How many kJ are required to convert 45.0 grams of water, initially at 25°C, to steam at 110°C?

This problem may be related to Figure 19.4 if point G is at a temperature of 25°C and point H is at 110°C. The 45.0 grams of water are to be heated from G to H. Three heat calculations are needed: from G to D, from D to E, and from E to H. Calculate first the energy used in heating the water from 25°C to 100°C.

$$45.0\,g \times \frac{4.18\ J}{g\,°C} \times (100 - 25)°C = 1.4 \times 10^4\ J = 14\ kJ \quad [3.38\ kcal]$$

The heat required to vaporize 45.0 grams of water may now be determined.

$$45.0\,g \times \frac{2.26\ kJ}{g} = 102\ kJ \quad [24.3\ kcal]$$

Now find out what is needed to raise the temperature to 110°C.

$$45.0\,g \times \frac{2.0\ J}{g\,°C} \times (110 - 100)°C = 0.90\ J \quad [0.22\ kcal]$$

You now have the energy required for each of the three separate steps. What is the total energy for the whole process?

$$Q = 14\ kJ + 102\ kJ + 0.90\ kJ = 117\ kJ \quad [27.9\ kcal]$$

EXAMPLE 19.9 Copper melts at 1083°C. Calculate the number of kcal that will be lost by 250 grams of molten copper, initially at 1343°C, as it cools to the melting point, solidifies, and then cools as a solid down to room temperature, 26°C.

The specific heat of liquid copper is 0.42 J/(g °C) [0.10 cal/(g °C)]. Other data may be found in Tables 19.1 and 19.2. Complete the problem.

Cooling liquid: $250\,\cancel{g} \times \dfrac{0.42\text{ J}}{\cancel{g}\,°\cancel{C}} \times (1083 - 1343)°\cancel{C} = -27\,300$ J $= -27$ kJ $[-6.5$ kcal$]$

Solidification: $250\text{ g} \times \dfrac{-205\text{ J}}{g} = -51\,250$ J $= -51.3$ kJ $[-12$ kcal$]$

Cooling solid: $250\,\cancel{g} \times \dfrac{0.38\text{ J}}{\cancel{g}\,°\cancel{C}} \times (26 - 1083)°\cancel{C} = -100\,000$ J $= -1.0 \times 10^2$ kJ $[-24$ kcal$]$

Total heat flow $= -27$ kJ $- 51.3$ kJ $- 1.0 \times 10^2$ kJ $= -1.8 \times 10^2$ kJ $[-43$ kcal$]$

The total heat *lost* is 1.8×10^2 kcal.

part iii: energy in chemical change

19.7 THERMOCHEMICAL EQUATIONS

PG 19 H Given a chemical equation, or information from which it may be written, and the heat (enthalpy) of reaction, write the thermochemical equation in two forms.

At the beginning of Chapter 8 we introduced the idea of chemical equations by examining the reaction when sodium is placed into water. It was noted that the solution formed became hot and released heat to the surroundings. The process was described by the equation

$$2\,Na(s) + 2\,H_2O(\ell) \rightarrow H_2(g) + 2\,NaOH(aq) + \text{heat}$$

It was then pointed out that nearly all chemical changes involve an energy transfer, usually in the form of heat. We will now consider quantitatively the heat effects of a chemical change.

Heat flow resulting from a chemical change is referred to as **heat of reaction** or **enthalpy of reaction. Enthalpy, designated H, may be thought of as "heat content," the amount of heat possessed by a chemical substance at a given temperature and pressure.** The enthalpy of individual substances cannot be measured directly, but a change in enthalpy can be measured in a calorimeter. "Heat of reaction" is therefore equivalent to "change of enthalpy," or ΔH.

If you burn two moles of ethane, C_2H_6, 3080 kilojoules are released. The ΔH of the reaction is -3080 kilojoules, the negative sign indicating that energy has been lost to the surroundings. There are two ways to express this in a chemical equation. One is to write the equation in the usual way, including the energy term as a "product":

$$2\,C_2H_6(g) + 7\,O_2(g) \rightarrow 4\,CO_2(g) + 6\,H_2O(\ell) + 3080\,kJ \qquad (19.8)$$

Alternatively, the enthalpy change is written separately, to the right of the conventional equation:

$$2\,C_2H_6(g) + 7\,O_2(g) \rightarrow 4\,CO_2(g) + 6\,H_2O(\ell) \quad \Delta H = -3080\,kJ \quad (19.9)$$

When writing thermochemical equations, state designations, i.e., (g), (ℓ) and (s), *must* be used. The equation is meaningless without them because the size of the enthalpy change depends upon the state of the reactants and products. If Equation 19.9 is written with water in the gaseous state, for example, $2\,C_2H_6(g) + 7\,O_2(g) \rightarrow 4\,CO_2(g) + 6\,H_2O(g)$, the value of ΔH is -2820 kJ.

The burning of ethane is, as might be expected, an exothermic reaction: heat is given off. The law of conservation of energy requires that the total heat content of the reactants be greater by 3080 kJ than the total heat content of the products. In other words,

$$H_{products} = H_{reactants} - 3080\,kJ$$

It therefore follows that

$$\Delta H = H_{products} - H_{reactants} = -3080\,kJ$$

A graphic representation of this relationship is helpful in understanding the sign of ΔH. In Figure 19.5A, enthalpy is plotted on the vertical axis. The enthalpy of the reactants (2 moles C_2H_6 + 7 moles O_2) is shown by the hori-

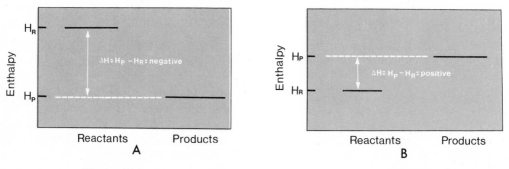

Figure 19.5
Enthalpy-reaction graphs for an exothermic change (A) and an endothermic change (B).

zontal line at the left. The line at the right gives the enthalpy of the products (4 moles of CO_2 + 6 moles of H_2O). The difference between them is ΔH, the enthalpy of reaction. For every exothermic reaction, $H_{products} < H_{reactants}$; total enthalpy has been reduced in the reaction. Therefore ΔH is always negative for an exothermic reaction. By similar reasoning, ΔH is always positive for an endothermic reaction, as shown by the enthalpy-reaction graph in Figure 19.5B. The thermal decomposition of potassium chlorate illustrates an endothermic reaction. The thermochemical equations are

$$2 \, KClO_3(s) \rightarrow 2 \, KCl(s) + 3 \, O_2(g) \quad \Delta H = +89.5 \, kJ$$

$$2 \, KClO_3(s) + 89.5 \, kJ \rightarrow 2 \, KCl(s) + 3 \, O_2(g)$$

EXAMPLE 19.10   The thermal decomposition of $CaCO_3(s)$ to $CaO(s)$ and $CO_2(g)$ is an endothermic reaction requiring 176 kJ per mole of $CaCO_3(s)$ decomposed. Write the thermochemical equation in two forms.

$$CaCO_3(s) \rightarrow CaO(s) + CO_2(g) \qquad \Delta H = 176 \, kJ$$

$$CaCO_3(s) + 176 \, kJ \rightarrow CaO(s) + CO_2(g)$$

19.8 THERMOCHEMICAL STOICHIOMETRY

PG 19 I   Given a thermochemical equation, or information from which it may be written, calculate the amount of heat evolved or absorbed for a given amount of reactant or product; alternately, calculate how many grams of a reactant are required to produce a given amount of heat.

The molar equivalence relationships that may be drawn from a chemical equation include energy in thermochemical equations. From

$$2 \, C_2H_6(g) + 7 \, O_2(g) \rightarrow 4 \, CO_2(g) + 6 \, H_2O(\ell) + 3080 \, kJ \qquad (19.8)$$

the following equivalences may be established:

$$2 \, mol \, C_2H_6 \simeq 7 \, mol \, O_2 \simeq 4 \, mol \, CO_2 \simeq 6 \, mol \, H_2O \simeq 3080 \, kJ \quad (19.10)$$

Heats of reaction may be related quantitatively to amounts of chemical species by changing the stoichiometry pattern (p. 183) to convert between moles of one species and kilojoules of energy.

EXAMPLE 19.11   How many kilojoules of heat are evolved by the burning of 84.0 grams of ethane, C_2H_6, according to Equation 19.8?

   This problem is a two-step conversion. The unit path is from given quantity to energy: grams ethane → moles ethane → kilojoules. The second conversion is based on the equivalence that may be drawn from Equation 19.10. Go all the way to the answer.

$$84.0 \text{ g } \cancel{C_2H_6} \times \frac{1 \text{ mole } \cancel{C_2H_6}}{30.0 \text{ g } \cancel{C_2H_6}} \times \frac{3080 \text{ kJ}}{2 \text{ moles } \cancel{C_2H_6}} = 4.31 \times 10^3 \text{ kJ}$$

The final conversion comes from the fact that 2 moles C_2H_6 are equivalent to 3080 kJ, according to Equations 19.8 and 19.10.

EXAMPLE 19.12   What mass of hexane, $C_6H_{14}(\ell)$, must be burned in order to provide 8.00×10^3 kJ of heat? The thermochemical equation is

$$2\,C_6H_{14}(\ell) + 19\,O_2(g) \rightarrow 12\,CO_2(g) + 14\,H_2O(\ell) \qquad \Delta H = 8280 \text{ kJ; or}$$

$$2\,C_6H_{14}(\ell) + 19\,O_2(g) \rightarrow 12\,CO_2(g) + 14\,H_2O(\ell) + 8280 \text{ kJ}$$

   In this problem the given quantity is kilojoules. The path is the reverse of that in the previous example. Carry it all the way to an answer.

$$8000 \text{ } \cancel{kJ} \times \frac{2 \text{ moles } \cancel{C_6H_{14}}}{8280 \text{ } \cancel{kJ}} \times \frac{86.0 \text{ g } C_6H_{14}}{1 \text{ mole } \cancel{C_6H_{14}}} = 166 \text{ grams } C_6H_{14}$$

A word about algebraic signs: In these examples we have been able to disregard the sign of ΔH because of the way the questions were worded. The question, "How much heat . . .?" is answered simply with a quantity, a number, of kilojoules. The wording of the question tells whether the heat is gained or lost. In Example 19.12, the question was "What mass of . . .?" Obviously that answer must be positive regardless of the sign of ΔH. However, whenever a question is of the form "What is the value of ΔH?" the algebraic sign is an essential part of the answer.

CHAPTER 19 IN REVIEW

PART I: THE NATURE OF ENERGY

19.1 PHYSICAL AND CHEMICAL ENERGY—
A REVIEW (444)

19.2 UNITS OF ENERGY

19 A State the relationship between the two principal scientific units for heat energy. Given one, calculate the other. (445)

PART II: ENERGY IN PHYSICAL CHANGE

19.3 ENERGY, TEMPERATURE, AND CHANGE
OF STATE (446)

19.4 ENERGY AND CHANGE OF TEMPERATURE:
SPECIFIC HEAT

19 B Given (1) the mass of a pure substance, (2) its specific heat, and (3) its temperature change, or initial and final temperatures, calculate the heat flow. (447)

19 C Given the heat flow to or from a known mass of a substance, and its temperature change, or initial and final temperatures, calculate the specific heat of the substance. (447)

19.5 ENERGY AND CHANGE OF STATE

19 D Given two of the following, calculate the third: (a) mass of a pure substance changing between the liquid and vapor states; (b) heat of vaporization or condensation; (c) heat flow. (451)

19 E Given two of the following, calculate the third: (a) mass of a pure substance changing between the solid and liquid states; (b) heat of fusion or solidification; (c) heat flow. (453)

19.6 CHANGE IN TEMPERATURE PLUS CHANGE
OF STATE

19 F Sketch, interpret, and/or identify regions in a graph of temperature versus energy for a pure substance over a temperature range from below the melting point to above the boiling point. (454)

19 G Given (1) the mass of a pure substance, (2) ΔH_{vap} and/or ΔH_{fus} of the substance, and (3) the average specific heat of the substance in the solid, liquid, and/or vapor state, calculate the total heat flow in going from one state and temperature to another state and temperature. (454)

PART III: ENERGY IN CHEMICAL CHANGE

19.7   THERMOCHEMICAL EQUATIONS

19 H Given a chemical equation, or information from which it may be written, and the heat (enthalpy) of reaction, write the thermochemical equation in two forms. (457)

19.8   THERMOCHEMICAL STOICHIOMETRY

19 I Given a thermochemical equation, or information from which it may be written, calculate the amount of heat evolved or absorbed for a given amount of reactant or product; alternately, calculate how many grams of a reactant are required to produce a given amount of heat. (459)

TERMS AND CONCEPTS

Joule, kilojoule (445)
Calorie, kilocalorie (445)
Specific heat (447)
Calorimeter (450)
Heat of vaporization (452)
Heat of condensation (452)
Heat of fusion (453)

Heat of solidification (454)
Heat of reaction (458)
Enthalpy of reaction, ΔH (458)
Enthalpy, H (458)
Thermochemical equation (458)
Thermochemical stoichiometry (459)

Most of these terms and many others are defined in the Glossary.

QUESTIONS AND PROBLEMS

An asterisk (*) identifies a question that is relatively difficult, or that extends beyond the performance goals of the chapter.

Section 19.2

19.1) The normal daily energy requirement for a person doing average work is about 2400 kilocalories. Express this energy in joules.

19.2)* It has been estimated that the heat reaching the earth from the sun is about 8.1 kJ/cm^2 each minute. How many kilocalories of radiant energy fall on a 40-square-foot roof panel of a solar heating system in one hour on a sunny day?

Section 19.4

19.3) A certain solution has a specific heat of 3.5 J/g °C. How many kilojoules must be removed from 325 grams of this solution to reduce its temperature from 22°C to −19°C?

19.28) Many common foods, including rice, bread, and potatoes, are about 75% carbohydrates. Calculate the kilocalories in the carbohydrate content of a one-pound loaf of bread if each gram of carbohydrate furnishes 17 kilojoules.

19.29)* The heating element of a toaster gives off 132 calories per second. Express this in kilojoules per minute.

19.30) How many kilojoules are required to warm 75 kilograms of an electroplating solution from 20°C to 35°C if its specific heat is 3.1 J/g °C?

19.4) In a carefully controlled calorimetry experiment 5.624 grams of an organic solid absorb 94.7 joules as the solid is warmed from 18.6°C to 32.4°C. Calculate the average specific heat over that temperature range.

19.5)* A piece of copper absorbs 245 joules while gaining 14.9°C in temperature. Calculate the mass of the copper.

19.6)* A student measures 100 grams of water into a calorimeter and finds that its temperature is 23.1°C. She then drops 39.0 grams of hot metal of unknown specific heat into the water, stirs, and records the maximum temperature reached, 29.6°C. If the metal was initially at 98.2°C, calculate its specific heat. Disregard any heat loss to the calorimeter or surroundings.

19.7) A hot cast iron hammer, initially at 70.0°C and with a specific heat of 0.498 J/g °C, weighs 666 grams. It is dropped into a can that contains 1110 grams of turpentine at 20.0°C. The final temperature of the hammer and turpentine is 27.0°C. How many kilojoules of heat were lost by the hammer?

Section 19.5

19.8) How many kilojoules are required to boil one gallon (3.785 kg) of water at 100°C?

19.9) Octane, C_8H_{18}, is a component of gasoline. Its heat of vaporization is 310 joules per gram. How much heat is absorbed in vaporizing 25.0 grams of octane in the process of ignition?

19.10) Using an electric immersion heater, it is found that 9.71 kJ of heat must be supplied to vaporize 50.0 grams of carbon tetrachloride, CCl_4, in a hood. Calculate the heat of vaporization of carbon tetrachloride in joules per gram.

19.11)* Ammonia is a widely used refrigerant. Its cooling effect arises from the heat absorbed from the surroundings as liquid ammonia is allowed to vaporize. Calculate the mass of ammonia that changes state in the transfer of 52.3 kJ in a small refrigerating unit. The molar heat of vaporization of ammonia is 21.8 kJ/mol.

19.12) If solid benzene is removed from a refrigerator and permitted to warm to 5°C, it begins to melt. How many kilojoules will be absorbed from the surroundings by 75.0 grams of benzene in melting if its heat of fusion is 137 J/g?

19.31) A 64.5-gram piece of a metal is cooled from 89.3°C to 31.8°C in a calorimeter that permits measurement of the heat lost. If the sample releases 2.17 kilojoules, calculate the specific heat of the metal.

19.32)* What temperature will be reached by a 185-gram pressing surface of an electric iron, initially at 23°C, if it absorbs 11.8 kJ of energy and its specific heat is 0.46 J/g °C?

19.33)* In an experiment similar to that described in Problem 19.6, a student uses toluene, $C_6H_5CH_3$, which has a specific heat of 1.8 J/g °C, instead of water in order to maximize temperature change of the liquid. 48.6 grams of a metal of unknown specific heat are heated to 91.9°C, and then dropped into 94.3 grams of toluene at 21.2°C. The entire system comes to equilibrium at 27.6°C. Calculate the specific heat of the metal.

19.34) From the information given in Problem 19.7, calculate the specific heat of turpentine.

19.35) Calculate the number of joules needed to vaporize 0.450 gram of metallic sodium at its boiling point.

19.36) Calculate the heat that must be removed from 50.0 grams of ethane, C_2H_6, when it is condensed at its normal boiling point of −87°C if $\Delta H_{vap} = 490$ J/g.

19.37) Sulfur dioxide, a gas at room temperature, boils at −10°C. In a laboratory experiment it is found that 9.33 kJ are absorbed when 42.0 grams of SO_2 liquid vaporize at − 10°C. Calculate the heat of vaporization in joules per gram.

19.38)* Components of gasoline are separated in the fractional distillation of petroleum in oil refineries. One such substance is hexane, C_6H_{14}, which has a molar heat of vaporization of 28.9 kJ/mol. How many grams of hexane condense when the process releases 34.9 kilojoules of heat?

19.39) Naphthalene, a compound commonly used for moth balls, melts at 80°C. Its heat of fusion is 151 J/g. Calculate the number of kilocalories required to melt 2.00×10^2 grams of naphthalene.

19.13) How many grams of silver at its melting point can be liquified by 6.28 kJ if the heat of fusion is 87.9 J/g?

19.14)* Toluene, C_7H_8, is one of the useful by-products recovered from the destructive distillation of coal to produce coke, a fuel used in making steel. Experimental data indicate that 23.9 kJ of heat must be furnished in order to boil 65.8 grams of toluene. Determine the *molar* heat of vaporization of toluene in kJ/mol.

19.40) 3.87 kilojoules are released by a sample of pure gold (Z = 79) when it solidifies at its freezing point. If the molar heat of fusion of gold is 13.2 kJ/mol, calculate the mass of the sample.

19.41)* A student determines in the laboratory that 1.83 kilojoules are required to melt 31.2 grams of tin at its melting point. Calculate the molar heat of fusion of tin.

Section 19.6

Figure 19.6 is a graph of energy versus temperature for a sample of a pure substance. Assume that letters J through P on the horizontal and vertical axes represent numbers, and that expressions such as R − S or X + Y + Z represent arithmetic operations to be performed with those numbers. The next five questions in each column are related to Figure 19.6.

19.15) What values are plotted, both horizontally and vertically?

19.16) Identify in Figure 19.6 all points on the curve where the substance is entirely liquid.

19.17) Identify all points on the curve in Figure 19.6 where the substance is partly liquid and partly gas.

19.18) Describe what happens physically as the energy represented by N − M is added to the sample.

19.19) Using letters from the graph, write the expression for the energy required to raise the temperature of the liquid from the freezing point to the boiling point.

19.20) Assume that the ice tray in your refrigerator holds 225 grams of water. How many kilojoules

19.42) Identify by letter the boiling and freezing points in Figure 19.6.

19.43) Identify all points on the curve in Figure 19.6 where the substance is entirely gas.

19.44) Identify in Figure 19.6 all points on the curve where the substance is partly solid and partly liquid.

19.45) Describe the physical changes that occur as energy N − P is removed from the sample.

19.46) Using letters from the graph, show how you would calculate the energy required to boil the liquid at its boiling point.

19.47) Some alloys, unlike most mixtures, have definite melting points. One such alloy melts at

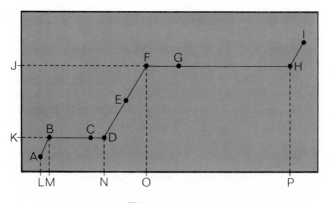

Figure 19.6

must your freezer remove from one tray of water initially at 16°C if the final temperature of the ice is −12°C? Disregard the heat that must be removed from the tray.

19.21) Nitrogen "boils" at −196°C. If 250 grams of liquid nitrogen at −209°C are poured from a thermos bottle into the air, it rises quickly to the boiling point, vaporizes, and then warms to room temperature, 23°C. If the average specific heat of liquid nitrogen is 1.99 J/g °C, the average specific heat of $N_2(g)$ is 1.00 J/g °C, and the heat of vaporization is 47.7 J/g, calculate the heat absorbed by 500 g N_2 in the process.

262°C, and has a heat of fusion of 21.7 joules per gram. As a solid this alloy has a specific heat of 0.120 J/g °C and, as a liquid, 0.146 J/g °C. Calculate the kilojoules of heat absorbed by an 8.55-kg slug of this metal as it comes to the 355°C temperature of a melting pot if the slug is dropped into the pot at room temperature, 30°C.

19.48) "Low pressure" steam leaves a turbine at 145°C, cools to the boiling point at atmospheric pressure, condenses, and then cools to 30°C at the point of discharge. Calculate the kilojoules of energy given up by each gallon (7.06 kg) of water in this process.

Section 19.7

Thermochemical equations may be written in two ways, one with a heat term as a part of the equation, and alternately with ΔH set apart from the regular equation. In the examples that follow, write the equations for the reactions described in both forms. Recall that state designations of all substances are essential in thermochemical equations.

19.22) When propane, $C_3H_8(g)$—a primary constituent in bottled gas used to heat trailers and rural homes—is burned to form gaseous carbon dioxide and liquid water, 2220 kilojoules of heat energy are released for every mole of butane consumed.

19.23) In "slaking" lime, CaO(s), by converting it to solid calcium hydroxide through reaction with water, 65.3 kilojoules of heat are released for each mole of calcium hydroxide formed.

19.24) The extraction of elemental aluminum from aluminum oxide is a highly endothermic electrolytic process. The reaction may be summarized by an equation in which the oxide reacts with carbon to yield aluminum and carbon dioxide gas. The energy requirement is 540 kJ/mol of aluminum metal produced. (Caution on the value of ΔH.)

19.49) 2820 kilojoules of energy are absorbed from sunlight in the photosynthesis reaction in which carbon dioxide and water combine to produce sugar, $C_6H_{12}O_6$, and release oxygen.

19.50) The electrolysis of water is an endothermic reaction, absorbing 286 kJ for each mole of liquid water decomposed to its elements.

19.51) The reaction in an oxyacetylene torch is highly exothermic, releasing 1310 kilojoules of heat for every mole of acetylene, $C_2H_2(g)$, burned. The end products are gaseous carbon dioxide and liquid water.

Section 19.8

19.25) How much heat energy will be released by burning 4.00×10^3 grams of butane, C_4H_{10}, one of the two principal fuels used in bottled gas? The thermochemical equation is $2 C_4H_{10}(g) + 13 O_2(g) \rightarrow 8 CO_2(g) + 10 H_2O(\ell) + 5770$ kJ.

19.52) The direct combination of powdered zinc with powdered sulfur is a spectacular reaction—though not one to be tried by students!—that yields bright light, flame and smoke:

$$Zn(s) + S(s) \rightarrow ZnS(s).$$

Calculate the energy released by the reaction of 9.63 grams of zinc if ΔH = −621 kJ/mol.

19.26) Calculate the number of grams of hexane, a component of gasoline, required to release 5.5×10^4 kJ if $\Delta H = -4140$ kJ for the reaction $C_6H_{14}(\ell) + \frac{19}{2} O_2(g) \rightarrow 6 CO_2(g) + 7 H_2O(\ell)$.

19.27) How many kilograms of aluminum can be processed by 1.82×10^7 kJ, according to the equation from Question 19.24?

19.53) Carbon monoxide is used as a fuel in many industrial processes:

$$2 CO(g) + O_2(g) \rightarrow 2 CO_2(g) + 565 \text{ kJ}.$$

How much carbon monoxide must be burned in a process requiring 7660 kJ of energy?

19.54) In slaking lime (see Question 19.23), how many kJ are released for every kilogram of CaO?

20

chemical equilibrium

20.1 PHYSICAL AND CHEMICAL EQUILIBRIA—A REVIEW

PG 20 A Identify the essential characteristics of an equilibrium system.

The first dynamic equilibrium in this text was between benzene, $C_6H_6(\ell)$, and its saturated vapor, $C_6H_6(g)$, in a closed (sealed) container (p. 317). The system was described by the reversible reaction equation

$$C_6H_6(\ell) \rightleftharpoons C_6H_6(g)$$

where the double arrow indicates two directions of change. Read from left to right, from liquid benzene to gaseous benzene, the equation describes the **forward reaction**; the change from right to left, or vapor to liquid, is the **reverse reaction**. This is a physical equilibrium, since the changes of state between liquid and vapor, evaporation and condensation, are physical changes. Using Figure 14.9, the development of the equilibrium was traced in terms of the rates of the forward and reverse changes. *When forward and reverse rates became equal, the system was at equilibrium.*

Another physical equilibrium is the formation of a saturated solution of an ionic solid in a liquid. This was described on page 334, leading ultimately to the equilibrium

$$NaCl(s) \rightleftharpoons Na^+(aq) + Cl^-(aq)$$

In any equilibrium between undissolved solute and its saturated solution, the rate of dissolving (forward reaction above) is equal to the rate of crystallization (reverse reaction).

In introducing the Brönsted-Lowry theory of acids and bases (Section 17.2), the equilibrium between ammonia and water was considered:

$$NH_3(aq) + HOH(\ell) \rightleftharpoons NH_4^+(aq) + OH^-(aq)$$

This is a chemical equilibrium because the changes that occur are chemical changes. *At equilibrium the rate at which ammonia and water molecules form ammonium and hydroxide ions (forward reaction) is equal to the rate at which the ions form the molecules (reverse reaction).*

Three features of these equilibria are true of any chemical equilibrium:

1. The equilibrium is a "closed" system—closed in the sense that substances on each side of the equation remain to react to form substances on the other side. There is no equilibrium if one or more substances enter or leave the system.

2. The equilibrium is dynamic. The reversible changes occur continuously, even though there is no appearance of change. By contrast, items in a static equilibrium are stationary—e.g., an object hanging on a spring.

3. The things that are *equal* in an equilibrium are the forward reaction rate (the rate from left to right) and the reverse reaction rate (from right to left). Specifically, amounts of substances present are *not* equal. If the substances on the right side of the equation are in greater abundance, the equilibrium is said to be **favored** in the forward direction; if there are more of the substances on the left, the reverse reaction is favored.

Because of the importance of rate of change in chemical equilibria, it is logical to begin our study by looking at the factors that affect reaction rate.

20.2 THE COLLISION THEORY OF CHEMICAL REACTIONS

PG 20 B Explain why, according to the collision theory of chemical reactions, some molecular collisions result in a chemical reaction and others do not.

If a chemical reaction is to occur between two molecules, it is reasonable to expect that the molecules must come into contact with each other. The view of a chemical reaction as the overall effect of a huge number of collisions between reacting particles is the **collision theory of chemical reactions.** We will examine an imaginary reaction in detail.

In order for the reaction $A_2 + B_2 \rightarrow 2\,AB$ to take place, a molecule of A_2 must collide with a molecule of B_2. What happens then depends upon the nature of the collision. If there is to be a reaction, the chemical bond between A atoms in A_2 must be broken; similarly, the bond in a B_2 molecule must also be broken. It takes energy to break these bonds. This energy comes from the kinetic energy of the molecules just before they collide. In other words, it must be a violent, bond-breaking collision. This is most likely to occur if the molecules are moving at high speeds and meet in a direction that not only shatters the bond but also causes each A atom to form a new bond with a B atom, yielding two molecules of AB. A collision of this kind is shown in Figure 20.1A.

What happens if the colliding molecules do not have enough kinetic energy to break existing bonds, or the proper orientation to form new bonds? In this case the molecules simply bounce off each other and separate with the same A_2 and B_2 identity they had before the collision. The collision is ineffective, as far as producing a reaction is concerned. Sometimes the orientation may be correct, but the molecules may not have enough kinetic energy to smash the existing bond. Figure 20.1B shows this kind of a collision. Other collisions may have enough kinetic energy, but the orientation may push atoms in the original molecules closer together, rather than tearing them apart. This is shown in Figure 20.1C. Moreover, this kind of collision does not favor the formation of an A—B bond between the outside A and B atoms.

In the study of equilibrium, the main concern is with the rates of opposing reactions. The collision theory suggests that reaction rate depends upon the *frequency* of effective collisions. The effectiveness of a collision depends in part upon kinetic energy. To understand the role of kinetic energy completely we must examine the energy changes that occur *during* the collision more closely.

20.3 ENERGY CHANGES DURING A MOLECULAR COLLISION

PG 20 C Sketch and/or interpret an enthalpy-reaction graph for either an exothermic or an endothermic reaction. Identify (a) the activated complex point; (b) activation energy; and (c) ΔH for the reaction.

In some ways what happens during a molecular collision is similar to what happens when a rubber ball bounces on the floor. For simplicity we will assume some sort of "superball" that bounces with no loss of energy—a perfectly elastic collision, as we pictured gas particles bouncing off the walls of their containers in the model of an ideal gas (p. 268). As the ball approaches the floor it has a certain amount of kinetic energy (Fig. 20.2A). As it hits the floor it flattens and begins to lose kinetic energy (Fig. 20.2B). The lost kinetic energy is replaced by potential energy in the ball, which is unstable in its deformed shape and tends to spring back to its natural round form. According to the Law of Conservation of Energy, the total kinetic energy plus potential energy remains constant throughout the collision. When the ball is at its lowest point (Fig. 20.2C) nearly all the kinetic energy has been converted to potential energy. As the ball

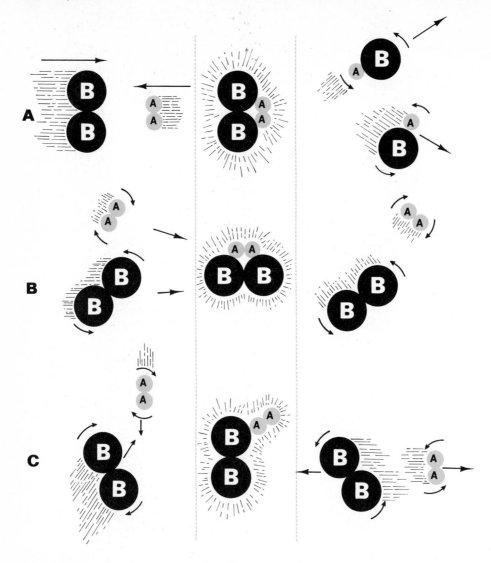

Figure 20.1

Molecular collisions and chemical reactions.

A, Two rotating molecules move toward each other on a collision course (left), collide or otherwise interact with sufficient impact and proper orientation to break existing bonds and form new ones (center), and separate as product molecules of the reaction that has occurred (right).

B, Two molecules approach slowly (left), collide with proper orientation but insufficient energy for a reaction (center), and separate as unreacted molecules, just as they were before collision (right).

C, Molecules with sufficient kinetic energy to produce a reaction approach (left), but collide with poor orientation (center), and therefore separate unreacted (right).

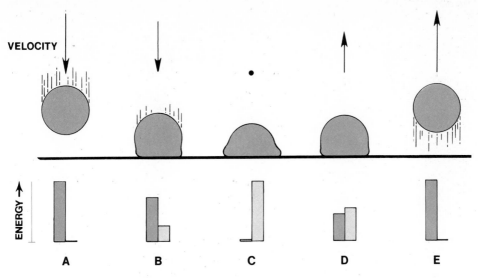

Figure 20.2
Kinetic and potential energy in a bouncing ball. If the collision is "perfectly elastic," the total kinetic energy plus potential energy is constant in each stage of the bounce. The distribution of energy among these forms is shown by plotting kinetic energy in color and potential energy in black beneath each illustration. See the text for a detailed description.

begins its upward bounce (Fig. 20.2D) the potential energy is converted back into kinetic energy. The conversion is complete when the ball leaves the floor (Fig. 20.2E).

It is believed that there is a similar conversion of kinetic energy to potential energy during a collision between molecules. This can be shown on an enthalpy-reaction graph, introduced on page 458 and repeated here as part of Figure 20.3. The enthalpies of the reactants and products are potential energies. The white line traces the increase in potential energy of the system during the collision. The hump in the curve is a **potential energy barrier** that must be overcome before a particle collision can result in a reaction. If the particles do not have enough kinetic energy to get over the hump they end up as bounce-off collisions. A mechanical system that suggests the same kind of potential energy barrier is shown in Figure 20.4.

The smallest amount of energy needed to get a reaction over the potential energy barrier is called **activation energy.** On the enthalpy-reaction graph the activation energy is the difference between the summit of the barrier and the enthalpy of the reactants. The unstable intermediate species formed as molecules come together during collision is called an **activated complex.** It has an extremely short life span. If the collision lacks the energy or orientation to produce a reaction, the activated complex separates as the original particles; if the collision has both the energy and orientation for a reaction it breaks immediately into two product molecules.

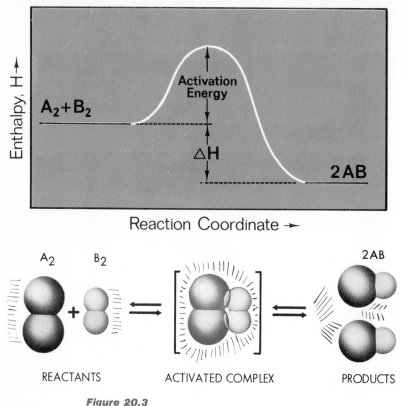

Figure 20.3
Potential energy diagram for the reaction $A_2 + B_2 \rightarrow 2AB$.

20.4 CONDITIONS THAT AFFECT THE RATE OF A CHEMICAL REACTION

THE EFFECT OF TEMPERATURE ON REACTION RATE

PG 20 D State and explain the relationship between reaction rate and temperature.

Chemical reactions are faster at higher temperatures. This is seen in two simple facts from the kitchen. The changes that occur in cooking are chemical reactions. One way to speed up cooking reactions that are performed in boiling water is to use a pressure cooker. Under pressure, the boiling point of water is higher, so the time required for a given cooking operation is reduced. The

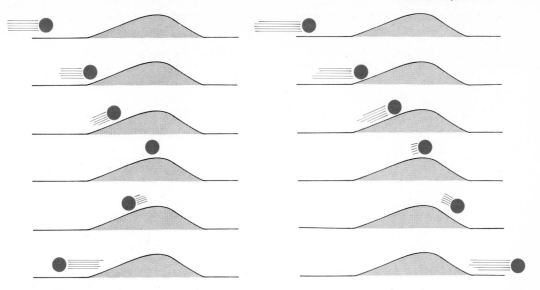

Figure 20.4
Potential energy barrier. As the ball rolls toward the small hill (potential energy barrier), its speed (kinetic energy factor) determines whether or not it will pass over the hill. In the left series of pictures the slow moving ball slows down even more as it begins to climb the hill in the third picture. In the fourth picture all of its kinetic energy has been converted to potential energy before it reaches the peak. In the fifth picture it is rolling back down the hill. In the sixth position the ball's potential energy from position four has been restored to kinetic energy as it rolls back to its starting point. In the right series the faster moving ball has more than enough kinetic energy to convert to potential energy at the top of the hill. It therefore passes over the potential energy barrier and rolls down the other side.

opposite effect is seen in open cooking at high altitudes, where reduced atmospheric pressure allows water to boil at lower temperatures. Here cooking is slower, the result of reduced reaction rates.

To understand the effect of temperature on reaction rates, recall that at any temperature the particles in a sample of matter have a wide range of kinetic energies. Some molecules move slowly, some very fast, and the majority fall in between. We have already seen that collisions between molecules with low kinetic energies are bounce-off collisions, and only high energy collisions yield reactions. This means that only a fraction of the sample, that fraction with high kinetic energy, can engage in reaction-producing collisions.

If the temperature of the reactants is increased, the molecules will have a higher average kinetic energy. It follows that a larger fraction of the sample will have enough kinetic energy to convert to potential energy and cross the potential energy barrier. Quite often a temperature increase of about 10°C will double the rate of a reaction, suggesting that the number of particles with sufficient kinetic energy to react has been doubled. Is this reasoning familiar? You have encountered it before; it is the same reasoning that explains the sharp increase in vapor pressure at higher temperatures (p. 319).

It can be argued that a higher temperature increases the frequency of all collisions, and that this alone would increase reaction rate. This is correct, but this explanation cannot by itself account for the increase in reaction rate of about 100% when the average absolute temperature is raised by about 3%. We must conclude that *the main reason a chemical change occurs more rapidly at higher temperatures is that a larger portion of the total number of particles has sufficient kinetic energy for an effective collision.*

THE EFFECT OF A CATALYST ON REACTION RATE

PG 20 E Explain how a catalyst affects reaction rate. Compare a catalyzed and an uncatalyzed reaction on an enthalpy-reaction graph.

In driving from one city to another you usually have a choice of two or more alternative routes. One of these would be the quickest, and in all probability your choice. If a superhighway were to be built between the cities you would have another route that would be faster than any of the earlier choices. If your purpose was to make the trip in the minimum time, you would no doubt turn to the superhighway for future trips.

A **catalyst** for a chemical reaction is similar to the highway in furnishing an alternative "route" for the reactants to change to products. It does this by changing the reaction path. Specifically, a lower energy activated complex is found, requiring a lower activation energy than the uncatalyzed reaction. Figure 20.5 compares the reaction paths for catalyzed and uncatalyzed reactions. E_a is the activation energy for the reaction without a catalyst, and E_a' represents the activation energy in the presence of a catalyst. Many molecules that don't have enough kinetic energy to overcome the E_a potential energy barrier are able to cross the lower E_a' hump. With a larger fraction of molecules engaging in reaction-producing collisions, the reaction is faster.

Catalysts exist in several different forms, and the precise function of many catalysts is not clearly understood. Some catalysts are mixed intimately in the reacting chemicals, while others seem to do no more than provide a surface upon which the reaction may occur. In either case, the catalyst is not permanently affected by the reaction. Many times the catalyst is only a bystander, although an important one from the standpoint of the reaction. In other cases, the catalyst may actually participate in the reaction and undergo chemical change, but eventually it will be regenerated in the same amount as at the start.

The catalytic reaction that appears most often in beginning chemistry laboratories is the making of oxygen by decomposing potassium chlorate in the presence of manganese dioxide as a catalyst. A well-known industrial process is the catalytic cracking of crude oil, in which large hydrocarbon molecules are broken down into simpler and more useful products in the presence of a catalyst. Biological reactions are controlled by catalysts called enzymes.

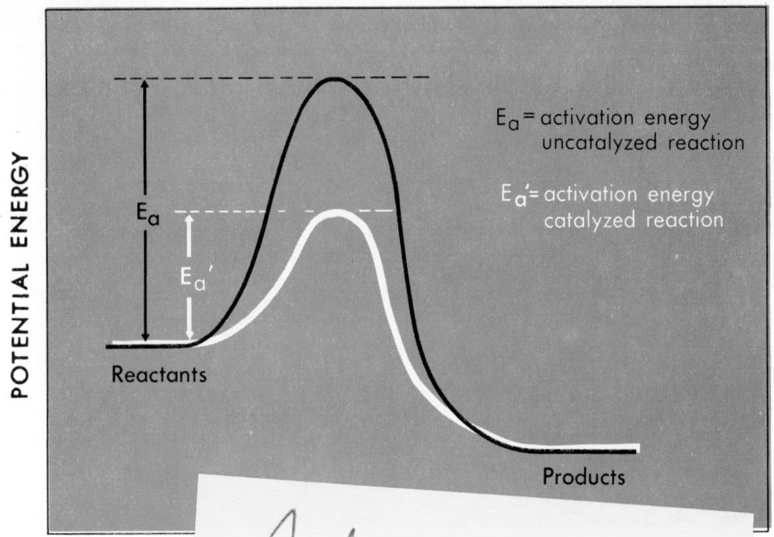

E_a = activation energy
uncatalyzed reaction

E_a' = activation energy
catalyzed reaction

POTENTIAL ENERGY

E_a

E_a'

Reactants

Products

Figure 20.5
Lowering of activa

Rob
Here's another
explanation. Sometimes
more than one book
helps.
Hang in there. I love
you. Dad

Some substances interfere with a
products, forcing the reaction to a high
Such substances are called **negative ca**
to control the rates of certain industria
can have disastrous results, as when
biological function of enzymes.

THE EFFECT OF CONCENTRATI

PG 20 F Explain the relationship between reactant concentration and reaction rate.

If a reaction rate depends upon frequency of effective collisions, the influence of concentration is readily predictable. The more particles there are in a
given space, the more frequently collisions will occur, and the more rapidly the
reaction will take place.

The effect of concentration on reaction rate is easily seen in the rate at
which objects burn in air compared to the rate of burning in an atmosphere of
pure oxygen. If a burning splint is thrust into pure oxygen the burning is
brighter, more vigorous, and much faster. In fact, the typical laboratory test for

oxygen is to ignite a splint, blow it out, and then, while there is still a faint glow, place it in oxygen. It immediately bursts back into full flame and burns vigorously. Charcoal, phosphorus, and other substances behave similarly.

* * * * *

Three factors that influence the rate of a chemical reaction have been identified. In relating these factors to equilibrium considerations, we will consider only concentration and temperature. These variables affect forward and reverse reaction rates differently. A catalyst, on the other hand, has the same effect on both forward and reverse rates, and therefore does not alter a chemical equilibrium. A catalyst does cause a system to reach equilibrium more quickly.

20.5 THE DEVELOPMENT OF A CHEMICAL EQUILIBRIUM

PG 20 G   Trace the changes in concentrations of reacting species that lead to a chemical equilibrium.

The role of concentration in chemical equilibrium may be illustrated by tracing the development of an equilibrium. We will do this in two ways, graphically and by means of the reversible reaction equation. For our purpose, the hypothetical reaction $A_2 + B_2 \rightarrow 2 AB$ will be assumed to take place by the simple collision of A_2 and B_2 molecules, which separate as two AB molecules (Fig. 20.1). Furthermore, the reverse reaction will be thought of as exactly the reverse process: two AB molecules collide and separate as one A_2 molecule and one B_2 molecule.

Figure 20.6 is a graph of forward and reverse reaction rates vs. time. Initially, at Time 0, pure A_2 and B_2 are introduced to the reaction chamber. At the initial concentrations of A_2 and B_2 the forward reaction begins at a certain rate, F_0. Initially there are no AB molecules present, so the reverse reaction cannot occur. At Time 0 the reverse reaction rate, R_0, is zero. These points are plotted on the graph.

As soon as the reaction begins, A_2 and B_2 are consumed, thereby reducing their concentrations in the reaction vessel. As these reactant concentrations decrease, the forward reaction rate declines. Consequently, at Time 1, the forward reaction rate will be F_1. During the same interval some AB molecules will be produced by the forward reaction, and the concentration of AB will be greater than zero. Therefore the reverse reaction begins with the reverse rate rising to R_1 at Time 1.

What happens during the next interval of time, beginning at Time 1? At that instant the forward rate is greater than the reverse rate. Therefore A_2 and B_2 are consumed by the forward reaction more rapidly than they are produced

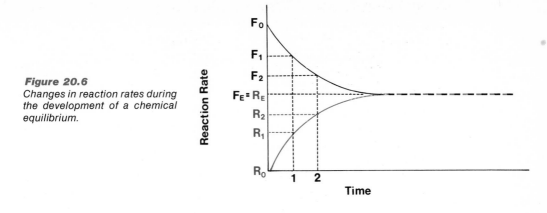

Figure 20.6
Changes in reaction rates during the development of a chemical equilibrium.

by the reverse reaction. The net change in the concentrations of A_2 and B_2 is therefore downward, causing a further reduction in the forward rate at Time 2. Conversely, the forward reaction produces AB more rapidly than the reverse reaction uses it. The net change in the concentration of AB is therefore an increase. This, in turn, raises the reverse reaction rate at Time 2.

Similar changes occur over successive intervals until the forward and reverse rates eventually become equal. At this point a dynamic equilibrium will have been established. From this analysis we may state the following generalization: *For any reversible reaction in a closed system, whenever the opposing reactions are occurring at different rates, the faster reaction will gradually become slower, and the slower reaction will become faster. Finally they become equal, and equilibrium is established.*

20.6 Le CHATELIER'S PRINCIPLE

In the last section you saw how concentrations and reaction rates jockey with each other until the rates become equal and equilibrium is reached. In this section we will start with a system already at equilibrium and see what happens to it when the equilibrium is upset. To "upset" an equilibrium you must somehow make the forward and reverse reactions unequal, at least temporarily. One way to do this is to add something to or remove something from the system. A gaseous equilibrium can often be upset simply by changing the volume of the container.

How an equilibrium responds to a disturbance can be predicted from the concentration and temperature effects already considered. The predictions may be summarized in **Le Chatelier's Principle**, which says that **if an equilibrium system is subjected to a change, processes occur that tend to partially counteract the initial change, thereby bringing the system to a new position of equilibrium.**

We will now apply this system and see why it works.

THE CONCENTRATION EFFECT

PG 20 H   Given the equation for a chemical equilibrium, predict the direction in which the equilibrium will shift because of a change in the concentration of one species.

The reaction of hydrogen and iodine to produce hydrogen iodide comes to equilibrium with hydrogen iodide as the favored species. The equal forward and reverse reaction rates are shown by the equal length arrows in the equation

$$H_2(g) + I_2(g) \rightleftharpoons 2\,HI(g)$$

The sizes of the formulas represent the relative concentrations of the different species in the reaction. If more HI is forced into the system, [HI] is increased. (Recall from page 407 that a formula enclosed in square brackets refers to the concentration of the substance in moles per liter.) This raises the rate of the reverse reaction, shown by the longer arrow in the equation, in which HI is a reactant:

$$H_2(g) + I_2(g) \rightleftharpoons 2\,HI(g)$$

The rates no longer being equal, the equilibrium is destroyed.

At this point the unequal reaction rates cause the system to "shift" in the direction of the faster rate—to the left, or in the reverse direction. As a result H_2 and I_2 are made by the reverse reaction faster than they are used by the forward reaction. Their concentrations increase, so the forward rate increases. Simultaneously HI is used faster than it is produced, reducing both [HI] and the reverse reaction rate. Eventually the rates become equal at an intermediate value (note arrow lengths) and a new equilibrium is reached:

$$H_2(g) + I_2(g) \rightleftharpoons 2\,HI(g)$$

The sizes of the formulas indicate that all three concentrations are larger than they were originally, although [HI] has come back from the maximum it reached just after it was added.

The above example shows *why* an equilibrium shifts when it is disturbed in terms of concentrations and reaction rates. Le Chatelier's Principle makes it possible to predict the direction of a shift, forward or reverse, without such a detailed analysis. Just remember that the shift is always in the direction that tries to return the substance disturbed to its original condition.

EXAMPLE 20.1   The system $N_2(g) + 3\,H_2(g) \rightleftharpoons 2\,NH_3(g)$ is at equilibrium. Use Le Chatelier's Principle to predict the direction in which the equilibrium will shift if ammonia is withdrawn from the reaction chamber.

The equilibrium disturbance is clearly stated: ammonia is withdrawn. In which direction must the reaction shift to *counteract* the removal of ammonia—that is, to *produce* more ammonia to replace some of what has been taken away?

The shift will be in the *forward* direction. Ammonia is the product of the forward reaction and therefore will be partially restored to its original concentration by a shift in the forward direction.

EXAMPLE 20.2 Predict the direction, forward or reverse, of a Le Chatelier shift in the equilibrium $CH_4(g) + 2 H_2S(g) \rightleftharpoons 4 H_2(g) + CS_2(g)$ caused by each of the following:
(a) increase $[H_2S]$;   (b) reduce $[CS_2]$;   (c) increase $[H_2]$.

(a) Forward, counteracting partially the increase in $[H_2S]$ by consuming some of it.

(b) Forward, restoring some of the CS_2 removed.

(c) Reverse, consuming some of the added H_2.

THE VOLUME EFFECT

PG 20 I Given the equation for a chemical equilibrium involving one or more gases, predict the direction in which the equilibrium will shift because of a change in the volume of the system.

A change in the volume of an equilibrium that includes one or more gases changes the concentrations of those gases. Usually—there is one exception, as you will see shortly—there is a Le Chatelier shift that partially offsets the initial change. Both the change and the adjustment involve the pressure exerted by the entire system.

From the ideal gas equation, $PV = nRT$, the pressure exerted by a gas at constant temperature is proportional to the concentration of the gas, n/V, measured in moles per liter:

$$P = RT \times \frac{n}{V} = \text{constant} \times \frac{n}{V}; \qquad P \propto \frac{n}{V} \qquad (20.1)$$

If volume is reduced, the denominator in n/V becomes smaller and the fraction becomes larger. Pressure increases. Le Chatelier's Principle calls for a shift that will partially counteract the change—to reduce pressure. What change can reduce the pressure of the system at the new volume? The numerator in Equation

20.1 must be reduced; there must be fewer gaseous molecules in the system. In general, *if a gaseous equilibrium is compressed, the increased pressure will be partially relieved by a shift in the direction of fewer gaseous molecules; if the system is expanded, the reduced pressure will be partially restored by a shift in the direction of more gaseous molecules.*

Realizing that the coefficients of gases in an equation are in the same proportion as the number of gaseous molecules, let's try some examples.

EXAMPLE 20.3   Predict the direction of shift resulting from an expansion in the volume of the equilibrium $2\,SO_2(g) + O_2(g) \rightleftharpoons 2\,SO_3(g)$.

First, will total pressure increase or decrease as a result of expansion?

_____   _____

Decrease. Pressure and volume are inversely related.

Next, will the decrease in pressure be counteracted by an increase or decrease in the number of molecules?

_____   _____

Increase. Pressure is directly related to number of molecules.

Finally, examining the equation, notice that a forward shift finds *three* reactant molecules, two SO_2 and one O_2, forming *two* SO_3 product molecules, whereas a reverse shift has *two* reactant molecules yielding *three* product molecules. In which direction will the reaction shift to increase the total number of molecules?

_____   _____

Reverse. Three product molecules replace two reactant molecules in a reverse shift. The increase in number of molecules partially restores the total pressure lost in expansion.

EXAMPLE 20.4   The volume occupied by the equilibrium

$$SiF_4(g) + 2\,H_2O(g) \rightleftharpoons SiO_2(s) + 4\,HF(g)$$

is reduced. Predict the direction of shift in the position of equilibrium.

Will the shift be in the direction of more gaseous molecules or fewer?

_____   _____

Fewer. If volume is reduced, pressure increases. Increased pressure will be counteracted by fewer molecules.

Now predict the direction of shift and justify your prediction by stating the numerical loss in molecules predicted by the equation.

_____   _____

The shift will be in the reverse direction. The reverse reaction has *four gaseous* molecules being reduced to three. Note the number is not five reduced to three, but four. Only the gaseous molecules are involved in pressure adjustments.

EXAMPLE 20.5 Returning to the familiar $H_2(g) + I_2(g) \rightleftharpoons 2\,HI(g)$, predict the direction of the shift that will occur because of a volume increase.

Take it all the way, but be careful.

- - - - - -　　　　　　　　　　　　　　　　　　　　- - - - - -

There will be no shift. In this case there are two gaseous molecules on *each* side of the equation. Therefore there is no gain or loss in number of molecules by a shift in either direction. When the number of gaseous molecules on the two sides of the equilibrium equation is the same, increasing or reducing volume has no effect on the equilibrium.

THE TEMPERATURE EFFECT

PG 20 J Given a thermochemical equation for a chemical equilibrium, predict the direction in which the equilibrium will shift because of a change in temperature.

A change in temperature of an equilibrium will change both forward and reverse reaction rates, but the rate changes are not equal. The equilibrium is therefore destroyed temporarily. The events that follow are again predictable by Le Chatelier's Principle.

EXAMPLE 20.6 If the temperature of the system

$$PCl_5(g) \rightleftharpoons PCl_3(g) + Cl_2(g) + 92.5 \text{ kJ}$$

is increased, predict the direction of the Le Chatelier shift.

In order to raise the temperature of something, it must be heated. We therefore interpret an increase in temperature as the "addition of heat," and a lowering of temperature as the "removal of heat." In applying Le Chatelier's Principle to a thermochemical equation, heat may be regarded in much the same manner as any chemical species in the equation. Accordingly, if heat is *added* to the equilibrium system shown, in which direction must it shift to *use up*, or *consume*, some of the heat that was added? (If chlorine were *added*, in which direction would the equilibrium shift to use up some of the chlorine?)

The equilibrium must shift in the *reverse* direction to use up some of the added heat. An endothermic reaction consumes heat. As the equation is written, the reverse reaction is endothermic.

EXAMPLE 20.7 The thermal decomposition of limestone reaches the following equilibrium: $CaCO_3(s) + 176$ kJ $\rightleftharpoons CaO(s) + CO_2(g)$. Predict the direction this equilibrium will shift if the temperature is reduced.

The shift would be in the *reverse* direction. Reduction in temperature is interpreted as the removal of heat. The reaction will respond to replace some of the heat removed—as an exothermic reaction. Heat is produced as the reaction proceeds in the reverse direction.

20.7 THE LAW OF CHEMICAL EQUILIBRIUM: THE EQUILIBRIUM CONSTANT

PG 20 K Given any chemical equilibrium equation, write the equilibrium constant expression.

GAS PHASE EQUILIBRIA

If 1.000 mole of $H_2(g)$ and 1.000 mole of $I_2(g)$ are introduced into a 1.000-liter reaction vessel under pressure, they will react to produce the following equilibrium: $H_2(g) + I_2(g) \rightleftharpoons 2$ HI(g). At a temperature of 440°C, analysis at equilibrium will show hydrogen and iodine concentrations of 0.218 mole per liter, and a hydrogen iodide concentration of 1.564 moles per liter. Development of these concentrations is shown in black in Figure 20.7.

Working in the opposite direction, if gaseous hydrogen iodide is introduced to a reaction chamber at 440°C at an initial concentration of 2.000 moles per liter, it will decompose by the reverse reaction. The system will eventually come to equilibrium with $[H_2] = [I_2] = 0.218$, and $[HI] = 1.564$, exactly the

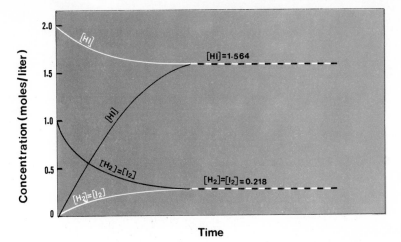

Figure 20.7
Changes in concentrations during the development of a chemical equilibrium. Black lines represent the development of the equilibrium from hydrogen and iodine each initially at 1.00 mole/liter, and no HI present. White lines show the system reaching the identical equilibrium from hydrogen iodide initially at 2.00 moles/liter, with no hydrogen or iodine present.

same equilibrium concentrations as in the first example. This illustrates the experimental fact that the position of an equilibrium is the same, regardless of the direction from which it is approached (Fig. 20.7, white lines).

As a third example, if the initial hydrogen iodide concentration is half as much, 1.000 mole per liter, the equilibrium concentration would be half what they are above: $[H_2] = [I_2] = 0.109$ and $[HI] = 0.782$.

If the experiment is repeated at the same temperature, starting this time with hydrogen at 1.000 mole per liter and iodine at 0.500 mole per liter, the reaction again proceeds to equilibrium. The equilibrium concentrations will be $[H_2] = 0.532$, $[I_2] = 0.032$, and $[HI] = 0.936$, as shown in Figure 20.8.

The data of these four equilibria are summarized in Table 20.1. Analysis of these data leads to the observation that, at equilibrium, the ratio of the

Figure 20.8
Changes in concentrations during the development of a chemical equilibrium. Hydrogen concentration initially at 1.00 mole/liter; iodine initially at 0.500 mole/liter; no HI present initially.

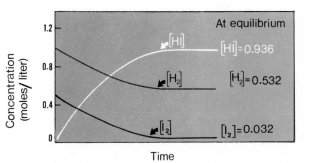

Table 20.1

Equi-librium	Initial			Equilibrium			$\dfrac{[HI]^2}{[H_2][I_2]}$
	$[H_2]$	$[I_2]$	$[HI]$	$[H_2]$	$[I_2]$	$[HI]$	
1	1.000	1.000	0	0.218	0.218	1.564	51.5
2	0	0	2.000	0.218	0.218	1.564	51.5
3	0	0	1.000	0.109	0.109	0.782	51.5
4	1.000	0.500	0	0.532	0.032	0.936	51.5

square of the hydrogen iodide concentration to the product of the hydrogen and iodide concentrations, a ratio that may be expressed mathematically as $\dfrac{[HI]^2}{[H_2][I_2]}$, has a value of 51.5 for all four equilibria. For that matter, this ratio of equilibrium concentrations is always the same regardless of the initial concentrations of hydrogen, iodine, and hydrogen iodide, as long as the temperature is held at 440°C.

Data from countless other equilibrium systems show a similar regularity. Expressed in words, it is known as the **Law of Chemical Equilibrium: For any equilibrium at a given temperature, the ratio of the product of the concentrations of the species on the right side of the equilibrium equation, each raised to a power equal to its coefficient in the equation, to the corresponding product of the concentrations of the species on the left side of the equation, is a constant.** The constant is called the **equilibrium constant, K.** Thus for the

$$H_2(g) + I_2(g) \rightleftharpoons 2\,HI(g)$$

equilibrium at 440°C,

$$K = \frac{[HI]^2}{[H_2][I_2]} = 51.5$$

The statement of the equilibrium law sets the procedure by which any equilibrium constant expression may be written. For the general equilibrium $aA + bB \rightleftharpoons cC + dD$, where A, B, C, and D are chemical formulas, and a, b, c, and d are their coefficients in the equilibrium equation,

1. Write in the *numerator* the concentration of each species on the *right* side of the equation.

$$K = \frac{[C][D]}{}$$

2. For each species on the right side of the equation, use its coefficient in the equation as an exponent.

$$K = \frac{[C]^c[D]^d}{}$$

3. Write in the *denominator* the concentration of each species on the *left* side of the equation.

$$K = \frac{[C]^c[D]^d}{[A][B]}$$

4. For each species on the left side of the equation, use its coefficient in the equation as an exponent.

$$K = \frac{[C]^c[D]^d}{[A]^a[B]^b} \qquad (20.2)$$

If the equation were written in reverse, $cC + dD \rightleftharpoons aA + bB$, the equilibrium constant would be

$$K' = \frac{[A]^a[B]^b}{[C]^c[D]^d},$$ (20.3)

which is the reciprocal of Equation 20.2. Every equilibrium constant value must be associated with a specific equilibrium equation.

EXAMPLE 20.8   Write equilibrium constant expressions for each of the following equilibria:

(a) $2 HI(g) \rightleftharpoons H_2(g) + I_2(g)$

(b) $HI(g) \rightleftharpoons \frac{1}{2} H_2(g) + \frac{1}{2} I_2(g)$

(c) $2 Cl_2(g) + 2 H_2O(g) \rightleftharpoons 4 HCl(g) + O_2(g)$

- -

(a) $K = \dfrac{[H_2][I_2]}{[HI]^2}$. Notice that this is the reciprocal of the equilibrium constant developed in the text, when the equation was written $H_2(g) + I_2(g) \rightleftharpoons 2 HI(g)$. Its numerical value at 440°C would be 1/51.5, or 0.0194.

(b) $K = \dfrac{[H_2]^{1/2}[I_2]^{1/2}}{[HI]}$. The value of this equilibrium constant is *not* the same as above, 0.0194. It is, in fact, the square root of 0.0194, or 0.139. This emphasizes the importance of associating any equilibrium constant expression with a specific chemical equation.

(c) $K = \dfrac{[HCl]^4[O_2]}{[Cl_2]^2[H_2O]^2}$. The procedure for writing the equilibrium constant expression is the same no matter how complex the equation is.

SOLIDS IN EQUILIBRIUM CONSTANT EXPRESSIONS

In Chapter 16 silver chloride was identified as an "insoluble" ionic compound. It is more accurately described as a very, very slightly soluble compound. Its solution becomes saturated at an extremely low concentration, roughly 0.000 01 mole per liter. The equilibrium equation is $AgCl(s) \rightleftharpoons Ag^+(aq) + Cl^-(aq)$.

The equilibrium constant expression, written according to the above procedure, is

$$K = \frac{[Ag^+][Cl^-]}{[AgCl]}$$

At this point we might ask, "What is the meaning of the denominator, $[AgCl]$, inasmuch as silver chloride is a solid? How do we express the 'concentration' of a solid in moles/liter?" The answer is that we don't. We could argue that the concentration of a solid is fixed by its density, but that has little if anything to do with reaction rates. Surface area does. The more surface area there is, the faster the reaction is—in *both* directions, forward and reverse. The effect cancels out. The fact is that, as long as there is undissolved solute in contact with the saturated solution, the *amount* of solute does not influence the ion concentrations. It is the indicated product of the ionic concentrations that is constant at equilibrium:

$$K_{sp} = [Ag^+][Cl^-]$$

The subscript in K_{sp} identifies a **solubility product constant,** the special name given to equilibrium constants for low solubility solids.

The same conclusion can be reached by more rigorous reasoning and, more importantly, it is confirmed in a large number of experiments. Accordingly, *when we write a "practical" equilibrium constant expression for an equilibrium that includes a solid as one of the reacting species, the concentration of the solid is omitted.* The resulting equilibrium constant expression is thereby restricted to substances that are variable in concentration.

EXAMPLE 20.9   Write the equilibrium constant expressions for each of the following equilibria:

(a) $CaCO_3(s) \rightleftharpoons CaO(s) + CO_2(g)$

(b) $Li_2CO_3(s) \rightleftharpoons 2\,Li^+(aq) + CO_3^{2-}(aq)$

(c) $4\,H_2O(g) + 3\,Fe(s) \rightleftharpoons 4\,H_2(g) + Fe_3O_4(s)$

--------   --------

(a) $K = [CO_2]$,   (b) $K = [Li^+]^2[CO_3^{2-}]$;   (c) $K = \dfrac{[H_2]^4}{[H_2O]^4}$

In all cases the equilibrium constant expression is written in the usual way, except that the solids are omitted.

LIQUID WATER IN EQUILIBRIUM
CONSTANT EXPRESSIONS

The equilibrium equation for the ionization of acetic acid, a weak acid, by reaction with water may be written

$$HC_2H_3O_2(aq) + H_2O(\ell) \rightleftharpoons H_3O^+(aq) + C_2H_3O_2^-(aq) \qquad (20.4)$$

The equilibrium constant expression is

$$K = \frac{[H_3O^+][C_2H_3O_2^-]}{[HC_2H_3O_2][H_2O]}$$

To two significant figures, $[H_2O]$ is constant in dilute aqueous solutions. Therefore multiplying both sides of the equation by $[H_2O]$ yields on the left the product of two constants, which is another constant, K_a. This is called the **acid constant** for the ionization of a weak acid:

$$K[H_2O] = K_a = \frac{[H_3O^+][C_2H_3O_2^-]}{[HC_2H_3O_2]}$$

The liquid water effect is the same in all dilute aqueous solution equilibria. Therefore, *when we write a "practical" equilibrium constant expression for a dilute aqueous solution equilibrium that includes liquid water as one of the reacting species, the concentration of water, $[H_2O]$, is omitted.* Again, the resulting expression includes only those species whose concentrations are variable.

EXAMPLE 20.10   Write the equilibrium constant expression beneath each equilibrium equation:

   (a) $HF(aq) + H_2O(\ell) \rightleftharpoons H_3O^+(aq) + F^-(aq)$

   (b) $NH_3(aq) + H_2O(\ell) \rightleftharpoons NH_4^+(aq) + OH^-(aq)$

- -

(a) $K = \dfrac{[H_3O^+][F^-]}{[HF]}$ (b) $K = \dfrac{[NH_4^+][OH^-]}{[NH_3]}$

20.8 THE SIGNIFICANCE OF THE VALUE OF K

PG 20 L   Given an equilibrium equation and the value of the equilibrium constant, identify which direction is favored.

By definition, an equilibrium constant is a ratio—a fraction. The numerical value of an equilibrium constant may be very large, very small, or any place in between. The value of 51.5 in the $H_2(g) + I_2(g) \rightleftharpoons 2\,HI(g)$ equilibrium is in the "in between" group. While there is no defined intermediate range, we might think of constants between 0.01 and 100 (10^{-2} to 10^2) as neither very large nor very small. Equilibria with such constants will have significant quantities of all species present at equilibrium.

To see what is meant by "very large" or "very small" K values, consider an equilibrium similar to the hydrogen iodide system we studied earlier. Substituting chlorine for iodine we have

$$H_2(g) + Cl_2(g) \rightleftharpoons 2\,HCl(g)$$

At 25°C,

$$K = \frac{[HCl]^2}{[H_2][Cl_2]} = 2.4 \times 10^{33},$$

a very large number. If hydrogen and chlorine, each initially at 1.00 mole per liter, were to react to reach equilibrium with HCl, 99.999 999 999 999 999 959% of the hydrogen and chlorine initially present would be converted to HCl in order to satisfy the value of K. Remaining unreacted would be 0.000 000 000 000 000 041% of the elements. This is quite a contrast to the 21.8% of the original elements that were present at equilibrium when we started with H_2 and I_2 concentrations of 1.000 mole/liter. For all practical purposes, the hydrogen-chlorine reaction "goes to completion."

A large equilibrium constant implies that at least one of the reacting species must be almost totally consumed, leaving a denominator concentration close to zero. Such a reaction goes essentially to completion in the forward direction. This is described as an equilibrium in which the forward reaction is favored.

By contrast, if the equilibrium constant is very small, the numerator must include some species at a very low concentration. In such a system the reaction goes nearly to completion in the reverse direction. For example, for the equilibrium $2\,N_2(g) + O_2(g) \rightleftharpoons 2\,N_2O(g)$,

$$K = \frac{[N_2O]^2}{[N_2]^2[O_2]} = 2.0 \times 10^{-37}$$

This very small value of K indicates that at equilibrium the reverse reaction is strongly favored, with $[N_2O]$ very small.

To summarize, if an equilibrium constant is very large, the forward reaction is favored; if the constant is very small, the reverse reaction is favored. If the constant is neither large nor small, appreciable quantities of all species will be present at equilibrium.

20.9 EQUILIBRIUM CONSTANT CALCULATIONS (OPTIONAL)

Calculations based on the equilibrium constant cover a wide range of problem types. Thorough understanding of them is essential to the comprehension of many chemical phenomena in the laboratory, in industry, and in living organisms. Only four representative examples will be presented here, leaving the greater number and variety of problems to your more advanced courses.*

If you know the initial concentrations of the reactants in an equilibrium, and the equilibrium concentration of at least one species, you can calculate the value of the equilibrium constant. A gas phase equilibrium may be used to illustrate the approach.

EXAMPLE 20.11 10.0 moles of NO and 6.0 moles of O_2 are placed in an empty 1.00-liter reaction vessel at a certain temperature. They react until equilibrium is established according to the equation $2\,NO(g) + O_2(g) \rightleftharpoons 2\,NO_2(g)$. At equilibrium the vessel contains 8.8 moles of NO_2. Determine the value of K at that temperature.

Solution: In this example only the equilibrium concentration of NO_2 is given. The other concentrations must be determined by stoichiometry. It is usually helpful to organize your work and your thinking along lines of a table indicating the initial concentration of each species, I; the number of moles per liter of each species reacting, R; and the equilibrium concentration of each species, E. For this problem

	$2\,NO(g)$	$+$	$O_2(g)$	$\rightleftharpoons$	$2\,NO_2(g)$
I	10.0		6.0		0
R					
E					8.8

is the table, filled out as far as possible from the information given in the example. The initial [NO], [O_2], and final [NO_2] are stated, and the word "empty" indicates zero NO_2 present at the beginning.

If there were no NO_2 present initially, and 8.8 moles of NO_2 are present after equilibrium is reached, it follows that 8.8 moles must have been produced in the reaction. This is the value of R for NO_2. Adding that figure to the table, it becomes

	$2\,NO(g)$	$+$	$O_2(g)$	$\rightleftharpoons$	$2\,NO_2(g)$
I	10.0		6.0		0
R					+8.8
E					8.8

The plus sign indicates that 8.8 moles of NO_2 were produced, and that its concentration increased by that amount.

Now that the number of moles of NO_2 formed is known, the moles of other species used or produced can be determined from the reaction equation. The equation tells us that 2 moles of NO are consumed while 2 moles of NO_2 are formed, or, more generally, the moles of NO_2 formed are equal to the moles of NO used. If 8.8 moles of NO_2 were formed, 8.8 moles of NO must have been used. Similarly, 1 mole of O_2 is used for every 2 moles of NO_2 produced. Therefore the production of 8.8 moles of NO_2 requires 4.4 moles of O_2. The table is now extended to include these reaction values:

	$2\,NO(g)$	$+$	$O_2(g)$	$\rightleftharpoons$	$2\,NO_2(g)$
I	10.0		6.0		0.0
R	−8.8		−4.4		+8.8
E					8.8

The negative signs indicate that these quantities are consumed, and are to be subtracted from the starting quantities. Beginning with 10.0 moles of NO, 8.8 moles react, leaving 1.2

---

*For a thorough consideration of equilibrium problems see Chapters 13 to 17 of Peters, *Problem Solving for Chemistry*, Second Edition, W. B. Saunders Company, 1976.

moles at equilibrium. Similarly 6.0 − 4.4 = 1.6 moles of O_2 left at equilibrium. The table may now be completed:

	2 NO(g)	+	O_2(g)	⇌	2 NO_2(g)
I	10.0		6.0		0.0
R	−8.8		−4.4		+8.8
E	1.2		1.6		8.8

The figures in the equilibrium concentration line may now be entered into the equilibrium constant expression and the numerical answer calculated:

$$K = \frac{[NO_2]^2}{[NO]^2[O_2]} = \frac{(8.8)^2}{(1.2)^2(1.6)} = 34$$

If the value of K and the starting concentrations of a weak acid are known, the hydrogen ion concentration—and all other concentrations, for that matter—can be calculated. Many chemical processes are very sensitive to hydrogen ion concentrations in aqueous solution. The example that follows is the beginning of the theory that leads to understanding and control of these processes.

EXAMPLE 20.12 Find $[H^+]$, $[C_2H_3O_2^-]$, and $[HC_2H_3O_2]$, at equilibrium in 0.10 M $HC_2H_3O_2$. The value of K_a is 1.8×10^{-5} for $HC_2H_3O_2$(aq) ⇌ H^+(aq) + $C_2H_3O_2^-$(aq).

Solution: This time you are given the initial concentration of $HC_2H_3O_2$. Assume that the concentrations of H^+ and $C_2H_3O_2^-$ are zero initially. The equilibrium equation shows that each mole of $HC_2H_3O_2$ that ionizes produces one mole of each ion, but there is no indication of the number of moles of acid ionized. Let y represent the moles per liter of acid ionized in reaching equilibrium. This means that y moles of H^+ and y moles of $C_2H_3O_2^-$ per liter will be produced; their equilibrium concentrations will be y. The concentration of acetic acid at equilibrium is the acid initially present minus that which ionized, 0.10 − y. Summarized in a table, this is

	$HC_2H_3O_2$(aq)	⇌	H^+(aq)	+	$C_2H_3O_2^-$(aq)
I	0.10		0		0
R	−y		+y		+y
E	0.10 − y		y		y

On substituting the equilibrium concentrations into the K expression and equating to the known value of K we find ourselves confronted with a not-very-welcome quadratic equation:

$$K_a = \frac{[H^+][C_2H_3O_2^-]}{[HC_2H_3O_2]} = \frac{y^2}{0.10 - y} = 1.8 \times 10^{-5}$$

This equation may be solved by rearranging and using the quadratic formula, but the process is long and usually not necessary. A better way is to take advantage of what we know about significant figures and weak acids. A weak acid ionizes to a very small extent. Is it possible that the amount that ionizes, y, will be negligible compared to the 0.10 mole per liter initially present? If so, application of significant figures rules to 0.10 − y will yield 0.10 as the difference. Assuming this to be the case, the denominator becomes simply 0.10, and the problem becomes

$$\frac{y^2}{0.10} = 1.8 \times 10^{-5}$$

$$y = 1.34 \times 10^{-3}$$

This result, rounded off to the two significant figures dictated by both the concentration and equilibrium constant, means that $[H^+] = [C_2H_3O_2^-] = 1.3 \times 10^{-3} = 0.0013$—*if* the assumption that y is negligible compared to 0.10 is valid. The validity can be checked according to the significant figure rule for subtraction (page 591):

$$\begin{array}{r} 0.10 \\ -0.0013 \\ \hline 0.0987 = 0.10 \end{array}$$

Accordingly, $[HC_2H_3O_2] = 0.10$ at equilibrium, just as it was initially.
As shown on page 408,

$$pH = -\log[H^+] = -\log(1.3 \times 10^{-3}) = 2.89$$

The use of the assumption that a term may be negligible in addition or subtraction (*never* in multiplication or division) simplifies many equilibrium calculations. Such an assumption is usually valid if the equilibrium constant is very small, usually 10^{-4} or less, and the molecular acid concentration is in the area of 0.1 M or more. Considering measurement uncertainties in K values and solution concentrations, the results are as accurate as those obtained by solving the quadratic formula, rounded off to the proper number of significant figures. Calculated by the quadratic formula, the value of y in Example 20.12 is again 1.3×10^{-3}.

The sensitivity of chemical processes to very small hydrogen ion concentrations was mentioned earlier. It is possible to **buffer** a solution, or to stabilize its pH at some value that is known to yield satisfactory results. Most buffers contain a weak acid that neutralizes any unexpected OH^- ions and a weak base that will react with stray H^+ ions that may appear. In the simplest kind of buffer they may be conjugates of each other. For example, a solution buffered by acetic acid and acetate ion includes the equilibrium

$$HC_2H_3O_2(aq) \rightleftharpoons H^+(aq) + C_2H_3O_2^-(aq);$$

$$K_a = \frac{[H^+][C_2H_3O_2^-]}{[HC_2H_3O_2]} = 1.8 \times 10^{-5}$$

Dividing both sides of the equilibrium constant equation by $[H^+]$ yields

$$\frac{[C_2H_3O_2^-]}{[HC_2H_3O_2]} = \frac{K_a}{[H^+]} \tag{20.5}$$

Equation 20.5 may be used to determine quantities of acetic acid and a source of acetate ion, probably sodium acetate, that will buffer a solution at a desired pH.

EXAMPLE 20.13 Calculate the $[C_2H_3O_2^-]/[HC_2H_3O_2]$ ratio that will buffer a solution at pH = 4.10. How many grams of $NaC_2H_3O_2$ must be added to the solution along with 500 grams of $HC_2H_3O_2$ to form the buffer?

Solution: First find the $[H^+]$ that corresponds with a pH of 4.10:

$$[H^+] = \text{antilog} - 4.10 = 10^{-4.10} = 7.9 \times 10^{-5}$$

This can now be substituted into Equation 20.5 to find the $[C_2H_3O_2^-]/[HC_2H_3O_2]$ ratio:

$$\frac{[C_2H_3O_2^-]}{[H_2CH_3O_2]} = \frac{K_a}{[H^+]} = \frac{1.8 \times 10^{-5}}{7.9 \times 10^{-5}} = 0.23$$

We now have a ratio of concentrations—a ratio of moles per liter values—but no volume is specified. It is not needed because the ratio of concentrations is the same as the ratio of moles. Both the acetate ion and the acetic acid are in the same solution, which can have only one volume. Therefore volume cancels:

$$\frac{\text{mol } C_2H_3O_2^-/L}{\text{mol } HC_2H_3O_2/L} = \frac{\text{mol } C_2H_3O_2^-}{\text{mol } HC_2H_3O_2} = 0.23$$

The final question comes down to how many grams of $NaC_2H_3O_2$ must be added to 500 grams of $HC_2H_3O_2$ if there is to be 0.23 mol $C_2H_3O_2^-$ for each mol $HC_2H_3O_2$ (0.23 mol $C_2H_3O_2^- \cong 1$ mol $HC_2H_3O_2$). By dimensional analysis,

$$500 \text{ g } HC_2H_3O_2 \times \frac{1 \text{ mol } HC_2H_3O_2}{60.0 \text{ g } HC_2H_3O_2} \times \frac{0.23 \text{ mol } C_2H_3O_2^-}{1 \text{ mol } HC_2H_3O_2}$$

$$\times \frac{82.0 \text{ g } NaC_2H_3O_2}{1 \text{ mol } C_2H_3O_2^-} = 157 \text{ g } NaC_2H_3O_2$$

160 grams $NaC_2H_3O_2$ (two significant figures) must be mixed with 500 grams $HC_2H_3O_2$ to buffer the solution at pH = 4.10.

Solubility equilibria are among the easier applications of equilibrium constants.

EXAMPLE 20.14   Calculate the solubility of silver chloride in moles per liter if $K_{sp} = 1.7 \times 10^{-10}$ for $AgCl(s) \rightleftharpoons Ag^+(aq) + Cl^-(aq)$.

Solution: From the equation, for each mole of AgCl that dissolves, 1 mole of Ag^+ and 1 mole of Cl^- appear in the solution. The concentrations of these ions are therefore equal to each other and equal to the solubility of silver chloride in moles per liter. Let $[Ag^+] = [Cl^-] = s$.

$$K_{sp} = [Ag^+][Cl^-] = 1.7 \times 10^{-10}$$
$$s^2 = 1.7 \times 10^{-10}$$
$$s = 1.3 \times 10^{-5}$$

The answer to Example 20.14 gives an indication of just how low the solubility of an "insoluble" salt may be. The above concentration represents about 0.002 gram per liter of solution, roughly a 0.0002% solution that is saturated with all the solute it can hold. Silver iodide is less soluble by three orders of magnitude,* about 10^{-8} mole per

---

*"Orders of magnitude" refers to multiples of ten. For example, quantities that differ by three orders of magnitude differ by a ratio of one thousand to one—$10 \times 10 \times 10 = 10^3 = 1000$.

liter, and silver sulfide has a solubility of about 10^{-13} mole per liter, which comes to about 0.000 000 000 01% by weight.

CHAPTER 20 IN REVIEW

20.1 PHYSICAL AND CHEMICAL EQUILIBRIA— A REVIEW

20 A Identify the essential characteristics of an equilibrium system. (467)

20.2 THE COLLISION THEORY OF CHEMICAL REACTIONS

20 B Explain why, according to the collision theory of chemical reactions, some molecular collisions result in a chemical reaction and others do not. (468)

20.3 ENERGY CHANGES DURING A MOLECULAR COLLISION

20 C Sketch and/or interpret an enthalpy-reaction graph for either an exothermic or an endothermic reaction. Identify (a) the activated complex point; (b) activation energy; and (c) ΔH for the reaction. (469)

20.4 CONDITIONS THAT AFFECT THE RATE OF A CHEMICAL REACTION

20 D State and explain the relationship between reaction rate and temperature. (472)

20 E Explain how a catalyst affects reaction rate. Compare a catalyzed and an uncatalyzed reaction on an enthalpy-reaction graph. (474)

20 F Explain the relationship between reactant concentration and reaction rate. (475)

20.5 THE DEVELOPMENT OF A CHEMICAL EQUILIBRIUM

20 G Trace the changes in concentrations of reacting species that lead to a chemical equilibrium. (476)

20.6 Le CHATELIER'S PRINCIPLE

20 H Given the equation for a chemical equilibrium, predict the direction in which the equilibrium will shift because of a change in the concentration of one species. (478)

20 I Given the equation for a chemical equilibrium involving one or more gases, predict the direction in which the equilibrium will shift because of a change in the volume of the system. (479)

20 J Given a thermochemical equation for a chemical equilibrium, predict the direction in which the equilibrium will shift because of a change in temperature. (481)

20.7 THE LAW OF CHEMICAL EQUILIBRIUM: THE EQUILIBRIUM CONSTANT

20 K Given any chemical equilibrium equation, write the equilibrium constant expression. (482)

20.8 THE SIGNIFICANCE OF THE VALUE OF K

20 L Given an equilibrium equation and the value of the equilibrium constant, identify which direction is favored. (487)

20.9 EQUILIBRIUM CONSTANT CALCULATIONS (OPTIONAL) (488)

TERMS AND CONCEPTS

Forward reaction, reverse reaction (467)
Dynamic equilibrium (468)
Static equilibrium (468)
Favored (direction of an equilibrium) (468)
Collision theory of chemical reactions (468)
Potential energy barrier (471)
Activation energy (471)
Activated complex (471)
Catalyst (474)

Negative catalyst, inhibitor (475)
Le Chatelier's Principle (477)
Shift (as in direction of a reversible reaction) (478)
Law of Chemical Equilibrium (484)
Equilibrium constant, K (484)
Solubility product constant, K_{sp} (486)
Acid constant, K_a (487)
Buffer (491)

QUESTIONS AND PROBLEMS

An asterisk (*) identifies a question that is relatively difficult, or that extends beyond the performance goals of the chapter.

Section 20.1

20.1) What things are equal in an equilibrium? Give an example.

20.2) What is meant by saying that an equilibrium is confined to a "closed system?" Is it possible to have an equilibrium in an open beaker?

20.3) With regard to the equilibrium

$$F^-(aq) + H_2O(\ell) \rightleftarrows HF(aq) + OH^-(aq)$$

what is meant by saying that the equilibrium is favored in the reverse direction?

20.36) What does it mean when an equilibrium is described as *dynamic*? Compare a dynamic equilibrium with one that is static.

20.37) Undissolved table salt is in contact with a saturated salt solution in (a) a sealed container and (b) an open beaker. Which system, if either, can reach equilibrium? Explain your answer.

20.38) The equilibrium equation for the ionization of formic acid is

$$HCHO_2(aq) \rightleftarrows H^+(aq) + CHO_2^-(aq)$$

and the equilibrium is favored in the reverse direction. If you were to write a net ionic equation in which formic acid is a reactant, would you show formic acid in its ionized form or molecular form? Why?

Section 20.2

20.4) According to the collision theory of chemical reactions, what two conditions must be satisfied if a molecular collision is to result in reaction?

20.39) Explain why a molecular collision can be sufficiently energetic to cause a reaction, yet no reaction occurs as a result of that collision.

Section 20.3

20.5) Is ΔH for an exothermic reaction positive or negative? Confirm your answer by reference to an enthalpy-reaction graph for an exothermic reaction.

20.40) Two reactions, A and B, are the reverse of each other, as M → N and N → M. They therefore have the same activated complex, and their ΔH values are the same in magnitude, but opposite in sign: $\Delta H_A = +X$, and $\Delta H_B = -X$. Sketch typical enthalpy-reaction graphs for these reactions. Which of your two graphs indicates the greater activation energy?

20.6) Sketch an enthalpy-reaction graph for a reaction in which $\Delta H = -182.4$ kJ. Show how the ΔH appears on the graph. Indicate also the activation energy, and the position of the activated complex.

20.41) Sketch an enthalpy-reaction graph for a reaction in which $\Delta H = +39.3$ kJ. Show how ΔH appears on the graph. Also indicate the activation energy and the activated complex point.

20.7) Explain the significance of activation energy. For two reactions that are identical in all respects except activation energy, identify the reaction that would have the higher rate and tell why.

20.42) What is an "activated complex"? Why is it that we cannot list the physical properties of the species represented as an activated complex?

Section 20.4

20.8) "At a given temperature, only a small fraction of the molecules in a sample has sufficient kinetic energy to engage in chemical reaction." What is the meaning of that statement?

20.43) State the effect of a temperature increase and a temperature decrease on the rate of a chemical reaction. Explain each effect.

20.9) What is a catalyst? Explain how a catalyst affects reaction rates.

20.44) Suppose that two substances are brought together under conditions that cause them to react and reach equilibrium. Suppose that in another vessel the same substances and a catalyst are brought together, and again equilibrium is reached. How are the processes alike, and how are they different?

20.10) For the hypothetical reaction A + B → C, what will happen to the rate of reaction if the concentration of A is increased? What will happen if the concentration of B is decreased? Explain why in both cases.

20.45) For the hypothetical reaction A + B → C, what will happen to the reaction rate if the concentration of A is increased *and* the concentration of B is decreased? Explain.

Section 20.5

If nitrogen and hydrogen are brought together at the proper temperature and pressure, they will react until they reach equilibrium:

$$N_2(g) + 3 H_2(g) \rightleftharpoons 2 NH_3(g)$$

Answer the following two questions in each column with regard to the establishment of that equilibrium:

20.11) When will the forward reaction rate be at a maximum: at the start of the reaction, after equi-

20.46) When will the reverse reaction rate be at a maximum: at the start of the reaction, after equilibrium has been reached, or at some point in between?

20.12) What happens to the concentration of each of the three species between the start of the reaction and the time equilibrium is reached?

Section 20.6

20.13) Predict the direction in which the equilibrium $SO_2(g) + NO_2(g) \rightleftharpoons SO_3(g) + NO(g)$ will shift if the concentration of NO is increased. Explain or justify your prediction.

20.14) The equilibrium system

$$COCl_2(g) \rightleftharpoons CO(g) + Cl_2(g)$$

has some of the chlorine removed. Predict the direction in which the equilibrium will shift, and explain your prediction.

20.15) In which direction would the equilibrium $HC_2H_3O_2(aq) \rightleftharpoons H^+(aq) + C_2H_3O_2^-(aq)$ shift if the concentration of the $C_2H_3O_2^-$ ion were increased? Explain your prediction.

20.16) Which direction will be favored if you reduce the volume of the equilibrium

$$N_2O_3(g) \rightleftharpoons N_2O(g) + O_2(g)?$$

Explain.

20.17) What shift will occur, according to Le Chatelier's Principle, to the equilibrium

$$C(s) + H_2O(g) \rightleftharpoons CO(g) + H_2(g)$$

if volume is increased? Explain.

20.18) The equilibrium $CO(g) + H_2O(g) \rightleftharpoons CO_2(g) + H_2(g) + 41.4\,kJ$ is heated. Predict the direction in which the equilibrium will be favored as a result, and justify your prediction in terms of Le Chatelier's Principle.

20.19) If you wished to increase the relative amount of HI in the equilibrium

$$H_2(g) + I_2(g) + 25.9\,kJ \rightleftharpoons 2\,HI(g)$$

would you heat or cool the system? Explain your decision.

20.20) Consider the equilibrium:

$$4\,NH_3(g) + 5\,O_2(g) \rightleftharpoons$$
$$4\,NO(g) + 6\,H_2O(g) + 905\,kJ$$

librium has been reached, or at some point in between?

20.47) On a single set of coordinate axes, sketch graphs of the forward reaction rate vs. time and the reverse reaction rate vs. time from the moment the reactants are mixed to a point beyond the establishment of equilibrium.

20.48) If the system $2\,SO_2(g) + O_2(g) \rightleftharpoons 2\,SO_3(g)$ is at equilibrium and the concentration of O_2 is reduced, predict the direction in which the equilibrium will shift. Justify or explain your prediction.

20.49) If additional oxygen is pumped into the equilibrium system

$$4\,NH_3(g) + 5\,O_2(g) \rightleftharpoons 4\,NO(g) + 6\,H_2O(g),$$

in which direction will the reaction shift? Justify your answer.

20.50) Predict the direction of shift for the equilibrium

$$Cu(NH_3)_4{}^{2+}(aq) \rightleftharpoons Cu^{2+}(aq) + 4\,NH_3(aq)$$

if the concentration of ammonia were reduced. Explain your prediction.

20.51) A container holding the equilibrium

$$4\,H_2(g) + CS_2(g) \rightleftharpoons CH_4(g) + 2\,H_2S(g)$$

is enlarged. Predict the direction of the Le Chatelier shift. Explain.

20.52) In what direction will

$$CO(g) + H_2O(g) \rightleftharpoons CO_2(g) + H_2(g)$$

shift as a result of a reduction in volume? Explain.

20.53) Which direction of the equilibrium

$$2\,NO_2(g) \rightleftharpoons N_2O_4(g) + 59.0\,kJ$$

will be favored if the system is cooled? Explain.

20.54) If your purpose were to increase the yield of SO_3 in the equilibrium

$$SO_2(g) + NO_2(g) \rightleftharpoons SO_3(g) + NO(g) + 41.8\,kJ$$

would you use the highest or lowest operating temperature possible? Explain.

20.55)* The solubility of calcium hydroxide is low; it reaches about $2.4 \times 10^{-2}\,M$ at saturation. In acid solutions, with many H^+ ions present, calcium hydroxide is quite soluble. Explain this fact

Determine the direction of the Le Chatelier shift, forward or reverse, for each of the following actions: (a) add ammonia; (b) raise temperature; (c) reduce volume; (d) remove $H_2O(g)$.

in terms of Le Chatelier's Principle. (Hint: Recall what you know of reactions in which molecular products are formed.)

Section 20.7

For each equilibrium equation shown, write the equilibrium constant expression.

20.21) $2 SO_2(g) + O_2(g) \rightleftharpoons 2 SO_3(g)$

20.22) $4 H_2(g) + CS_2(g) \rightleftharpoons CH_4(g) + 2 H_2S(g)$

20.23) $Cd(OH)_2(s) \rightleftharpoons Cd^{2+}(aq) + 2 OH^-(aq)$

20.24) $HNO_2(aq) \rightleftharpoons H^+(aq) + NO_2^-(aq)$

20.25) $Ag(CN)_2^-(aq) \rightleftharpoons Ag^+(aq) + 2 CN^-(aq)$

20.26) "The equilibrium constant expression for a given reaction depends upon how the equilibrium equation is written." Explain the meaning of that statement. You may, if you wish, use the equilibrium equation $N_2(g) + 3 H_2(g) \rightleftharpoons 2 NH_3(g)$ to illustrate your explanation.

20.56) $CO(g) + H_2O(g) \rightleftharpoons CO_2(g) + H_2(g)$

20.57) $C(s) + H_2O(g) \rightleftharpoons CO(g) + H_2(g)$

20.58) $Zn_3(PO_4)_2(s) \rightleftharpoons 3 Zn^{2+}(aq) + 2 PO_4^{3-}(aq)$

20.59) $HNO_2(aq) + H_2O(\ell) \rightleftharpoons$
$$H_3O^+(aq) + NO_2^-(aq)$$

20.60) $Cu(NH_3)_4^{2+}(aq) \rightleftharpoons Cu^{2+}(aq) + 4 NH_3(aq)$

20.61) The equilibrium between nitrogen monoxide, oxygen, and nitrogen dioxide may be expressed in the equation

$$2 NO(g) + O_2(g) \rightleftharpoons 2 NO_2(g)$$

Write the equilibrium constant expression for this equation. Then express the same equilibrium in at least two other ways, and write the equilibrium constant expression for each. Are the constants numerically equal? Cite some evidence to support your yes or no answer.

Section 20.8

20.27) The equilibrium constant is 2.3×10^{-8} for the equilibrium

$$HCO_3^-(aq) + HOH(\ell) \rightleftharpoons$$
$$H_2CO_3(aq) + OH^-(aq)$$

In which direction is the reaction favored at equilibrium? State the basis for your answer.

20.28) Acetic acid, $HC_2H_3O_2$, is a soluble weak acid. When placed in water it partially ionizes and reaches equilibrium. Write the equilibrium equation for the ionization. Will the equilibrium constant be large or small? Justify your answer.

20.62) If sodium cyanide solution is added to silver nitrate solution the following equilibrium will be reached:

$$Ag^+(aq) + 2 CN^-(aq) \rightleftharpoons Ag(CN)_2^-(aq)$$

For this equilibrium $K = 5.6 \times 10^{18}$. In which direction is the equilibrium favored? Justify your answer.

20.63) A certain equilibrium has a very small equilibrium constant. In which direction, forward or reverse, is the equilibrium favored? Explain.

In Chapter 16 you learned how to write solution inventories and net ionic equations, based on the solubility of ionic compounds, the strength of acids, and the stability of certain ion combinations. Use these ideas to predict the favored direction of each equilibrium below. In each case, state whether you expect the equilibrium constant to be large or small.

20.29)* (a) $HCl(aq) \rightleftharpoons H^+(aq) + Cl^-(aq)$
(b) $BaSO_4(s) \rightleftharpoons Ba^{2+}(aq) + SO_4^{2-}(aq)$

20.64)* (a) $H_2SO_3(aq) \rightleftharpoons H_2O(\ell) + SO_2(aq)$
(b) $H^+(aq) + C_2H_3O_2^-(aq) \rightleftharpoons$
$$HC_2H_3O_2(aq)$$

Section 20.9

20.30) At 1350°K the system

$$2 SO_2(g) + O_2(g) \rightleftharpoons 2 SO_3(g)$$

reaches equilibrium with $[SO_2] = 0.90$, $[O_2] = 0.45$, and $[SO_3] = 0.60$. Find K at that temperature.

20.31) $[H^+] = 0.0265$ in 1.0 M HF. Calculate K_a for the ionization equilibrium

$$HF(aq) \rightleftharpoons H^+(aq) + F^-(aq).$$

20.32) 3.00 moles of HI are introduced into a 1.00-liter reaction chamber at 440°C, the temperature at which $K = 51.5$ for the reaction

$$H_2(g) + I_2 \rightleftharpoons 2 HI(g).$$

The compound decomposes until equilibrium is reached. Find the equilibrium concentrations of all species.

20.33) At what pH will a solution be buffered if it contains 11.8 g $HC_3H_5O_2$ (propanoic acid, $K_a = 1.4 \times 10^{-5}$) and 32.5 g $KC_3H_5O_2$ (potassium propionate) in 500 mL solution? in 1000 mL?

20.34) Calculate the solubility of calcium carbonate in moles per liter if $K_{sp} = 8.7 \times 10^{-9}$.

20.35) $K_{sp} = 6.5 \times 10^{-9}$ for magnesium fluoride, MgF_2. Calculate its solubility in moles per liter.

20.65) Calculate the equilibrium constant for the system $Ni(CO)_4(g) \rightleftharpoons Ni(s) + 4 CO(g)$ at 400 K if $[Ni(CO)_4] = 1.80$ and $[CO] = 0.800$.

20.66) 0.50 M $HCHO_2$ (formic acid) is 2.0% ionized at equilibrium. Calculate the value of the acid constant, K_a, for

$$HCHO_2(aq) \rightleftharpoons H^+(aq) + CHO_2^-(aq).$$

20.67) Calculate $[HC_7H_5O_2]$, $[H^+]$, and $[C_7H_5O_2^-]$ in 0.20 M $HC_7H_5O_2$ (benzoic acid) if $K_a = 6.6 \times 10^{-5}$.

20.68) How many grams of $KC_3H_5O_2$ (112 g/mol) must be dissolved in 250 mL 0.500 M $HC_3H_5O_2$ to produce a buffer in which pH = 4.50? (See Question 20.33 for data.)

20.69) How many grams of calcium sulfate are dissolved in 10.0 liters of a saturated solution if $K_{sp} = 1.9 \times 10^{-4}$?

20.70) What is the solubility (moles per liter) of silver chloride in a salt solution in which $[Cl^-]$ is fixed at 0.010? $K_{sp} = 1.7 \times 10^{-10}$ (Hint: The solubility of AgCl is equal to $[Ag^+]$ at equilibrium.)

nuclear chemistry

21.1 THE DAWN OF NUCLEAR CHEMISTRY

Serendipity. This pleasant sounding word refers to finding valuable things you are not looking for—an accidental discovery, in other words. Serendipity has played an important part in many scientific discoveries, but rarely has an unplanned observation led to a discovery as far reaching as that of Henri Becquerel. What he stumbled across now affects the lives of you and me, and potentially of every creature on this planet. Becquerel discovered nuclear chemistry.

Becquerel became interested in X-rays soon after they were discovered, also accidentally, by Wilhelm Roentgen in 1895. Becquerel was also interested in fluorescence and wondered if the two were related. To produce fluorescence you shine one color of light on an object, and it later glows with another color. X-rays are like light. They are part of the electromagnetic spectrum (p. 74), but they are outside the visible range. They are high energy radiations that are able to go through human tissue, black paper, and many other solids that visible light cannot penetrate.

Becquerel's plan was to put some fluorescent material, a uranium salt, on top of unexposed photographic film wrapped in black paper and place them in the sunlight. The sunlight could not expose the film by itself, but if the film were exposed it would show that fluorescent rays from the uranium salt had passed through the paper like X-rays.

The day Becquerel picked for his experiment was cloudy. After waiting in vain for the sun to come out, he put the uranium salt and wrapped film into a drawer, to be used the next sunny day. The weather continued to be dull for several days. Becquerel finally decided to develop the film to see if any fluorescent light had gotten through, even on the cloudy day. To his amazement, the film was strongly exposed. The only possible conclusion was that rays were being emitted from the salt all the time in the darkness of the drawer, and that they had nothing whatever to do with sunlight. It was later shown that, although their penetrating ability is like that of X-rays, the emissions were not X-rays but a form of energy that had never been observed before.

This is how Becquerel discovered **radioactivity**, a natural, spontaneous process that has been going on in the nuclei of atoms since the beginning of time.

Once the door to nuclear chemistry was opened, progress was rapid. While working with a uranium-bearing mineral called *pitchblende*, Marie and Pierre Curie discovered that a second element, thorium, gave off penetrating radiations. They also found two new radioactive elements: polonium, which is about 400 times more radioactive than uranium, and radium, which is again many times more radioactive than polonium.

21.2 NATURAL RADIOACTIVITY

PG 21 A Define radioactivity. Name, identify from a description, or describe the three types of natural radioactive emission.

Radioactivity is the spontaneous emission of rays resulting from the decay, or breaking up, of an atomic nucleus. Three kinds of rays can be identified. If a beam consisting of all three rays is directed into an electric field, as in Figure 21.1, the individual rays are separated. One, called an alpha ray, or α-ray,* is attracted toward the negatively charged plate, indicating that it has a positive charge. The α-ray has little penetrating power; it can be stopped by the outer layer of skin or a few sheets of paper (Fig. 21.2). Alpha rays are now known to be nuclei of helium atoms, having the nuclear symbol $_2^4$He.† They are commonly called alpha particles, or α particles.

The second kind of ray also turns out to be a beam of particles, but they are negatively charged and are therefore attracted to the positively charged plate (Fig. 21.1). Called beta rays, or β-rays, they have been identified as electrons. The nuclear symbol for a β particle is $_{-1}^0$e, indicating zero atomic mass and a -1 charge emerging from the nucleus. β particles have considerably more

---

*α, β, and γ are the Greek letters alpha, beta, and gamma.

†Recall from page 59 that the nuclear symbol is $_Z^A$Sy, where Sy is the symbol of the element, Z is the nuclear charge, or atomic number, and A is the mass number, the total number of protons plus neutrons in the nucleus of the isotope. It follows that the number of neutrons is A − Z.

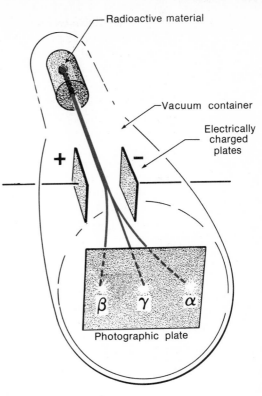

Figure 21.1
Paths of alpha, beta, and gamma emissions in an electric field. Alpha rays are attracted toward the negative plate, indicating that they have a positive charge. Beta rays are attracted toward the positive plate, indicating a negative charge. Gamma rays are unaffected, showing that they have no electrical charge.

penetrating power than α particles, but they can be stopped by a sheet of aluminum about 4 millimeters thick (Fig. 21.2).

The third kind of radiation is the gamma ray, or γ-ray. Gamma rays are not particles, but very high energy electromagnetic rays, similar to X-rays. Because of their high energy, gamma rays have high penetrating power. They can be stopped only by thick layers of lead or heavy concrete walls, as shown in Figure 21.2. Not having an electric charge, gamma rays are not deflected by an electric field (Fig. 21.2).

Figure 21.2
Penetration power of radioactive emissions. A few sheets of paper are unable to stop beta or gamma rays, but they will stop alpha emissions. Gamma rays will pass through an aluminum sheet 4 mm thick, but beta rays will not. A heavy layer of concrete or lead is required to stop gamma radiations.

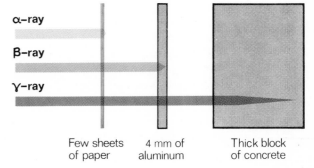

21.3 INTERACTION OF RADIOACTIVE EMISSIONS WITH MATTER

PG 21 B Describe how emissions from radioactive substances affect gases and living organisms.

When α, β, or γ radiation collides with an atom or molecule, some of its energy is given to the target particle. The collision changes the electron arrangement in the species hit. An electron may be excited to a higher energy level, leading to electromagnetic radiation as the electron drops back to its ground state level. The electron may be knocked all the way out of an atom or molecule, ionizing it. Air, or any gas, can be ionized by a radioactive substance. If the radiation strikes chemically bonded atoms, it often breaks those bonds, thereby causing a chemical reaction.

The ionization of molecules in air by radioactive emission has a present-day application in one kind of home smoke detector. These "ionization" detectors generally use a tiny chip of americium-241, a radioactive isotope of an element not found in nature. The ionization of air causes a small current to flow through the air inside the detector. When smoke enters, it breaks the circuit and sets off the alarm.

It is the ability of radiation to start chemical change that is responsible for its effect on body tissue, leading to illness and even death in extreme cases. These effects have been studied among the survivors of the atomic bombs dropped at Hiroshima and Nagasaki at the close of World War II, and among workers in nuclear plants and laboratories who have been overly exposed to radiation. People working in such plants wear a small device called a dosimeter that monitors the amount of radiation that falls upon them. By this and other safety measures, the danger to workers associated with industrial radiation has been minimized and controlled.

21.4 DETECTION OF EMISSIONS FROM RADIOACTIVE SUBSTANCES

PG 21 C Define or describe a cloud chamber. Explain how it operates.
21 D Define or describe a Geiger counter. Explain how it operates.

There are several ways to detect radioactivity. Perhaps the most obvious, but not necessarily the most convenient, is through its effect on photographic film, the property that caused it to be discovered in the first place. Another is based on its ability to produce visible emissions in fluorescent materials. Luminous watch dials, for example, are painted with a mixture including zinc sulfide and a tiny trace of radium sulfate. Radiations from the radium cause fluorescence in the zinc sulfide, making it visible in the dark. A device called a *scintillation counter* may be used to measure radiant energy by fluorescence.

Perhaps the best visible evidence of radiation is found in a **cloud chamber** (Fig. 21.3). Radioactive emissions are directed through an air space that is supersaturated with some vapor, often water. As the molecules in the air are ionized, the vapor condenses on the ions. The condensate is visible, and leaves a growing track behind the radiation as it travels through the space, much as a vapor trail forms behind a high-flying jet aircraft on a clear day. Cloud chamber tracks may be photographed; one historically significant photograph is shown as Figure 21.4.

The **Geiger counter** is probably the best known instrument for detecting and measuring radiation. It consists of a tube filled with a gas, as shown in Figure 21.5. The gas is ionized by radiation entering the tube through a window, permitting a surge of electrical current between two electrodes. The current is related to the quantity of radiation received, and can be read on a meter. Many Geiger counters indicate radiation by a clicking sound.

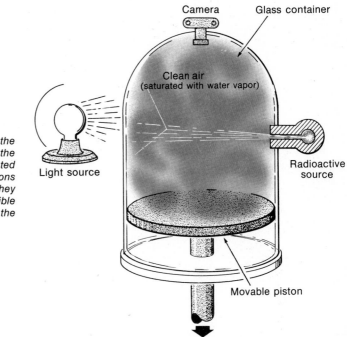

Figure 21.3
Wilson cloud chamber. Moving the piston down causes the air in the chamber to become supersaturated in water vapor. Radioactive emissions ionize the air through which they pass, causing condensation of visible droplets that leave a trail behind the emitted ray.

Camera   Glass container

Clean air
(saturated with water vapor)

Light source

Radioactive source

Movable piston

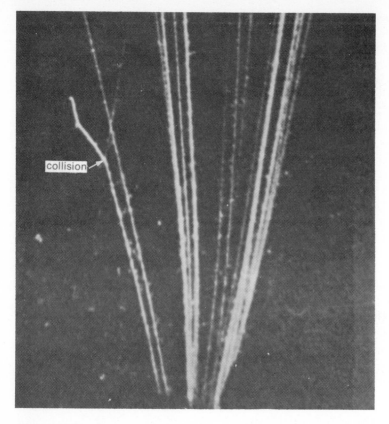

Figure 21.4
Alpha particle tracks as they appear in a cloud chamber. This particular photograph is historically significant. At the point marked "collision," an alpha particle struck a nitrogen nucleus, producing a proton and an oxygen-17 nucleus. This was the first artificial nuclear transformation.

21.5 NATURAL RADIOACTIVE DECAY SERIES— NUCLEAR EQUATIONS

PG 21 E Describe a natural radioactive decay series.

21 F Given the identity of a radioactive isotope and the particle it emits, write a nuclear equation for the emission.

When a nucleus emits an α or β particle, there is a change in the make-up of the nucleus. It has a new identity directly related to the particle emitted. The emission of one or more gamma rays does not, by itself, change the compo-

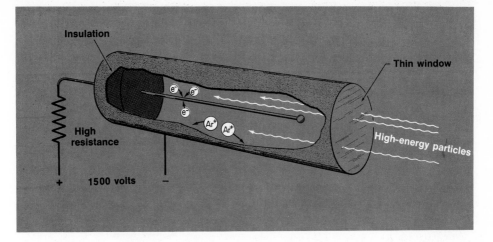

Figure 21.5
*Schematic drawing of a Geiger counter. The tube contains argon at low pressure. A high electrical
potential is established between a positively charged wire in the center and the negatively charged
case. When radioactive emissions enter the tube window, electrons are knocked out of the argon
atoms, ionizing the gas. Electrons are accelerated toward the positively charged wire, and argon
atoms move toward the negatively charged case, producing more ions en route. The buildup of ions
is such that a sharp electrical discharge occurs, producing an audible "click." These discharges
may be counted and recorded, indicating the intensity of the radioactivity.*

sition of the nucleus. For the balance of the chapter, therefore, we will consider
only alpha and beta particles in nuclear reactions.

When a radioactive nucleus emits an alpha or beta particle, there is a
transmutation of an element, or a change from one element to another. Recall
that the elemental identity of an atom depends upon the number of protons in
the nucleus. If an atom changes from one element to another, there must be a
change in the number of protons. A change of this kind is represented by a
nuclear equation showing the nuclear symbols of the reactant and product
isotopes. The emission observed by Becquerel was an alpha particle emission,
also called an **alpha decay reaction.** The nuclear product remaining after a $^{238}_{92}U$
nucleus emits an α particle is a thorium nucleus, $^{234}_{90}Th$. In other words, a $^{238}_{92}U$
nucleus has *disintegrated*, or *decayed*, into a $^{4}_{2}He$ nucleus and a $^{234}_{90}Th$ nucleus.
This may be shown in a nuclear equation:

$$^{238}_{92}U \rightarrow {}^{4}_{2}He + {}^{234}_{90}Th \qquad (21.1)$$

Notice that the above equation is "balanced" in both neutrons and protons. The total number of neutrons and protons is 238, the mass number of the
uranium isotope. The total mass number of the two products is $234 + 4$, again
238. In terms of protons, the 92 in a uranium nucleus are accounted for by 90 in
the thorium nucleus plus 2 in the helium nucleus. A nuclear equation is balanced if the sums of the mass numbers on the two sides of the equation are
equal, and if the sums of the atomic numbers are equal.

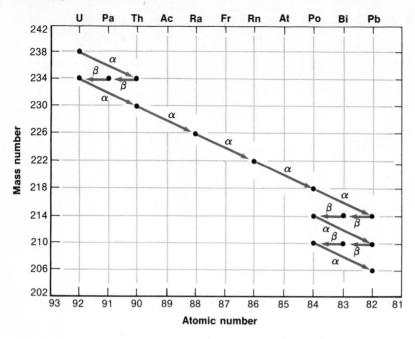

Figure 21.6
Radioactive decay series. This series begins with $^{238}_{92}U$, and after eight alpha emissions and six beta emissions, produces $^{206}_{82}Pb$ as a stable end product.

The $^{234}_{90}Th$ nucleus resulting from the disintegration of uranium-238 is also radioactive. It is a **beta decay reaction;** it emits a beta particle, $_{-1}^{0}e$, and produces an isotope of protactinium, $^{234}_{91}Pa$:

$$^{234}_{90}Th \rightarrow {}^{234}_{91}Pa + {}^{0}_{-1}e \qquad (21.2)$$

In a β particle emission the mass numbers of the reactant and product isotopes are the same, while the atomic number increases by 1. Though the actual process is more complex, it appears as if a neutron divides into a proton and an electron, and the electron is ejected.

The two radioactive disintegration steps described by Equations 21.1 and 21.2 are only the first two of fourteen steps that begin with $^{238}_{92}U$. There are eight α particle emissions and six β particle emissions, leading ultimately to a stable isotope of lead, $^{206}_{82}Pb$. This entire **natural radioactive decay series** is described in Figure 21.6. There are two other natural disintegration series, as they are also called. One begins with $^{232}_{90}Th$ and ends with $^{208}_{82}Pb$, and the other passes from $^{235}_{92}U$ to $^{207}_{82}Pb$.

EXAMPLE 21.1 Write the nuclear equation for the changes that occur in the uranium-238 disintegration series when $^{226}_{88}Ra$ ejects an α particle. Ra is the symbol for radium, one of the elements discovered by Pierre and Marie Curie in their study of radioactivity.

In writing a nuclear equation, one product will be the particle ejected. The mass number of the other product will be such that, when added to the mass number of the ejected particle, the total will be the mass number of the original isotope. What is the mass number of the second product of the emission of an alpha particle from a $^{226}_{88}$Ra nucleus?

------   ------

222. The reactant isotope has a mass number of 226. It emits a particle having a mass number of 4. This leaves $226 - 4 = 222$ as the mass number of the remaining particle.

Now find the atomic number of the second product particle. The atomic number of the starting isotope is 88. It emitted a particle having two protons. How many protons are left in the nucleus of the other product?

------   ------

86. If two protons are emitted from a nucleus having 88 protons, 86 remain.

You now know the mass number and the atomic number of the second product of an alpha particle emission from $^{226}_{88}$Ra. Using a periodic table you can find the elemental symbol of this product, and assemble all three symbols into the required nuclear equation.

------   ------

$$^{226}_{88}\text{Ra} \rightarrow {}^{4}_{2}\text{He} + {}^{222}_{86}\text{Rn}$$

The second product is a radioactive isotope of the noble gas radon, whose atomic number is 86. This isotope continues the natural emission series by emitting another α particle.

EXAMPLE 21.2   Write the nuclear equation for the emission of a β particle from $^{210}_{83}$Bi. The method is the same. Remember the beta particle, $_{-1}^{0}$e, has zero mass number, and an effective atomic number of -1. Both mass number and atomic number must be conserved in the equation.

------   ------

$$^{210}_{83}\text{Bi} \rightarrow {}^{0}_{-1}\text{e} + {}^{210}_{84}\text{Po}$$

In the emission of a β particle, which has effectively no mass, the mass number of the radioactive isotope and the product isotope are the same. The product isotope has an atomic number greater by one than the radioactive isotope, an increase of one proton. Po is the symbol that corresponds to atomic number 84. The element is polonium, the other element discovered by the Curies in their investigation of radioactivity. The name of the element was selected to honor Mme. Curie's native Poland.

21.6 HALF-LIFE

PG 21 G   Describe or illustrate what is meant by the half-life of a radioactive isotope.

    21 H   Given the starting quantity of a radioactive isotope and two of the following, calculate the third: half-life; elapsed time; quantity of isotope remaining.

The rate at which a single step in nuclear disintegration occurs is measured by its **half-life, the time required for the disintegration of one half of the radioactive atoms in a sample.** Each radioactive isotope has its own unique half-life, commonly written $t_{1/2}$. The half-lives of some isotopes are very long; the half-life of $^{238}_{92}U$ is 4.5 billion years. Other half-lives are very short. The half-life of $^{234}_{90}Th$ (Equation 21.1) is a comfortable 24.1 days, but the half-life of the particle that appears after $^{234}_{90}Th$ emits a β particle, $^{234}_{91}Pa$, is only 1.18 minutes. The half-life of one species in the same radioactive disintegration series, $^{214}_{84}Po$, is only 0.000 16 second. Needless to say, people who work with radioactive substances with such short half-lives are rather pressed for time.

Figure 21.7 illustrates the meaning of half-life. Shown in black along the vertical axis is the fraction of the original sample that is left at the end of any number of half-lives, plotted horizontally in black. One half of the material present at the beginning of each half-life cycle will remain at the end of that cycle. At the end of the first cycle, $\frac{1}{2}$ of the original sample remains. At the end of the second half-life, $\frac{1}{2}$ of $\frac{1}{2}$, or $\frac{1}{4}$ remains. This is the same as saying that the fraction of the original sample remaining after *two* half-lives is $(\frac{1}{2})^2$. At the end of *three* half-lives, the amount remaining will be $\frac{1}{2}$ of $\frac{1}{4}$, or $\frac{1}{8}$, which is the same as $(\frac{1}{2})^3$. At the end of n half-lives the fraction of the original sample that remains is $(\frac{1}{2})^n$. If S is the starting quantity and R is the amount that remains after n half-lives,

$$R = S \times (\tfrac{1}{2})^n \qquad (21.3)$$

Using logarithms, this can be solved for n:*

$$n = \frac{\log (S/R)}{0.301} = \frac{\log S - \log R}{0.301} \qquad (21.4)$$

To further illustrate the meaning of a half-life cycle, the decay of a sample of $^{210}_{83}Bi$, the radioactive isotope used in Example 21.2, is shown by colored numbers in Figure 21.7. The half-life of that isotope is 5.0 days. If you begin

*Multiplying both sides by $2^n/R$,

$$2^n = S/R$$

Taking logarithms of both sides, and using $\log x^n = n \log x$ and $\log a/b = \log a - \log b$,

$$n \log 2 = 0.301 \ n = \log (S/R) = \log S - \log R$$

Divide by 0.301 to get Equation 21.4.

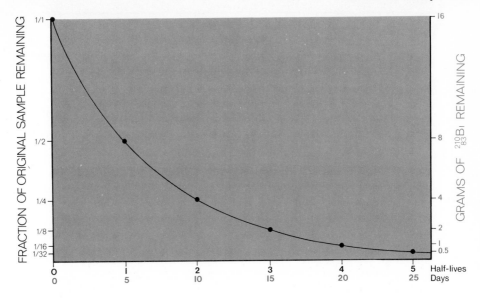

Figure 21.7
Half-life decay curve for a radioactive substance. During each half-life period, half of the material present at the beginning of the period undergoes radioactive disintegration. This is shown by the black numbers. Colored numbers trace the decay of 16 grams of $^{210}_{83}Bi$ through five half-lives of five days each. The fraction of a sample remaining after n half-lives is $(\frac{1}{2})^n$.

with 16 grams on the first day of April, only 8 grams will remain on April 5. By April 10 your sample will be reduced to 4 grams, and on April 15, the middle of the month, it will be down to 2 grams. One gram will still be there five days later, on April 20, dropping to 0.5 gram on April 25. At the end of April you will have only 0.25 gram of your original 16 to carry into the month of May. The same result for six half-life cycles can be reached with Equation 21.3:

$$R = S \times (\tfrac{1}{2})^n = 16\,g \times (\tfrac{1}{2})^6 = 0.25\,g$$

To solve the problem with a calculator you must punch in 0.5 for $\frac{1}{2}$, and use the y^x key to raise it to the sixth power. The methods for a TI-30 II and HP-32E are:

TI-30 II			HP-32E	
Press	*Display*		*Press*	*Display*
.5	0.5		.5	0.5
y^x	0.5		ENTER	0.5000
6	6		6	6
$\times$	0.015625		y^x	0.0156
16	16		16	16
=	0.25		$\times$	0.2500

EXAMPLE 21.3   The half-life of $^{45}_{19}$K is 20 minutes. If you have a sample containing 2.1×10^3 micrograms (μg) of this isotope at noon, how many micrograms will remain at 3 o'clock in the afternoon?

   First of all, through how many half-lives will the decay process pass between noon and 3 o'clock, at 20 minutes for each half-life?

———————   ———————

$$3 \text{ hours} \times \frac{60 \text{ minutes}}{1 \text{ hour}} \times \frac{1 \text{ half-life}}{20 \text{ minutes}} = 9 \text{ half-lives}$$

From here the solution comes by direct substitution into Equation 21.3.

———————   ———————

$$R = 2.1 \times 10^3 \ \mu g \times (\tfrac{1}{2})^9 = 4.1 \ \mu g$$

   To find the half-life of a radioactive isotope you must determine initial and remaining quantities over a measured period of time. The number of half-lives in that period is found from Equation 21.4. Dividing time by number of half-lives gives $t_{1/2}$.

EXAMPLE 21.4 $^{125}_{51}$Sb has a longer half-life than many man-made radioisotopes. The mass of a sample is found to be 8.623 grams. The sample is set aside for 73 days—a convenient period equal to 0.20 year—and then weighed again. The mass is 8.192 grams. Find $t_{1/2}$.

   The problem gives all the information needed to find n, the number of half-lives in 0.20 year. Substitute the numbers, but do not calculate the answer.

———————   ———————

$$n = \frac{\log \ (S/R)}{0.301} = \frac{\log \ (8.623/8.192)}{0.301} =$$

   One way to use a calculator to solve the above problem is to find first the quotient of 8.623/8.192. With that number in the display, press the "log" key and you will have the numerator. Divide by 0.301 to complete the problem. Calculator sequences are given below, but see if you can complete the problem before looking at them.

———————   ———————

$$n = \frac{\log \ (S/R)}{0.301} = \frac{\log \ (8.623/8.192)}{0.301} = 0.0740 \text{ half-life}$$

	TI-30			HP-32E	
Press	Display		Press	Display	
8.623	*8.623*		8.623	*8.623*	
÷	*8.623*		ENTER	*8.6230*	
8.192	*8.192*		8.192	*8.192*	
=	*1.052 6 12 3*		÷	*1.0526*	
log	*0.022 268 44*		gold f	*1.0526*	
÷	*0.022 268 44*		log	*0.0223*	
.301	*0.30 1*		.301	*0.30 1*	
=	*0.073 98 1 54*		÷	*0.0740*	

You now know that 0.20 year is 0.0740 half-life. Calculate years per half-life.

--------   --------

$$\frac{0.20 \text{ year}}{0.0740 \text{ half-life}} = 2.7 \text{ years/half-life}$$

This half-life rate of decay of radioactive substances is the basis of **radio-carbon dating,** or simply "carbon dating." Carbon is the principal chemical element in all living organisms, both plant and animal. Most carbon atoms are carbon-12; but a small portion of the carbon in atmospheric carbon dioxide is carbon-14, a radioactive isotope with a half-life of 5.72×10^3 years. When a plant or animal is alive, it takes in this isotope from the atmosphere, while the same isotope in the organism is disappearing by nuclear disintegration. A "steady state" situation exists while the system lives, maintaining a constant ratio of carbon-14 to carbon-12. When the organism dies, the disintegration of carbon-14 continues, but its intake stops. This leads to a gradual reduction in the ratio of $^{14}_{6}C$ to $^{12}_{6}C$. By measuring the ratio and the amount of $^{14}_{6}C$ now present in a sample it is possible to calculate the $^{14}_{6}C$ present when the organism died. Equation 21.4 is then used to calculate the age of the sample.

EXAMPLE 21.5 An archaeologist analyzes an organic fossil and finds that for every 14 units of $^{14}_{6}C$ present in the sample at death, 9.3 remain today. Calculate the age of the sample.

The quantities 14 and 9.3 refer to radioactive disintegrations detected by a Geiger counter or some other instrument. They express initial and final amounts, just as if they were expressed in grams. Use these quantities to calculate the number of half-lives by direct substitution into Equation 21.4.

--------   --------

$$n = \frac{\log (14/9.3)}{0.301} = 0.59 \text{ half-life}$$

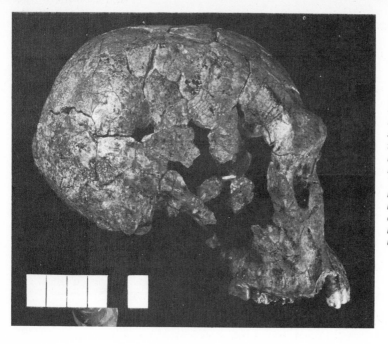

Figure 21.8
Skull of Australopithecus africanus, a small-brained, small-toothed remote ancestor of Man, discovered by Richard Leakey in the Koobi Fora Formation near Lake Turkana in Kenya's Rift Valley. This skull is believed to be two to three million years old. (Courtesy of Richard Leakey and National Museums of Kenya.)

If each half-life is 5.72×10^3 years, find the age of the fossil.

$$0.59 \text{ half-life} \times \frac{5.72 \times 10^3}{1 \text{ half-life}} = 3.4 \times 10^3 \text{ years}$$

Carbon dating has produced evidence of modern man's presence on earth as long ago as 14 000 to 15 000 years, and recent findings indicate that *Homo sapiens* may date back as many as 100 000 years (Fig. 21.8). Similar dating techniques are also applied to mineral deposits. Analyses of geological deposits have yielded rocks with an estimated age of 3.0 to 4.5 billion years, the latter figure being the scientist's estimate of the age of the earth. The oldest moon rocks analyzed to date indicate an age of about 3.5 billion years.

21.7 NUCLEAR REACTIONS AND ORDINARY CHEMICAL REACTIONS COMPARED

PG 21 I List or identify four ways in which nuclear reactions differ from ordinary chemical reactions.

Now that you have seen the nature of a nuclear change and the type of equation by which it is described, we will pause to compare nuclear reactions with the others you have studied. There are four areas of comparison:

1. In ordinary chemical reactions, the chemical properties of an element depend only upon the electrons outside the nucleus, and the properties are essentially the same for all isotopes of the element. The nuclear properties of the various isotopes of an element are quite different, however. In the radioactive decay series beginning with uranium-238, $^{234}_{90}$Th emits a β particle, whereas a bit farther down the line $^{230}_{90}$Th ejects an α particle. Both $^{214}_{82}$Pb and $^{210}_{82}$Pb are β particle emitters toward the end of the series, while the final product, $^{206}_{82}$Pb, has a stable nucleus, emitting neither alpha nor beta particles, nor gamma rays.

2. Radioactivity is independent of the state of chemical combination of the radioactive isotope. The reaction of $^{210}_{83}$Bi occurs for atoms of that particular isotope whether they are in pure elemental bismuth, combined in bismuth chloride, $BiCl_3$, bismuth sulfate, $Bi_2(SO_4)_3$, or any other bismuth compound, or if they happen to be present in the low melting bismuth alloy used for fire protection in sprinkler systems in large buildings.

3. Nuclear reactions usually result in the formation of different elements because of changes in the number of protons in the nucleus of an atom. In ordinary chemical reactions the atoms keep their identity while changing from one compound as a reactant to another as a product.

4. Both nuclear and ordinary chemical changes involve energy, but the amount of energy for a given amount of reactant in a nuclear change is enormous—greater by several orders of magnitude, or multiples of ten—compared to the energies of ordinary chemical reactions. Further comment appears in Sections 21.11 and 21.13.

21.8 NUCLEAR BOMBARDMENT AND ARTIFICIAL RADIOACTIVITY

PG 21 J   Define or identify nuclear bombardment, reactions.

   21 K   Distinguish natural radioactivity from artificial radioactivity produced by bombardment reactions.

   21 L   Define or identify transuranium elements.

In natural radioactive decay, we find an example of the alchemist's get-rich-quick dream of converting one element to another. But the natural process for uranium did not yield the gold coveted by the alchemist; rather, it produced the element lead, with which the dreamer wanted to begin his transmutation. The question was still present after the discovery of radioactivity: can man cause the transmutation of one element into another?

In 1919, Rutherford produced a "Yes" answer to that question. He found that he could "bombard" the nucleus of a nitrogen atom with a beam of alpha particles from a radioactive source, knocking a proton out of the nucleus and producing an atom of oxygen-17:

$$^{14}_{7}\text{N} + ^{4}_{2}\text{He} \rightarrow ^{17}_{8}\text{O} + ^{1}_{1}\text{H}$$

The oxygen isotope produced is stable; the experiment did not yield any man-made radioactive isotopes. Similar experiments were conducted with other elements, using high speed alpha particles as atomic "bullets." It was found that most of the elements up to potassium can be changed to other elements by nuclear bombardment. None of the isotopes produced were radioactive.

One experiment during this period was first thought to yield a nuclear particle that emitted some sort of high energy radiation, perhaps a gamma ray. In 1932 James Chadwick correctly interpreted the experiment and, in doing so, he became the first person to identify the neutron. The reaction comes from bombarding a beryllium atom with a high-energy α particle:

$$^{9}_{4}\text{Be} + ^{4}_{2}\text{He} \rightarrow ^{12}_{6}\text{C} + ^{1}_{0}\text{n}$$

$^{1}_{0}\text{n}$ is the nuclear symbol for the neutron, with a mass number of 1 and zero charge.

Two years later, in 1934, Irene Curie, the daughter of Pierre and Marie Curie, and her husband, Frederic Joliot, used high energy α particles to produce the first man-made radioactive isotope. Their target was boron-5; the product was a radioactive nitrogen nucleus not found in nature:

$$^{10}_{5}\text{B} + ^{4}_{2}\text{He} \rightarrow ^{13}_{7}\text{N} + ^{1}_{0}\text{n}$$

When $^{13}_{7}\text{N}$ decays, it emits a particle having the mass of an electron and a charge equal to that of an electron, except that it is positive. This "positive electron" is called a **positron**, and it is represented by the symbol $^{0}_{1}\text{e}$. The decay equation is

$$^{13}_{7}\text{N} \rightarrow ^{13}_{6}\text{C} + ^{0}_{1}\text{e}$$

Today hundreds of **radioisotopes** have been produced in laboratories all over the world, and they find broad use in medicine, industry, and research. Many of these isotopes have been made in different kinds of *particle accelerators*, which use electric fields to increase the kinetic energy of the charged particles that bombard nuclei. Among the earliest and best known accelerators is the cyclotron (Fig. 21.9), designed by E. O. Lawrence at the University of California, Berkeley. Other more powerful accelerators are approximately circular in shape (Fig. 21.10), and running almost unnoticed beneath a busy freeway in the foothills just west of the campus of Stanford University is a two-mile long linear accelerator.

One of the more exciting areas of research with bombardment reactions has been the production of elements that do not exist in nature. Except for trace quantities, no natural elements having atomic numbers greater than 92 have ever been discovered. In 1940, however, it was found that $^{238}_{92}\text{U}$ is capable of capturing a neutron:

$$^{238}_{92}\text{U} + ^{1}_{0}\text{n} \rightarrow ^{239}_{92}\text{U} \tag{21.5}$$

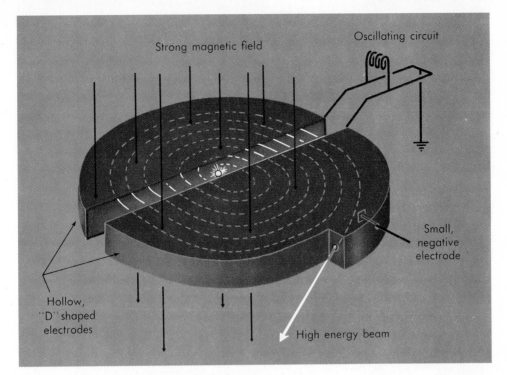

Figure 21.9
The cyclotron consists of two oppositely charged, evacuated "dees" placed between the poles of a powerful electromagnet. Positive ions, originating at the center, enter the upper dee, which is originally at a negative potential. They pass through this dee in a curved path. At the instant they reenter the central corridor, the polarity of the dees is reversed, and the particles enter the lower dee at an increased velocity. This procedure is repeated over and over; the particles move at higher and higher velocities in paths of greater and greater radius. Eventually they are deflected from the periphery of one of the dees to strike the target.

The newly formed isotope is unstable, progressing through two successive β particle emissions, yielding isotopes of the elements having atomic numbers 93 and 94:

$$^{239}_{92}\text{U} \rightarrow {}^{0}_{-1}\text{e} + {}^{239}_{93}\text{Np} \ (\text{neptunium}) \tag{21.6}$$

$$^{239}_{93}\text{Np} \rightarrow {}^{0}_{-1}\text{e} + {}^{239}_{94}\text{Pu} \ (\text{plutonium}) \tag{21.7}$$

Neptunium, plutonium, and all the other man-made elements having atomic numbers greater than 92 are called the **transuranium elements**. To the time of this writing, all elements up to atomic number 106 have been identified. All the transuranium isotopes are radioactive, and a few have been isolated only in isotopes with very short half-lives and in extremely small quantities. Some of the bombardments yielding transuranium products use relatively heavy isotopes as bullets. For example, einsteinium-247 is produced by bombarding uranium-238 with ordinary nitrogen nuclei:

$$^{238}_{92}\text{U} + {}^{14}_{7}\text{N} \rightarrow {}^{247}_{99}\text{Es} + 5\,{}^{1}_{0}\text{n}$$

A

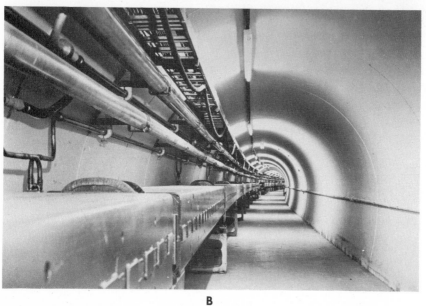

B

Figure 21.10
Particle accelerator at the Fermi National Laboratory near Batavia, Illinois. A, Aerial view of main accelerator. B, Magnets in interior of main accelerator, which is 4 miles in circumference and 1.27 miles in diameter. (Fermi Lab Photo.)

21.9 USES OF RADIOISOTOPES

Shortly after radioactivity was discovered, it was thought that the radiations had certain curative powers. Radium compounds were made, and radium solutions were bottled and sold for drinking and bathing, before the harmful effects of radiation exposure were well understood. Today's medical practitioners are much wiser, and they have devised sophisticated ways to examine their patients, diagnose their illnesses, and treat their disorders, all using man-made radioisotopes.

One means of radioisotope examination involves injecting a radioactive sodium isotope into the bloodstream, and then tracing its progress through the body with a suitable detector, perhaps a computer-controlled body scan technique. If some portion of the body shows a low radiation count, it is an indication of a circulatory problem in that area. The normal absorption of iodine by the thyroid glands is checked by adding a radioactive iodine isotope to drinking water, followed by monitoring the radiation of that isotope. The same isotope is used to treat cancer of the thyroid glands. Radioactive isotopes of cobalt, phosphorus, radium, yttrium, strontium, and cesium are used to destroy cancerous tissue in different forms of radiation therapy.

Industrial applications of radioisotopes include studies of piston wear and corrosion resistance. Petroleum companies use radioisotopes to monitor the progress of certain oils through pipelines. The thickness of thin sheets of metal, plastic, and paper is subject to continuous production control by using a Geiger counter to measure the amount of radiation that passes through the sheet; the thinner the sheet, the more radiation that will be detected by the counter. Quality control laboratories can detect small traces of radioactive elements in a metal part.

Scientific research is another major application of radioisotopes. Chemists use "tagged" atoms as *radioactive tracers* to study the mechanism, or series of individual steps, in complicated reactions. For example, by using water containing radioactive oxygen it has been determined that the oxygen in the glycose, $C_6H_{12}O_6$, formed in photosynthesis,

$$6\,CO_2(g) + 6\,H_2O(\ell) \rightarrow C_6H_{12}O_6(s) + 6\,O_2(g)$$

comes entirely from the carbon dioxide, and all oxygen from water is released as oxygen gas. Archaeologists use neutron bombardment to produce radioactive isotopes in an artifact, which makes it possible to analyze the item without destroying it. Biologists employ radioactive tracers in the water absorbed by the roots of plants to study the rate at which the water is distributed throughout the plant system. These are but a few of the many ingenious applications that have been devised for this useful tool of science.

21.10 NUCLEAR FISSION

PG 21 M   Define or identify a nuclear fission reaction.
   21 N   Define or identify a chain reaction.

In 1938, during the period when Nazi Germany was moving steadily toward war, dramatic and far-reaching events were taking place in her laboratories.

A team made up of Otto Hahn, Fritz Strassman, and Lise Meitner was working with neutron bombardment of uranium. In the products of the reaction they were finding, surprisingly, atoms of barium and krypton, and other elements far removed in both atomic mass and atomic number from the uranium atoms and neutrons used to produce them. The only explanation was, at that time, unbelievable: the nucleus must be splitting into two nuclei of smaller mass. This kind of reaction is called **nuclear fission.**

Only 0.7% of all naturally occurring uranium is $^{235}_{92}U$, one isotope that is capable of the **chain reaction** required to keep a fission reaction going. In the fission of Uranium-235 there are many products; it is not possible to write a single equation to show what happens. A representative equation is

$$^{235}_{92}U + ^1_0n \rightarrow ^{94}_{38}Sr + ^{139}_{54}Xe + 3\,^1_0n$$

Notice that it takes a neutron to initiate the reaction. Notice also that the reaction produces *three* neutrons. If one or two of these collide with other fissionable uranium nuclei, there is the possibility of another fission or two. And the neutrons from those reactions can trigger others, repeatedly, as long as the supply of nuclei lasts. This is what is meant by a *chain reaction* (Fig. 21.11), in which a nuclear product of the reaction becomes a nuclear reactant in the next step, thereby continuing the process.

The number of neutrons produced in the fission of $^{235}_{92}U$ varies with each reaction. Some reactions yield two neutrons per uranium atom, others, like the above, yield three, and still others produce four or more. The average is about 2.5. If the quantity of uranium, or any other fissionable isotope, is large enough that most of the neutrons produced are captured within the sample, rather than escaping to the surroundings, the chain reaction will continue. The minimum quantity required for this purpose is called the **critical mass.**

Because uranium-235 makes up less than 1% of all naturally occurring uranium, it is not a very satisfactory source of nuclear fuel. An alternate is the plutonium isotope, $^{239}_{94}Pu$, produced from $^{238}_{92}U$, the more abundant uranium isotope (Equations 21.5 to 21.7). $^{234}_{94}Pu$ has a long half-life (24 360 years) and is fissionable. It has been used in the production of atomic bombs and is also used in some nuclear power plants to generate electrical energy. It is made in a **breeder reactor,** the name given to a device whose purpose is to produce fissionable fuel from nonfissionable isotopes. The future of breeder reactors is uncertain, however, because of unsolved safety problems.

21.11 NUCLEAR ENERGY

PG 21 0 Explain the source of energy in a nuclear reaction.

The term *atomic energy* is widely used; a better term is *nuclear energy*, for it is the nucleus that is the source of the enormous energy associated with a

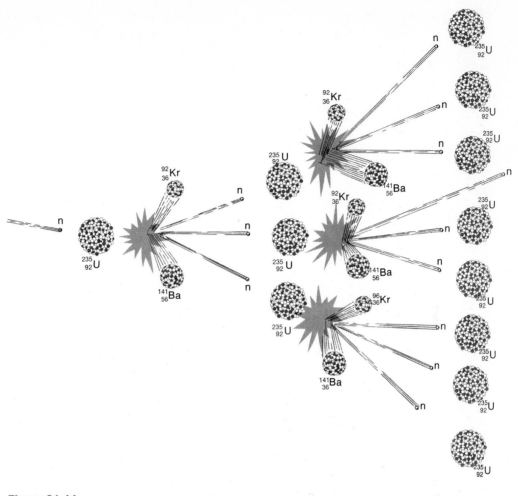

Figure 21.11
Illustration of a chain reaction initiated by capture of a stray neutron. (Many pairs of different isotopes are produced but only one kind of pair is shown.)

nuclear reaction. In Chapter 2 it was pointed out that the conservation of mass and energy laws must be combined to account for nuclear energy, in which matter is converted to energy. The conversion is expressed by Einstein's equation, $\Delta E = \Delta mc^2$, in which ΔE is the change in energy, Δm is the change in mass, and c is the speed of light. The amount of energy for a given quantity of matter is huge. For example, it can be shown that the energy released by changing a given mass of matter to energy is about 2.5 *billion* times greater than the energy released by burning the same mass of coal (see Fig. 21.12).

An atom of carbon-12 can be used to illustrate the mass-energy relationship. By the definition of an atomic mass unit, a carbon-12 atom has a mass of exactly 12 atomic mass units. The atom consists of a nucleus and six electrons. If

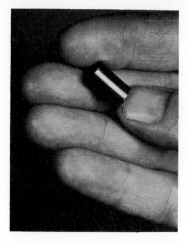

Figure 21.12
This uranium pellet, which provides nuclear fuel for a reactor, produces the energy equivalent of one ton of coal. (Courtesy of Duke Power Company.)

the mass of the electrons (0.000 549 amu each) is subtracted from the mass of the atom, the difference is the mass of the nucleus:

Mass of atom:	12.000 000 amu
Mass of electrons (6 × 0.000 549):	−0.003 294 amu
Mass of nucleus:	11.996 706 amu

The nucleus is made up of six protons (1.007 27 amu each) and six neutrons (1.008 67 amu each). The sum of the masses of these component parts is

Mass of protons (6 × 1.007 28):	6.043 68 amu
Mass of neutrons (6 × 1.008 67):	+6.052 02 amu
Total	12.095 70 amu

The difference between the total mass of the *parts* of the nucleus and the actual mass of the *whole* nucleus is called the **mass defect.** For carbon-12 it is 12.095 70 − 11.996 706 = 0.098 99 amu. This 0.098 99 amu of mass is lost—converted to energy—when six protons and six neutrons combine to form a nucleus of $^{12}_{6}C$. The energy represented by this difference is called the **binding energy;** it is the energy that holds the nucleus together against the tremendous repulsion forces between the tightly packed positively charged protons.

In nuclear changes some of this binding energy is lost as one nucleus is destroyed, and another quantity of energy is absorbed as new nuclei form. The differences between these binding energies for all atoms involved represent the energy of a nuclear reaction.

21.12 ELECTRICAL ENERGY FROM NUCLEAR FISSION

Aside from hydroelectric plants located on major rivers, most of the electrical energy consumed in the world comes from generators driven by steam. Traditionally the steam comes from boilers fueled by oil, gas, or coal. The fast dwindling supplies of these fossil fuels, and the uncertainties surrounding the availability and cost of petroleum from the countries where it is so abundant, have turned attention to nuclear fission as an alternative energy source.

A diagram of a nuclear power plant is shown in Figure 21.13. The turbine, generator, and condenser are similar to those found in any fuel-burning power plant. The nuclear fission reactor has three main components: the fuel elements, control rods, and moderator. The fuel elements are simply long trays that hold fissionable material in the reactor. As the fission reaction proceeds, fast-moving neutrons are released. These neutrons are slowed down by a moderator, which is water in the reactor illustrated. When the slower neutrons collide with more fissionable material the reaction is continued. The reaction rate is governed by cadmium control rods, which absorb excess neutrons. At times of peak power demand, the control rods are largely withdrawn from the reactor, permitting as many neutrons as necessary to find fissionable nuclei. When demand drops, the control rods are pushed in, absorbing neutrons and limiting the reaction.

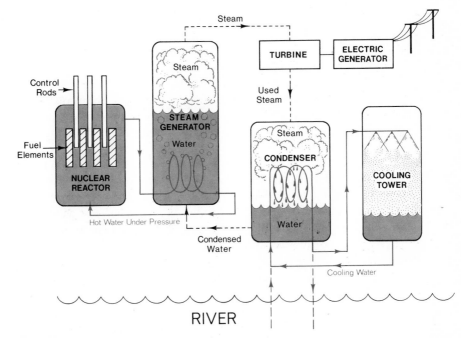

Figure 21.13
Schematic illustration of a nuclear power plant. Nuclear fission occurs in the reactor. Control rods are largely withdrawn during period of high power demand, permitting much energy to be released in the fission process. Inserting control rods during low demand periods limits nuclear reaction and energy release. Fission energy is used to convert water to steam in the steam generator. High pressure steam is delivered to the turbine, where it drives the generator that produces electrical energy, which is distributed to the user. Spent steam from the turbine is converted to liquid water in the condenser, and then recycled back to the steam generator. Cooling water for the condenser comes from a cooling tower, to which it is recycled. Makeup cooling water, and sometimes the cooling water itself, is drawn from a river, lake or ocean.

The energy released in a nuclear reactor appears as heat. This heat is transferred to the water, which is in the liquid state under a pressure of 50 atmospheres or more. The water from the reactor enters a heat exchanger at about 300°C. After giving up some of its heat in the heat exchanger, it is returned to the reactor, as shown.

Low pressure water that runs through a coil in the heat exchanger absorbs the heat coming from the high temperature reactor water. The low pressure water forms steam, which is delivered to the turbines that drive the generators. Spent steam from the turbines is sent through a condenser where residual heat is removed and the steam condenses to water. The water is then cycled through the reactor heat exchanger again.

The building, and even the continued use, of nuclear power plants faces stiff opposition from both individuals and public groups. The threat of an accident that might release large amounts of radiation over a densely populated area is the major concern. To the date of this writing, the safety record of nuclear power plants is unblemished, but there have been a number of "close calls." The best known of these is the Three Mile Island accident in 1979. These incidents point out the need for absolutely foolproof safeguards in the design, construction, inspection, and operation of nuclear power facilities.

Even if a nuclear accident never occurs, there is still the problem of how and where to dispose of the dangerous radioactive wastes from nuclear reactors. One method is to collect them in large containers that may be buried in the earth, as in Figure 21.14. People who live and work near such disposal sites are seldom enthusiastic about this solution.

Finally there is fear that some irresponsible government may use nuclear fuel for the manufacture of nuclear weapons, spreading the threat of atomic warfare. Almost more frightening is the possibility that some terrorist group might steal the materials needed to build a bomb. While these threats cannot be removed from the earth today, perhaps they would be minimized if the large scale production of nuclear fuel for electric power were eliminated.

On the other side of all these concerns is, of course, the question, "If we do not build and operate nuclear power plants, how else will the energy needs of the coming decades be met?" Perhaps the next section offers one answer, if it can only be reached in time.

21.13 NUCLEAR FUSION

PG 21 P   Define or identify a nuclear fusion reaction.

There is nothing new about nuclear energy. Man didn't invent it. In fact, without knowing it, man has been enjoying its benefits since the beginning of recorded time, and before. In its common form, though, we do not call it nuclear energy. We call it solar energy. It is the energy that comes from the sun.

The energy the earth derives from the sun comes from another type of

Figure 21.14
Burial of nuclear waste. Land burial trench at the Oak Ridge National Laboratory reservation. Each day's accumulation of waste containers is buried by three or more feet of earth. This is one of several partial solutions to the problem of disposal of radioactive waste.

nuclear reaction called **nuclear fusion, in which two small nuclei combine to form a larger nucleus.** The smaller nuclei are "fused" together, you might say. The typical fusion reaction believed to be responsible for the heat energy radiated by the sun is represented by the equation

$$\,_1^2H + \,_1^3H \rightarrow \,_2^4He + \,_0^1n$$

Fusion processes are, in general, more energetic than fission reactions. The fusion of one gram of hydrogen in the above reaction yields about four times as much energy as the fission of an equal mass of uranium-235. So far man has been able to produce only one kind of fusion reaction, and that has been the explosion of a hydrogen bomb (Fig. 21.15).

Much research effort is being made to develop nuclear fusion as a source of useful energy. It has several advantages over a fission reactor. It presents more energy per given quantity of fuel. The isotopes required for fusion are far more abundant than those needed for fission. Perhaps the biggest advantage is that fusion yields no radioactive waste, removing both the need for extensive dis-

Figure 21.15
A hydrogen bomb explosion. Note that the giant battleships are dwarfed by the cloud. (Photography by H. Armstrong Roberts.)

posal systems and the danger of accidental release of radiation to the atmosphere.

The main obstacle to be overcome before energy can be obtained from fusion is the extremely high temperature required to start and sustain the reaction. This is no problem on the sun, where temperatures are more than one million degrees. On earth, no substance known can hold the reactants at the required temperature. Experiments are now in progress on "magnetic containment," in which the fuel is suspended in a magnetic field. The only way now known to reach the necessary temperature in a hydrogen bomb is to explode a small fission bomb first. New research is investigating "energy pellets" that react when "hit" by a laser or ion beam.

Even if the technological obstacles to energy from fusion are overcome, time remains a serious problem. Only the most optimistic predictions forsee an operating plant in this century, and it will probably be well into the next before a significant portion of our energy needs can be supplied by fusion.

CHAPTER 21 IN REVIEW

21.1   THE DAWN OF NUCLEAR CHEMISTRY (499)

21.2   NATURAL RADIOACTIVITY

21 A Define radioactivity. Name, identify from a description, or describe the three types of natural radioactive emission. (500)

21.3 INTERACTION OF RADIOACTIVE EMISSIONS WITH MATTER

21 B Describe how emissions from radioactive substances affect gases and living organisms. (502)

21.4 DETECTION OF EMISSIONS FROM RADIOACTIVE SUBSTANCES

21 C Define or describe a cloud chamber. Explain how it operates. (502)

21 D Define or describe a Geiger counter. Explain how it operates. (502)

21.5 NATURAL RADIOACTIVE DECAY SERIES—NUCLEAR EQUATIONS

21 E Describe a natural radioactive decay series. (504)

21 F Given the identity of a radioactive isotope and the particle it emits, write a nuclear equation for the emission. (504)

21.6 HALF-LIFE

21 G Describe or illustrate what is meant by the half-life of a radioactive isotope. (508)

21 H Given the starting quantity of a radioactive isotope and two of the following, calculate the third: half-life; elapsed time; quantity of isotope remaining. (508)

21.7 NUCLEAR REACTIONS AND ORDINARY CHEMICAL REACTIONS COMPARED

21 I List or identify four ways in which nuclear reactions differ from ordinary chemical reactions. (512)

21.8 NUCLEAR BOMBARDMENT AND ARTIFICIAL RADIOACTIVITY

21 J Define or identify nuclear bombardment reactions. (513)

21 K Distinguish natural radioactivity from artificial radioactivity produced by bombardment reactions. (513)

21 L Define or identify transuranium elements. (513)

21.9 USES OF RADIOISOTOPES (516)

21.10 NUCLEAR FISSION

21 M Define or identify a nuclear fission reaction. (517)

21 N Define or identify a chain reaction. (517)

21.11 NUCLEAR ENERGY

21 O Explain the source of energy in a nuclear reaction. (518)

21.12 ELECTRICAL ENERGY FROM NUCLEAR FISSION (520)

21.13 NUCLEAR FUSION

21 P Define or identify a nuclear fusion reaction. (522)

TERMS AND CONCEPTS

Radioactivity (500)
Alpha particle, α particle (500)
Beta particle, β particle (500)
Gamma ray, γ-ray (501)
Cloud chamber (503)
Geiger counter (503)
Transmutation (505)
Alpha decay reaction; beta decay reaction (505)
Natural radioactive decay series (506)
Half-life (508)
Radiocarbon dating (511)
Nuclear bombardment (514)

Radioisotope (514)
Particle accelerator (514)
Transuranium elements (515)
Radioactive tracer (517)
Nuclear fission (518)
Chain reaction (518)
Critical mass (518)
Breeder reactor (518)
Mass defect (520)
Binding energy (520)
Nuclear fusion (523)

QUESTIONS AND PROBLEMS

An asterisk (*) identifies a question that is relatively difficult, or that extends beyond the performance goals of the chapter.

Section 21.2

21.1) What is meant by radioactivity?

21.2) Identify the three types of emission normally associated with radioactive decay. What is each type of emission made of; that is, what is its "structure?" Write the nuclear symbol, if any, for each kind of emission.

21.29) What is the meaning of the terms *disintegration* and *decay* in relation to radioactivity?

21.30) Compare the three principal forms of radioactive emission in terms of mass, electrical charge, and penetrating power.

Section 21.3

21.3) When an emission from a radioactive substance passes through air or any other gas, what effect does it have?

21.31) Explain how rays from radioactive substances can cause injury or damage to internal tissue in a living organism.

Section 21.4

21.4) What is a cloud chamber? By what process do "tracks" of radioactive emission particles appear?

21.32) What is a Geiger counter? How does it work?

Section 21.5

21.5) What is meant by the expression *transmutation of an element?*

21.6) Describe the change in mass number and atomic number that accompanies an alpha particle emission from a radioactive nucleus.

21.7) Write nuclear equations for beta emissions from $^{212}_{82}Pb$ and from $^{231}_{90}Th$.

21.8) Write nuclear equations for the ejection of an alpha particle from $^{228}_{90}Th$ and from $^{222}_{86}Rn$.

Section 21.6

21.9) What is meant by the half-life of a radioactive substance? What fraction of an original sample remains after the passage of six half-lives?

21.10) The half-life of $^{210}_{82}Pb$ is 22 years. How many grams of that isotope will remain after 66 years if the original sample had a mass of 100 grams?

21.11)* One of the more hazardous radioactive isotopes in the fallout of atomic bombs is strontium-90, $^{90}_{38}Sr$, for which $t_{1/2} = 28$ years. If 600 grams of Sr-90 descended on a family farm on the day a child was born in 1964, how many grams will still be on that farm when his granddaughter is born in 2014?

21.12) The half-life of $^{13}_{7}N$ is 10.0 minutes. How long will it take for the mass of that particular isotope in a sample of ammonium chloride to be reduced to 0.120 gram if the sample originally contained 0.960 gram?

21.13)* Analysis of charcoal taken from an ancient campfire yielded a $^{14}_{6}C/^{12}_{6}C$ ratio that is 0.135 times the ratio in trees living today. If $t_{1/2} = 5720$ years for carbon-14, estimate the age of the sample.

21.14) $^{228}_{89}Ac$ is an isotope in the radioactive decay series that begins with $^{232}_{90}Th$. An 8.64-gram sample of $^{228}_{89}Ac$ was weighed and the time noted. It was supposed to have been weighed again exactly

21.33) What is a *natural radioactive decay series?* How many such series have been found?

21.34) If a radioactive nucleus emits a beta particle, what change will occur in the atomic number and mass number of the nucleus that remains?

21.35) Write nuclear equations for beta emissions from $^{228}_{89}Ac$ and from $^{214}_{83}Bi$.

21.36) Write nuclear equations for alpha decay of $^{216}_{84}Po$; of $^{234}_{92}U$.

21.37) Suggest some of the difficulties that might surround the determination of physical properties of the radioactive isotope of an element that has a short half-life. What fraction of the sample would remain after the passage of four half-lives?

21.38) $^{222}_{86}Rn$ has a half-life of roughly half a week. If you had 50 milligrams of that isotope, how many milligrams would be left after you returned from a three week vacation?

21.39)* A radiochemistry research laboratory ends its workday at 5 P.M., and resumes at 9 A.M. the next morning. If they had in storage 4.8 centigrams of $^{24}_{11}Na$ ($t_{1/2} = 15$ hours) at the close of a work day, how much would they have at the beginning of the next work day? What would be the beginning supply on the Tuesday morning after a three-day weekend that began at 2 P.M. the previous Friday, if they left 24.0 centigrams in their vault at the start of the holiday?

21.40) The half-life of iodine-131, an isotope used in radiotherapy, is eight days. If a hospital wishes to have available at all times material containing no less than 10 micrograms of the isotope, how soon after the inventory shows 160 micrograms will they have to restock, assuming none of the supply is used during the period?

21.41)* The full decay process for $^{238}_{92}U$ may be expressed by the overall equation

$$^{238}_{92}U \rightarrow ^{206}_{82}Pb + 8\,^{4}_{2}He + 6\,^{0}_{-1}e$$

The half-life for the process is 4.5×10^9 years. Assuming that all the lead is from the uranium originally present, how old is a rock sample that contains three times as much $^{238}_{92}U$ as $^{206}_{82}Pb$?

21.42) $^{212}_{83}Bi$ is an isotope in the radioactive decay series starting with $^{232}_{90}Th$. A sample of $^{212}_{83}Bi$ was weighed in a laboratory at 10:00 A.M.; its mass was

12 hours later, but the night shift didn't get around to it until 12 hours and 15 minutes had elapsed. It then weighed 2.16 grams. What is the half-life of $^{228}_{89}Ac$?

21.15)* $^{137}_{55}Cs$ is one of the hazardous isotopes found in nuclear waste. In studying the decay of this material, a sample was weighed, buried in a deep vault for 1.00 year, dug up, and weighed again. The second weighing showed that 97.72% of the original mass was still present. Calculate the half-life of $^{137}_{55}Cs$.

Section 21.7

21.16) Explain why the chemical properties of an element are the same for all isotopes in an ordinary chemical change, but depend upon the particular isotope for a nuclear change.

21.17)* If the uranium in pure UCl_4 and UBr_4 has all isotopes in their normal percentage distribution in nature, which will exhibit the greatest amount of radioactivity, 100 grams of UCl_4 or 100 grams of UBr_4? How about 0.10 mole of UCl_4 or 0.10 mole of UBr_4? Explain both answers.

21.18) A fundamental idea in Dalton's atomic theory is that atoms of an element can be neither created nor destroyed. How, then, can you account for the fact that the number of lead atoms in the world is constantly increasing?

Section 21.8

21.19) What is the meaning of the expression *nuclear bombardment*?

21.20)* Particles used for nuclear bombardment reactions frequently do not have sufficient kinetic energy when obtained from their natural sources. Identify some of the "particle accelerators" that have been developed to increase their kinetic energy.

21.21) What are transuranium elements? Where are they located on the periodic table?

214 grams. At 2:02 P.M. the mass was only 13.4 grams. Find the half-life of $^{212}_{83}Bi$.

21.43)* $^{214}_{82}Pb$ is the first of three lead isotopes in the natural decay series of uranium-238. In an experiment to find its half-life, a 15.4 gram sample was used. After 5.00 minutes it weighed only 13.5 grams. How many half-lives are represented by the 5.00 minutes? What is the half-life of the isotope?

21.44) Two bottles of the same lead compound are carelessly left exposed. Explain the circumstances under which one of these bottles might be hazardous, but not the other.

21.45) An ore sample containing a certain quantity of radioactive uranium disintegrates at 7000 counts per minute, a way of expressing rate of radioactive decay when measured with a Geiger counter. If all the uranium in the sample is extracted and isolated as a pure element, would you expect the rate of decay to remain at 7000 counts per minute, would it be less than 7000, or would it be more? (Disregard any loss of the radioactive isotope because of disintegration during the extraction process.) Explain your answer.

21.46) A radiochemical laboratory prepares a sample of pure KCl containing a measurable amount of $^{43}_{19}K$, a radioactive isotope that emits a beta particle and has a half-life of 22.4 hours. The compound is securely stored overnight in an inert atmosphere. The next day the compound will no longer be pure. Why? With what element would you expect it to be contaminated?

21.47) How does artificial radioactivity differ from natural radioactivity?

21.48)* Describe the principle by which a particle accelerator increases the kinetic energy of a particle used in nuclear bombardment. What major nuclear particle cannot be accelerated? Why?

21.49)* The first eleven transuranium elements were discovered at the Lawrence Radiation Laboratory, named after the inventor of the cyclotron, at the University of California, Berkeley. (Other laboratories participated with the Berkeley group in the discovery of some of the elements.) The names of the elements represent the continuation of a series already stated with uranium, some nationalism, some state and school loyalties, and the desire to honor men who have made major contributions to chemistry, mostly in the area of nuclear chemistry. You will find it interesting to review those names to see if you can connect them with their sources, most of which are identified in this chapter.

Complete each of the following nuclear bombardment equations by supplying the nuclear symbol for the missing species:

21.22) $^{98}_{42}Mo + ^{2}_{1}H \rightarrow ? + ^{1}_{0}n$

21.23) $^{238}_{92}U + ^{4}_{2}He \rightarrow 3\,^{1}_{0}n + ?$

21.24) $? + ^{2}_{1}H \rightarrow ^{60}_{27}Co + ^{1}_{1}H$

21.50) $^{44}_{20}Ca + ^{1}_{1}H \rightarrow ? + ^{1}_{0}n$

21.51) $^{252}_{98}Cf + ^{10}_{5}B \rightarrow 5\,^{1}_{0}n + ?$

21.52) $^{106}_{46}Pd + ^{4}_{2}He \rightarrow ^{109}_{47}Ag + ?$

Section 21.10

21.25) Explain what is meant by a *fission* reaction. You will find the significance of this term easier to remember if you associate it with another science course you probably have taken. Even without the course, looking the word up in the dictionary should fix its significance in your thought.

21.53) What is a *chain reaction?* What essential feature must be present in a reaction before it can become a chain reaction?

Section 21.11

21.26) What is the source of the enormous energy released in a nuclear reaction?

21.54) For all practical purposes, the Law of Conservation of Mass is obeyed in ordinary chemical reactions, but not in nuclear reactions. Would the mass of the products of an atomic bomb explosion be more than or less than the mass of the reactants? Explain your answer.

Section 21.12

21.27) List some of the principal advantages that are associated with nuclear power plants as a source of electrical energy.

21.55)* In January, 1981, when this question was written, there were serious concerns about using nuclear reactors as a source of electrical energy. List those concerns. In the period since 1981, have there been changes that have removed or reduced the earlier objections to nuclear energy? If so, identify them. Has anything happened over the same period to show that the worries of 1981 were justified, and that continuing to build nuclear power plants is an unwise procedure? If so, identify the events. How do you feel about nuclear power sources today?

Section 21.13

21.28) What is a *fusion* reaction? How does it differ from a fission reaction?

21.56) Why is nuclear fusion more promising as a source of electrical energy than nuclear fission? What major obstacle prevents us from building nuclear fusion power plants?

22

introduction to organic chemistry

Chapter 22 is a brief survey of organic chemistry. It is unlike the first twenty-one chapters, in which we could expect to find mastery of specific chemical concepts. While we might seek similar achievement in selected areas of this chapter, most topics are presented so briefly it is unrealistic to set for them the kind of performance goals used earlier. In their place the following chapter-wide performance goals are offered for your guidance:

1. Distinguish between organic and inorganic chemistry.
2. Define a hydrocarbon.
3. Distinguish between saturated and unsaturated hydrocarbons.
4. Write, recognize, or otherwise identify (1) the structural unit, or functional group, (2) the general formula, and (3) molecular or structural formulas and/or names of specific examples of the following classes of organic compounds: alkanes, alkenes, alkynes, aromatic hydrocarbons, alcohols, ethers, aldehydes, ketones, carboxylic acids, and esters.
5. Define and give examples of isomerism.

22.1 THE NATURE OF ORGANIC CHEMISTRY

In the early development of chemistry, the logical starting point was a study of substances that occur in nature. As in the organization of any body of knowledge, substances were grouped by certain common characteristics. One system assigns substances to groups labeled animal, vegetable, or mineral. So far in this text, attention has been directed almost entirely to minerals and the compounds that may be derived from them. These substances make up that area of chemistry commonly known as *inorganic chemistry*.

Animal and vegetable substances are, or at one time were, composed of living organisms, a distinction that sets them apart from inorganic substances. **Organic chemistry** was originally defined as the chemistry of living organisms, including those compounds directly derived from living organisms by natural processes of decay. It was learned, however, that compounds called "organic" can be produced from inorganic chemicals. Ultimately it was recognized that all organic compounds contain the element **carbon.** We now consider organic chemistry to be **the chemistry of carbon compounds.** Carbonates, cyanides, oxides of carbon, and a few other compounds are exceptions that are still classified as inorganic.

The only really unique feature of organic chemistry is the huge and rapidly growing number of organic compounds—many times more than the total number of known compounds that do not contain carbon. All of the chemical principles we have studied for inorganic chemicals, such as bonding, reaction rates, equilibrium and others, apply equally to organic compounds. In particular, a clear picture of bonding and the structure of molecules is the cornerstone of all that we understand about organic chemistry, as the remaining pages will show.

Some science fiction writers have speculated about the possibility that "life" can be based on some element other than carbon. Silicon is a possibility. The particular value of carbon to life on earth is due to its ability to form long chains of carbon atoms. Without these chains, none of the complex proteins that comprise living organisms could be formed. No other element can form such chains on earth. However, perhaps somewhere in the universe, a form of life based on completely different chemical structures has found an environment suitable for it.

22.2 THE MOLECULAR STRUCTURE OF ORGANIC COMPOUNDS

Table 22.1 summarizes the covalent bonding properties of carbon, hydrogen, oxygen, nitrogen, and the halogens, the elements most frequently found in organic compounds. All the bond geometries are indicated—the linear, planar, or three-dimensional shapes and, where constant, the actual bond angles. Of particular significance is the number of covalent bonds that atoms of the different elements can form, when each atom contributes one electron to the bond. This is determined by the electron configuration of the atom. The bonding relationships of these elements are basic to your understanding of the structure of organic compounds.

When they are all single bonds, the geometry of the four bonds of carbon limits the accuracy by which molecular shape may be pictured in a book. The directional orientation of the bonds around the carbon atom is tetrahedral (Fig. 22.1), a shape that cannot be shown correctly on a two-dimensional page. If, in the figure, bond axis number 1 rising from the top of the atom is *in* the plane of the paper, then bond axis number 2 is coming *out* of the plane of the paper

Table 22.1
Bonding in Organic Compounds

Element	Number of Bonds*	Bond Geometry			
		Single Bond	*Double Bond*	*Double Bonds*	*Triple Bond*
Carbon	4	$\overset{\mid}{\underset{/\,\diagdown}{C}}$ Tetrahedral: 109.5° angles	$\diagup\!\!\!^{\diagdown}\!C=$ Planar: 120° angles	$=C=$ Linear: 180° angle	$-C\equiv$ Linear: 180° angle
Hydrogen	1	H—			
Halogens	1	$:\!\ddot{X}\!-$			
Oxygen	2	$\overset{..}{\underset{/\,\diagdown}{\ddot{O}}}$ Bent structure	$\ddot{O}=$		
Nitrogen	3	$\overset{..}{\underset{/\,\diagdown}{N}}$ Pyramidal structure	$\underset{\diagdown}{\overset{..}{\underset{/}{N}}}$ Bent structure		$:N\equiv$

*Number of bonds to which an atom of the element shown can contribute *one* electron.

toward you, while bond axes 3 and 4 are extending *behind* the plane of the paper.

Rather than wrestle with these difficulties in accurate illustrations, the chemist usually represents the four bonds radiating from a carbon atom thus:

$$-\overset{\mid}{\underset{\mid}{C}}-$$

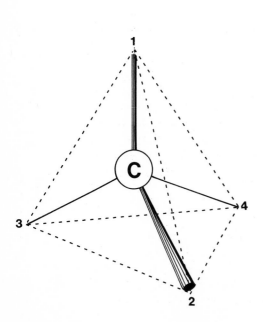

Figure 22.1
Tetrahedral bond orientation about a carbon atom.

These bonds are identical from the standpoint of directional orientation. Thus, in a three-carbon chain,

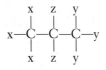

all three x positions are geometrically equal, three y positions are equal, and two z positions are equal. It follows that

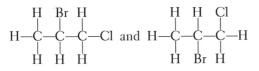

are identical compounds.

HYDROCARBONS

The simplest type of organic compound is the hydrocarbon. As the name suggests, **hydrocarbons** consist of two elements, hydrogen and carbon. A hydrocarbon may be classified into one of several categories based upon its structure: (1) alkanes, (2) alkenes, (3) alkynes, and (4) aromatic hydrocarbons. The first three of these are sometimes grouped together as the *aliphatic* hydrocarbons, in which the carbon atoms are arranged in chains. Some aliphatic hydrocarbons are shown in Table 22.2; the aromatic hydrocarbons will be considered separately.

22.3 SATURATED HYDROCARBONS: THE ALKANES

The **alkanes** are known as **saturated hydrocarbons.** This means that each carbon atom is bonded by four single covalent bonds to four other atoms, the maximum number possible according to the octet rule (p. 220). Structural models of the first three alkane molecules, methane, ethane, and propane, are shown in Figure 22.2. Notice the tetrahedral orientation of atoms bonded to carbon in all three molecules. Because of this tetrahedral arrangement any continuous chain of three or more carbon atoms through a saturated hydrocarbon has a crooked path. Furthermore there is free rotation around single carbon-carbon bonds. The shape of the molecule varies. At one moment the molecule can curl up so the end carbons are close to each other, and later the molecule is stretched out with the end carbons far apart. The straight lines we draw on paper are not true representations of carbon chains.

THE GENERAL FORMULA FOR THE ALKANES— A HOMOLOGOUS SERIES

If you examine the molecular and structural formulas of methane, ethane and propane, you will find a pattern. Each additional carbon atom is accompanied

Table 22.2

Aliphatic Hydrocarbons

Number of Carbon Atoms	Alkane	Alkene	Alkyne
1	H | H—C—H | H CH_4 Methane		
2	H H | | H—C—C—H | | H H C_2H_6 Ethane	H H | | H—C=C—H C_2H_4 Ethylene ethene	H—C≡C—H C_2H_2 Acetylene ethyne
3	H H H | | | H—C—C—C—H | | | H H H C_3H_8 Propane	H H H | | | H—C=C—C—H | H C_3H_6 Propylene propene	H | H—C≡C—C—H | H C_3H_4 Propyne
Structural Unit or Functional Group	—C—	>C=C<	—C≡C—

by two more hydrogen atoms. The alkane with four carbon atoms is butane, C_4H_{10}, five carbon atoms yield pentane, C_5H_{12}, and so forth, with each additional step extending the chain by a —CH_2— structural unit.

A series of compounds in which each member differs from the members before and after it by the same structural unit is called a **homologous series.** The alkane series may be represented by the general formula C_nH_{2n+2}, where n is the number of carbon atoms in the molecule. With this general formula you can produce the molecular formula for any member of the series. For octane, the alkane with 8 carbon atoms, n = 8. The number of hydrogen atoms is $2(8) + 2 = 18$. The formula of octane is therefore C_8H_{18}.

The formulas and names of the first ten alkanes are shown in Table 22.3. Also shown are the melting and boiling points of the so-called **normal** alkanes, those in which the carbon atoms form a continuous chain. You will recall from Chapter 14 that intermolecular forces between nonpolar molecules increase with increasing molecular size, and that stronger intermolecular attractions yield higher boiling points. Accordingly, alkanes having fewer than five carbons have the weakest intermolecular attractions, low boiling points, and are gases at normal temperatures. All are used as fuels for stoves. Methane is the main constituent of "natural gas." It is also found in large quantities in the forbidding atmosphere of planets like Jupiter and Saturn.

Intermolecular forces are stronger between the larger alkane molecules

Figure 22.2
Ball-and-stick and space-filling models of the first three members of the alkane hydrocarbon series. (Photograph by Janice Peters.)

from C_5H_{12} to $C_{17}H_{35}$. These higher boiling compounds are liquids at room temperature. Several of the lower molecular weight liquid alkanes, notably octane, are present in gasoline. Diesel fuel and lubricating oils are made up largely of higher molecular weight liquid alkanes. Alkanes with molecular weights greater than 250 are normally solids at room temperature.

Table 22.3
The Alkane Series

Molecular Formula	Name	Number of Carbon Atoms	Prefix	Melting Point (°C)	Boiling Point (°C)
CH_4	Methane	1	Meth-	−183	−162
C_2H_6	Ethane	2	Eth-	−172	−89
C_3H_8	Propane	3	Prop-	−187	−42
C_4H_{10}	Butane	4	But-	−138	0
C_5H_{12}	Pentane	5	Pent-	−130	36
C_6H_{14}	Hexane	6	Hex-	−95	69
C_7H_{16}	Heptane	7	Hept-	−91	98
C_8H_{18}	Octane	8	Oct-	−57	126
C_9H_{20}	Nonane	9	Non-	−54	151
$C_{10}H_{22}$	Decane	10	Dec-	−30	174

Because of isomerism (see below), a molecular formula does not identify a compound adequately. Structural formulas, on the other hand, are complex, although frequently the only satisfactory representation of a compound. A compromise sometimes used is the **condensed formula**, or **line formula**, in which the structure is indicated by repeating the groups it contains. Line formulas for a few sample alkanes are shown below:

Ethane, C_2H_6 CH_3CH_3
Propane, C_3H_8 $CH_3CH_2CH_3$
Butane, C_4H_{10} $CH_3CH_2CH_2CH_3$
Octane, C_8H_{18} $CH_3CH_2CH_2CH_2CH_2CH_2CH_2CH_3$

When line formulas become long, as in the case of octane, they are sometimes shortened by grouping the CH_2 units: $CH_3(CH_2)_6CH_3$.

The alkane hydrocarbons also serve to introduce organic nomenclature. Table 22.3 illustrates the system for the first ten alkanes. Each alkane is named by combining a prefix and a suffix. The prefix indicates the number of carbons in the chain. The first ten prefixes used in this nomenclature system appear in the fourth column of Table 22.3. The suffix identifying an alk*ane* is *-ane*. Thus the name of methane comes from combining the prefix *meth-*, indicating one carbon atom, with the suffix *-ane* indicating an alkane. We will see shortly that these prefixes are used in naming other organic compounds and groups as well.

ISOMERISM IN THE ALKANE SERIES

Not all alkanes have their carbons bonded in a continuous chain; some have branches. The smallest alkane in which this is possible is butane, which has two possible structures:*

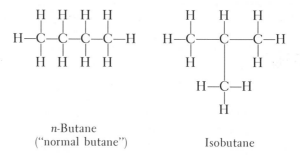

n-Butane
("normal butane")   Isobutane

You will observe that in the compound at the left, called normal butane, or *n*-butane, the four carbons are in a single chain. In isobutane, the structure at the right, there are only three carbons in the chain, with the fourth carbon branching off the middle carbon of the three. Both compounds have the same molecular formula, C_4H_{10}. These compounds are **isomers: two compounds having the same molecular formula but different structures are called isomers.** It

---

*The length of lines representing bonds has no significance. Different lengths are used only to separate on paper those parts of the molecule that are not bonded to each other.

should be realized that isomers are distinctly different compounds, having different physical and chemical properties.

The number of isomers that are possible increases rapidly with the number of carbon atoms in the compound. There are three different pentanes; their structures, showing only the carbon skeletons to make the diagrams less "cluttered," are

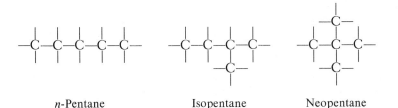

n-Pentane   Isopentane   Neopentane

There are five isomeric hexanes, nine heptanes, and seventy-five possible decanes. It is possible to draw over 300 000 isomeric structures for $C_{20}H_{42}$, and more than one hundred million for $C_{30}H_{62}$. Obviously not all of them have been prepared and identified! This does give us some idea, though, why there are so many organic compounds.

(You might like to try your hand at writing isomers of the alkanes. See if you can sketch the five isomers of hexane. In doing so, be sure they are all different. You would be wise to start with the longest chain possible, then shorten it by one, and again shorten it by one, drawing all possible structures each time until all combinations are exhausted. The correct diagrams are shown on page 540.)

Distinguishing between isomeric structures requires some extension of nomenclature rules. As noted earlier, a continuous straight chain alkane is called a normal alkane—hence the name "normal butane," written *n*-butane. In a normal alkane each carbon atom is bonded to no more than two other carbon atoms. The branched chain isomer of butane is called isobutane. In it, one carbon atom is bonded to three other carbon atoms. The *normal* and *iso*- terminology is carried forth to the isomers of pentane, and is expanded to *neopentane* to provide for the third isomer in which one carbon atom is bonded to four other carbon atoms.

Beyond pentane the number of isomers becomes so large it is necessary to develop a systematic nomenclature. In fact, many chemists prefer to drop the *iso*- and *neo*- prefixes and use a single system for all hydrocarbons. Several different systems exist, but the one most widely adopted is the IUPAC* system. But before we can consider this system we must define an *alkyl group*.

ALKYL GROUPS

In inorganic chemistry we found it convenient to assign names to certain groups of atoms, such as sulfate, nitrate, and ammonium ions, that function as units in forming chemical compounds. Similarly, in organic chemistry, it is convenient

---

*International Union of Pure and Applied Chemistry (p. 244).

to identify **alkyl groups** that may be derived from an alkane. If, on paper, we remove a hydrogen atom from methane, CH_4, we get —CH_3. This —CH_3 group, appearing in the structural formula of a compound, is called a **methyl group**, the term being made up of the prefix *meth-* for one carbon (Table 22.3) and the suffix *-yl* applied to all alkyl groups. If we compare two compounds,

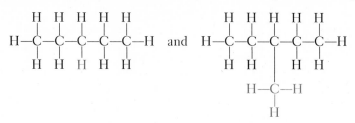

we see that the colored H in the first compound has been replaced by a —CH_3 group, or methyl group, in the second. If the replacement group has two carbon atoms,

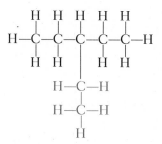

it is an ethyl group, —C_2H_5, one hydrogen short of ethane, C_2H_6. All the alkyl groups are similarly named.

Frequently we wish to show a bonding situation in which *any* alkyl group may appear. The letter R is used for this purpose. Thus R—OH could be CH_3OH, C_2H_5OH, C_3H_7OH, or any other alkyl group attached to an —OH group.

Some chemists consider alkyl groups as **functional groups** also. A functional group is **an atom or group of atoms that both establishes the identity of a class of compounds and determines its chemical properties.** The structural units in the bottom row of Table 22.2 may be considered as the functional groups of the aliphatic hydrocarbons. Other functional groups will appear later in the chapter.

NAMING THE ALKANES BY THE IUPAC SYSTEM

We are now ready to describe the IUPAC system of naming isomers of the alkanes, as well as other compounds we will encounter shortly. The system follows a set of rules:

1. **Identify as the parent alkane the longest continuous chain.** For example, in the compound having the structure

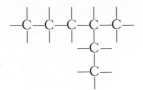

the longest chain is six carbons long, not five as you might first expect. This is readily apparent if we number the carbon atoms in the original representation of the structure and an equivalent layout:

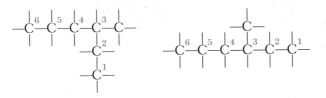

2. **Identify by number the carbon atom to which the alkyl group (or other species) is bonded to the chain.** In the example compound this is the *third* carbon, as shown. Notice that counting always begins at that end of the chain that places the branch on the *lowest* number carbon atom possible.
3. **Identify the branched group (or other species).** In this example the branch is a methyl group, $-CH_3$, shown in color.

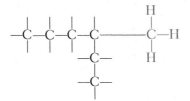

    These three items of information are combined to produce the name of the compound, 3-*methylhexane*. The 3 comes from the third carbon (Step 2); methyl comes from the branch group (Step 3); and hexane is the parent alkane (Step 1).

    Sometimes the same branch appears more than once in a single compound. This situation is governed by the following rule:

4. **If the same alkyl group, or other species, appears more than once, indicate the number of appearances by di-, tri-, tetra-, etc., and show the location of each branch by number.** For example,

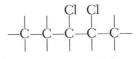

would be 2,3-dichloropentane. In the other direction, to write the structural formula for 1,1,5-tribromohexane, we would establish a six-carbon skeleton and attach bromines as required, two to the first carbon and one to the fifth:

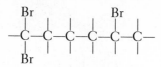

5. **If two or more different alkyl groups, or other species, are attached to the parent chain, they are named in order of increasing group size or in alphabetical order.** By this rule the compound

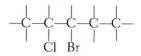

would be named 3-bromo-2-chloropentane. The formula for 2,2-dibromo-4-chloroheptane would be

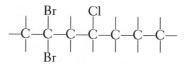

The five isomers of hexane, models of which are pictured in Figure 22.3, further illustrate the application of these rules:

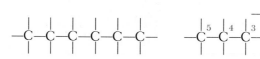

n-Hexane   2-Methylpentane

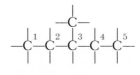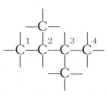

3-Methylpentane   2,3-Dimethylbutane

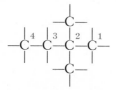

2,2-Dimethylbutane

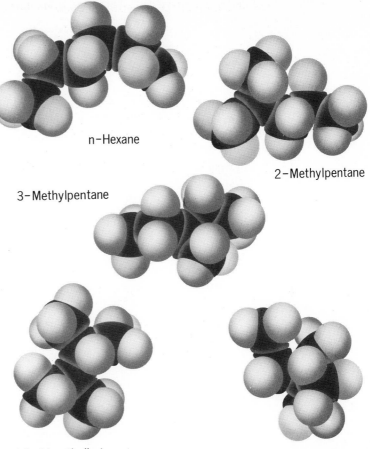

n-Hexane

2-Methylpentane

3-Methylpentane

2,3-Dimethylbutane

2,2-Dimethylbutane

Figure 22.3
Space-filling models of the isomeric hexanes.

22.4 UNSATURATED HYDROCARBONS: THE ALKENES AND THE ALKYNES

STRUCTURE AND NOMENCLATURE

On page 531 a saturated hydrocarbon was identified as one in which each carbon atom is bonded to four other atoms. **Hydrocarbons in which two or more carbon atoms are (1) connected by a double or triple bond and (2) bonded to fewer than four other atoms are said to be unsaturated.**

If one hydrogen atom, complete with its electron, is removed from each of two adjacent carbon atoms in an alkane (A below), each carbon is left with a single unpaired electron (B). These electrons may then form a second bond between the two carbon atoms (C):

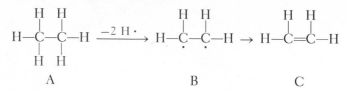

$$
\underset{A}{
\begin{array}{c}
\text{H}\quad\text{H} \\
| \quad\;\; | \\
\text{H}-\text{C}-\text{C}-\text{H} \\
| \quad\;\; | \\
\text{H}\quad\text{H}
\end{array}}
\xrightarrow{-2\,\text{H}\cdot}
\underset{B}{
\begin{array}{c}
\text{H}\quad\text{H} \\
| \quad\;\; | \\
\text{H}-\overset{..}{\text{C}}-\overset{..}{\text{C}}-\text{H}
\end{array}}
\rightarrow
\underset{C}{
\begin{array}{c}
\text{H}\quad\text{H} \\
| \quad\;\; | \\
\text{H}-\text{C}=\text{C}-\text{H}
\end{array}}
$$

Each carbon atom is now bonded to three other atoms. **An aliphatic hydrocarbon in which two carbon atoms are bonded to three other atoms and double-bonded to each other is called an alkene.** Figure 22.4 shows two models of the simplest alkene.

Removal of another hydrogen atom from each of the double-bonded carbon atoms in an alkene yields a triple bond:

$$
\begin{array}{c}
\text{H}\quad\text{H} \\
| \quad\;\; | \\
\text{H}-\text{C}=\text{C}-\text{H}
\end{array}
\xrightarrow{-2\,\text{H}\cdot}
\text{H}-\overset{..}{\text{C}}=\overset{..}{\text{C}}-\text{H}
\rightarrow
\text{H}-\text{C}\equiv\text{C}-\text{H}
$$

Each carbon atom is now bonded to two other atoms. **An aliphatic hydrocarbon in which two carbon atoms are bonded to two other atoms and triple-bonded to each other is called an alkyne.** Models of acetylene, the most common alkyne, are shown in Figure 22.4.

Both the alkenes and the alkynes make up a new homologous series. Just as with the alkanes, each series may be extended by adding —CH_2— units. Longer chains may have more than one multiple bond, but we will not consider such compounds in this text. The general formula for an alkene is C_nH_{2n}, and for an alkyne, C_nH_{2n-2}.

Figure 22.4
Ball-and-stick and space-filling models of the first members of the alkene and alkyne hydrocarbon series. (Photograph by Janice Peters.)

Table 22.4
Unsaturated Hydrocarbons

Hydrocarbon Series	n	Formulas		Names	
		Molecular	*Structural*	*IUPAC*	*Common*
Alkene, C_nH_{2n}	2	C_2H_4	$H_2C{=}CH_2$	Ethene	Ethylene
	3	C_3H_6	$H_2C{=}CH{-}CH_3$	Propene	Propylene
	4	C_4H_8	$H_2C{=}CH{-}CH_2{-}CH_3$	Butene	Butylene
Alkynes, C_nH_{2n-2}	2	C_2H_2	$H{-}C{\equiv}C{-}H$	Ethyne	Acetylene
	3	C_3H_4	$H{-}C{\equiv}C{-}CH_3$	Propyne	—

Table 22.4 gives the names and formulas of some of the simpler unsaturated hydrocarbons. The IUPAC nomenclature system for the alkenes matches that of the alkanes. The suffix designating the alkene hydrocarbon series is *-ene*, just as *-ane* identifies an alkane. The same prefixes are used to show the total number of carbon atoms in the molecule. For example, pentene is C_5H_{10}, hexene is C_6H_{12}, and octene is C_8H_{16}. The common names for the alkenes are produced similarly, except that the suffix is *-ylene*. These names are firmly entrenched in reference to the lower alkenes: C_2H_4 is almost always called ethylene, C_3H_6 is propylene and C_4H_8 is butylene.

Acetylene, C_2H_2, the first member of the alkyne series, is always called by its common name. The next alkynes are sometimes named as derivatives of acetylene, as methyl acetylene, C_3H_4, and ethyl acetylene, C_4H_6. The IUPAC system is more often employed for all alkynes except acetylene. The same prefixes are used, and the alkyne suffix is *-yne*. Thus, for the alkynes with two, three, and four carbon atoms, the formal names are ethyne, propyne and butyne, respectively.

ISOMERISM AMONG THE UNSATURATED HYDROCARBONS

All possibilities for isomerism among the alkanes are duplicated in the alkenes and alkynes. Moreover, the unsaturated hydrocarbons introduce a second form of isomerism, and the alkenes even a third. The unique isomerism they both have concerns the location of the multiple bond, which can be anywhere in the chain. Double and triple bonds are handled similarly. In the simplest example, butene may have either of these two structures:

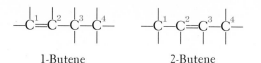

1-Butene   2-Butene

The number appearing before the name is the lowest number possible to identify the carbon atom to which the double bond is attached. The compound

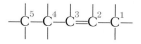

is 2-pentene, because the double bond is attached to the *second* carbon atom counting from the right.

 A form of isomerism shown only by alkenes arises because a double bond allows no rotation about its axis. This leads to two possible arrangements around the double bond. The first alkene in which these options appear is butene:

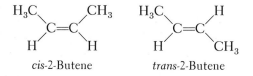

cis-2-Butene   *trans*-2-Butene

The two methyl groups attached to the double-bonded carbons can be on the *same* side of the double bond, as in *cis*-2-butene, or on *opposite* sides, as in *trans*-2-butene. *Cis*- and *trans*- are prefixes meaning, respectively, *on this side* and *across*. The latter is perhaps most easily remembered by association with such words as transcontinental, suggesting across a continent.

22.5 SOURCES AND PREPARATION OF ALIPHATIC HYDROCARBONS

PETROLEUM PRODUCTS

Alkanes and alkenes are natural products which have resulted from the decay of organic compounds from plants and animals that lived millions of years ago. They are found today as petroleum, mixtures of hydrocarbons containing up to 30 to 40 carbon atoms in the molecule. Different components of petroleum may be separated by fractional distillation (Fig. 22.5). This process separates "fractions" that boil at different temperatures. The lower alkanes and alkenes, up to four or five carbons per molecule, may be obtained in pure form by this method.

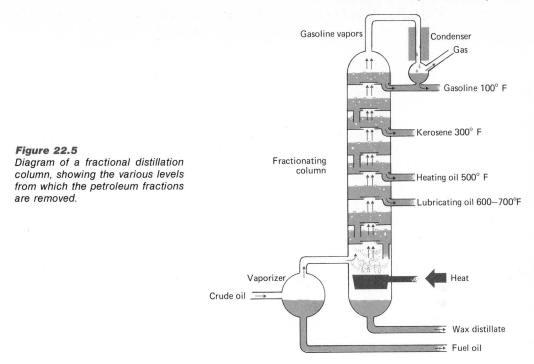

Figure 22.5
Diagram of a fractional distillation column, showing the various levels from which the petroleum fractions are removed.

The boiling points of larger compounds are too close for their complete separation, so chemical methods must be employed to obtain pure samples.

PREPARATION OF ALKENES

Two ways alkenes are produced are the *dehydration* of alcohols and the *dehydrohalogenation* of an alkyl halide. These two impressive terms describe very similar processes which are quite simple, at least in principle if not in practice. Dehydration is removal of water; dehydrohalogenation is removal of a hydrogen and a halogen. As an example, propanol is an alcohol (Section 22.9). Its formula is C_3H_7OH. Under certain conditions a molecule of water may be separated from a molecule of the alcohol, producing propene:

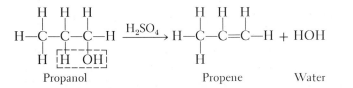

An alkyl halide is an alkane in which a halogen atom has been substituted for a hydrogen atom; or, viewed in another way, an alkyl halide is an alkyl group bonded to a halogen. The molecule is attacked with a base in the presence of an alcohol.

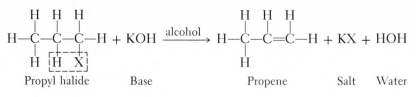

| Propyl halide | Base | | Propene | Salt | Water |

PREPARATION OF ALKANES

There are several industrial and laboratory methods by which alkanes may be prepared. One of the more important is the catalytic **hydrogenation** of an alkene. Hydrogenation is the reaction of a substance with hydrogen. The general reaction for the hydrogenation of an alkene is

$$C_nH_{2n} + H_2 \xrightarrow{\text{catalyst}} C_nH_{2n+2}$$

PREPARATION OF ACETYLENE

One alkyne is of major importance—acetylene. It is produced commercially in a two-step process in which calcium oxide reacts with coke (carbon) at high temperatures to produce calcium carbide and carbon monoxide:

$$CaO(s) + 3\ C(s) \rightarrow CaC_2(s) + CO(g)$$

Calcium carbide then reacts with water to produce acetylene:

$$CaC_2(s) + 2\ H_2O(\ell) \rightarrow C_2H_2(g) + Ca(OH)_2(s)$$

22.6 CHEMICAL PROPERTIES OF ALIPHATIC HYDROCARBONS

The combustibility—ability to burn in air—of the hydrocarbons is probably one of the most important of all chemical reactions to modern man. As components of liquid and gaseous fuels, hydrocarbons are among the most heavily processed and distributed chemical products in the world. When burned in an excess of air, the end products are water and carbon dioxide (p. 163).

One major distinction separates the chemical properties of saturated hydrocarbons from the unsaturated hydrocarbons. By opening a multiple bond in an alkene or alkyne, the compound is capable of reacting by **addition,** simply by adding atoms of some element to the molecule. By contrast, an alkane molecule is literally saturated; there is no room for an atom to join the molecule without first removing a hydrogen atom. A reaction in which a hydrogen atom in an alkane is replaced by an atom of another element is called a **substitution** reaction.

Both alkanes and alkenes undergo *halogenation* reactions—reaction with a halogen. These reactions serve to show the difference between addition and substitution reactions:

Addition reaction:

Propene   Chlorine   1,2-Dichloropropane

Substitution reaction:

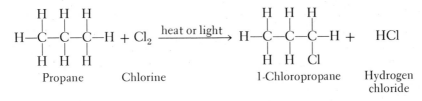

Propane   Chlorine   1-Chloropropane   Hydrogen
chloride

The substituted chlorine atom may appear on either an end carbon atom or the middle carbon; the actual product is usually a mixture of 1-chloropropane and 2-chloropropane.

Normally, addition reactions are more readily accomplished than substitution reactions. This is hinted in the reaction conditions specified above. The addition of a halogen to an alkene will occur easily at room temperature, whereas the substitution of a halogen for a hydrogen in an alkane requires either high temperature or ultraviolet light. This shows that unsaturated hydrocarbons are more reactive than saturated hydrocarbons.

Hydrogenation is also an addition reaction. We have already indicated that the hydrogenation of an alkene may be used to produce an alkane. Hydrogenation of an alkyne is a stepwise process, which may often be controlled to give the intermediate alkene as a product:

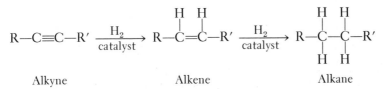

Alkyne   Alkene   Alkane

A particularly interesting addition reaction is the addition of ethylene to itself. It is an example of **polymerization.** Polymerization is the process whereby small molecular units called **monomers** join together to form giant molecules called **polymers.** Ethylene polymerizes as follows:

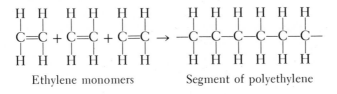

Ethylene monomers   Segment of polyethylene

In this reaction the double bond of each ethylene molecule opens, and each carbon atom joins to a carbon atom of another molecule, producing the chain shown. The chains formed yield molecules of molecular weights in the area of 20 000. Most plastics are polymers: the example above is the familiar polyethylene used for squeeze bottles, toys, packaging, etc.

Pyrolysis is a decomposition by heat. The pyrolysis of petroleum, consisting primarily of high molecular weight alkanes, is called *cracking*, and is usually done in the presence of a catalyst. Pyrolysis is conducted at temperatures in the range of 400 to 600°C. The usual products are alkanes of fewer carbon atoms, alkenes, and hydrogen.

Cracking is one of the major operations in extracting gasoline from raw petroleum. It increases both the yield and quality of gasoline. Yield is increased because some of the longer chain hydrocarbons are reduced to acceptable length for use as gasoline (five to ten carbon atoms per molecule). Quality is improved because the alkenes resulting from the reaction have good antiknock properties.*

22.7 THE AROMATIC HYDROCARBONS

Historically the term **aromatic** was associated with a series of compounds found in such pleasant-smelling substances as oil of cloves, vanilla, wintergreen, cinnamon, and others. Ultimately it was found that the key structure in these compounds is the benzene ring.

Benzene has been studied thoroughly in an attempt to determine its structure. Its molecular formula is C_6H_6. It is also known to be a ring compound. How this structure is to be represented in print is a problem, and a universally agreed upon answer has yet to be found. Two common forms are:

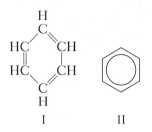

I   II

Structure I is perhaps the most complete representation in that it shows all the carbon and hydrogen atoms, as well as alternating single and double bonds that would satisfy the requirements of the octet rule. This structure is not in agreement with experimental fact, however; among other things, all carbon-carbon bonds are known to be alike, rather than some being single, some double. Struc-

*"Knocking" in an internal combustion engine is a sharp detonation of the fuel-air mixture, rather than a smooth explosion, producing the familiar knocking sound heard in automobiles when accelerating too quickly, or when climbing a hill. Knocking is rated on an arbitrary scale in which *n*-heptane is given an **octane number** of zero, and 2,2,4-trimethyl pentane ("isooctane") is rated at 100 in octane number.

ture II is a compromise that suggests equality among these bonds. It is also understood that each "corner" of the hexagon represents a carbon atom that forms the equivalent of four covalent bonds, three within the benzene ring, and one without. If the fourth bonded atom is not shown, it is understood to be hydrogen.

An alkyl group, halogen, or other species may replace a hydrogen in the benzene ring.

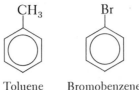

Toluene   Bromobenzene

If two bromines substitute for hydrogens on the same ring we must consider three possible isomers:

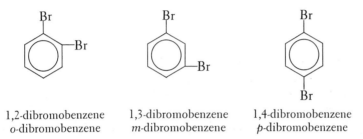

1,2-dibromobenzene   1,3-dibromobenzene   1,4-dibromobenzene
o-dibromobenzene   m-dibromobenzene   p-dibromobenzene

Two names are given for each isomer. The number system, which counts locations around the ring beginning at the substituted position that yields the lowest numbers, is more formal and serves any number of substituents. The other names are pronounced *ortho*-dibromobenzene, *meta*-dibromobenzene, and *para*-dibromobenzene. *Ortho*-, *meta*-, and *para*- are prefixes commonly used when two hydrogens have been replaced from the benzene ring. Relative to position X, the other positions are shown:

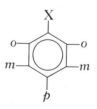

The physical properties of benzene and its derivatives are quite similar to those of other hydrocarbons. The compounds are nonpolar, insoluble in polar solvents such as water, but generally soluble in nonpolar solvents. In fact, benzene is widely used as the solvent for many nonpolar organic compounds. Like other hydrocarbons of comparable molecular weight, benzene is a liquid at room temperature. Members of the homologous series increase in boiling point in the usual manner as the number of carbon atoms increases.

The principal industrial source for benzene has been as a byproduct of the preparation of coke from coal. More recently, commercial methods have been developed by which certain petroleum products are converted to aromatic hydrocarbons. For example, toluene may be prepared from *n*-heptane:

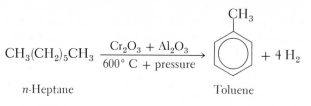

$$CH_3(CH_2)_5CH_3 \xrightarrow[\text{600}°\,C + \text{pressure}]{\text{Cr}_2\text{O}_3 + \text{Al}_2\text{O}_3} \text{(Toluene)} + 4\,H_2$$

n-Heptane   Toluene

Perhaps the most significant—and surprising—chemical property of benzene is that, despite its high degree of unsaturation, it does not normally engage in addition reactions. The most important reaction of benzene itself is the substitution reaction in which one hydrogen is displaced from the benzene ring. Several substances may be used for substitution, including the halogens:

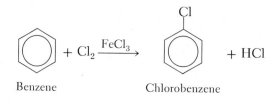

$$\text{(Benzene)} + Cl_2 \xrightarrow{\text{FeCl}_3} \text{(Chlorobenzene)} + HCl$$

Benzene   Chlorobenzene

Substitutions with nitric and sulfuric acids yield, respectively, nitrobenzene and benzenesulfonic acid. Second substitutions on the same ring are possible, though more difficult to bring about. Substitution reactions may also be performed on benzene derivatives, such as toluene, yielding isomers of nitrotoluene, for example. A triple nitro substitution produces 2,4,6-trinitrotoluene, better known simply as TNT.

22.8 SUMMARY OF THE HYDROCARBONS

Four types of hydrocarbons we have considered are summarized in Table 22.5:

Table 22.5
Hydrocarbons

Type	Name	Formula	Saturation	Structure
Aliphatic Open Chain	Alkane	C_nH_{2n+2}	Saturated	$-\overset{\textstyle\mid}{\underset{\textstyle\mid}{C}}-$
	Alkene	C_nH_{2n}	Unsaturated	$>C=C<$
	Alkyne	C_nH_{2n-2}	Unsaturated	$-C\equiv C-$
Aromatic	—	—	Unsaturated	

ORGANIC COMPOUNDS WITH OXYGEN

Thus far we have considered only the hydrocarbons and their derivatives. The third element most commonly found in organic chemicals is oxygen. Capable of forming two bonds (Table 22.1), oxygen serves as a connecting link between two other elements or, double-bonded, usually to carbon, it is a terminal point in a functional group. We will now examine functional groups that contain oxygen: the alcohols, ethers, aldehydes and ketones, acids and esters.

22.9 THE ALCOHOLS

THE STRUCTURE OF ALCOHOLS

Alcohol is the name given to a large class of compounds containing the **hydroxyl** group, —OH.* This functional group is not to be confused with the hydroxide ion of inorganic chemistry, which exists as an entity in ionic compounds and solutions. In alcohols the hydroxyl group is covalently bonded to an alkyl or other hydrocarbon group. Thus the general formula for an alcohol is R—OH, where R represents the alkyl group.

As shown in Table 22.1, and also in the two models of ethyl alcohol, C_2H_5OH in Figure 22.6, the bond angle around an oxygen atom is close to the tetrahedral angle—about 105°. Thus the alcohol molecule is a water molecule in which one hydrogen has been replaced by an alkyl group:

Water	Functional group	Alcohol

This structural similarity correctly suggests similar intermolecular forces and therefore similar physical properties. The lower alcohols (one to three carbon atoms) are liquids with boiling points ranging from 65°C to 97°C, comparable with water but well above the boiling points of alkanes of about the same

*Some chemists refer to the hydroxyl group as the *hydroxy group.*

Figure 22.6
Ball-and-stick and space-filling models of ethanol (ethyl alcohol), C_2H_5OH. (Photograph by Janice Peters.)

molecular weight. This is largely because of hydrogen bonding, very much in evidence in the lower alcohols (Fig. 22.7). Hydrogen bonding also accounts for the complete miscibility (solubility) between lower alcohols and water. As usual, boiling points rise with increasing molecular weight. Solubility drops off sharply as the alkyl chain lengthens and the molecule assumes more the character of the parent alkane.

When the hydroxyl group is attached to the end carbon in a chain, the compound is a *primary* alcohol. If the hydroxyl group is bonded to a carbon that is bonded to two other carbons, it is a *secondary* alcohol. Isopropyl alcohol (see below) is a secondary alcohol. A *tertiary* alcohol has the hydroxyl group attached to a carbon that is bonded to three other carbon atoms.

NAMES OF THE ALCOHOLS

Alcohols are best known by their common names, which originate in the name of the alkyl group to which the hydroxyl group is bonded. This system names the alkyl group, followed by "alcohol." Thus CH_3OH is *methyl alcohol* and C_2H_5OH is *ethyl alcohol*. When we reach propyl alcohol we are confronted with two isomers;

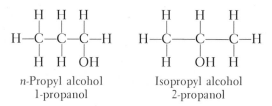

n-Propyl alcohol   Isopropyl alcohol
1-propanol   2-propanol

Under IUPAC nomenclature rules for alcohols, the *e* at the end of the corresponding alkane is replaced with the suffix *-ol* and the result is the name of the alcohol. Thus methyl alcohol becomes *methanol*, and ethyl alcohol is formally *ethanol*. The propane isomers are distinguished by stating the number of the carbon atom to which the hydroxyl group is bonded. Accordingly, n-propyl alcohol becomes *1-propanol* and isopropyl alcohol is designated *2-propanol*.

SOURCES AND PREPARATION OF ALCOHOLS

Hydration of alkenes. The major industrial source of several of our most important alcohols is the hydration of alkenes obtained from the cracking of petro-

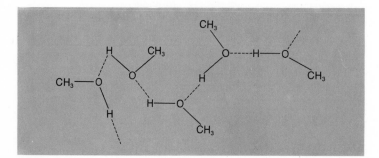

Figure 22.7
Hydrogen bonding in methanol.

leum. Beginning with ethylene, for example, the reaction may be summarized

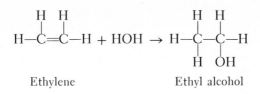

Ethylene   Ethyl alcohol

Fermentation of carbohydrates. Making ethyl alcohol by the fermentation of sugars in the presence of yeast is probably the oldest synthetic chemical process known:

$$C_6H_{12}O_6 \xrightarrow{\text{yeast}} 2\,CO_2 + 2\,C_2H_5OH$$

Glucose (sugar)   Ethyl alcohol

A solution that is 95% ethyl alcohol (190 proof) may be obtained from the final mixture by fractional distillation. The mixture also yields two other products that are of commercial importance today: *n*-butyl alcohol and acetone (Section 22.11).

Synthesis of methyl alcohol. Methyl alcohol is sometimes called wood alcohol because it was once made by the destructive distillation (heating in absence of air) of wood. It is now produced by the catalytic hydrogenation of carbon monoxide at high pressure and temperature:

$$CO + 2\,H_2 \xrightarrow[\text{250 atm} + 300°C]{ZnO + Cr_2O_3} CH_3OH$$

CHEMICAL PROPERTIES OF ALCOHOLS

The chemical properties of alcohols are essentially the chemical properties of the functional group, —OH. In some reactions the C—OH bond is broken, separating the entire hydroxyl group. This is true in the dehydration of alcohols to form alkenes (p. 545). In other reactions of alcohols the O—H bond within the hydroxyl group is broken. We will postpone discussion of these reactions until the structures of the various products—aldehydes, ketones, carboxylic acids, and esters—have been examined.

SOME COMMON ALCOHOLS

Methyl alcohol is an important industrial chemical with production measured in the billions of pounds annually. It is a raw material for the production of many chemicals, particularly formaldehyde, which is widely used in the plastics industry. It is also used in antifreezes, commercial solvents, and as a denaturant, or additive to ethyl alcohol to make it unfit for human consumption. Taken internally, methyl alcohol is a deadly poison, frequently causing blindness in less than lethal doses.

In addition to its uses in beverages, ethyl alcohol is used in organic solvents and in the preparation of various organic compounds such as chloroform and ether. Its production is also measured in the billions of pounds annually.

Other widely used alcohols include isopropyl alcohol, which is sold as rubbing alcohol, and *n*-butyl alcohol, used in lacquers in the automobile industry. Alcohols containing more than one hydroxyl group are also common. Permanent antifreeze in automobiles is ethylene glycol, which has two hydroxyl groups in the molecule. Glycerine, or glycerol, a trihydroxyl alcohol, has many uses in the manufacture of drugs, cosmetics, explosives, and other chemicals.

22.10 THE ETHERS

In the previous section we pointed out that structurally an alcohol might be considered as a water molecule in which one hydrogen has been replaced by an alkyl group. An **ether** may be similarly considered, except that *both* hydrogens are replaced by an alkyl group. The functional group that identifies an ether is simply the oxygen atom bonded to two alkyl groups:

Water   Functional group   Ether

The R′ indicates that the functional groups may or may not be identical. For example, methyl ethyl ether and ethyl ether have the structures

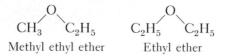

Methyl ethyl ether   Ethyl ether

Figure 22.8 shows models of methyl ether.

Ether molecules are less polar than alcohol molecules, and there is no opportunity for hydrogen bonding between them. Intermolecular attractions are therefore lower, as are the dependent boiling points. Up to three carbons, ethers are gases at room conditions, and the familiar ethyl ether is a volatile

Figure 22.8
Ball-and-stick and space-filling models of methyl ether, CH_3OCH_3. (Photograph by Janice Peters.)

liquid that boils at 35°C. The solubility of an ether in water is about the same as the solubility of its isomeric alcohol, primarily because of hydrogen bonding between the ether molecule and water molecule:

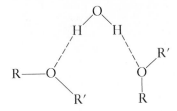

All ethers are called "ether," and identified specifically by naming first the two alkyl groups that are bonded to the functional group. If the groups are identical, the prefix *di-* may be used, as in diethyl ether.

Under properly controlled conditions, ethers can be prepared by dehydrating alcohols. At 140°C, and with constant alcohol addition to replace the ether as it distills from the mixture, ethyl ether is formed from two molecules of ethanol:

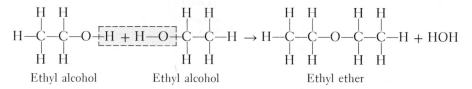

Aside from combustion, ethers are relatively unreactive compounds, being quite resistant to attack by active metals, strong bases and oxidizing agents. They are, however, highly flammable and must be handled cautiously in the laboratory.

The isolated word "ether" generally prompts one to think of the anesthetic that is so identified. This compound is ethyl ether; its line formula is $C_2H_5-O-C_2H_5$. Recently its isomer, methyl propyl ether, $CH_3-O-C_3H_7$, has been gaining popularity as a substitute in this use; it has fewer objectionable after-effects than ethyl ether. Ethyl ether is also used as a solvent for fats from foods and animal tissue in the laboratory.

22.11 THE ALDEHYDES AND KETONES

Aldehydes and **ketones** are characterized by the **carbonyl functional group,**

$$\overset{\displaystyle O}{\underset{\displaystyle /C\backslash}{\|}}$$

If at least one hydrogen atom is bonded to the carbonyl carbon, the compound is an aldehyde, RCHO; if two alkyl groups are attached, the compound is a ketone, R—CO—R'.

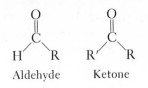

Aldehyde   Ketone

The simplest carbonyl compound is formaldehyde, HCHO, which has two hydrogen atoms bonded to the carbonyl carbon. If a methyl group replaces one of the hydrogens of formaldehyde, the result is acetaldehyde, CH_3CHO (Fig. 22.9). Replacement of both formaldehyde hydrogens with methyl groups yields acetone.

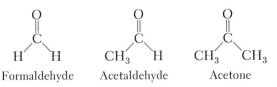

Formaldehyde   Acetaldehyde   Acetone

The carbonyl group is polar, thereby making ketone and aldehyde molecules polar, although not as polar as alcohols. Only formaldehyde is definitely a gas at room temperature (boiling point $-21°C$); acetaldehyde boils at $20°C$. Aldehydes and ketones of up to about five carbons enjoy some solubility in water, no doubt because of the polarity of both the solvent and solute molecules and hydrogen bonding. The liquid "formaldehyde" we encounter in the laboratory is actually a water solution, sold under the trade name "Formalin."

Figure 22.9
Ball-and-stick and space-filling models of acetaldehyde and acetone. (Photograph by Janice Peters.)

The lower aldehydes are best known by their common names. The IUPAC nomenclature system for aldehydes employs the name of the parent hydrocarbon, substituting the suffix -al for the final e to identify the compound as an aldehyde. Thus the IUPAC name for formaldehyde is methanal, for acetaldehyde, ethanal, and so forth.

Ketones are named by one of two systems. The first duplicates the method of naming ethers: identify each alkyl group attached to the carbonyl group, followed by the class name, ketone. Accordingly, methyl ethyl ketone has the structure

Under the IUPAC system the number of carbons in the longest chain carrying the carbonyl carbon establishes the hydrocarbon base, which is followed by -one to identify the ketone as the class of compound. Methyl ethyl ketone, having four carbons, would be called *butanone*. Two isomers of pentanone would be *2-pentanone* and *3-pentanone*, the number being used to designate the carbonyl carbon:

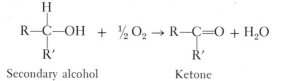

2-Pentanone   3-Pentanone

Aldehydes and ketones may be prepared by oxidation of alcohols. If the product is to be a ketone, the alcohol must be a *secondary* alcohol, in which the hydroxyl group is bonded to an interior carbon:

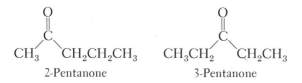

Secondary alcohol   Ketone

Care must be taken not to overoxidize aldehyde preparations, since aldehydes are easily oxidized to carboxylic acids (see next section).

Aldehydes and ketones may also be produced by the hydration of alkynes. If the triple bond is on an end carbon, an aldehyde is produced; if between internal carbons, the result is a ketone. A typical reaction is the commercial preparation of acetaldehyde:

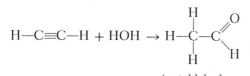

Acetaldehyde

The double bond of the carbonyl group can engage in addition reactions, just like the double bond in the alkenes. One such reaction is the catalytic hydrogenation of ketones to secondary alcohols, in which the hydroxyl group is bonded to a carbon atom *within* the chain:

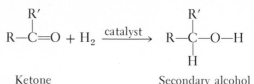

Ketone   Secondary alcohol

Oxidation reactions occur quite readily with aldehydes, but are resisted by ketones. When an aldehyde is oxidized, the product is a carboxylic acid:

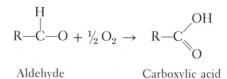

Aldehyde   Carboxylic acid

Formaldehyde is probably the best known carbonyl compound. It is widely used as a preservative. Large quantities are consumed in the manufacture of resins and in the preparation of numerous organic compounds. Acetaldehyde finds use in the manufacture of acetic acid, ethyl acetate, and other organic products.

Acetone is the most important ketone. It is a solvent for many organic chemicals, including cellulose derivatives, varnish, lacquer, plastics, and resins. Methyl ethyl ketone finds application in the petroleum industry, and it is also familiar as an ingredient of fingernail polish remover.

22.12 CARBOXYLIC ACIDS AND ESTERS

In the last section we saw that oxidation of an aldehyde produces a **carboxylic acid,** the general formula of which is frequently represented as RCOOH. The functional group, —COOH, shown at the left below, is a combination of a carbonyl group and a hydroxyl group, appropriately called the **carboxyl group.** You can probably pick out the carboxyl group in the formic and acetic acid models in Figure 22.10. In an **ester,** the carboxyl carbon may be bonded to a hydrogen atom or an alkyl group, and the carboxyl hydrogen is replaced by another alkyl group, as shown:

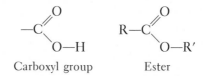

Carboxyl group   Ester

The geometry of the carboxyl group results in strong dipole attractions and hydrogen bonding between molecules. As a consequence, boiling points

Figure 22.10
Ball-and-stick and space-filling models of formic and acetic acids. (Photograph by Janice Peters.)

tend to be high compared to compounds of similar molecular weight. Formic acid, for example, boils at 100.5°C. Lower acids are completely miscible in water, but solubility drops off as the aliphatic chain lengthens and the molecule behaves more like a hydrocarbon.

Common names continue to be used for most acids. Formic acid, HCOOH, with only a hydrogen attached to the carboxyl group, is the simplest of the carboxylic acids. Next in the series is acetic acid, in which the methyl group is bonded to the carboxyl group: CH_3COOH. The names of many acids come from their sources or some physical property associated with them. Butyric acid, C_3H_7COOH, for example, is responsible for the odor of rancid butter, for which the Latin word is *butyrum*. The IUPAC system for naming carboxylic acids drops the *e* from the alkane of the same number of carbon atoms, and replaces it with *-oic*. Thus HCOOH is *methanoic acid*; CH_3COOH is *ethanoic acid*; C_2H_5COOH is *propanoic acid*, and so forth.

Formic acid is prepared commercially from sodium formate, produced by the reaction of carbon monoxide and sodium hydroxide. In a typical molecular product reaction (p. 000), sodium formate reacts with hydrochloric acid to yield formic acid and sodium chloride:

$$HCOONa + HCl \rightarrow HCOOH + NaCl$$

Acetic acid, by far the most important of the carboxylic acids, is produced by the stepwise oxidation of ethanol, first to acetaldehyde and then to acetic acid:

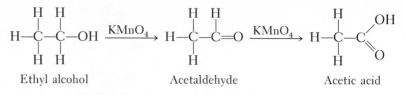

Ethyl alcohol   Acetaldehyde   Acetic acid

Carboxylic acids are weak acids that release a proton from the carboxyl group on ionization.* Acetic acid, for example, ionizes in water as follows:

$$CH_3COOH(aq) \rightleftharpoons CH_3COO^-(aq) + H^+(aq)$$

The ionization takes place but slightly; only about 1% of the acetic acid molecules ionize. The solution consists primarily of molecular CH_3COOH. This notwithstanding, acetic acid participates in typical acid reactions such as neutralization,

$$CH_3COOH(aq) + OH^-(aq) \rightarrow HOH(\ell) + CH_3COO^-(aq)$$

and the release of hydrogen on reaction with a metal,

$$2\ CH_3COOH(aq) + Ca(s) \rightarrow 2\ CH_3COO^-(aq) + Ca^{2+}(aq) + H_2(g)$$

Metal acetate salts may be obtained by evaporating the resulting solutions to dryness.

The reaction between an acid and an alcohol is called **esterification.** The products of the reaction are an ester and water. A typical esterification reaction is

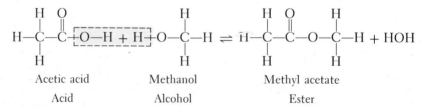

Acetic acid   Methanol   Methyl acetate

Acid   Alcohol   Ester

Notice how the water molecule is formed: the *acid contributes the entire hydroxyl group,* while the *alcohol furnishes only the hydrogen.*

The names of esters are derived from the parent alcohol and acid. The first term is the alkyl group associated with the alcohol; the second term is the name of the anion derived from the acid. In the example above, methyl alcohol (methanol) yields methyl as the first term, and acetic acid yields acetate as the second term.

Carboxylic acids engage in typical proton transfer acid-base type reactions with ammonia to produce salts. The ammonium salt so produced may then be heated, which causes it to lose a water molecule. The resulting product is called

---

*In more advanced consideration of organic reactions, the term *acid* is also used in reference to Lewis acids (p. 401). This is why the adjective *carboxylic* is used to identify an organic acid containing the carboxyl group.

an *amide*. Compared to the original acid, an amide substitutes an —NH_2 group for the —OH group of the acid (Section 22.14).

$$R-\overset{\overset{\displaystyle O}{\|}}{C}-O-H + NH_3 \rightarrow \quad R-\overset{\overset{\displaystyle O}{\|}}{C}-O^- \ NH_4^+ \xrightarrow{\Delta} R-\overset{\overset{\displaystyle O}{\|}}{C}-NH_2 + H_2O$$

$$\text{Acid} \qquad\qquad\qquad \text{Salt} \qquad\qquad \text{Amide}$$

Formic acid and acetic acid are the two most important carboxylic acids. Formic acid is the source of irritation in the bite of ants and other insects, or the scratch of nettles. A liquid with a sharp, irritating odor, it is used in manufacturing esters, salts, plastics, and other chemicals. Acetic acid is present to about 4 to 5% in vinegar, and is responsible for its odor and taste. Acetic acid is among the least expensive organic acids, and is therefore a raw material in many commerical processes that require a carboxylic acid. Sodium acetate is one of several important salts of carboxylic acids. It is used to control the acidity of chemical processes and in the preparation of soaps and pharmaceutical agents.

Ethyl acetate and butyl acetate are two of the relatively few esters produced in large quantity. Both are used as solvents, particularly in the manufacture of lacquers. Other esters are involved in the plastics industry, and some find application in the medicinal fields. Esters are responsible for the odor of most fruit and flowers, leading to their use in the food and perfume industries.

ORGANIC COMPOUNDS WITH NITROGEN

Nitrogen is one of the important elements found in living organisms, so we might expect to find it in many organic compounds. We will mention briefly two classes of nitrogen-bearing organic compounds, the amines and the amides.

22.13 AMINES

Amines are organic derivatives of ammonia. In much the same way that an alcohol can be viewed structurally as a water molecule in which a hydrogen atom has been replaced by an alkyl group, and ether molecules are water molecules with both hydrogen atoms replaced by alkyl groups, an amine is an ammonia molecule with one, two, or three hydrogen atoms replaced by alkyl groups. The number of hydrogens so replaced distinguishes between a primary, secondary, and tertiary amine. Amines are named by identifying the alkyl groups that are bonded to the nitrogen atom, using appropriate prefixes if two or three identical groups are present, followed by the suffix *-amine*. Illustrative examples follow:

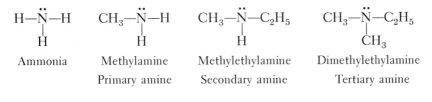

Ammonia

Methylamine
Primary amine

Methylethylamine
Secondary amine

Dimethylethylamine
Tertiary amine

As ammonia is polar and capable of forming hydrogen bonds, so are the primary and secondary amines, but to a lesser extent. Tertiary amines are essentially nonpolar. These structural features contribute in the usual way to the physical properties of the amines. All three methylamines and ethylamines are gases, with boiling points in the range of $-6°C$ to $11°C$. Other amines are liquids with boiling points that increase with molecular weight and more complex structure. Amines can form hydrogen bonds with water molecules; therefore lower amines—particularly primary and secondary amines—are very soluble in water, and less soluble in nonpolar solvents.

Because of the unshared electron pair in the nitrogen atom, amines behave as Brönsted-Lowry or Lewis bases. A typical reaction is

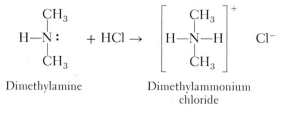

Dimethylamine   Dimethylammonium chloride

By reaction first with nitrous acid and then hydrogen, dimethylamine is made into dimethyl hydrazine, $(CH_3)_2NNH_2$, which is used as a rocket propellant. Dimethylamine and trimethylamine are both used in making anion exchange resins. Dyes, drugs, herbicides, fungicides, soaps, insecticides, and photographic developers are among the chemical products made from amines. Aniline, or phenylamine, an aromatic amine, is among the more important materials used in dye making.

22.14 AMIDES

On page 561, an **amide** was shown to be a derivative of a carboxylic acid in which the hydroxyl part of the carboxyl group is replaced by an NH_2 group. For example,

Acetic acid   Acetamide

by substitution of the $-NH_2$ in place of the $-OH$, as shown. An amide is named by replacing the -*ic acid* name of the acid with *amide*.

Amides are polar compounds that are capable of strong hydrogen bonding between the electronegative oxygen of the carboxyl group of one molecule and the electropositive amide hydrogen of the next. As a consequence the amides as a group have higher melting and boiling points than otherwise similar compounds. Only formamide, $HCONH_2$, is a liquid at room temperature; all higher amides are solids. Polarity and hydrogen bonding predict accurately the solubility of the lower amides in water.

The amide structure appears in an important biochemical system, protein, as a connecting link between amino acids. The linkage is commonly called a *peptide linkage*. This linkage has the form

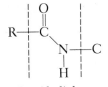

Peptide linkage

An *amino acid* is an acid in which an amine group is substituted for a hydrogen atom in the molecule. The amino acids involved in the protein structure have the general formula

in which the amine and carboxyl groups are bonded to the same carbon atom. The peptide linkage is formed when the carboxyl group of one amino acid and the amine group of another combine by removing a water molecule:

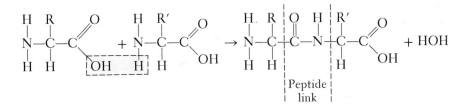

Proteins are chains of such links between as many as 18 different amino acids, producing huge molecules with molar weights ranging from about 34 500 to 50 000 000.

22.15 SUMMARY OF THE ORGANIC COMPOUNDS OF CARBON, HYDROGEN, OXYGEN, AND NITROGEN

The eight types of organic compounds of carbon, hydrogen, oxygen and nitrogen we have considered are summarized in Table 22.6.

Table 22.6
Classes of Organic Compounds

Compound Class	General Formula	Functional Group	Names*
Alcohol	R—OH	—OH	Alkyl group + *alcohol;* methyl alcohol Alkane prefix + *-ol:* methanol
Ether	R—O—R′	(structure)	Name both alkyl groups + *ether:* methyl ethyl ether Alkyl group + *-oxy-* + alkane: meth-oxyethane
Aldehyde	R—CHO	(structure)	Common prefix + *-aldehyde:* formal-dehyde Alkane prefix + *-al:* methanal
Ketone	R—CO—R′	(structure)	Name both alkyl groups + *ketone:* methyl ethyl ketone; methyl n-propyl ketone (Number carbonyl carbon) + alkane prefix + *-one:* butanone; 2-penta-none
Acid	R—COOH	(structure)	Common name + acid: formic acid Alkane prefix + *-oic* + *acid:* meth-anoic acid
Ester	R—CO—OR′	(structure)	Alcohol alkyl group + acid anion: methyl acetate Alcohol alkyl group + acid alkane prefix + *-oate;* methyl ethanoate
Amine	RNH_2 R_2NH R_3N	—N—	Name alkyl group(s) + *-amine:* meth ylamine *Amino-* + alkane: aminomethane
Amide	$R—CONH_2$	(structure)	Common acid prefix + *-amide:* for-mamide Alkane prefix + *-amide:* methan-amide

*Common name followed by IUPAC name.

TERMS AND CONCEPTS

Organic chemistry (531)
Hydrocarbon (533)
Aliphatic hydrocarbon (533)
Saturated hydrocarbon (533)
Alkane (533)
Homologous series (534)
Normal alkane (534)
Condensed (line) formula (536)
Isomer; isomerism (536)
Alkyl group (538)
Functional group (538)
Unsaturated hydrocarbon (541)
Alkene (542)
Alkyne (542)

Cis-, *trans-* isomers (544)
Fractional distillation (544)
Dehydration(545)
Dehydrohalogenation (545)
Hydrogenation (546)
Addition reaction (546)
Substitution reaction (546)
Halogenation (547)
Polymerization (polymer, monomer) (547)
Pyrolysis (548)
Cracking (548)
Aromatic hydrocarbon (548)
Alcohol (551)
Hydroxyl group (551)

Primary, secondary tertiary alcohol (552)
Ether (554)
Aldehyde (555)
Ketone (555)
Carbonyl group (555)
Carboxylic acid (558)

Carboxyl group (558)
Ester (558)
Esterification (560)
Amine (561)
Primary, secondary, tertiary amine (561)
Amide (562)

QUESTIONS AND PROBLEMS

Section 22.1

22.1) Distinguish between organic chemistry and inorganic chemistry. Define organic chemistry.

Section 22.3

22.2) Define hydrocarbon.

22.3) Define an alkane, and explain how it is an example of a homologous series.

22.4) What is a condensed formula, or line formula? What advantage does it have over a molecular formula?

22.5) Write both the structural formula and condensed (line) formula for the normal alkane having nine carbon atoms in its molecules.

22.6) Define isomerism. From the following list of molecular and condensed formulas, select two that are isomers and explain why they may be so classified:
  (a) $CH_3CH(CH_3)CH_2CH_2CH(CH_3)CH_2CH_3$
  (b) $CH_3CH_2C(CH_3)_2CH_2CH_3$
  (c) $CH_3(CH_2)_8CH_3$
  (d) C_6H_{14}
  (e) C_9H_{20}

22.7) What is meant by an alkyl group? How are alkyl groups named? Give three examples of alkyl groups, both name and formula.

22.8)

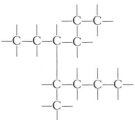

is the skeleton structure for an alkane. Identify

22.43) Name some of the classes of carbon-bearing compounds that are not included in organic chemistry.

22.44) Distinguish between saturated and unsaturated hydrocarbons.

22.45) Write the molecular formula for the alkane having 19 carbon atoms in its molecules. Explain how you determined this formula.

22.46) Write a structural formula for the following compound: $CH_3(CH_2)_4CH_3$.

22.47) Write the name of

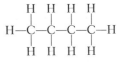

22.48) Draw structural diagrams for all the isomers of heptane.

22.49) What is the significance of the letter R in a formula such as R—Cl? Write two structural formulas that might be described by R—Cl.

22.50) Redraw the skeleton structure of the alkane shown in question 22.8 so its identification will be more evident.

the alkane on which its IUPAC name would be based (e.g., propane, butane, etc.). Justify your choice.

Write the names of those compounds in Problems 22.9 to 22.16 and 22.51 to 22.58 whose structural formulas are given, and the structural formulas for the compounds whose names are given.

22.9) 3-methylheptane

22.10) 2,4-dimethyloctane

22.11)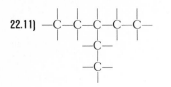

22.12)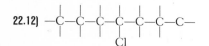

22.13)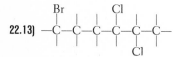

22.14) 1,1-dibromoethane

22.15) 1,1,4,5-tetrabromopentane

22.16) 1,2,2-tribromo-1-chlorobutane

22.51) 3-ethylhexane

22.52) 2,2,4-trimethyl-3,6-diethyloctane

22.53)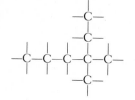

22.54)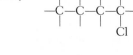

22.55)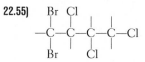

22.56) 1,2,3-trichloropropane

22.57) 2,3-dibromo-4-chlorohexane

22.58) 1,1,1-tribromo-3-chloropentane

Section 22.4

22.17) What structural feature identifies an alkene? Write the general formula for an alkene.

22.18) Write structural formulas and names for the first three alkynes.

22.19) Explain in words the difference among 1-hexene, 2-hexene, and 3-hexene.

22.20) What are *cis-*, *trans-* isomers?

22.21) What is the name of

22.59) What structural feature identifies an alkyne? Write the general formula for an alkyne.

22.60) Write structural formulas and names for the first three alkenes.

22.61) Write the structural formula for 3-hexyne.

22.62) Write structural formulas for the *cis-* and *trans-* isomers of 3-hexene, and state which is which.

22.63) What is the name of

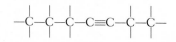

Section 22.5

22.22) Write the formula and name of the alkene that can be prepared by the dehydration of the following alcohol:

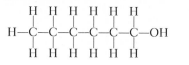

22.23) Write the structural formula and name of the alkene that may be prepared by the dehydrohalogenation of 1-chloropentane.

22.64) Which of the two alcohol structures shown could be used to produce 2-butene? Explain.

(a)

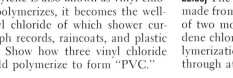

(b)

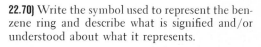

22.65) Write the structural formula and name of two isomeric alkenes that could be formed by the dehydrohalogenation of 2-bromobutane.

Section 22.6

22.24) Distinguish between an addition reaction and a substitution reaction.

22.25) Write an equation for the hydrogenation of propene.

22.26) What is polymerization?

22.27) Chloroethylene is also known as vinyl chloride. When it polymerizes, it becomes the well-known polyvinyl chloride of which shower curtains, phonograph records, raincoats, and plastic pipe are made. Show how three vinyl chloride monomers would polymerize to form "PVC."

22.66) Explain why the chlorination of ethane yields chloroethane, whereas the chlorination of ethylene produces 1,2-dichloroethane.

22.67) Write an equation for the hydrogenation of propyne.

22.68) Distinguish between monomers and polymers.

22.69) The popular kitchen material Saran Wrap is made from the copolymerization (polymerization of two monomers) of chloroethylene and vinylidene chloride, $CH_2{=}CCl_2$. Show how this copolymerization occurs, and write the structure through at least eight carbon atoms.

Section 22.7

22.28) How do aromatic hydrocarbons differ from aliphatic hydrocarbons?

22.29) Name the compounds represented by these three formulas:

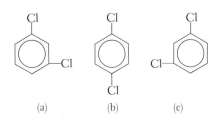

(a)   (b)   (c)

22.70) Write the symbol used to represent the benzene ring and describe what is signified and/or understood about what it represents.

22.71) Draw a structural formula for 1,3,5-trichlorobenzene.

Section 22.9

22.30) What is the functional group of an alcohol? Write both its name and formula. Write the general formula of an alcohol.

22.31) Write the structural formula for 1-pentanol.

22.32) There must be at least three carbon atoms in a secondary alcohol. Explain why this is so. Write the structural formulas of two different secondary alcohols that have five carbon atoms.

22.72) Account for the high boiling points and water solubility of alcohols compared to the same properties of alkanes of comparable molecular weight.

22.73) Give the common and IUPAC name of

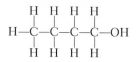

22.74) Write the structural formula of the tertiary alcohol that contains the smallest possible number of carbon atoms.

Section 22.10

22.33) Show how water, alcohols, and ethers are structurally related.

22.34) Write the structural formula for methyl propyl ether.

22.75) Write a structural equation that shows how methyl ether might be prepared from an alcohol.

22.76) "Each ether with two or more carbon atoms has an alcohol with which it is isomeric." Show that this statement is true.

Section 22.11

22.35) Use structural formulas to show how an aldehyde is formed by oxidizing an alcohol.

22.36) How are aldehydes and ketones alike, and how do they differ?

22.77) Show why a ketone rather than an aldehyde is produced when a secondary alcohol is oxidized.

22.78) Write structural formulas for butanal and butanone.

Section 22.12

22.37) Write the general formula for a carboxylic acid, as well as the formula of the functional group.

22.38) Account for the high boiling points of carboxylic acids compared to the boiling points of other compounds of comparable molecular weight.

22.39) Demonstrate by equation the acid character of the carboxylic group.

22.79) Write the structural formula for hexanoic acid. What is the name of C_4H_9COOH?

22.80) Account for the high solubility of the lower carboxylic acids.

22.81) Write the equation for the reaction between formic acid, $HCOOH$, and propanol, and name the ester formed.

Section 22.13

22.40) Explain the relationship between ammonia and the amines.

22.82) "Amines are bases in many chemical reactions." Explain why this is so.

22.41) What is the name of

$$CH_3-\underset{\underset{H}{|}}{N}-C_3H_7$$

22.83) Write the structural formula of dibutyl-amine.

Section 22.14

22.42) Compare the functional groups of carboxylic acids and amides.

22.84) Explain by structural formula how a peptide link is formed.

appendices

appendix I:

chemical calculations

If you plan to take a chemistry course after this one, one of the main purposes of the present course is to build problem-solving skills. These skills fall into two groups. First, you must learn the chemical principles and thought processes by which you set up a problem. These are fully described throughout your text. Second, you must be able to calculate the answer.

Calculation skills are learned gradually through years of mathematics courses, and it is assumed that beginning chemistry students have these skills at their fingertips, ready for immediate use. This assumption is often wrong. Many students who once learned these skills, perhaps well, perhaps not, but have not used them regularly, have forgotten them. A mathematical refresher, or at least a handy reference to the exact procedures used in chemistry, is important for these students. The first section of this Appendix meets that need.

PART A: NUMBERS

When we count objects, the result is given in a **number.** One, two, three, etc., are **counting numbers,** usually written as **numerals** 1, 2, 3. . . . Counting numbers are also called **whole numbers, integers,** and **integral numbers.** Zero is an integer. Numbers may be **positive** or **negative.** Numbers that lie between integers are called **fractions.** Fractions are sometimes written as a **ratio** of integers, such as $\frac{1}{2}$, $\frac{3}{4}$, and $\frac{5}{3}$. The number above the line is the **numerator,** and the number below the line is the **denominator.** In a **proper fraction** the numerator is smaller than the denominator; if the

numerator is larger than the denominator it is an **improper fraction.** Any fraction in which the numerator is the same as or equal to the denominator has a value of 1.

In chemistry, fractions are usually written as **decimal fractions.** Whole numbers appear to the left of the decimal point, and fractional parts on the right. The positions to the left and right of the decimal are **place values** or **digit values.** Reading left from the decimal point, the place values show, in order, the number of ones or units, tens, hundreds, etc.; from the decimal to the right are the tenths, hundredths, thousandths, etc. Each place value is ten times larger than the place value to its right, and is said to be larger by one **order of magnitude** or a **multiple of 10**. For numbers less than 1, orders of magnitude are called **submultiples of 10**.

It has long been customary in the United States to write commas after every three orders of magnitude in a large number. One billion, for example, is written 1,000,000,000. Commas are not used to separate groups of three orders of magnitude in numbers less than 1; one millionth is 0.000001. Other nations use different "punctuation" in writing numbers. There is, at present, an international effort to standardize number writing by placing a space between each three orders of magnitude on both sides of the decimal point. By this procedure the number 32,845,702.83785 is written 32 845 702.837 85, and the number 0.04237986 is written 0.042 379 86. The space is often omitted for numbers that begin or end four spaces to the left or right of the decimal. Thus 4 968 is written 4968, and 0.589 3 is written 0.5893. This type of spacing in large and small numbers is used throughout this book.

Another standard practice in scientific writing is to place a zero before the decimal point of a decimal fraction. Thus .5893 is written 0.5893.

PART B: THE HAND CALCULATOR

GENERAL DESCRIPTION

With the number of problems to be solved in a chemistry course, it is almost impossible to get by without a calculator. If you are willing to do some pencil and paper work on problems with exponential notation and logarithms, you can solve the problems in this book with a calculator that is able to add, subtract, multiply, and divide. Most students have calculators that can solve exponential notation and logarithm problems too. If you are going to buy a calculator, you should include these functions. As a bonus you will no doubt find keys for reciprocals, squares, square root, raising numbers to a power, and trigonometric functions.

There are many fine calculators on the market, and it is not our intention to recommend one over any other. We will select two, however, to illustrate how calculators are used. We have chosen two that are widely used by beginning chemistry students, the Texas Instrument TI-30 II shown in Figure AP-1, and the Hewlett-Packard HP-32E, shown in Figure AP-2. Two are necessary because there are two *logic* systems, each with its own order of operations for solving a problem. Texas Instrument calculators use what is called the Algebraic Operating System (AOS), and Hewlett-Packard uses Reverse Polish Notation (RPN). Other brands use one or the other of these systems, and for normal operations the instructions given here will apply to them also. If

Figure AP-1
Texas Instrument TI-30II calculator.

Figure AP-2
Hewlett-Packard HP-32E calculator.

you find minor variations in procedure, check the instruction book supplied with your calculator.

This section describes only the calculator operations needed to solve the problems in this book. It will not mention other functions your calculator may be able to perform. Particular attention is given to methods that you are apt to overlook if you learn from a manufacturer's instruction book alone.

A calculator should not be used as a substitute for thinking. If a problem is simple and can be solved mentally, do it "in your head." You will make fewer mistakes. If you use your calculator, *think* your way through each problem and estimate the answer mentally. Suggestions on approximating answers are given on page 586. If the calculator answer appears reasonable, round it off properly and write it down. Then run through the

calculation again to be sure you haven't made a keyboard error. Your calculator is an obedient and faithful servant that will do exactly what you tell it to do, but it is not responsible for the mistakes you make in your instructions.

It was suggested above that you round off your answer properly before recording it. This is because calculator answers to many problems are limited in length only by the display. For example, $273 \div 45.6$ equals 5.986 842 1 on a TI-30 II. Some calculators can show more numbers, and therefore do. Usually only the first three or four digits have meaning, and the others should be discarded. Procedures for deciding how many digits to write are given in the section on significant figures, which begins on page 587. In this section we will record all the numbers shown on the displays of the TI-30 II and HP-32E. If you use a different calculator you may expect minor differences in the last digits of the display.

Consult your instruction book to learn how to turn your calculator on and off, and how to clear an incorrect entry or to clear the entire unit. If you have a Hewlett-Packard calculator you also have a "display control" by which you may choose the number of decimal places in the display. Without instructions it shows four. Sometimes this throws unnecessary zeros into an answer, and sometimes needed digits in the answer do not appear. The extra zeros can be ignored. The missing digits can be brought to the display temporarily by pressing first the blue g key, and then MANT (appears in blue on front of ENTER key), holding it down. All decimals will be displayed until MANT is released. Suggestions for the display control will be made as needed.

The instructions that follow are based on TI-30 II and HP-32E calculators being sold in January, 1981. They are subject to design changes made after that date.

RECIPROCALS, SQUARE ROOTS, AND SQUARES

Finding the reciprocal, square root, or square of a number are called one-number functions because only one number must be keyed into the calculator. The procedures are as follows:

PROBLEM	TI-30 II		HP-32E	
	Press	Display	Press	Display
$\dfrac{1}{12.34}$	12.34 1/x	*12.34* *0.08 103728*	12.34 1/x	*12.34* *0.08 10*
$\sqrt{12.34}$	12.34 $\sqrt{x}$	*12.34* *3.5 128336*	12.34 $\sqrt{x}$	*12.34* *3.5 128*
12.34^2	12.34 x^2	*12.34* *152.2756*	12.34 blue g x^2	*12.34* *12.3400* *152.2756*

ADDITION, SUBTRACTION, MULTIPLICATION, AND DIVISION

The differences between AOS and RPN logics appear in ordinary arithmetic operations. Follow the procedure for your calculator.

AOS LOGIC	RPN LOGIC
The procedure for a common arithmetic operation is identical to the arithmetic equation for the same calculation. If you wish to add X to Y, the equation is X + Y = . The calculator procedure is:	The procedure for a common arithmetic operation is to key in *both* numbers and then tell the calculator what to do with them. If you wish to add X to Y, you enter X, key in Y, and instruct the calculator to add. The procedure is:
1. Key in X.	1. Key in X.
2. Press the function key (+ for addition).	2. Press ENTER.
3. Key in Y.	3. Key in Y.
4. Press = .	4. Press the function key (+ for addition).
The display will show the calculated result.	The display will show the calculated result.

EXAMPLE: Solve 12.34 + 0.0567* = ? (AOS)
Solution:

Press	Display
12.34	*12.34*
+	*12.34*
.0567	*0.0567*
=	*12.3967*

The numbers in the PRESS column are, in order, Steps 1, 2, 3, and 4 of the procedure. In Step 2, the function key is − for subtraction, × for multiplication, and ÷ for division.

EXAMPLE: Solve 12.34 + 0.0567* = ? (RPN)
Solution:

Press	Display
12.34	*12.34*
ENTER	*12.3400*
.0567	*0.0567*
+	*12.3967*

The numbers in the PRESS column are, in order, Steps 1, 2, 3, and 4 of the procedure. In Step 4, the function key is − for subtraction, × for multiplication, and ÷ for division.

---

*In scientific writing, a decimal fraction less than 1 is shown with a zero to the left of the decimal point. Most calculators show these zeros, but it is not necessary to key them into the calculator.

You may wish to confirm the following results on your calculator:

12.34 − 0.0567 = 12.2833;
12.34 × 0.0567 = 0.699 678;
12.34 ÷ 0.0567 = 217.636 68.

(An HP-32E calculator will round off all answers to four decimals. In ordinary arithmetic, particularly when answers are 0.0005 or less, it is advisable to key in additional decimals at the start. The procedure is to press first the gold f key, and then FIX, which is the $\sqrt{x}$ key—notice that FIX is printed in gold above the key—followed by whatever number of decimals you want from the number keys. If you select 9 you will have nine digits, many of which are apt to be tail-end zeros. They may be ignored unless needed for significant figures, as shown on page 590.)

CHAIN CALCULATIONS

A "chain calculation" is a series of two or more operations performed on three or more numbers. To the calculator, the sequence is a series of two-number operations in which the first number is always the result of all calculations completed to that point. For example, in X + Y − Z, the calculator first finds X + Y = A. The quantity A is already in the calculator. All that needs to be done is to subtract Z from it. In the following example we deliberately begin with a negative number to illustrate the way such a number is introduced to the calculator. You simply press in the number, followed by the key that changes the sign: +/− on a TI-30, and CHS on an HP-32E. The example: −2.45 + 18.7 + 0.309 − 24.6 = ?

AOS LOGIC

Press	Display
2.45	2.45
+/−	−2.45
+	−2.45
18.7	18.7
+	16.25
.309	0.309
−	16.559
24.6	24.6
=	−8.041

RPN LOGIC

Press	Display
2.45	2.45
CHS	−2.45
ENTER	−2.4500
18.7	18.7
+	16.2500
.309	0.309
+	16.5590
24.6	24.6
−	−8.0410

Notice that it is not necessary to press the = key after each step *in a chain calculation involving only addition and/or subtraction.*

This display shows four decimals, as it will unless you choose otherwise.

Combinations of multiplication and division are handled the same way. To solve 9.87 × 0.0654 ÷ 3.21:

AOS LOGIC

Press	Display
9.87	9.87
×	9.87
.0654	0.0654
÷	0.645498
3.21	3.21
=	.20108972

RPN LOGIC

Press	Display
9.87	9.87
ENTER	9.8700
.0654	0.0654
×	0.6455
3.21	3.21
÷	.2011

Notice that it is not necessary to press the = key after each step *in a chain calculation involving only multiplication and/or division.*

This display rounds off to four decimals, as it will unless you choose otherwise. Additional decimal numbers may be seen by pressing blue g and holding down MANT.

Combination multiplication/division problems similar to the one above usually appear in the form of fractions in which all multipliers are in the numerator and all divisors are in the denominator. Thus $9.87 \times 0.0654 \div 3.21$ is the same as $\frac{9.87 \times 0.0654}{3.21}$. In chemistry there are often several numerator factors and several denominator factors. A simple calculation that is easily completed "in your head" brings out some important facts about using calculators for chain calculations. Mentally, right now, calculate $\frac{9 \times 4}{2 \times 6} = ?$

There are several ways to get the answer. The most probable one, if you do it mentally, is to multiply $9 \times 4 = 36$ in the numerator, and then multiply $2 \times 6 = 12$ in the denominator. This changes the problem to $\frac{36}{12}$. Dividing 36 by 12 gives 3 for the answer. This perfectly correct approach is often followed by the beginning calculator user when faced with numbers that cannot be multiplied and divided mentally. It is not the best method, however. It is longer and there is greater probability of error than is necessary.

In solving a problem such as $\frac{9 \times 4}{2 \times 6}$ you can begin with any number and perform the required operations with other numbers in any order. Logically you begin with one of the numerator factors. That gives you 12 different calculation sequences that yield the correct answer. They are

$9 \times 4 \div 2 \div 6$	$4 \times 9 \div 2 \div 6$
$9 \times 4 \div 6 \div 2$	$4 \times 9 \div 6 \div 2$
$9 \div 2 \div 6 \times 4$	$4 \div 2 \div 6 \times 9$
$9 \div 2 \times 4 \div 6$	$4 \div 2 \times 9 \div 6$
$9 \div 6 \times 4 \div 2$	$4 \div 6 \times 9 \div 2$
$9 \div 6 \div 2 \times 4$	$4 \div 6 \div 2 \times 9$

Practice a few of these sequences on your calculator to see how freely you may choose.

There is a common error in chain calculations that you should avoid. This is to interpret the above problem as 9×4 divided by 2×6, which is correct, but then punch it into the calculator as $9 \times 4 \div 2 \times 6$, which is not correct. The calculator interprets these instructions as $9 \times 4 = 36$; $36 \div 2 = 18$; $18 \times 6 = 108$. The last step should be $18 \div 6 = 3$, as in the first setup in the list above. *In chain calculations you must always divide by each factor in the denominator.*

There are over a hundred different sequences by which $\frac{7.83 \times 86.4 \times 291}{445 \times 807 \times 0.302}$ can be calculated. Practice some of them and see if you can duplicate the answer, 1.815 214 7.

You have seen that in multiplication and division you can take the factors in any order. This is possible in addition and subtraction too, provided that you keep each positive and negative sign with the number that follows it and treat the problem as an algebraic addition of signed numbers. When you mix addition/subtraction with multiplication/division, however, you must obey the rules that govern the order in which arithmetic operations are performed. Briefly, these rules are:

1. Simplify all expressions enclosed in parentheses.

2. Complete all multiplications and divisions.

3. Complete all additions and subtractions.

If your calculator is able to store and recall numbers, it can solve problems with a very complex order of operations. In this book you will find no such problems, but only those that require the simplest application of the first rule. Our comments will be limited to that application, and we will not use the storage capacity of your calculator, as the instruction book would probably recommend.

A typical calculation is
$$6.02 \times (22.1 - 48.6) \times 0.134.$$
Recalling that factors in a multiplication problem may be taken in any order, you rearrange the numbers so the enclosed factor appears first:
$$(22.1 - 48.6) \times 6.02 \times 0.134.$$
You may then perform the calculation in the order in which the numbers appear:

AOS LOGIC	
Press	Display
22.1	22.1
−	22.1
48.6	48.6
=	−26.5
×	−26.5
6.02	6.02
×	−159.53
.134	0.134
=	−21.37702

RPN LOGIC	
Press	Display
22.1	22.1
ENTER	22.1000
48.6	48.6
−	−26.5000
6.02	6.02
×	−159.5300
.134	0.134
×	−21.3770

Notice that, in a chain calculation involving *both* addition/subtraction *and* multiplication/division, *it is necessary to press the = key after each addition or subtraction sequence, before you proceed to a multiplication/division.*

Sometimes a factor in parentheses appears in the denominator of a fraction, where it is not easily taken as the first factor to be entered into the calculator. Again such problems in this book are relatively simple. They may be solved by working the problem upside down, and at the end using the 1/x key to turn it right side up.

The process is demonstrated in calculating

$$\frac{13.3}{2.59(88.4 - 27.2)}.$$

The procedure is to calculate $\frac{2.59(88.4 - 27.2)}{13.3}$ and find the reciprocal of the result.

AOS LOGIC	
Press	Display
88.4	88.4
−	88.4
27.2	27.2
=	61.2
×	61.2
2.59	2.59
÷	158.508
13.3	13.3
=	11.917895
1/x	.08390744

RPN LOGIC	
Press	Display
88.4	88.4
ENTER	88.4000
27.2	27.2
−	61.2000
2.59	2.59
×	158.5080
13.3	13.3
÷	11.9179
1/x	0.0839

EXPONENTIAL NOTATION

Modern calculators use exponential notation (p. 584) for very large or very small numbers. Ordinarily, if numbers are entered as decimal numbers, the answer appears as a decimal number. If the answer is too large or too small to be displayed, it "overflows" or "underflows" into exponential notation automatically. If you want the answer in exponential notation, even if the calculator can display it as a decimal number, you instruct the machine accordingly. On a TI-30 II you simply introduce the numbers in exponential notation. With an HP-32E you press, in order, gold f, SCI, and the number of decimals you want in the answer, and then proceed in the usual way. (SCI refers to scientific notation, which is the same as exponential notation.)

A number shown in exponential notation has a space in front of the last two digits, which are at the right side of the display. These last two digits are the exponent. Thus 4.68×10^{14} is displayed as 4.68 14; 2.39×10^6 is 2.39 06. If the exponent is negative, a minus sign is present; 4.68×10^{-14} is 4.68 − 14.

To key 4.68×10^{14} and 4.68×10^{-14} into a calculator, proceed as follows:

TI-30 II		HP-32E	
Press	Display	Press	Display
		gold f	*0.0000*
		SCI	*0.0000 00*
		2	*0.00 00*
4.68	*4.68*	4.68	*4.68*
EE	*4.68 00*	EEX	*4.68 00*
14	*4.68 14*	14	*4.68 14*
The calculator display now shows 4.68×10^{14}. Use the next step only if you wish to change to a negative exponent, 4.68×10^{-14}.			
+/−	*4.68 -14*	CHS	*4.68 -14*
Function key: +, −, ×, or ÷	*4.68 14* or *4.68 -14*	ENTER	*4.68 14* or *4.68 -14*

The calculator is now ready for the next number to be keyed in, as in earlier examples.

In the section on exponential notation (p. 584) you will find specific examples showing how to use your calculator to solve problems with these numbers. In Chapter 17 you will combine exponential notation with logarithms.

PART C: ARITHMETIC AND ALGEBRA

We present here a brief review of arithmetic and algebra to the point that it is used or assumed in this text. Formal mathematical statement and development is avoided. The only purpose of this section is to refresh your memory in those areas where it may be needed.

1. ADDITION. a + b. Example: $2 + 3 = 5$. The result of an addition is a **sum**.

2. SUBTRACTION. a − b. Example: $5 − 3 = 2$. Subtraction may be thought of as the addition of a negative number. In that sense, $a − b = a + (−b)$. Example:
$$5 − 3 = 5 + (−3) = 2.$$
The result of a subtraction is a **difference**.

3. MULTIPLICATION. $a \times b = ab =$ $a \cdot b = a(b) = (a)(b) = b \times a = ba = b \cdot a = b(a)$. The foregoing all mean that **factor** a is to be multiplied by factor b. Reversing the sequence of the factors, $a \times b = b \times a$, indicates that factors may be taken in any order when two or more are multiplied together. Examples:
$$2 \times 3 = 2 \cdot 3 = 2(3) = (2)(3) =$$
$$3 \times 2 = 3 \cdot 2 = 3(2) = 6.$$
The result of a multiplication is a **product**.

Grouping of factors: $(a)(b)(c) = (ab)(c) = (a)(bc)$. Factors may be grouped in any way in multiplication. Example:
$$(2)(3)(4) = (2 \times 3)(4) = (2)(3 \times 4) = 24.$$

Multiplication by 1: $n \times 1 = n$. If any number is multiplied by 1, the product is the original number. Examples:

$$6 \times 1 = 6; \qquad 3.25 \times 1 = 3.25.$$

Multiplication of fractions:

$$\frac{a}{b} \times \frac{c}{d} \times \frac{e}{f} = \frac{ace}{bdf}$$

If two or more fractions are to be multiplied, the product is equal to the product of the numerators divided by the product of the denominators. Example:

$$4 \times \frac{9}{2} \times \frac{1}{6} = \frac{4}{1} \times \frac{9}{2} \times \frac{1}{6}$$
$$= \frac{4 \times 9 \times 1}{1 \times 2 \times 6} = \frac{36}{12} = 3$$

4. DIVISION. $a \div b = a/b = \frac{a}{b}$. The foregoing all mean that a is to be divided by b. Example: $12 \div 4 = 12/4 = \frac{12}{4} = 3$. The result of a division is a **quotient.**

Special case: If any number is divided by the same or an equal number, the quotient is equal to 1. Examples: $\frac{4}{4} = 1$; $\frac{8-3}{4+1} = \frac{5}{5} = 1$; $\frac{n}{n} = 1$.

Division by 1. $\frac{n}{1} = n$. If any number is divided by 1, the quotient is the original number. Examples: $\frac{6}{1} = 6$; $\frac{3.25}{1} = 3.25$. From this it follows that any number may be expressed as a fraction having 1 as the denominator. Examples: $4 = \frac{4}{1}$; $9.12 = \frac{9.12}{1}$; $m = \frac{m}{1}$.

5. RECIPROCALS. If n is any number, the reciprocal of n is $\frac{1}{n}$; if $\frac{a}{b}$ is any fraction, the reciprocal of $\frac{a}{b}$ is $\frac{b}{a}$. The first part of the foregoing sentence is actually a special case of the second part: If n is any number, it is equal to $\frac{n}{1}$. Its reciprocal is therefore $\frac{1}{n}$.

A reciprocal is sometimes referred to as the **inverse** (more specifically, the multiplicative inverse) of a number. This is because the product of any number multiplied by its reciprocal equals 1. Examples:

$$2 \times \frac{1}{2} = \frac{2}{2} = 1 \qquad n \times \frac{1}{n} = \frac{n}{n} = 1$$
$$\frac{4}{3} \times \frac{3}{4} = \frac{12}{12} = 1 \qquad \frac{m}{n} \times \frac{n}{m} = \frac{mn}{mn} = 1$$

Division may be regarded as multiplication by a reciprocal:

$$a \div b = \frac{a}{b} = a \times \frac{1}{b}.$$

Example: $6 \div 2 = \frac{6}{2} = 6 \times \frac{1}{2} = 3$.

$$a \div b/c = \frac{a}{b/c} = a \times \frac{c}{b}.$$

Example: $6 \div \frac{2}{3} = \frac{6}{2/3} = 6 \times \frac{3}{2} = 9$.

6. SUBSTITUTION. If $d = b + c$, then $a(b + c) = ad$. Any number or expression may be substituted for its equal in any other expression. Example: $7 = 3 + 4$. Therefore $2(3 + 4) = 2 \times 7$.

7. "CANCELLATION". $\frac{ab}{ca} = \frac{\cancel{a}b}{c\cancel{a}} = \frac{b}{c}$. The process commonly called **cancellation** is actually a combination of grouping of factors (see 3), substitution (see 6) of 1 for a number divided by itself (see 4), and multiplication by 1 (see 3). Note the steps in the following examples:

$$\frac{xy}{yz} = \frac{yx}{yx} = \left(\frac{y}{y}\right)\left(\frac{x}{z}\right) = 1 \cdot \frac{x}{z} = \frac{x}{z}$$
$$\frac{24}{18} = \frac{6 \times 4}{6 \times 3} = \frac{6}{6} \times \frac{4}{3} = 1 \times \frac{4}{3} = \frac{4}{3}$$

Note that only *factors*, or *multipliers*, can be canceled. There is no cancellation in $\frac{a+b}{a+c}$.

8. EXPONENTIALS. An exponential has the form B^p, where B is the **base** and p is the **power** or **exponent**. An exponential indicates the number of times the base is used as a factor in multiplication. For example, 10^3 means 10 is to be used as a factor 3 times:

$$10^3 = 10 \times 10 \times 10 = 1000.$$

Raising a product to a power: $(ab)^n = a^n \times b^n$. When the product of two or more factors is raised to some power, each factor is raised to that power. Example:

$$(2 \times 5y)^3 = 2^3 \times 5^3 \times y^3$$
$$= 8 \times 125 \times y^3 = 1000y^3$$

Raising a fraction to a power: $\left(\dfrac{a}{b}\right)^n = \dfrac{a^n}{b^n}$.
When a fraction is raised to some power, the numerator and denominator are both raised to that power. Example:

$$\left(\frac{2x}{5}\right)^3 = \frac{2^3x^3}{5^3} = \frac{8x^3}{125} = 0.064x^3$$

Multiplication of exponentials having the same base: $a^m \times a^n = a^{m+n}$. To multiply exponentials, add the exponents. Example:
$$10^3 \times 10^4 = 10^7.$$

Division of exponentials having the same base: $a^m \div a^n = \dfrac{a^m}{a^n} = a^{m-n}$. To divide exponentials, subtract the denominator exponent from the numerator exponent. Example:

$$10^7 \div 10^4 = \frac{10^7}{10^4} = 10^{7-4} = 10^3.$$

Zero power: $a^0 = 1$. Any base raised to the zero power equals 1. The fraction $\dfrac{a^m}{a^m} = 1$ because the numerator is the same as the denominator. By division of exponentials, $\dfrac{a^m}{a^m} = a^{m-m} = a^0$.

Square root of exponentials: $\sqrt{a^{2n}} = a^n$. To find the square root of an exponential, divide the exponent by 2. Example: $\sqrt{10^6} = 10^3$. If the exponent is odd, see below.

Square root of a product: $\sqrt{ab} = \sqrt{a} \times \sqrt{b}$. The square root of the product of two numbers equals the product of the square roots of the numbers. Example:
$$\sqrt{9 \times 10^{-6}} = \sqrt{9} \times \sqrt{10^{-6}} = 3 \times 10^{-3}.$$

Using this principle, by adjusting a decimal point you may take the square root of an exponential having an odd exponent. Example:

$$\sqrt{10^5} = \sqrt{10 \times 10^4} = \sqrt{10} \times \sqrt{10^4}$$
$$= 3.16 \times 10^2$$

The same technique may be used in taking the square root of a number expressed in exponential notation. Example:

$$\sqrt{1.8 \times 10^{-5}} = \sqrt{18 \times 10^{-6}}$$
$$= \sqrt{18} \times \sqrt{10^{-6}}$$
$$= 4.2 \times 10^{-3}$$

9. SOLVING AN EQUATION FOR AN UNKNOWN QUANTITY: Most problems in this book can be solved by dimensional analysis methods. There are times, however, when algebra

must be used, particularly in relation to the gas laws. Solving an equation for an unknown involves rearranging the equation so that the unknown is the only item on one side and only known quantities are on the other. "Rearranging" an equation may be done in several ways, but the important thing is that *whatever is done to one side of the equation must also be done to the other*. The resulting relationship remains an equality, a true equation. Among the operations that may be performed on both sides of an equation are addition, subtraction, multiplication, division, and taking square root.

In the following examples, a, b, and c represent known quantities, and x is the unknown. The object in each case is to solve the equation for x. The steps of the algebraic solution are shown, as well as the operation performed on both sides of the equation. Each example is accompanied by a practice problem that is solved by the same method. You should be able to solve the problem, even if you have not yet reached that point in the book where such a problem is likely to appear. Answers to these practice problems may be found on page 593.

(1) $\quad x + a \quad = b$

$\quad\quad x + a - a = b - a \quad$ Subtract a.

$\quad\quad\quad\quad x \quad = b - a \quad$ Simplify.

PRACTICE: 1) If $P = p_{O_2} + p_{H_2O}$, find p_{O_2}, when $P = 748$ torr and $p_{H_2O} = 24$ torr.

(2) $\quad ax = b$

$\quad\quad \dfrac{\cancel{a}x}{\cancel{a}} = \dfrac{b}{a} \quad$ Divide by a which is called the **coefficient** of x.

$\quad\quad x = \dfrac{b}{a} \quad$ Simplify.

PRACTICE: 2) At a certain temperature, $PV = k$. If $P = 1.23$ atm and $k = 1.62$ L atm, find V.

(3) $\quad \dfrac{x}{a} = b$

$\quad\quad \dfrac{\cancel{a}x}{\cancel{a}} = ba \quad$ Multiply by a.

$\quad\quad x = ba \quad$ Simplify.

PRACTICE: 3) In a fixed volume, $\dfrac{P}{T} = k$. Find P if $k = \dfrac{2.4\ torr}{K}$ and $T = 300\ K$.

(4) $\dfrac{a}{x} = b$

$\dfrac{x a}{\not x} = bx$ Multiply by x.

$a = bx$ Simplify.

$x = \dfrac{a}{b}$ Divide by b.

PRACTICE: 4) *At what value of T will P =
760 torr when* $k = \dfrac{2.4\ torr}{K}$ *and* $\dfrac{P}{T} = k$?

(5) $\dfrac{a}{x} = \dfrac{b}{c}$

$\dfrac{cx}{b} \cdot \dfrac{a}{\not x} = \dfrac{\not b}{\not c} \cdot \dfrac{\not c x}{\not b}$ Multiply by $\dfrac{cx}{b}$.

$\dfrac{ac}{b} = x$ Simplify.

PRACTICE: 5) *For gases at constant volume,*
$\dfrac{P_1}{T_1} = \dfrac{P_2}{T_2}$. *If* $P_1 = 0.80\ atm$ *at* $T_1 = 320\ K$, *at
what value of* T_2 *will* $P_2 = 1.00\ atm$?

(6) $\dfrac{a}{b + x} = c$

$(b + x)\dfrac{a}{(b + x)} = c(b + x)$ Multiply by
$(b + x)$.

$a = c(b + x)$ Simplify.

$\dfrac{a}{c} = \dfrac{\not c(b + x)}{\not c}$ Divide by c.

$\dfrac{a}{c} = b + x$ Simplify.

$\dfrac{a}{c} - b = x$ Subtract b.

PRACTICE: 6) *In how many grams of water
must you dissolve 20.0 g of salt to make a 25%
solution? The formula is*

$$\dfrac{g\ salt}{g\ salt\ +\ g\ water} \times 100 = \%;\ or$$

$$\dfrac{g\ salt}{g\ salt\ +\ g\ water} = \dfrac{\%}{100}$$

PART D: PROPORTIONALITIES

It is quite common, both in the laboratory and
out, to find two quantities that are related to each
other in a constant way. The weight of milk you
carry out of a grocery store depends upon the
number of half-gallon cartons you buy. Four cartons weigh twice as much as two, and three weigh
half as much as six. Mathematically we say that
the weight is **directly proportional** to the number
of cartons. Using W for weight, C for number of
cartons, and the symbol $\propto$ for "is proportional
to," the proportionality may be written

$$W \propto C \qquad (AP.1)$$

A proportionality may be changed to an equation
by introducing a **proportionality constant**, k:

$$W = kC \qquad (AP.2)$$

Quite often a proportionality constant has
units and physical significance. It does here. Solving Equation AP.2 for k gives

$$k = \dfrac{W}{c} \qquad (AP.3)$$

Equation AP.3 shows the "constant way" in
which weight is related to the number of cartons:
the ratio between them is constant. If three cartons weigh 14.4 pounds, five cartons weigh 24.0
pounds, and eight cartons weigh 38.4 pounds, the
ratios of pounds to cartons is

$$\dfrac{38.4\ lb}{8\ cartons} = \dfrac{24.0\ lb}{5\ cartons} = \dfrac{14.4\ lb}{3\ cartons},$$

which all reduce to $\dfrac{4.8\ lb}{1\ carton}$. Thus the units of
k and its physical significance are 4.8 pounds per
carton.

Because the ratio of weight to number of
cartons is constant, the ratios can be set equal to
each other and the equation can be used to find
the weight of any number of cartons. If you want
to know the weight of 7 cartons, you can set any

known ratio equal to $\frac{x}{7}$, where x is the required weight, and solve. Using $\frac{24.0 \text{ lb}}{5 \text{ cartons}}$,

$$\frac{24.0 \text{ lb}}{5 \text{ cartons}} = \frac{x \text{ lb}}{7 \text{ cartons}}$$

$$x \text{ lb} = \frac{24.0 \text{ lb} \times 7 \text{ cartons}}{5 \text{ cartons}} = 33.6 \text{ pounds}$$

This result is easily checked by reason. If each carton weighs 4.8 pounds, then 7 cartons must weigh $7 \times 4.8 = 33.6$ pounds.

Problems set up and solved this way are called "ratio and proportion" problems. This approach can be used in chemistry, and once was quite popular. Now dimensional analysis (p. 31) is much more common because most students find it easier to understand and to set up. Complete with units, the dimensional analysis setup of the above problem is identical to that reached by ratio and proportion. Dimensional analysis is emphasized in this book.

It can be shown that whenever one variable is proportional to two others, it is also proportional to their product. The number of acres, A, in a rectangular field is proportional to both the length, L, and width, W, of the field:

$$A \propto L \quad \text{and} \quad A \propto W; \quad \text{therefore,} \quad A \propto L \times W$$

Introducing a proportionality constant, k',

$$A = k'LW$$

Solving for k',

$$k' = \frac{A}{LW}$$

which has the physical meaning of acres per square foot. The value of k' is $\frac{1}{43\,560}$, which is the inverse of the number of square feet in an acre.

In a direct proportionality, if one variable rises, the other rises too. An **inverse proportionality** is exactly the opposite; as one variable increases, the other decreases. A common example is the time it takes to drive from one point to another at different speeds. At 40 miles per hour (mph) it will take half as long as at 20 mph. As the speed is doubled, the time is cut in half. At 60

mph—three times as fast as 20 mph—the time is one-third. If r is speed and t is time, the inverse proportionality is written

$$r \propto \frac{1}{t}$$

This time we will use d for the proportionality constant:

$$r = d\left(\frac{1}{t}\right)$$

Solving for d gives

$$d = rt \qquad \text{(AP.4)}$$

which you may recognize as the familiar distance = rate × time equation. The physical significance of d in Equation AP.4 is distance. Notice that in an inverse proportionality it is the product of the variables that is constant.

Not all proportionality constants have units and significance. The universal gas constant, R (p. 280), is made up of a combination of units that have no apparent meaning. A very familiar proportionality constant comes from the proportionality between the circumference and diameter of a circle: $c \propto d$. The constant is π, as in the equation $c = \pi d$. Solving for π gives a ratio of lengths. The units cancel, leaving a dimensionless, or pure, number.

Summary. Most of what appears here is background information to help you understand proportionality relationships as they appear in the text. The important things to recognize are:

1. A mathematical statement of proportionality may be changed to an equation by introducing a proportionality constant.

2. A proportionality constant may have units and physical meaning, which can be determined by solving the equation for the constant.

3. If one variable is proportional to two other variables, it is proportional to the product of those variables.

4. If A is *directly* proportional to B, then the ratio A/B is constant. If A increases, B increases; if A decreases, B decreases.

5. If A is *inversely* proportional to B, then the product A × B is constant. If A increases, B decreases; if A decreases, B increases.

PART E: PER AND PERCENT

If you drive 100 miles in 4 hours, what is your average speed? Twenty-five miles per hour, you say. Fine. But how did you get it? Was it by *dividing* 100 by 4? In other words, 100 miles *per* 4 hours is the same as 25 miles *per* 1 hour, or $100 \div 4 = 25$, or $\frac{100}{4} = \frac{25}{1}$. The results of many division problems are expressed as some quantity *per unit*—per *one*—or something else. Thus the term *per* means *division*. Finally, *per* represents an equivalence. As long as speed is 25 miles per hour, 25 miles $\simeq$ 1 hour.

A familiar use of *per* is in **percent**. *Cent-* is an expression that refers to 100: there are 100 *centimeters* in a meter, and 100 *cents* in a dollar. Percent is commonly used to state the amount of one part of a mixture in 100 total parts of the mixture—parts per 100 parts total. If 78 of 100 cars in a parking lot are made in the United States, 78% are American made.

Suppose a laboratory analyzes 761 grams of sea water and finds that it contains 21.3 grams of salt. What percentage is salt? According to the data, there are 21.3 grams of salt per 761 grams of sea water, or $\frac{21.3 \text{ g salt}}{761 \text{ g sea water}}$. If X is the grams of salt in 100 grams of sea water—X grams of salt per 100 grams of sea water, or $\frac{X \text{ g salt}}{100 \text{ g sea water}}$ then, by definition, X is the percentage of sea water that is salt. The ratio of salt to sea water is the same regardless of the size of the sample, so the two ratios can be set equal to each other:

$$\frac{X \text{ g salt}}{100 \text{ g sea water}} = \frac{21.3 \text{ g salt}}{761 \text{ g sea water}} \qquad (AP.5)$$

Solving this "ratio and proportion" problem (p. 582) for X gives

$$X = \frac{21.3 \text{ g salt}}{761 \text{ g sea water}} \times 100 \text{ g sea water} \qquad (AP.6)$$
$$= 2.80\% \text{ salt}$$

The form of this equation shows how the percentage of one part, A, of any mixture may be found:

$$\% \text{ of A} = \frac{\text{parts of A}}{100 \text{ parts total}} \times 100 \qquad (AP.7)$$

PRACTICE: 7) During a certain period, 89 Chinook salmon, 14 steelhead trout, and 21 fish of other kinds passed through a counting station in a fish ladder bypass around a hydro-electric dam. Calculate the percentage of Chinook salmon.*

In addition to methods you have already learned, percentage problems can be solved by dimensional analysis. Percentage establishes an equivalence between a given number of one part in a mixture and 100 total parts in the mixture. If sea water is 2.80% salt, then

$$2.80 \text{ grams salt} \simeq 100 \text{ grams sea water} \qquad (AP.8)$$

This relationship can be used to calculate the number of grams of salt in 5000 grams of sea water:

$$5000 \text{ g sea water} \times \frac{2.80 \text{ g salt}}{100 \text{ g sea water}}$$
$$= 140 \text{ g salt}$$

The usual way to solve this problem is to calculate 2.80% of 5000. The percentage is changed to a decimal fraction by dividing by 100 (moving the decimal point two places left) and then multiplying by 5000: $0.028 \times 5000 = 140$.

PRACTICE: 8) 84% of the registered voters in a certain precinct cast ballots in a recent election. If there are 1482 registered voters in the district, how many ballots were cast?

The dimensional analysis approach to percentage problems is even more useful in the reverse direction. For example, a state-supported university establishes a quota such that 27% of the freshmen admitted in a certain year will be from another state. If 2891 incoming freshmen were from out of state, what was the total number of freshmen enrolled that year? Dimensional analysis interprets 27% as 27 out of state students (OSS) $\simeq$ 100 students. Applied to the given quantity,

$$2891 \text{ OSS} \times \frac{100 \text{ students}}{27 \text{ OSS}} = 10707 \text{ students}$$

*Answers to practice problems are on page 593.

The algebraic approach to this problem is

$$27\% \text{ of total students} = \text{OSS}$$

$$0.27 \times \text{total students} = 2891$$

$$\text{total students} = \frac{2891}{0.27} = 10\,707$$

PRACTICE: 9) The total number of pages in a monthly magazine is planned on the number of advertising pages sold. If 58 pages of advertising sold for a forthcoming issue are to represent 46% of the pages in the magazine, what will be the total number of pages?

PART F: EXPONENTIAL NOTATION

EXPRESSING NUMBERS IN EXPONENTIAL NOTATION

Problems in chemistry sometimes involve extremely large or very small numbers. For example, the mass of a single atom of hydrogen is 0.000 000 000 000 000 000 000 000 001 66 gram. In one liter of hydrogen, measured at one atmosphere pressure and 0°C, there are 53 770 000-000 000 000 000 000 hydrogen atoms.

Very large and very small numbers can be written in the form $C \times 10^n$, where C is called the **coefficient** and 10^n is an exponential. The exponent n may be a positive or negative integer. This form is called **exponential notation** or **scientific notation**. In **standard** exponential notation the coefficient is always equal to or greater than 1, but less than 10.* Unless there is reason for doing otherwise, C is written in that range.

To change an ordinary number to exponential notation, you first count the number of places, n, that the decimal point must be moved so it will appear immediately after the first nonzero digit. This number is the size of the exponent. If the decimal point in the original number is moved to the left, the exponent is positive; if the decimal moves right, the exponent is negative. The rule, left-positive, right-negative, is easily learned, but just as easily reversed in one's memory. A more certain way is to think of the original number multiplied by 1 in the form of 10^0. Two examples show the thought process:

$$1234 = 1234 \times 10^0$$
$$= 1.234 \times 10^3$$

$$0.000\,567\,8 = 0.000\,567\,8 \times 10^0$$
$$= 5.678 \times 10^{-4}$$

In the first case, the coefficient became smaller (1234 became 1.234), while the exponent became larger, or *more positive* (0 became 3). In the second case, the coefficient was made larger (0.000 567 8 became 5.678), while the exponent was made smaller, or *more negative* (0 became −4). One factor always becomes larger, and the other becomes smaller.

*PRACTICE: 10) Write each of the following numbers in exponential notation:**

10a) 3 762 199   *10c)* 0.000 098   *10e)* 0.004 60
10b) 198.75   *10d)* 10.080

In converting a number expressed in exponential notation to an ordinary decimal number, the **absolute value** of the exponent (its *numerical value, regardless of sign*) tells you how many places the decimal point must be moved. If the exponent is positive, the number is larger than 1, so the decimal is moved to the right; if the exponent is negative, the number is smaller than 1, so the decimal goes to the left. For example, in 7.89×10^5 the exponent is positive. This indicates a large number, so the decimal is moved five

---

*Some books use the term *exponential* notation if C is not in the range from 1 to less than 10, and *scientific* notation if C is in the standard range.

*Answers to practice problems are on page 594.

places to the right, the same number of places as the value of the exponent: 789 000. In 5.37×10^{-4} the negative exponent identifies a small number, so the decimal point is moved four places to the left: 0.000 537.

MULTIPLICATION WITH EXPONENTIAL NOTATION

In multiplying numbers expressed in exponential notation, the factors are rearranged, placing all the coefficients in one group and all the exponentials in a second group. The two groups are multiplied separately. The decimal result and the exponential result are then combined as the exponential notation expression of the product. For example:

$(3.96 \times 10^4)(5.19 \times 10^{-7})$
$$= 3.96 \times 10^4 \times 5.19 \times 10^{-7}$$
$$= 3.96 \times 5.19 \times 10^4 \times 10^{-7}$$
$$= (3.96 \times 5.19)(10^4 \times 10^{-7})$$
$$= 20.6 \times 10^{4-7}$$
$$= 20.6 \times 10^{-3}$$

To make the coefficient between 1 and 10, the decimal point must be moved one place to the left. The coefficient becomes smaller (20.6 becomes 2.06), so the exponent must become larger, or more positive. The exponent must increase from -3 to -2. Be careful when changing negative exponents. Although 3 is larger than 2, *minus* 3 is smaller than *minus* 2. The more positive—or the less negative—the exponent, the larger the exponential.

If your calculator works in exponential notation, you will no doubt use it to solve problems like the one above. As noted in the section on calculators, if you have an HP-32E calculator, you must instruct it to display the answer in exponential notation and select the number of decimals to be shown. You do this by pressing, in order, gold f, SCI, and the number. We will assume that you have pressed 4. The steps for solving the problem $(3.96 \times 10^4)(5.19 \times 10^{-7})$ are:

AOS LOGIC			RPN LOGIC		
Press	Display		Press	Display	
3.96	*3.96*		3.96	*3.96*	
EE	*3.96 00*		EEX	*3.96 00*	
4	*3.96 04*		4	*3.96 04*	
×	*3.96 04*		ENTER	*3.96 00 04*	
5.19	*5.19*		5.19	*5.19*	
EE	*5.19 00*		EEX	*5.19 00*	
7	*5.19 07*		7	*5.19 07*	
+/−	*5.19−07*		CHS	*5.19−07*	
=	*2.0552−02*		×	*2.0552−02*	

PRACTICE: 12) *Perform each of the following calculations and express the result in exponential notation:*

a) $(3.26 \times 10^4)(1.54 \times 10^6) =$
b) $(8.39 \times 10^{-7})(4.53 \times 10^9) =$
c) $(6.73 \times 10^{-3})(9.11 \times 10^{-3}) =$
d) $(2.93 \times 10^5)(4.85 \times 10^6)(5.58 \times 10^{-3}) =$

e) $(35\ 780)(761.9) =$
f) $(0.000\ 91)(235.6) =$
g) $(0.0890)(0.000\ 000\ 726) =$

DIVISION WITH EXPONENTIAL NOTATION

Division of numbers expressed in exponential notation is done the same way as multiplication. The coefficients and exponential factors are separated into two groups; the value of each group is calculated separately; and finally the decimal result is combined with the exponential result as the exponential notation form of the quotient. To illustrate,

$$3.96 \times 10^4 \div 5.19 \times 10^{-7} = \frac{3.96 \times 10^4}{5.19 \times 10^{-7}}$$

$$= \left(\frac{3.96}{5.19}\right)\left(\frac{10^4}{10^{-7}}\right)$$

$$= 0.763 \times 10^{11}$$

$$= 7.63 \times 10^{10}$$

PRACTICE: 13) Perform each of the following calculations and express the result in exponential notation.

a) $\dfrac{8.94 \times 10^6}{4.35 \times 10^4} =$

b) $\dfrac{5.08 \times 10^{-3}}{7.23 \times 10^{-5}} =$

c) $\dfrac{(3.05 \times 10^{-6})(2.19 \times 10^{-3})}{5.48 \times 10^{-5}} =$

d) $\dfrac{(867)(7.2 \times 10^{-7})}{0.000\,000\,000\,927} =$

e) $\dfrac{(7180)(9\,420\,000)}{(258\,000)(0.000\,618)} =$

ADDITION AND SUBTRACTION IN EXPONENTIAL NOTATION

As in arithmetic, addition and subtraction of exponentials require that digit values (hundreds, units, tenths, etc.) be aligned vertically. This may be achieved by adjusting coefficients and exponents so the exponents of all numbers are the same, and then proceeding as in ordinary arithmetic. (This adjustment is not necessary on a calculator. The numbers are keyed in just as they appear in the problem.) Shown below is the addition of $6.44 \times 10^{-7} + 1.3900 \times 10^{-5}$ in three ways: with exponents of 10^{-5}, with exponents of 10^{-7}, and as decimal numerals.

$$\begin{aligned}1.3900 &\times 10^{-5} \\ 0.0644 &\times 10^{-5} \\ \hline 1.4544 &\times 10^{-5}\end{aligned}$$

$$\begin{aligned}139.00 &\times 10^{-7} \\ 6.44 &\times 10^{-7} \\ \hline 145.44 &\times 10^{-7}\end{aligned}$$

$$\begin{aligned}0.000\,0139\,00 \\ 0.000\,0006\,44 \\ \hline 0.000\,0145\,44\end{aligned}$$

PRACTICE: 14) Add or subtract the following numbers:

a) $5.2 \times 10^5 + 3.98 \times 10^4 =$
b) $3.971 \times 10^2 + 1.98 \times 10^{-1} =$
c) $5.63 \times 10^4 + 2.93 \times 10^3 + 5.42 \times 10^5 =$
d) $1.05 \times 10^{-4} - 9.7 \times 10^{-5} =$
e) $\dfrac{9.02 \times 10^7}{265.9} - \dfrac{7.89 \times 10^3}{2.37 \times 10^{-2}} =$

PART G: ESTIMATING CALCULATION RESULTS

On page 573 it was suggested that you should estimate calculation results to be sure they are reasonable before writing them down. There is no single "right" way to estimate an answer. As your mathematical skills grow, you will develop techniques that are best for you. You will also find that one method works best on one kind of problem, and another method on another problem.

The ideas that follow should help you get started in this important practice.

In general, estimating a calculated result involves rounding off the given numbers and calculating the answer mentally. For example, if you multiply 325 by 8.36 on your calculator, you will get 2717. To see if this answer is reasonable, you might round off 325 to 300, and 8.36 to 8. The problem then becomes 300×8, which you can calculate mentally to 2400. This is reasonably close to your calculator answer. Even so, you should run your calculator again to be sure you haven't made a small "typing" error.

If your calculator answer for the above problem had been 1254.5, or 38.975 598, or 27 170, your estimated 2400 would signal that an error was made. These three numbers represent common calculation errors. The first comes from a mistake that may arise any time a number is transferred from one position (your paper) to another (your calculator). It is called transposition, and appears when two numerals are changed in position. In this case, 325×3.86 (instead of 8.36) $= 1254.5$.

The answer 38.875 598 comes from pressing the wrong function key: $325 \div 8.36 = 38.875\ 598$. This answer is so unreasonable—$325 \times 8.36 =$ about 38!—that no mental arithmetic should be necessary to tell you it is wrong. But, like molten idols, calculators speak to some students with a mystic authority they would never dare to challenge. On one occasion neither student nor teacher could figure out why or how, on a test, the student used a calculator to divide 428 by 0.01, and then wrote down 7290 for the answer, when all he had to do was move the decimal two places!

The answer 27 170 is a decimal error, as might arise from transposing the decimal point and a number, or putting an incorrect number of zeros in numbers like 0.001 23 or 123 000. Decimal errors are also apt to appear through an incorrect use of exponential notation, either with or without a calculator.

Exponential notation is a valuable aid in estimating results. For example, in calculating $41\ 300 \times 0.0524$, you can regard both numbers as falling between 1 and 10 for a quick calculation of the coefficient: $4 \times 5 = 20$. Then, thinking of the exponents, changing 41 300 to 4 moves the decimal four places left, so the exponential is 10^4. Changing 0.0524 to 5 has the decimal moving two places right, so the exponential is 10^{-2}. Adding exponents gives $4 + (-2) = 2$. The estimated answer is 20×10^2, or 2000. On the calculator it comes out to 2164.12.

Another "trick" that can be used is to move decimals in such a way that the moves cancel and at the same time the problem is simplified. In $41\ 300 \times 0.0524$, the decimal in the first factor can be moved two places left (divide by 100), and in the second factor two places right (multiply by 100). Dividing and multiplying by 100 is the same as multiplying by $\frac{100}{100}$, which is equal to 1. The problem simplifies to 413×5.24, which is easily estimated as $400 \times 5 = 2000$. The same technique can be used to simplify fractions, too. $\frac{371\ 000}{6240}$ can be simplified to $\frac{371}{6.24}$ by moving the decimal three places left in both numerator and denominator, which is the same as multiplying by 1 in the form $\frac{0.001}{0.001}$. An estimated $\frac{360}{6}$ give 60 as the approximate answer. The calculator answer is 59.455 128. Similarly, $\frac{0.000\ 406}{0.000\ 839}$ becomes about $\frac{4}{8}$, or 0.5; by calculator, the answer is 0.483 904 2.

On pages 585–586 there are some practice problems (12 and 13) for exponential notation. Try estimating the answers to some of those problems, and then compare them with the calculated answers on page 594. Problems 12d, 13c, and 13e are good for this purpose, and use others if they will be helpful.

PART H: SIGNIFICANT FIGURES

UNCERTAINTY IN MEASUREMENT

Every physical measurement has some uncertainty associated with it. This is evident as you attempt to measure the length of a board with a series of meter sticks that are graduated in smaller and smaller subdivisions, as shown in Figure AP-3. In Figure AP-3A there are no graduations. Looking at the top illustration alone, the best that can be done is to estimate the length of the

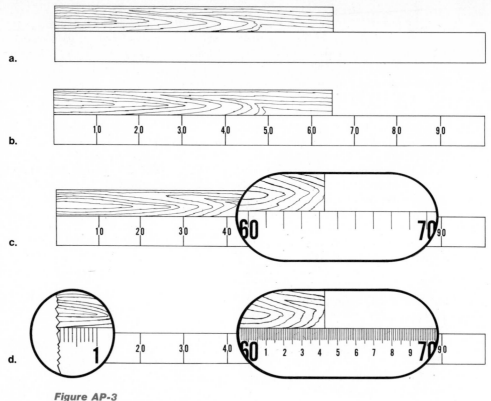

Figure AP-3
Significant figures in a measurement.

board as 0.6 or 0.7 meters. The number of tenths of a meter is uncertain. Uncertainty is often added to a measurement as a "plus or minus." In this case the length of the board might be recorded as 0.6 ± 0.1 meter, or 0.7 ± 0.1 meter.

In Figure AP-3B the meter stick is graduated in tenths of a meter (decimeters), but numbered in centimeters. It is clear that the board is between 0.6 and 0.7 meters, or between 60 and 70 centimeters. Can you estimate the next digit? The end of the board is almost halfway between 60 and 70 cm, but it is hard to tell if it is closer to 64 or 65. Either number is reasonable, so we recognize that the number of centimeters is uncertain. The reading might be recorded as 64 ± 1 cm or 65 ± 1 cm. If the length is recorded in meters, the numerals are the same, but the decimal is two places to the left: 0.64 ± 0.01 m or 0.65 ± 0.01 m. The uncertain digit, the 4 or 5 in either expression of the length, is sometimes called the **doubtful digit**.

If the measuring stick is graduated in centimeters, as in Figure AP-3C, the length of the board may be estimated to the next decimal place. 64.2 ± 0.1 or 64.3 ± 0.1 cm, or 0.642 ± 0.001 or 0.643 ± 0.001 m, are acceptable measurements. The doubtful digit is now the 2 or 3. Extending graduations to millimeters in Figure AP-3D lets us select 64.3 cm as the better reading, but roughness at the ends of the board and meter stick plus the thickness of the graduation lines make it impractical to estimate tenths of a millimeter. To record such a distance would be misleading, suggesting measuring precision that simply is not present.

It is important in scientific work to make accurate measurements and to record them correctly. It is also important to know how reliable a measurement is—that is, how large an uncertainty is present. One way to accomplish this, and the way used in this book, is to record each measurement so the last digit written is the first doubtful

Table AP-1

FIGURE AP-3	READING	DOUBTFUL DIGIT VALUE	SIGNIFICANT FIGURES
A	0.6 ± 0.1 m	Tenths	1
B	0.64 ± 0.01 m	Hundredths	2
	64 ± 1 cm	Units	2
C & D	0.643 ± 0.001 m	Thousandths	3
	64.3 ± 0.1 cm	Tenths	3

digit—the uncertain digit. This was done in each of the above measurements, which are summarized in Table AP-1. In Figure AP-3A the uncertain digit is *tenths* of a meter, and the length is recorded as six *tenths* of a meter. In Figure AP-3B, uncertainty is one centimeter, and the last digit shown is the number of whole centimeters.

In Figure AP-3C and AP-3D, where uncertainty is in tenths of a centimeter, the recorded measurement ends in tenths of a centimeter. If this rule is applied consistently, it is not necessary to write the uncertainty when recording a measurement. It is simply understood that *all measurements are doubtful in the last digit shown.*

COUNTING SIGNIFICANT FIGURES

The name that is given to this method of recording measurements is **significant figures.** We speak of the number of significant figures in a measurement. This number is set by the following rules:

1. *The number of significant figures in any quantity is the number of digits that are known accurately plus one that is doubtful.*

2. *The number of significant figures in a quantity is determined by uncertainty in measurement processes, not by the units in which the quantity is expressed.*

3. *If a quantity is properly expressed in terms of significant figures, the doubtful digit is the last digit shown.*

4. *To count the number of significant figures in a quantity, begin with the first nonzero digit and end with the doubtful digit—the last digit shown.*

   Using the above rules when recording measurements and counting significant figures sometimes becomes confusing when zeros are concerned. For example, if you measure a distance with a meter stick and find it to be 75.0 ± 0.1 cm, how do you write the number? There is a strong temptation to write 75 cm because it is the same thing as 75.0 cm. But 75 is not

correct. It takes 75.0 to indicate that the measurement is accurate to the closest 0.1 cm, whereas 75 suggests a less precise measuring method with an uncertainty of a whole centimeter. When a tail-end zero to the right of the decimal is the doubtful digit, that zero *must be written.* Always, ". . . the doubtful digit is the last digit shown" (Rule 3, above).

   The measurement process illustrated in Figures AP-3C and AP-3D yielded answers of 0.643 ± 0.001 m and 64.3 ± 0.1 cm. Both are three significant figure numbers. In both cases, counting begins with the first nonzero digit, 6, and ends with 3, the doubtful digit. The same measurement expressed in kilometers is 0.000 643 ± 0.000 001. This is still a three significant figure number. It must be; changing the unit in which the measurement is expressed does not change the measuring process (Rule 2 above). This illustrates an important fact: *the location of a decimal point has nothing to do with significant figures.* In particular, begin counting significant figures at the first nonzero digit, *not at the decimal point.*

   If 0.643 ± 0.001 m were to be expressed in micrometers the number would be 643 000 ± 1000 μm. If you drop this uncertainty and show 643 000 μm, Rule 3 indicates that the measurement is accurate to ±1 μm. This obviously is not correct. In fact, if you simply look at 643 000 μm, there is no way to tell if the uncertainty lies in thousands, hundreds, tens, or units. The way to get around this is to write the number in exponential notation (p. 584). This puts the doubtful digit to the right of the decimal. Thus, 643 000 to three, four, five, and six significant figures becomes.

. . . 6.43 × 10⁵—doubtful digit in thousands
(three significant figures)

. . . 6.430 × 10⁵—doubtful digit in hundreds
(four significant figures)

. . . 6.4300 × 10⁵—doubtful digit in tens
(five significant figures)

. . . 6.430 00 × 10⁵—doubtful digit in ones
(six significant figures)

If you find the significant figure rules for large and small numbers confusing—and many students do—use exponential notation for both. You will never be tempted to begin counting at the decimal point of a small number, and you will not have too many zeros on a large number. If the doubtful digit happens to be zero, be sure to write it even though it is to the right of the decimal.

Let's see how much of all this you have mastered. Write the number of significant figures

in each of the following measurements. While doing so, cover up the answers, which follow immediately. When the list is completed, check your answers.

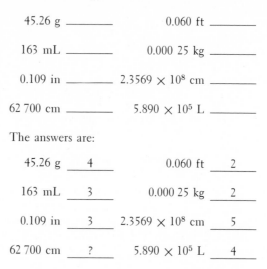

The answers are:

45.26 g	4	0.060 ft	2
163 mL	3	0.000 25 kg	2
0.109 in	3	2.3569 × 10⁸ cm	5
62 700 cm	?	5.890 × 10⁵ L	4

There is no way to tell whether the doubtful digit is in hundreds, tens, or ones in 62 700 cm. Exponential notation is required. The zero at the end of 5.890 × 10⁵ L identifies it as the doubtful digit and therefore significant. The equivalent value, 589 000, would be questionable; the 9 or any 0 could be the doubtful digit.

EXACT NUMBERS

Notice that everything that has been said about significant figures is related to *measured* quantities. Uncertainty in numbers appears only in measurements. Numbers found by counting are exact. A bicycle has exactly two wheels, and at any instant there may be exactly five apples in a fruit basket. Numbers in definitions are also exact. There are exactly 16 ounces in 1 pound, 1000 meters in 1 kilometer, and 12 eggs in 1 dozen eggs. Exact numbers are infinitely significant.

ROUNDING OFF

Sometimes, when experimentally measured quantities are added, subtracted, multiplied, or divided, the answer contains figures that are not significant. When this happens the result must be rounded off. Rules for rounding off are as follows:

1. If the first digit to be dropped is less than 5, leave the digit before it unchanged.

2. If the first digit to be dropped is 5 or more, increase the digit before it by 1.

Other rounding off rules vary in their handling of 5 if it is the first digit to be dropped. This may change the final digit by 1 in some cases. This is acceptable, however, because by any method only the doubtful digit is affected.

PRACTICE: 15) *Round off each of the following quantities to three significant figures:**

a) 1.427 52 g/cm³ _____

b) 643.349 cm² _____

c) 0.007 456 2 kg _____

d) 2.103 × 10⁴ cm _____

e) 45 853 cm _____

f) 0.039 449 8 m _____

g) 3.605 × 10⁻⁷ cm _____

h) 3.5000 sec _____

ADDITION AND SUBTRACTION

The **significant figure rule for addition and subtraction** is

> *Round off a sum or a difference to the first column that has a doubtful digit.*

Suppose a student weighs four different chemicals into a preweighed beaker and wishes to find their sum. The individual weights and the total are as follows:

Beaker	319.542	grams
Chemical A	20.460	grams
Chemical B	0.0639	grams
Chemical C	38.2	grams
Chemical D	4.173	grams
Total	382.4389	grams

The total shown is what would appear in the display of a calculator. This sum must now be rounded off to the proper number of significant figures. According to the rule above, this means rounding off to the first column that has a doubtful digit. Chemical C has its doubtful digit in the tenths column, compared to the other measurements that are doubtful in the thousandths and ten-thousandths columns. To round off to tenths we drop the hundredth and all smaller columns. The number of hundredths is 3, which is less than 5, so it is dropped without changing the 4 in the tenths column. The correct answer is 382.4 grams.

Another way to solve the above problem is to round off all numbers before adding. Rounding off everything to tenths gives

$$\begin{array}{r} 319.5 \\ 20.5 \\ 0.1 \\ 38.2 \\ 4.2 \\ \hline 382.5 \end{array}$$

Both methods yield uncertainty in the tenths digit, so both are technically acceptable. The recommended procedure is to use all the data in the calculation and then round off the final answer.

This example may be used to justify the rule for addition and subtraction. A sum or difference digit must be doubtful if any number entering into that sum or difference is doubtful or unknown. In the left addition below all doubtful digits are shown in color, and all digits to the right of a colored digit are simply unknown.

In the left addition the 4 in the tenths column is clearly the first doubtful digit. The addition at the right shows a mechanical way to locate the first doubtful digit in a sum. Draw a vertical line after the last column that has every space occupied and the doubtful digit in the sum will be just left of that line. The result must be rounded off to the line.

PRACTICE: 16) *In an experiment in which oxygen is produced by heating potassium chlorate in the presence of a catalyst, a student assembled and weighed a test tube, test tube holder, and catalyst. The sum was 24.05 grams. He then added potassium chlorate to the test tube and weighed the assembly again, getting 26.255 grams. The weight of potassium chlorate is the difference between these numbers. Calculate that weight and express the result in the proper number of significant figures.*

*Answers to practice problems are on page 594.

MULTIPLICATION AND DIVISION

The **significant figure rule for multiplication and division** is

> *Round off a product or a quotient to the same* number *of significant figures as the smallest number of significant figures in any factor.*

To calculate the mass of a given volume of gas of known density, you multiply the volume by the density. If the density of a certain gas is 1.436 grams per liter, the mass of 0.0573 liter of that gas is

$$0.0573 \text{ liter } \times \frac{1.436 \text{ grams}}{1 \text{ liter}} = 0.082\ 282\ 8 \text{ grams,}$$

according to a calculator. The answer must be rounded off to the same number of significant figures as the *smallest number* of significant figures in either factor. This number is three; there are four significant figures in 1.436, but only three in 0.0573. The product should therefore be written 0.0823 grams.

Again the example may be used to justify the rule for multiplication and division. Each

product number that comes from a doubtful multiplier is doubtful or unknown. Colored numbers indicate the doubtful or unknown digits in the detailed multiplication.

$$
\begin{array}{r}
1.436 \\
\times 0.0573 \\
\hline
4308 \\
10052 \\
7180 \\
\hline
0.0822828 = 0.0823 \text{ grams}
\end{array}
$$

PRACTICE: 17) Assuming that the numbers are from experimental measurements, solve

$$\frac{(2.86 \times 10^4)(3.163 \times 10^{-2})}{1.8} =$$

Express the answer in the correct number of significant figures.

PRACTICE: 18) How many millimeters are in 0.6294 centimeters? Express the answer in the correct number of significant figures. (1 centimeter = 10 millimeters)

SIGNIFICANT FIGURES AND THIS BOOK

Most modern calculators report answers in all the digits they are able to display, usually eight or more. Rarely does a chemistry problem call for that many significant digits in a calculated result. Such an answer is simply unrealistic and should never be used. You should assume that all data in a problem are based on measurements expressed

to the proper number of significant figures, and write your calculated answers to the number of significant figures justified by those data. This is the procedure followed in this book. Most problems have been written to give results in three significant figures.

ADDITIONAL PRACTICE PROBLEMS

The following problems are given for additional practice. Answers for the problems in the left column are on page 574. Problems in the right column are not answered in the book.

Problems 19 to 21: Indicate the number of significant figures in each of the following quantities:

19) 4.060 liters

20) 1.6×10^{-4} gram

21) 312°C

19a) 10.40 kilograms

20a) 0.0026 liter

21a) 19 623.0 milligrams

Problems 22 to 24: Round off each of the following quantities to three significant figures:

22) 8.3562 g/cm³

22a) 21.369 ml

23) 124.563 g

23a) 0.000 549 841 kg

24) 5.0635 × 10⁻⁴ liter

24a) 4200 mg

Problems 25 to 28: Solve the problems from the data given. Express the answer in the proper number of significant figures.

25) The following amounts of a precious metal compound are drawn from a stock bottle for four different purposes: 0.475 gram, 3.40 grams, 1.8 grams, 12.92 grams. Find the total number of grams used.

25a) Over a given period of time a farmer's five cows produce the following quantities of milk: 34.3 gallons, 41.07 gallons, 38 gallons, 36.294 gallons, and 35.0 gallons. What is the total volume of milk?

26) The density of a substance is found by dividing the mass of a sample by its volume. In an experiment it is found that 48.85 cubic centimeters of magnesium have a mass of 85.0 grams. What is the density of magnesium?

26a) What is the weight in pounds of a rock that has a mass of 1862 grams? (454 grams = 1 pound)

27) A certain liquid dispensing unit delivers 10.36 mL each time it is used. If it is used twice each hour for 12 hours, how many milliliters of solution will it deliver?

27a) A pump in a chemical plant delivers 14.6 liters of hydrochloric acid each hour to a continuous industrial process. If the pump operates 518.6 hours in one working month, what will be the total volume of acid consumed?

28) An empty graduated cylinder has a mass of 62.9 grams. When liquid is added the total mass is 113 grams. What is the mass of the liquid?

28a) An empty 55-gallon drum weighs 19 pounds. When filled with machine oil the total weight of the drum and its contents is 424.6 pounds. How much does the oil weigh?

ANSWERS TO PRACTICE PROBLEMS IN APPENDIX I

1) $P_{O_2} = P - P_{H_2O} = 748 - 24 = 724$ torr

2) $V = \dfrac{k}{P} = \dfrac{1.62 \text{ L·atm}}{1.23 \text{ atm}} = 1.32$ L

3) $P = kT = \dfrac{2.4 \text{ torr}}{K} \times 300 \, K = 720$ torr

4) $T = \dfrac{P}{k} = 760 \text{ torr} \times \dfrac{K}{2.4 \text{ torr}} = 317$ K

5) $T_2 = \dfrac{T_1 P_2}{P_1} = \dfrac{(320 \text{ K})(1.00 \text{ atm})}{0.80 \text{ atm}} = 400$ K

6) Because of the complexity of this problem, it is easier to substitute the given values into the original equation and then solve for the unknown. The steps in the solution correspond to those on page 581.

$$\frac{\text{g salt}}{\text{g salt} + \text{g water}} = \frac{\%}{100}$$

$$\frac{20.0}{20.0 + \text{g water}} = \frac{25}{100}$$

$$20.0 = 0.25(20.0 + \text{g water})$$

$$= 5.0 + 0.25(\text{g water})$$

$$20.0 - 5.0 = 0.25(\text{g water})$$

$$\text{g water} = \frac{15.0}{0.25} = 60 \text{ g water}$$

7) total fish = 89 Chinook + 14 steelhead + 21 others = 124 fish

$\dfrac{89 \text{ Chinook}}{124 \text{ total}} \times 100 = 71.8\%$ (arbitrarily rounded off to three significant figures)

8) 1482 registered $\times \dfrac{84 \text{ voters}}{100 \text{ registered}} = 1245$ voters;

or 1482 × 0.84 = 1245

9) $58 \ \cancel{p \ sold} \times \dfrac{100 \text{ pages total}}{46 \ \cancel{\text{pages sold}}} = 126 \text{ pages total};$

$$\text{or } 0.46 \times \text{total} = 58$$
$$\text{total} = 126$$

10) a) $3.762\,199 \times 10^6$ d) 1.0080×10^1
    b) 1.9875×10^2 e) 4.60×10^{-3}
    c) 9.8×10^{-5}

11) a) $0.000\,000\,000\,034\,9$ c) 10
    b) $51\,600$ d) 3.75

12) *Answers are adjusted for significant figures.*
    a) 5.02×10^{10}
    b) $38.0 \times 10^2 = 3.80 \times 10^3$
    c) $61.3 \times 10^{-6} = 6.13 \times 10^{-5}$
    d) $79.3 \times 10^8 = 7.93 \times 10^9$
    e) $27.26 \times 10^6 = 2.726 \times 10^7$
    f) $21 \times 10^{-2} = 2.1 \times 10^{-1}$
    g) $64.6 \times 10^{-9} = 6.46 \times 10^{-8}$

13) *Answers are adjusted for significant figures.*
    a) 2.06×10^2
    b) $0.703 \times 10^2 = 7.03 \times 10^1$
    c) 1.22×10^{-4}
    d) 6.7×10^5
    e) 4.24×10^8

14) *Answers are adjusted for significant figures.*
    a) $\begin{array}{r} 5.2 \quad \times 10^5 \\ 0.398 \times 10^5 \\ \hline 5.6 \quad \times 10^5 \end{array}$

    b) $\begin{array}{r} 3.971 \quad \times 10^2 \\ 0.00198 \times 10^2 \\ \hline 3.973 \quad \times 10^2 \end{array}$

    c) $\begin{array}{r} 0.563 \quad \times 10^5 \\ 0.0293 \times 10^5 \\ 5.42 \quad \times 10^5 \\ \hline 6.01 \quad \times 10^5 \end{array}$

    d) $\begin{array}{r} 10.5 \times 10^{-5} \\ - \ \ 9.7 \times 10^{-5} \\ \hline 0.8 \times 10^{-5} = 8 \times 10^{-6} \end{array}$

    e) $\dfrac{9.02 \times 10^7}{265.9} = 3.39 \times 10^5$

    $\dfrac{7.89 \times 10^3}{2.37 \times 10^{-2}} = 3.33 \times 10^5$

$\begin{array}{r} 3.39 \times 10^5 \\ -3.33 \times 10^5 \\ \hline 0.06 \times 10^5 = 6 \times 10^3 \end{array}$

15) a) 1.43 g/cm^3
    b) 643 cm^2
    c) $0.007\,46$ or 7.46×10^{-3} kg
    d) 2.10×10^4 cm
    e) $4.59 \times 10^4 \text{ cm}^3$
    f) 0.0394 or 3.94×10^{-2} m
    g) 3.61×10^{-7} cm
    h) 3.50 sec

16) 26.255 grams test tube, test tube holder, catalyst + potassium chlorate
    24.05 grams test tube, test tube holder, + catalyst
    $\overline{}$
    $2.205 = 2.21$ grams potassium chlorate

17) 5.0×10^2. 500 is not correct, as it fails to show two significant figures, based on 1.8.

18) 6.294 millimeters. There are *exactly* 10 millimeters in a centimeter, by definition.

19) 4   20) 2   21) 3

22) 8.36 g/cm^3 23) 125 g

24) 5.06×10^{-4} L

25) $\begin{array}{r} 0.475 \text{ g} \\ 3.40 \ \text{ g} \\ 1.8 \ \text{ g} \\ 12.92 \ \text{ g} \\ \hline 18.595 \text{ g} = 18.6 \text{ g} \end{array}$

26) $\dfrac{85.0 \text{ g}}{48.85 \text{ cm}^3} = 1.74 \text{ g/cm}^3$

27) $12 \ \cancel{hr} \times \dfrac{2 \ \cancel{uses}}{1 \ \cancel{hr}} \times \dfrac{10.36 \text{ mL}}{1 \ \cancel{use}} = 248.6 \text{ mL}$
    (12 and 2 are exact numbers.)

28) $\begin{array}{r} 113 \ \ \text{ g cylinder + liquid} \\ 62.9 \text{ g cylinder} \\ \hline 50.1 = 50 = 5.0 \times 10^1 \text{ g liquid} \end{array}$

appendix II:

the si system of units

BASE UNITS

The International System of Units or *Système International* (SI), which represents an extension of the metric system, was adopted by the 11th General Conference of Weights and Measures in 1960. It is constructed from seven base units, each of which represents a particular physical quantity (Table AP-2).

Table AP-2
SI Base Units

PHYSICAL QUANTITY	NAME OF UNIT	SYMBOL
1. Length	meter	m
2. Mass	kilogram	kg
3. Time	second	s
4. Temperature	kelvin	K
5. Amount of substance	mole	mol
6. Electric current	ampere	A
7. Luminous intensity	candela	cd

Of the seven units listed in Table AP-2, the first five are particularly useful in general chemistry. They are defined as follows:

1. The *meter* was redefined in 1960 to be equal to 1 650 763.73 wavelengths of a certain line in the orange-red region of the emission spectrum of krypton-86.

2. The *kilogram* represents the mass of a platinum-iridium block kept at the International Bureau of Weights and Measures at Sevres, France.

3. The *second* was redefined in 1967 as the duration of 9 192 631 770 periods of a certain line in the microwave spectrum of cesium-133.

4. The *kelvin* is 1/273.16 of the temperature interval between the absolute zero and the triple point of water ($0.01\,°C = 273.16$ K).

5. The *mole* is the amount of substance which contains as many entities as there are atoms in exactly 0.012 kg of carbon-12.

PREFIXES USED WITH SI UNITS

Decimal fractions and multiples of SI units are designated by using the prefixes listed in Table AP-3. Those which are most commonly used in general chemistry are underlined.

Table AP-3
SI Prefixes

FACTOR	PREFIX	SYMBOL	FACTOR	PREFIX	SYMBOL
10^{12}	tera-	T	10^{-1}	deci-	d
10^{9}	giga-	G	10^{-2}	centi-	c
10^{6}	mega-	M	10^{-3}	milli-	m
10^{3}	kilo-	k	10^{-6}	micro-	μ
10^{2}	hecto-	h	10^{-9}	nano-	n
10^{1}	deca-	da	10^{-12}	pico-	p
			10^{-15}	femto-	f
			10^{-18}	atto-	a

DERIVED UNITS

In the International System of Units, all physical quantities are expressed in combinations of the base units listed in Table AP-2. For example, the density of a substance is found by dividing the mass of a sample in kilograms by its volume in cubic meters. The resulting units are kilograms per cubic meter, or kg/m^3. Some of the derived units used in chemistry are given in Table AP-4.

If you have not studied physics the SI units of force, pressure, and energy are probably new to you. Force is related to acceleration, which has to do with changing the speed (more correctly, *velocity*) of an object. One **newton** is the force that, when applied for one second, will change the straight line speed of a 1-kilogram object by 1 meter per second.

A **pascal** is defined as a pressure of one newton acting on an area of one square meter. A pascal is a small unit, so pressures are commonly expressed in kilopascals (kPa), 1000 times larger than the pascal.

A **joule** (pronounced j͞ool, as in pool) is defined as the work done when a force of one newton acts through a distance of one meter. *Work* and *energy* have the same units. When speaking of large amounts of energy, it is often expressed in kilojoules, 1000 times larger than the joule.

Table AP-4
SI Derived Units

PHYSICAL QUANTITY	NAME OF UNIT	SYMBOL	DEFINITION
Area	square meter	m^2	
Volume	cubic meter	m^3	
Density	kilogram per cubic meter	kg/m^3	
Force	newton	N	$kg \cdot m/s^2$
Pressure	pascal	Pa	N/m^2
Energy	joule	J	$N \cdot m$

SOME CHOICES

It is difficult to predict the extent to which SI units will replace traditional metric units in the coming years. This makes it difficult to select and use the units that will be most helpful to the readers of this textbook. Add to that the author's joy that, after nearly 200 years, the United States has finally begun to adopt the metric system, including some units that the SI system would eliminate, and his deep desire to encourage rather than complicate the use of metrics in his native land. With particular apologies to Canadian readers, who are more familiar with SI units than Americans are, we list the areas in which this book does not follow SI recommendations:

1. The SI unit of length is the *metre*, spelled in a way that corresponds to its French pronunciation. In America, and in this book, it is written *meter*, which matches the English pronunciation. "What's in a name?"

2. The SI volume unit, *cubic meter*, is huge for most everyday uses. The *cubic decimeter* is 1/1000 as large, and much more practical. When referring to liquids it is customary to replace this six-syllable name with the two-syllable *liter*—or *litre*, for the French spelling. (Which would you rather buy at the grocery store, two cubic decimeters of milk, or two liters?) In the laboratory the common units are again 1/1000 as large, the *cubic centimeter* for solids and the *milliliter* for liquids.

3. The *millimeter of mercury* has an advantage over the *pascal* or *kilopascal* as a pressure unit because the common laboratory instrument for "measuring" pressure literally measures millimeters of mercury. We lose some of the advantage of this "natural" pressure unit by using its other name, *torr*. Reducing eight syllables to one is worth the sacrifice. Again, "What's in a name?" For large pressures we continue to use the traditional *atmosphere*, which is 760 torr.

CONVERSIONS BETWEEN ENGLISH AND METRIC UNITS

Table AP-5 lists the conversion relationships between metric units and some of the more common English units.

Table AP-5
Conversion Table—Metric and English Units

MASS	LENGTH	CAPACITY
454 grams = 1 pound	2.54 centimeters = 1 inch*	1 liter = 1.06 quarts
28.3 grams = 1 ounce	30.5 centimeters = 1 foot	3.785 liters = 1 gallon
1 kilogram = 2.2 pounds	1 meter = 39.4 inches	4.546 liters = 1 imperial gallon
	1 meter = 1.09 yards	
	1.61 kilometers = 1 mile	

PRESSURE	ENERGY
760.0 torr 760.0 mm mercury 76.00 cm mercury 101.3 kilopascals $\Big\}$ = 1 atm = $\begin{cases} 14.69 \text{ lb/in}^2 \\ 29.92 \text{ in mercury} \end{cases}$	4.184 joules = 1 calorie = 0.003 97 British thermal unit

*Exactly, by definition.

common names of chemicals

COMMON NAME	CHEMICAL NAME	FORMULA
Alumina	Aluminum oxide	Al_2O_3
Baking soda	Sodium hydrogen carbonate	$NaHCO_3$
Bluestone	Copper(II) sulfate pentahydrate	$CuSO_4 \cdot 5\,H_2O$
Borax	Sodium tetraborate decahydrate	$Na_2B_4O_7 \cdot 10\,H_2O$
Brimstone	Sulfur	S
Carbon tetrachloride	Tetrachloromethane	CCl_4
Chloroform	Trichloromethane	$CHCl_3$
Cream of tartar	Potassium hydrogen tartrate	$KHC_4H_4O_6$
Diamond	Carbon	C
Dolomite	Calcium magnesium carbonate	$CaCO_3 \cdot MgCO_3$
Epsom salts	Magnesium sulfate heptahydrate	$MgSO_4 \cdot 7\,H_2O$
Freon (refrigerant)	Dichlorodifluoromethane	CCl_2F_2
Galena	Lead(II) sulfide	PbS
Grain alcohol	Ethyl alcohol; ethanol	C_2H_5OH
Graphite	Carbon	C
Gypsum	Calcium sulfate dihydrate	$CaSO_4 \cdot 2H_2O$
Hypo	Sodium thiosulfate	$Na_2S_2O_3$
Laughing gas	Dinitrogen oxide	N_2O
Lime	Calcium oxide	CaO
Limestone	Calcium carbonate	$CaCO_3$
Lye	Sodium hydroxide	$NaOH$
Marble	Calcium carbonate	$CaCO_3$
MEK	Methyl ethyl ketone	$CH_3COC_3H_5$
Milk of magnesia	Magnesium hydroxide	$Mg(OH)_2$
Muriatic acid	Hydrochloric acid	HCl
Oil of vitriol	Sulfuric acid (conc.)	H_2SO_4
Plaster of Paris	Calcium sulfate hemihydrate	$CaSO_4 \cdot \frac{1}{2}\,H_2O$
Potash	Potassium carbonate	K_2CO_3
Pyrites (fool's gold)	Iron disulfide	FeS_2
Quartz	Silicon dioxide	SiO_2
Quicksilver	Mercury	Hg
Rubbing alcohol	Isopropyl alcohol	$(CH_3)_2CHOH$
Sal ammoniac	Ammonium chloride	NH_4Cl
Salt	Sodium chloride	$NaCl$
Saltpeter	Sodium nitrate	$NaNO_3$
Slaked lime	Calcium hydroxide	$Ca(OH)_2$
Sugar	Sucrose	$C_{12}H_{22}O_{11}$
Washing soda	Sodium carbonate decahydrate	$Na_2CO_3 \cdot 10\,H_2O$
Wood alcohol	Methyl alcohol; methanol	CH_3OH

appendix IV

Table AP-6
Greek Prefixes Used in Chemical Nomenclature

NUMBER	PREFIX	NUMBER	PREFIX
1	mono-	7	hepta-
2	di-	8	octa-
3	tri-	9	nona-
4	tetra-	10	deca-
5	penta-	11	hendeca-
6	hexa-	12	dodeca-

appendix V

Table AP-7
Water Vapor Pressure

TEMPERATURE (°C)	VAPOR PRESSURE (torr)	TEMPERATURE (°C)	VAPOR PRESSURE (torr)
0	4.6	28	28.3
5	6.5	29	30.0
10	9.2	30	31.8
15	12.8	31	33.7
16	13.6	32	35.7
17	14.5	33	37.7
18	15.5	34	40.0
19	16.5	35	42.2
20	17.5	40	55.3
21	18.6	45	71.9
22	19.8	50	92.5
23	21.1	60	149.4
24	22.4	70	233.7
25	23.8	80	355.1
26	25.2	90	525.8
27	26.7	100	760.0

glossary

A

absolute zero—the temperature predicted by extrapolation of experimental data where translational kinetic energy theoretically becomes zero; the zero of the absolute temperature scale, which is equivalent to $-273.15°C$.

acid—a substance that yields hydrogen (hydronium) ions in aqueous solution (Arrhenius definition); a substance that donates protons in chemical reaction (Brönsted-Lowry definition); a substance that forms covalent bonds by accepting a pair of electrons (Lewis definition).

acidic solution—an aqueous solution in which the hydrogen ion concentration is greater than the hydroxide ion concentration; a solution in which the pH is less than 7.

actinides—elements 90 (Th) through 103 (Lr).

activated complex—an intermediate molecular species presumed to be formed during the interaction (collision) of reacting molecules in a chemical change.

activation energy—the energy barrier that must be overcome to start a chemical reaction.

alcohol—an organic compound consisting of an alkyl group and at least one hydroxyl group, having the general formula ROH.

aldehyde—a compound consisting of a carbonyl group bonded to a hydrogen on one side, and a hydrogen, alkyl, or aryl group on the other, having the general formula RCHO.

aliphatic hydrocarbon—an alkane, alkene or alkyne.

alkaline—basic; having pH greater than 7.

alkaline earth metal—a metal from Group 2A of the periodic table.

alkali metal—a metal from Group 1A of the periodic table.

alkane—a saturated hydrocarbon containing only single bonds, in which each carbon atom is bonded to four other atoms.

alkene—an unsaturated hydrocarbon containing a double bond, and each carbon atom that is double-bonded is bonded to a maximum of three atoms.

alkyl group—an alkane hydrocarbon group lacking one hydrogen atom, having the general formula C_nH_{2n+1}, and frequently symbolized by the letter R.

alkyne—an unsaturated hydrocarbon containing a triple bond, in which each carbon atom that is triple-bonded is bonded to a total of two atoms.

alpha (α) particle—the nucleus of a helium atom, often emitted in nuclear disintegration.

amide—a derivative of a carboxylic acid in which the hydroxyl group is replaced by a $-NH_2$ group, and having the general formula $RCONH_2$.

amine—an ammonia derivative in which one or more hydrogens are replaced by an alkyl group.

amorphous—a substance that is without definite structure or shape.

amphiprotic, amphoteric—a substance that can act as an acid or a base.

angstrom—a length unit equal to 10^{-10} m.

anhydride (anhydrous)—a substance that is without water, or from which water has been removed.

anion—a negatively charged ion.

aqueous—pertaining to water.

aromatic hydrocarbon—a hydrocarbon containing a benzene ring.

atmosphere (pressure unit)—a unit of pressure based on atmospheric pressure at sea level, and capable of supporting a mercury column 760 mm high.

atom—the smallest particle of an element that can combine with atoms of other elements in forming chemical compounds.

atomic mass—*see atomic weight.*

atomic mass unit (amu)—a unit of mass that is exactly $\frac{1}{12}$ of the mass of an atom of carbon-12.

atomic number (Z)—the number of protons in an atom of an element.

atomic weight—the number that expresses the average mass of the atoms of an element compared to the mass of an atom of carbon-12 at a value of exactly 12; the average mass of the atoms of an element expressed in atomic mass units.

Avogadro's number—the number of carbon atoms in exactly 12 grams of carbon-12; the number of units in 1 mole (6.02×10^{23}).

B

barometer—laboratory device for measuring atmospheric pressure.

base—a substance that yields hydroxide ions in aqueous solution (Arrhenius definition); a substance that accepts protons in chemical reaction (Brönsted-Lowry definition); a substance that forms covalent bonds by donating a pair of electrons (Lewis definition).

basic solution—an aqueous solution in which the hydroxide ion concentration is greater than the hydrogen ion concentration; a solution in which the pH is greater than 7.

beta (β) particle—a high energy electron, often emitted in nuclear disintegration.

binary compound—a compound consisting of two elements.

boiling point—the temperature at which vapor pressure becomes equal to the pressure above a liquid; the temperature at which vapor bubbles form spontaneously anyplace within a liquid.

boiling point elevation—the difference between the boiling point of a solution and the boiling point of the pure solvent.

bombardment (nuclear)—the striking of a target nucleus by an atomic particle, causing a nuclear change.

bond—*see chemical bond.*

bond angle—the angle formed by the bonds between two atoms that are bonded to a common central atom.

bonding electrons—the electrons transferred or shared in forming chemical bonds; valence electrons.

buffer—a solution that resists a change in pH.

C

calorie—a unit of heat equal to 4.184 joules.

calorimeter—laboratory device for measuring heat flow.

carbonyl group—an organic functional group,

$>$C$=$O , characteristic of aldehydes and ketones.

carboxyl group—an organic functional group,

, characteristic of carboxylic acids.

carboxylic acid—an organic acid containing the carboxyl group, having the general formula RCOOH.

catalyst—a substance that increases the rate of a chemical reaction by lowering activation energy. The catalyst is either a nonparticipant in the reaction, or it is regenerated. *See inhibitor.*

cathode—the negative electrode in a cathode ray tube; the electrode at which reduction occurs in an electrochemical cell.

cation—a positively charged ion.

chain reaction—a reaction that has, as a product, one of its own reactants; that product becomes a reactant, thereby allowing the original reaction to continue.

charge cloud—*see electron cloud.*

chemical bond—a general term that sometimes includes all of the electrostatic attractions among atoms, molecules, and ions, but more often refers to covalent and ionic bonds. *See covalent bond, ionic bond.*

chemical change—a change in which one or more substances disappear and one or more new substances are formed.

chemical family—a group of elements having similar chemical properties because of similar valence electron configuration, appearing in the same column of the periodic table.

chemical properties—the types of chemical change a substance is able to experience.

cloud chamber—a device in which condensation tracks form behind radioactive emissions as they travel through a supersaturated vapor.

colligative properties—physical properties of mixtures that depend upon concentration of particles irrespective of their identity.

colloid—a nonsettling dispersion of aggregated ions or molecules intermediate in size between the particles in a true solution and those in a suspension.

combustion—the process of burning.

compound—a pure substance that can be broken down into two or more other pure substances by a chemical change.

condense—to change from a vapor to a liquid or solid.

condensation—the act of condensing.

conjugate acid-base pair—a Brönsted-Lowry acid and the base derived from it when it loses a proton; or a Brönsted-Lowry base and the acid developed from it when it accepts a proton.

coulomb—a unit of electrical charge.

covalent bond—the chemical bond between two atoms that share a pair of electrons.

crystalline solid—a solid in which the ions and/or molecules are arranged in a definite geometric pattern.

D

decompose—to change chemically into simpler substances.

density—the mass of a substance per unit volume.

diatomic—that which has two atoms.

dilute—that which is relatively weak in concentration.

dipole—a polar molecule.

diprotic acid—an acid capable of yielding two protons per molecule in complete ionization.

dispersion forces—weak electrical attractions between molecules, temporarily produced by the shifting of electrons within molecules.

dissolve—to pass into solution.

distillation—the process of separating components of a mixture by boiling off and condensing the more volatile component.

dynamic equilibrium—a state in which opposing changes occur at equal rates, resulting in zero net change over a period of time.

E

electrode—conductor by which electric charge enters or leaves an electrolyte.

electrolysis—passage of electric charge through an electrolyte.

electrolyte—a substance which, when dissolved, yields a solution that conducts electricity; a solution or other medium that conducts electricity by ionic movement.

electron—subatomic particle carrying a unit negative charge and having a mass of 9.1×10^{-28} gram, or 1/1837 of the mass of a hydrogen nucleus, found outside the nucleus of the atom.

electron (charge) cloud—region of space around or between atoms that is occupied by electrons.

electron configuration—the orbital arrangement of electrons in ions or atoms.

electron-dot diagram (structure)—see Lewis diagram.

electronegativity—a scale of the relative ability of an atom of one element to attract the electron pair that forms a single covalent bond with an atom of another element.

electron orbit—the circular or elliptical path supposedly followed by an electron around an atomic nucleus, according to the Bohr theory of the atom.

electron orbital—a mathematically described region in space within an atom in which there is a high probability that an electron will be found.

electron pair geometry—a description of the distribution of bonding and unshared electron pairs around a bonded atom.

electron pair repulsion—the principle that electron pair geometry is the result of repulsion between electron pairs around a bonded atom, causing them to be as far apart as possible.

electrostatic force—force of attraction or repulsion between electrically charged objects.

element—a pure substance that cannot be decomposed into other pure substances by ordinary chemical means.

empirical formula—a formula that represents the lowest integral ratio of atoms of the elements in a compound.

endothermic—a change that absorbs energy from the surroundings, having a positive ΔH, an increase in enthalpy.

energy—the ability to do work.

enthalpy—the heat content of a chemical system.

enthalpy of reaction—see heat of reaction.

equilibrium—see dynamic equilibrium.

equilibrium constant—with reference to an equilibrium equation, the ratio in which the numerator is the product of concentrations of the species on the right side of the equation, each raised to a power corresponding to its coefficient in the equation, and the denominator is the corresponding product of the species on the left side of the equation; symbol: K, K_c, or K_{eq}.

equilibrium vapor pressure—see vapor pressure.

equivalent—that quantity of an acid (or base) that yields or reacts with one mole of H^+ (or OH^-) in a chemical reaction; that quantity of a substance that gains or loses one mole of electrons in a redox reaction.

ester—an organic compound formed by the reaction between a carboxylic acid and an alcohol, having the general formula R—CO—OR'.

ether—an organic compound in which two alkyl groups are bonded to the same oxygen, having the general formula R—O—R'.

excited state—the state of an atom in which one

or more electrons have absorbed energy—become "excited"—to raise them to energy levels above ground state.

exothermic reaction—a reaction that gives off energy to its surroundings.

F

family—*see chemical family.*

fission—a nuclear reaction in which a large nucleus splits into two smaller nuclei.

formula, chemical—a combination of chemical symbols and subscript numbers that represent the elements in a pure substance and the ratio in which the atoms of the different elements appear.

formula unit—a real (molecular) or hypothetical (ionic) unit particle represented by a chemical formula.

formula weight—the weight in amu of one formula unit of a substance; the molar weight of formula units of a substance.

fractional distillation—separation of a mixture into fractions whose components boil over a given temperature range.

freezing point depression—the difference between the freezing point of a solution and the freezing point of the pure solvent.

fusion—the process of melting; also, a nuclear reaction in which two small nuclei combine to form a larger nucleus.

G

gamma (γ) ray—a high energy photon emission in radioactive disintegration.

Geiger counter—an electrical device for detecting and measuring the intensity of radioactive emission.

ground state—the state of an atom in which all electrons occupy the lowest possible energy levels.

group (periodic table)—the elements comprising a vertical column in the periodic table.

H

half-life ($t_{1/2}$)—the time required for the disintegration of one half of the radioactive atoms in a sample.

half-reaction—the oxidation or reduction half of an oxidation-reduction reaction.

halide ion—F^-, Cl^-, Br^-, or I^-.

halogen—the name of the chemical family consisting of fluorine, chlorine, bromine, and iodine; any member of the halogen family.

heat of fusion (solidification)—the heat flow

when one gram of a substance changes between a solid and a liquid at constant pressure and temperature.

heat of reaction—change of enthalpy in a chemical reaction.

heat of vaporization (condensation)—the heat flow when one gram of a substance changes between a liquid and a vapor at constant pressure and temperature.

heterogeneous matter—matter having a nonuniform composition, usually with visibly different parts or phases.

homogeneous matter—matter having a uniform appearance and uniform properties throughout.

homologous series—a series of compounds in which each member differs from the one next to it by the same structural unit.

hydrate—a crystalline solid that contains water of hydration.

hydrocarbon—an organic compound consisting of carbon and hydrogen.

hydrogen bond—an intermolecular bond (attraction) between a hydrogen atom in one molecule and a highly electronegative atom (fluorine, oxygen, or nitrogen) of another polar molecule; the polar molecule may be of the same substance containing the hydrogen, or a different substance.

hydronium ion—a hydrated hydrogen ion, H_3O^+.

hydroxyl group—an organic functional group, —OH, characteristic of alcohols.

I

ideal gas—a hypothetical gas that behaves according to the ideal gas model over all ranges of temperature and pressure.

ideal gas equation—the equation $PV = nRT$ that relates quantitatively the pressure, volume, quantity, and temperature of an ideal gas.

immiscible—insoluble (usually used only in reference to liquids).

indicator—a substance that changes from one color to another, used to signal the end of a titration.

inhibitor—a substance added to a chemical reaction to retard its rate; sometimes called a negative catalyst.

ion—an atom or group of covalently bonded atoms that is electrically charged because of an excess or deficiency of electrons.

ion combination reaction—when two solutions are combined, the formation of a precipitate or molecular compound by a cation from one solution and an anion from the second solution.

ionic bond—the chemical bond arising from the attraction forces between oppositely charged ions in an ionic compound.

ionic compound—a compound in which ions are held by ionic bonds.

ionic equation—a chemical equation in which dissociated ionic compounds are shown in ionic form.

ionization—the formation of an ion from a molecule or atom.

ionization energy—the energy required to remove an electron from an atom or ion.

isoelectronic—having the same electron configuration.

isomers—two compounds having the same molecular formulas, but different structural formulas and different physical and chemical properties.

isotopes—two or more atoms of the same element that have different atomic masses because of different numbers of neutrons.

IUPAC—International Union of Pure and Applied Chemistry.

J

joule—the SI energy unit, defined as a force of one newton applied over a distance of one meter; 1 joule = 0.239 calorie.

K

K—the symbol for the kelvin, the absolute temperature unit; the symbol for an equilibrium constant. K_a is the constant for the ionization of a weak acid; K_{sp} is the constant for the equilibrium between a slightly soluble ionic compound and a saturated solution of its ions; K_w is the constant for the ionization of water.

Kelvin temperature scale—an absolute temperature scale on which the degrees are the same size as Celsius degrees, with 0 K at absolute zero, or $-273.16°C$.

ketone—a compound consisting of a carbonyl group bonded on each side to an alkyl group, having the general formula R—CO—R′.

kinetic energy—energy of motion; translational kinetic energy is equal to $\frac{1}{2}$ mass × (velocity)^2.

kinetic molecular theory—the general theory that all matter consists of particles in constant motion, with different degrees of freedom distinguishing among solids, liquids, and gases.

kinetic theory of gases—that portion of the kinetic molecular theory that describes gases and from which the model of an ideal gas is developed.

L

lanthanides—elements 58(Ce) through 71(Lu).

Le Chatelier's Principle—if an equilibrium system is subjected to a change, processes occur that tend to counteract partially the initial change, thereby bringing the system to a new position of equilibrium.

Lewis diagram, structure, or symbol—a diagram representing the valence electrons and covalent bonds in an atomic or molecular species.

limiting reagent—the reactant first totally consumed in a reaction, thereby determining the maximum yield possible.

line spectrum—the spectral lines that appear when light emitted from a sample is analyzed in a spectroscope.

M

macromolecular crystal—a crystal made up of a large but indefinite number of atoms covalently bonded to each other to form a huge molecule.

manometer—a laboratory device for measuring gas pressure.

mass—a property reflecting the quantity of matter in a sample.

mass number—the total number of protons plus neutrons in the nucleus of an atom.

mass spectroscope—a laboratory device whereby a flow of gaseous ions may be analyzed in regard to their charge and/or mass.

matter—that which occupies space and has mass.

metal—a substance that possesses metallic properties, such as luster, ductility, malleability, good conductivity of heat, and electricity; an element that loses electrons to form monatomic cations.

miscible—soluble (usually used only in reference to liquids).

mixture—a sample of matter containing two or more pure substances.

molality—solution concentration expressed in moles of solute per kilogram of solvent.

molar heat of fusion (solidification)—the heat flow when one mole of a substance changes between a solid and a liquid at its normal melting point.

molar heat of vaporization (condensation)—the heat flow when one mole of a substance changes between a liquid and a vapor at its normal boiling point.

molarity—solution concentration expressed in moles of solute per liter of solution.

molar volume—the volume occupied by one mole, usually of a gas.

molar weight—the mass of one mole of any substance.

mole—that quantity of any species that contains the same number of units as the number of atoms in exactly 12 grams of carbon-12.

molecular compound—a compound whose fundamental particles are molecules rather than ions.

molecular crystal—a molecular solid in which the molecules are arranged according to a definite geometric pattern.

molecular geometry—a description of the shape of a molecule.

molecular weight (mass)—the number that expresses the average mass of the molecules of a compound compared to the mass of an atom of carbon-12 at a value of exactly 12; the average mass of the molecules of a compound expressed in atomic mass units.

molecule—the smallest unit particle of a pure substance that can exist independently and possess the identity of the substance.

monatomic—that which has only one atom.

monomer—the individual chemical structural unit from which a polymer may be developed.

monoprotic acid—an acid capable of yielding one proton per molecule in complete ionization.

N

negative catalyst—*see inhibitor.*

net ionic equation—an ionic equation from which all spectators have been removed.

neutralization—the reaction between an acid and a base to form a salt and water; any reaction between an acid and a base.

neutron—an electrically neutral subatomic particle having a mass of 1.7×10^{-24} gram, approximately equal to the mass of a proton, or 1 atomic mass unit, found in the nucleus of the atom.

noble gas—the name of the chemical family of relatively unreactive elemental gases appearing in Group 0 of the periodic table.

nonelectrolyte—a substance which, when dissolved, yields a solution that is a nonconductor of electricity; a solution or other fluid that does not conduct electricity by ionic movement.

nonpolar—pertaining to a bond or molecule having a symmetrical distribution of electric charge.

normal boiling point—the temperature at which a substance boils in an open vessel at one atmosphere pressure.

normality—solution concentration in equivalents per liter.

nucleus—the extremely dense central portion of the atom that contains the neutrons and protons which constitute nearly all the mass of the atom and all of the positive charge.

O

octet rule—the general rule that atoms tend to form stable bonds by sharing or transferring electrons until the atom is surrounded by a total of eight electrons.

orbit—*see electron orbit.*

orbital—*see electron orbital.*

organic chemistry—the chemistry of carbon compounds other than carbonates, cyanides, carbon monoxide, and carbon dioxide.

oxidation—chemical reaction with oxygen; a chemical change in which the oxidation number (state) of an element is increased; also, the loss of electrons in a redox reaction.

oxidation number—a number assigned to each element in a compound, ion, or elemental species by an arbitrary set of rules. Its two main functions are to organize and simplify the study of oxidation-reduction reactions and to serve as a base for one branch of chemical nomenclature.

oxidation state—*see oxidation number.*

oxidizer, oxidizing agent—the substance that takes electrons from another species, thereby oxidizing it.

oxyacid—an acid that contains oxygen.

oxyanion—an anion that contains oxygen.

P

partial pressure—the pressure one component of a mixture of gases would exert if it alone occupied the same volume as the mixture at the same temperature.

Pauli exclusion principle—the principle that says, in effect, that no more than two electrons can occupy the same orbital.

period (periodic table)—a horizontal row of the periodic table.

pH—a way of expressing hydrogen ion concentration; the negative of the logarithm of the hydrogen ion concentration.

phase—a visibly distinct part of a heterogeneous sample of matter.

physical change—a change in the physical form of a substance without changing its chemical identity.

physical properties—properties of a substance that can be observed and measured without changing the substance chemically.

pOH—a way of expressing hydroxide ion concentration; the negative logarithm of the hydroxide ion concentration.

polar—pertaining to a bond or molecule having an unsymmetrical distribution of electric charge.

polyatomic—pertaining to a species consisting of two or more atoms; usually said of polyatomic ions.

polymer—a chemical compound formed by bonding two or more monomers; frequently, in plastics, a huge macromolecule.

polymerization—the reaction in which monomers combine to form polymers.

polyprotic acid—an acid capable of yielding more than one proton per molecule on complete ionization.

potential energy—energy possessed by a body by virtue of its position in an attractive or repulsive force field.

precipitate—a solid that forms when two solutions are mixed.

pressure—force per unit area.

principal energy level(s)—the main energy levels within the electron arrangement in an atom. They are quantized by a set of integers beginning at $n = 1$ for the lowest level, $n = 2$ for the next, and so forth; also called the principal quantum number.

proton—a subatomic particle carrying a unit positive charge and having a mass of 1.7×10^{-24} gram, almost the same as the mass of a neutron, found in the nucleus of the atom.

pure substance—a sample consisting of only one kind of matter, either compound or element.

Q

quantization of energy—the existence of certain discrete electron energy levels within an atom such that electrons may have any one of these energies, but no energy between two such levels.

quantum mechanical model of the atom—an atomic concept that recognizes four quantum numbers by which electron energy levels may be described.

R

R—a symbol used to designate any alkyl group; the ideal gas constant, having a value of 0.0821 L atm/mol K.

radioactivity—spontaneous emission of rays and/or particles from an atomic nucleus.

redox—a term coined from REDuction-OXidation to refer to oxidation-reduction reactions.

reducer, reducing agent—the substance that loses electrons to another species, thereby reducing it.

reduction—a chemical change in which the oxidation number (state) of an element is reduced; also, the gain of electrons in a redox reaction.

reversible reaction—a chemical reaction in which the products may react to re-form the original reactants.

S

salt—the product of a neutralization reaction other than water; an ionic compound containing neither the hydrogen ion, H^+, oxide ion, O^{2-}, nor hydroxide ion, OH^-.

saturated hydrocarbon—a hydrocarbon that contains only single bonds, in which each carbon atom is bonded to four other atoms.

saturated solution—a solution of such concentration that it is or would be in a state of equilibrium with excess solute present.

SI unit—a unit associated with the International System of Units.

significant figures—the digits in a measurement that are known to be accurate plus one doubtful digit.

soluble—a substance that will dissolve in a suitable solvent.

solubility—the quantity of solute that will dissolve in a given quantity of solvent, or in a given quantity of solution, at a specified temperature, to establish an equilibrium between the solution and excess solute; frequently expressed in grams of solute per 100 grams of solvent.

solubility product constant—*under K, see K_{sp}.*

solute—the substance dissolved in the solvent; sometimes not clearly distinguishable from the solvent (see below), but usually the lesser of the two.

solution—a homogeneous mixture of two or more substances of molecular or ionic particle size, the concentration of which may be varied, usually within certain limits.

solution inventory—a precise identification of

the chemical species present in a solution, in contrast with the solute from which they may have come; i.e., sodium ions and chloride ions, rather than sodium chloride.

solvent—the medium in which the solute is dissolved; *see solute.*

specific gravity—the ratio of the density of a substance to the density of some standard, usually water at 4°C.

specific heat—the quantity of heat required to raise the temperature of one gram of a substance one degree Celsius.

spectator (ion)—a species present at the scene of a reaction but not a participant in it.

spectroscope—a laboratory instrument used to analyze spectra.

spectrum (*plural:* **spectra**)—the result of a dispersion of a beam of light into its component colors; also the result of a dispersion of a beam of gaseous ions into its component particles, distinguished by mass and electric charge.

spontaneous—a change that appears to take place by itself, without outside influence.

stable—that which does not change spontaneously.

standard temperature and pressure (STP)—arbitrarily defined conditions of temperature (0°C) and pressure (1 atmosphere) at which gas volumes and quantities are frequently measured and/or compared.

stoichiometry—the quantitative relationships between the substances involved in a chemical reaction, established by the equation for the reaction.

STP—abbreviation for standard temperature and pressure (see above).

strong acid—an acid that ionizes almost completely in aqueous solution; an acid that loses its protons readily.

strong base—a base that dissociates almost completely in aqueous solution; a base that has a strong attraction for protons.

strong electrolyte—a substance which, when dissolved, yields a solution that is a good conductor of electricity because of nearly complete ionization or dissociation.

strong oxidizer (oxidizing agent)—an oxidizer that has a strong attraction for electrons.

strong reducer (reducing agent)—a reducer that releases electrons readily.

sublevel—the levels into which the principal energy levels are divided according to the quantum mechanical model of the atom; usually specified s, p, d, and f.

supersaturated—a state of solution concentration which is greater than the equilibrium concentration (solubility) at a given temperature and/or pressure.

suspension—a mixture that gradually separates by settling.

T

tetrahedral—related to a tetrahedron; usually used in reference to the orientation of four covalent bonds radiating from a central atom toward the vertices of a tetrahedron, or to the 109°28' angle formed by any two corners of the tetrahedron and the central atom as its vertex.

tetrahedron—a regular four-sided solid, having congruent equilateral triangles as its four faces.

thermal—having to do with heat.

thermochemical equation—a chemical equation which includes an energy term, or for which ΔH is indicated.

thermochemical stoichiometry—stoichiometry expanded to include the energy involved in a chemical reaction, as defined by the thermochemical equation.

titration—the controlled and measured addition of one solution into another.

torr—a unit of pressure equal to the pressure unit millimeter of mercury.

transition element; transition metal—an element from one of the B groups or Group 8 of the periodic table.

transmutation—conversion of an atom from one element to another by means of a nuclear change.

transuranium elements—man-made elements whose atomic numbers are greater than 92.

triprotic acid—an acid capable of yielding three protons in complete ionization.

U

unsaturated hydrocarbon—a hydrocarbon that contains one or more multiple bonds.

V

valence electrons—the highest energy s and p electrons in an atom that determine the bonding characteristics of an element.

van der Waals forces—a general term for all kinds of weak intermolecular attractions.

vapor—a gas

vaporize, vaporization—changing from a solid or liquid to a gas.

vapor pressure—the partial pressure exerted by a specific vapor component in a mixture of gases. Frequently refers to the partial pressure of a vapor that is in equilibrium with its liquid state at a given temperature.

volatile—that which vaporizes easily.

W

water of crystallization, water of hydration—water molecules that are included as structural parts of crystals formed from aqueous solutions.

weak acid—an acid that ionizes only slightly in aqueous solution; an acid that does not donate protons readily.

weak base—a base that dissociates only slightly in aqueous solution; a base that has a weak attraction for protons.

weak electrolyte—a substance which, when dissolved, yields a solution that is a poor conductor of electricity because of limited ionization or dissociation.

weak oxidizer (oxidizing agent)—an oxidizer that has a weak attraction for electrons.

weak reducer (reducing agent)—a reducer that does not release electrons readily.

weight—a measure of the force of gravitational attraction of a body to the earth.

Y

yield—the amount of product from a chemical reaction.

Z

Z—atomic number.

answers

answers to questions and problems

CHAPTER 2

2.1)   (a) P; (b) C; (c) C; (d) P; (e) P.

2.2)   (a) C; (b) P; (c) P; (d) C; (e) P.

2.3)   Place the mixture in water. Salt will dissolve, sand will not. Solubility is generally considered a physical property.

2.4)   Gas: volume and shape variable; particle movement totally random within entire volume of closed container. Liquid: volume fixed, shape variable; particle movement random within fluid volume at bottom of container. Solid: volume and shape fixed; particle movement limited to vibration in fixed position.

2.5)   Particles of solid are rigidly fixed in position, giving definite shape that turns corner as a unit. Liquid particles, not fixed in position, flow ahead while bowl turns corner.

2.6)   "Fluid" refers to substance in which particles are not in fixed position, and therefore flow. Applies to liquids and gases only.

2.7)   A pure substance is a single kind of matter, unlike any other kind. A mixture consists of two or more pure substances.

2.8)   Properties of a pure substance are fixed; properties of a mixture vary with composition. Liquid is a mixture because it has variable boiling point, a physical property.

2.9)   Pure substance. Density constant in situation that would have resulted in varying concentration and density if liquid were a mixture.

2.10)   Element cannot be decomposed into simpler pure substances, while a compound can.

2.11)   Elements: tin, nitrogen, iron; compounds: carbon dioxide, baking soda.

2.12)   Homogeneous refers to sameness in appearance and behavior. The composition of a homogeneous substance is the same throughout.

2.13)   (a) and (c) heterogeneous; (b), (d), and (e) homogeneous.

2.14)   Mass is conserved in a chemical change. The combined mass of all reactants (carbon and oxygen) is equal to the combined mass of all products (carbon dioxide).

2.15)   Electrostatic force is force of attraction or repulsion between charged bodies. Gravitational force: attraction only.

2.16)   Exothermic, heat evolved, or given off; endothermic, heat absorbed. Boiling water is endothermic, as you must add heat to the water to boil it.

2.17)   (a) K; (b) P; (c) K; (d) K; (e) P; (f) P; (g) P.

2.18)   "Chemical (potential) energy" is released as atoms in the fuel are rearranged in electrical force fields.

CHAPTER 3

3.1) $\quad 48.5 \text{ gal} \times \dfrac{4 \text{ qt}}{1 \text{ gal}} = 194 \text{ qt}$

3.2) $\quad 0.35 \text{ ton} \times \dfrac{2000 \text{ lbs}}{1 \text{ ton}} \times \dfrac{16 \text{ oz}}{1 \text{ lb}} = 11\,200 \text{ oz}$

3.3) $\quad 5 \text{ month} \times \dfrac{1500 \text{ bikes}}{1 \text{ month}} \times \dfrac{2 \text{ wheels}}{1 \text{ bike}} = 15\,000 \text{ wheels}$

3.4) $\quad 2500 \text{ ft} \times \dfrac{1 \text{ mile}}{5280 \text{ ft}} \times \dfrac{1 \text{ hour}}{35 \text{ miles}} \times \dfrac{60 \text{ min}}{1 \text{ hour}} = 0.81 \text{ minute}$

3.5) $\quad 2.0 \text{ lbs} \times \dfrac{1 \text{ batch}}{75 \text{ lbs}} \times \dfrac{12 \text{ min}}{1 \text{ batch}} \times \dfrac{1 \text{ hr}}{60 \text{ min}} \times \dfrac{\$28}{1 \text{ hr}} \times \dfrac{100\cent}{\$} = 14.9\cent$

3.6) $\quad 3.6 \text{ yd}^3 \times \dfrac{3^3 \text{ ft}^3}{1 \text{ yd}^3} = 97.2 \text{ ft}^3$

3.7) $\quad 53.3 \text{ yd} \times 100 \text{ yd} \times \dfrac{3^2 \text{ ft}^2}{1 \text{ yd}^2} = 48\,000 \text{ ft}^2$ (three significant figures)

3.8) $\quad 28.0 \text{ in.} \times \dfrac{2.54 \text{ cm}}{1 \text{ in.}} = 71.1 \text{ cm}$

3.9) $\quad 148 \text{ km} \times \dfrac{1 \text{ mile}}{1.61 \text{ km}} = 91.9 \text{ miles}$

3.10) $\quad 39.8 \text{ yds} \times \dfrac{1 \text{ meter}}{1.09 \text{ yds}} = 36.5 \text{ meters}$

3.11) $\quad 5 \times 10$. As delivered, a "two-by-four" does not measure $2'' \times 4''$, but is closer to $1.5'' \times 3.5''$. The more accurate metric dimensions are 4 cm $\times$ 9 cm.

3.12) $\quad \dfrac{1 \text{ in.}}{8} \times \dfrac{2.54 \text{ cm}}{1 \text{ in.}} \times \dfrac{10 \text{ mm}}{1 \text{ cm}} = 3.175 \text{ mm}$

3.13) $\quad \dfrac{5}{8} \text{ in.} = 0.625 \text{ in.} \qquad 3.625 \text{ in.} \times \dfrac{2.54 \text{ cm}}{1 \text{ in.}} \times \dfrac{10 \text{ mm}}{1 \text{ cm}} = 92.075 \text{ mm}$

3.14) $\quad 0.786 \text{ m} \times \dfrac{1000 \text{ mm}}{1 \text{ m}} = 786 \text{ mm}$

3.15) $\quad 5.9 \times 10^3 \text{ Å} \times \dfrac{1 \text{ m}}{10^{10} \text{ Å}} \times \dfrac{10^2 \text{ cm}}{1 \text{ m}} = 5.9 \times 10^{-5} \text{ cm}$

3.16) $\quad 2.50 \text{ yd}^3 \times \dfrac{1^3 \text{ m}^3}{1.09^3 \text{ yd}^3} = 1.93 \text{ m}^3$

3.17) $\quad 14.2 \text{ m}^3 \times \dfrac{100^3 \text{ m}^3}{1^3 \text{ m}^3} = 1.42 \times 10^7 \text{ cm}^3$

3.18) $\quad 5.00 \text{ gal} \times \dfrac{3.785 \text{ L}}{1 \text{ gal}} = 18.9 \text{ L}$

3.19) $\quad$ "Weightlessness" refers to absence of a gravitational field. Mass is related to the amount of matter in a sample, regardless of a gravitational field. An astronaut is therefore never "massless."

3.20) $\quad 4.80 \text{ oz} \times \dfrac{28.3 \text{ g}}{1 \text{ oz}} = 136 \text{ g}$

3.21) $\quad 3.52 \text{ kg} \times \dfrac{1000 \text{ g}}{1 \text{ kg}} = 3.52 \times 10^3 \text{ g}$

3.22) $0.782 \text{ mg} \times \dfrac{1 \text{ g}}{1000 \text{ mg}} = 7.82 \times 10^{-4} \text{ mg}$

3.23) $0.218 \text{ g} \times \dfrac{1000 \text{ mg}}{1 \text{ g}} \times \dfrac{1 \text{ carat}}{200 \text{ mg}} = 1.09 \text{ carats}$

3.24) $\dfrac{93 \text{ ¢}}{1 \text{ lb}} \times \dfrac{2.20 \text{ lb}}{1 \text{ kg}} \times \dfrac{1 \text{ dollar}}{100 \text{ ¢}} = \$2.05/\text{kg}$

3.25) 0.89 g/cm^3. The density of the substance is 0.89 as great as the density of water.

3.26) $50.0 \text{ mL} \times \dfrac{1.60 \text{ g}}{1 \text{ mL}} = 80.0 \text{ g}$

3.27) $68.3 \text{ g} \times \dfrac{1 \text{ cm}^3}{19.3 \text{ g}} = 3.54 \text{ cm}^3$

3.28) $\dfrac{2.04 \times 10^3 \text{ g}}{150 \text{ mL}} = 13.6 \text{ g/mL};$ Sp. g. = 13.6

3.29) $1 \text{ qt} \times \dfrac{1 \text{ L}}{1.06 \text{ qts}} \times \dfrac{13.6 \text{ g}}{0.001 \text{ L}} \times \dfrac{1 \text{ lb}}{454 \text{ g}} = 28.3 \text{ lbs}$

3.30) $32 + 1.8(805) = 1481°\text{F}$ 3.31) $120 - 32 = 1.8(°\text{C})$ $°\text{C} = 48.9$

3.32) $-196°\text{C}; 96.1°\text{C}; 1349°\text{C}$ 3.33) $-407°\text{F}; 115°\text{F}; 2768°\text{F}$

CHAPTER 4

4.1) See page 53 of text.

4.2) 46 grams is a fixed mass of sodium. Oxygen masses that combine with 46 grams of sodium are 16 and 32 grams. $\frac{16}{32} = \frac{1}{2}$, a ratio of small whole numbers.

4.3) Atoms are not indivisible, as Dalton proposed.

4.4) Most of an atom is empty space through which the alpha particle can pass.

4.5) Atoms consist of extremely tiny and extremely dense positively charged nuclei, surrounded by "open space" that is thinly populated with electrons of very small mass.

4.6) The Rutherford experiment showed that massive atomic particles were in the nucleus. When found, the proton and neutron proved to be heavy particles.

4.7) If an atom is neutral, the number of protons (positive charges) must equal the number of electrons (negative charges). The number of neutrons may or may not be equal to the number of protons or electrons.

4.8) Atomic number, 3; mass number, 6; number of neutrons, 3.

4.9) Electrons and protons, 33; neutrons, 37; Z = 33; mass number = 70.

4.10) The name of the isotope is chromium-52; the mass number is 52; the symbol is $^{52}_{24}\text{Cr}$.

4.11) 26; 31; 26; $^{57}_{26}\text{Fe}$; iron-57.

4.12) An atomic mass unit is exactly $\frac{1}{12}$ of the mass of an atom of carbon-12.

4.13) Boron occurs in nature as a mixture of atoms that have different masses. If you determined the masses of a large number of atoms and calculated the average mass it would be 10.81 amu.

4.14) $0.7553 \times 34.968\ 85$ amu $= 26.41$ amu
$\underline{0.2447 \times 36.965\ 90}$ amu $= \underline{\ 9.046}$ amu
35.46 amu

4.15) 0.5182×106.9041 amu $= 55.40$ amu
0.4818×108.9047 amu $= \underline{52.47}$ amu
107.87 amu

4.16) 0.5725×120.9038 amu $= 69.22$ amu
0.4275×122.9041 amu $= \underline{52.54}$ amu
121.76 amu

4.17) $0.001\ 93 \times 135.907$ amu $= \quad 0.262$ amu
$0.002\ 50 \times 137.9057$ amu $= \quad 0.345$ amu
$0.8848 \quad \times 139.9053$ amu $= 123.8 \quad$ amu
$0.1107 \quad \times 141.9090$ amu $= \underline{15.71}$ amu
$140.1 \quad$ amu

4.18) Be, Mg, Ca, Sr, Ba, Ra; 11 to 18.

4.19) (a) and (d), same period; (b) and (c), same family.

4.20) $Z = 24$: 52.00; $Z = 50$: 118.7; $Z = 77$: 192.2.

4.21) Potassium, 39.10; sulfur, 32.06.

CHAPTER 5

5.1) In a continuous light spectrum one color blends into another, as in a rainbow. In a discrete line spectrum there is a series of sharp lines, separated by black areas.

5.2) Something that behaves "like a wave" has properties normally associated with waves. Some of these are listed in the caption under Figure 5.3.

5.3) Something that is quantized may, at any instant, have one of several distinct values, but never any value between those distinct values. Something that is not quantized may have any value.

5.4) At any instant an electron may have one of several allowable energies, but never any other energy.

5.5) The spectrum would be continuous; one color would blend gradually into another. Not being quantized, the electron could drop from any energy level to any other, thereby releasing light of any wavelength, or color. With billions upon billions of electrons, all colors would be produced.

5.6) An atom in the ground state has all electrons in the lowest possible energy levels. An atom in an excited state has one or more electrons at energy levels above the ground state.

5.7) The atom is described as a dense nucleus with electrons moving around it in circular orbits. Add a nucleus to the center of Figure 5.6 for the sketch.

5.8) Hydrogen atoms are the only neutral atoms that fit the Bohr model. The model also fails to account for the energy radiations that would accompany an electron moving in a circular orbit.

5.9) The principal energy levels of an atom are the quantized energy levels recognized by Bohr. "Principal" distinguishes them from sublevels not known to Bohr, but which were introduced with the quantum mechanical model of the atom.

5.10) Sublevels are quantized energy levels within the principal energy level. They are identified by the letters s, p, d, and f.

5.11) *Orbits* are circular or elliptical paths that electrons travel around the nucleus in the Bohr model. *Orbitals* are mathematically defined regions of space about the nucleus where there is a high probability of locating an electron. See Figure 5.7 for shapes.

5.12) The statements are incorrect. At any value of n, the first sublevel is always an s sublevel. At n = 2 or more, the second sublevel is a p sublevel, and at n = 3 or more, the third sublevel is a d sublevel. There is never more than one p sublevel at any value of n.

5.13) An orbital is the region in space where there is a high probability of finding an electron. This implies that there is a low probability of finding it outside that region. When drawn to scale, orbitals are frequently drawn for 90 or 95% probability that the electron is within them, which leaves 10 or 5% probability that the electron will be outside the figure drawn.

5.14) The Pauli exclusion principle limits the electron population of an orbital to two.

5.15) The statement is incorrect. First, the orbital is three-dimensional, and only looks like a figure 8 when sketched in two dimension. Second, the orbital represents a region in space, not a path. We do not and theoretically cannot know the path of an electron in an atom.

5.16) If n = 3 or more, the *actual* (not just the maximum) number of d orbitals is 5.

5.17) The models are consistent in identifying quantized energy levels. The quantum mechanical model substitutes orbitals (regions in space) for orbits (electron paths) in the Bohr model. The quantum model goes beyond the Bohr model in identifying energy sublevels.

5.18) $3p^4$ means that there are four electrons in the 3p sublevel.

5.19) Adding superscripts—the number of electrons, which is equal to the number of protons in a neutral atom—gives 9 for the atomic number. This is fluorine, which is in Period 2 and Group 7A.

5.20) [Ne] represents the "neon core," or the electron configuration of a neon atom: $1s^2 2s^2 2p^6$.

5.21) Group 4A.

5.22) N: $1s^2 2s^2 2p^3$; Ti: $1s^2 2s^2 2p^6 3s^2 3p^6 4s^2 3d^2$.

5.23) K: $1s^2 2s^2 2p^6 3s^2 3p^6 4s^1$; Mn: $1s^2 2s^2 2p^6 3s^2 3p^6 4s^2 3d^5$.

5.24) Ti: $[Ar]4s^2 3d^2$; K: $[Ar]4s^1$; Mn: $[Ar]4s^2 3d^5$.

5.25) Ge: $[Ar]4s^2 3d^{10} 4p^2$.

5.26) Ionization energy is the energy required to remove an electron from an atom. When an electron is taken from an atom, the ion that is left has a positive charge. Energy is required to overcome the attraction between the negatively charged electron and positively charged ion.

5.27) Aluminum has a lower ionization energy than chlorine. In each element, an atom is ionized by removing a 3d electron. The electron is closer to the nucleus in the smaller chlorine atom, and the 17 protons in the chlorine nucleus hold the electron more strongly than the 13 protons in the nucleus of an aluminum atom.

5.28) Two elements in the same family have similar chemical properties.

5.29) ns^1.

5.30)   Iodine, halogens. Rubidium, alkali metals.

5.31)   Many chemical properties are established by the number of electrons in the highest energy level. Both magnesium ($3s^2$) and calcium ($4s^2$) have two electrons in their highest occupied energy levels.

5.32)   The representative elements are in the "A" columns of the periodic table. Those to the left of the stair-step line that begins next to boron are metals; those to the right are nonmetals.

5.33)   Atoms of antimony ($Z = 51$) and bismuth ($Z = 83$) are larger than arsenic atoms, and atoms of nitrogen and phosphorus are smaller.

5.34)   As you go down a column of the periodic table, the outermost electrons are in increasingly higher energy levels, and therefore farther from the nucleus. "Farther from the nucleus" means a larger atom.

5.35)   See answer to Question 5.34. The number of occupied energy levels is apparently more important in determining atomic size than nuclear charge.

5.36)   Metal atoms tend to lose electrons easily to become positively charged ions.

5.37)   (a) Q M;   (b) D E.   5.38)   J L W.   5.39)   E D G.

CHAPTER 6

6.1)   Check alphabetical list of elements inside back cover.

6.2)   Two nitrogen atoms, one oxygen atom.

6.3)   One sulfur atom, two oxygen atoms, two chlorine atoms.

6.4) SiH_4.   6.5) C_3H_8O.

6.6)   An atom is the fundamental structural unit of an element. A molecule is the smallest stable particle of any pure substance, element, or compound. Molecules of noble gases are individual atoms.

6.7)   A monatomic substance has one atom per fundamental particle; a diatomic substance has two.

6.8)   It is a diatomic molecule.

6.9)   The structural unit of a molecular compound is a molecule, rather than the ions that make up an ionic compound.

6.10)   Water; carbon dioxide.

6.11)   Monatomic ions are made up of one atom; polyatomic ions have two or more atoms.

6.12)   The anion has a negative charge, and the cation a positive charge.

6.13)   A monatomic cation is formed when an atom loses one or more electrons.

6.14)

ATOMIC NUMBER	ION NAME	ION SYMBOL	ANION OR CATION
11	Sodium	Na^+	Cation
53	Iodide	I^-	Anion
3	Lithium	Li^+	Cation
20	Calcium	Ca^{2+}	Cation
16	Sulfide	S^{2-}	Anion

6.15)   Selenium forms the selenide ion, Se^{2-}. It is an anion.
Rubidium forms the rubidium ion, Rb^{2+}. It is a cation.

6.16) An acid is a solution that contains hydrogen ions.

6.17) HCl.   6.18) HNO_3; NO_3^-; nitrate ion.

6.19) Carbonic acid; H_2CO_3; CO_3^{2-}.   6.20) Hydroxide ion.

6.21) LiCl; calcium sulfide.   6.22) NH_4NO_3; barium carbonate.

6.23) $BaBr_2$; potassium phosphate.   6.24) $Mg_3(PO_4)_2$; ammonium sulfate.

6.25) A hydrate is a solid compound that includes water molecules in its crystal structure. When the water has been removed it is an anhydrous compound.

6.26) Two. Calcium chloride dihydrate, calcium chloride 2-water, or calcium chloride 2-hydrate.

6.27) $Ba(OH)_2 \cdot 8 H_2O$; barium hydroxide octahydrate, barium hydroxide 8-water, or barium hydroxide 8-hydrate.

ANSWERS TO FORMULA WRITING AND NOMENCLATURE EXERCISE NUMBER 1

1) LiBr, lithium bromide
2) $MgBr_2$, magnesium bromide
3) NH_4Br, ammonium bromide
4) $AlBr_3$, aluminum bromide
5) NaBr, sodium bromide
6) $BaBr_2$, barium bromide
7) KBr, potassium bromide
8) $CaBr_2$, calcium bromide
9) Li_2SO_4, lithium sulfate
10) $MgSO_4$, magnesium sulfate
11) $(NH_4)_2SO_4$, ammonium sulfate
12) $Al_2(SO_4)_3$, aluminum sulfate
13) Na_2SO_4, sodium sulfate
14) $BaSO_4$, barium sulfate
15) K_2SO_4, potassium sulfate
16) $CaSO_4$, calcium sulfate
17) LiOH, lithium hydroxide
18) $Mg(OH)_2$, magnesium hydroxide
19) NH_4OH, ammonium hydroxide
20) $Al(OH)_3$, aluminum hydroxide
21) NaOH, sodium hydroxide
22) $Ba(OH)_2$, barium hydroxide
23) KOH, potassium hydroxide
24) $Ca(OH)_2$, calcium hydroxide
25) LiF, lithium fluoride
26) MgF_2, magnesium fluoride
27) NH_4F, ammonium fluoride
28) AlF_3, aluminum fluoride
29) NaF, sodium fluoride
30) BaF_2, barium fluoride
31) KF, potassium fluoride
32) CaF_2, calcium fluoride
33) Li_2O, lithium oxide
34) MgO, magnesium oxide
35) $(NH_4)_2O$, ammonium oxide
36) Al_2O_3, aluminum oxide
37) Na_2O, sodium oxide
38) BaO, barium oxide
39) K_2O, potassium oxide
40) CaO, calcium oxide
41) $LiNO_3$, lithium nitrate
42) $Mg(NO_3)_2$, magnesium nitrate
43) NH_4NO_3, ammonium nitrate
44) $Al(NO_3)_3$, aluminum nitrate
45) $NaNO_3$, sodium nitrate
46) $Ba(NO_3)_2$, barium nitrate
47) KNO_3, potassium nitrate
48) $Ca(NO_3)_2$, calcium nitrate
49) Li_3PO_4, lithium phosphate
50) $Mg_3(PO_4)_2$, magnesium phosphate
51) $(NH_4)_3PO_4$, ammonium phosphate
52) $AlPO_4$, aluminum phosphate
53) Na_3PO_4, sodium phosphate
54) $Ba_3(PO_4)_2$, barium phosphate
55) K_3PO_4, potassium phosphate
56) $Ca_3(PO_4)_2$, calcium phosphate
57) LiCl, lithium chloride
58) $MgCl_2$, magnesium chloride
59) NH_4Cl, ammonium chloride
60) $AlCl_3$, aluminum chloride
61) NaCl, sodium chloride
62) $BaCl_2$, barium chloride
63) KCl, potassium chloride
64) $CaCl_2$, calcium chloride
65) Li_2S, lithium sulfide
66) MgS, magnesium sulfide
67) $(NH_4)_2S$, ammonium sulfide
68) Al_2S_3, aluminum sulfide
69) Na_2S, sodium sulfide
70) BaS, barium sulfide
71) K_2S, potassium sulfide
72) CaS, calcium sulfide
73) LiI, lithium iodide
74) MgI_2, magnesium iodide

75) NH_4I, ammonium iodide
76) AlI_3, aluminum iodide
77) NaI, sodium iodide
78) BaI_2, barium iodide
79) KI, potassium iodide
80) CaI_2, calcium iodide
81) Li_2CO_3, lithium carbonate

82) $MgCO_3$, magnesium carbonate
83) $(NH_4)_2CO_3$, ammonium carbonate
84) $Al_2(CO_3)_3$, aluminum carbonate
85) Na_2CO_3, sodium carbonate
86) $BaCO_3$, barium carbonate
87) K_2CO_3, potassium carbonate
88) $CaCO_3$, calcium carbonate

ANSWERS TO FORMULA WRITING EXERCISE NUMBER 2

1) $AlBr_3$
2) $Al_2(CO_3)_3$
3) $AlCl_3$
4) AlF_3
5) $Al(OH)_3$
6) AlI_3
7) $Al(NO_3)_3$
8) Al_2O_3
9) $AlPO_4$
10) $Al_2(SO_4)_3$
11) Al_2S_3
12) NH_4Br
13) $(NH_4)_2CO_3$
14) NH_4Cl
15) NH_4F
16) NH_4OH
17) NH_4I
18) NH_4NO_3
19) $(NH_4)_2O$
20) $(NH_4)_3PO_4$
21) $(NH_4)_2SO_4$
22) $(NH_4)_2S$

23) $BaBr_2$
24) $BaCO_3$
25) $BaCl_2$
26) BaF_2
27) $Ba(OH)_2$
28) BaI_2
29) $Ba(NO_3)_2$
30) BaO
31) $Ba_3(PO_4)_2$
32) $BaSO_4$
33) BaS
34) $CaBr_2$
35) $CaCO_3$
36) $CaCl_2$
37) CaF_2
38) $Ca(OH)_2$
39) CaI_2
40) $Ca(NO_3)_2$
41) CaO
42) $Ca_3(PO_4)_2$
43) $CaSO_4$
44) CaS

45) $LiBr$
46) Li_2CO_3
47) $LiCl$
48) LiF
49) $LiOH$
50) LiI
51) $LiNO_3$
52) Li_2O
53) Li_3PO_4
54) Li_2SO_4
55) Li_2S
56) $MgBr_2$
57) $MgCO_3$
58) $MgCl_2$
59) MgF_2
60) $Mg(OH)_2$
61) MgI_2
62) $Mg(NO_3)_2$
63) MgO
64) $Mg_3(PO_4)_2$
65) $MgSO_4$
66) MgS

67) KBr
68) K_2CO_3
69) KCl
70) KF
71) KOH
72) KI
73) KNO_3
74) K_2O
75) K_3PO_4
76) K_2SO_4
77) K_2S
78) $NaBr$
79) Na_2CO_3
80) $NaCl$
81) NaF
82) $NaOH$
83) NaI
84) $NaNO_3$
85) Na_2O
86) Na_3PO_4
87) Na_2SO_4
88) Na_2S

CHAPTER 7

7.1) Sodium chloride is a compound, not an element.

7.2) Atomic weights refer to elements, molecular weights to molecular compounds, and formula weights to ionic compounds.

7.3) Mg, atomic weight; Br_2 and CO_2, molecular weight; Al_2O_3 and NH_4Cl, formula weight.

7.4) Atomic mass units. One atomic mass unit is exactly $\frac{1}{12}$ of the mass of one atom of carbon-12.

7.5) A *mole* is that amount of any substance that contains the same number of units as the number of atoms in exactly 12 grams of carbon-12. In chemistry, the mole is the basis by which atoms, molecules, and formula units may be "counted" by weighing. The units are too small to count by any other method.

7.6) Avogadro's number, 6.02×10^{23}.

7.7) 64.2 grams of sulfur. In 40.4 grams of neon there are 2 moles of atoms: 40.4 grams/20.2 grams per mole = 2 moles. Two moles of sulfur atoms have a mass of 2×32.1 grams per mole, or 64.2 grams.

7.8) "Molar weight of hydrogen" does not specify hydrogen molecules, H_2, or hydrogen atoms, H.

7.9) LiBr: $6.9 + 79.9 = 86.8$ g/mol

7.10) Cl_2: $2(35.5) = 71.0$ g/mol

7.11) $CaSO_4$: $40.1 + 32.1 + 4(16.0) = 136.2 = 136$ g/mol

7.12) NH_4NO_3: $14.0 + 4(1.0) + 14.0 + 3(16.0) = 80.0$ g/mol

7.13) C_6H_6: $6(12.0) + 6(1.0) = 78.0$ g/mol

7.14) $Na_2CO_3 \cdot 10\ H_2O$: $2(23.0) + 12.0 + 3(16.0) + 20(1.0) + 10(16.0) = 286$ g/mol

7.15) CrO_3: $52.0 + 3(16.0) = 100$ g/mol

$$65.4 \text{ g } \cancel{Cr} \times \frac{100 \text{ g } CrO_3}{52.0 \text{ g } \cancel{Cr}} = 126 \text{ g } CrO_3$$

7.16) NaCl: $23.0 + 35.5 = 58.5$ g/mol

$$426 \text{ g } \cancel{Na} \times \frac{35.5 \text{ g chlorine}}{23.0 \text{ g } \cancel{Na}} = 658 \text{ g chlorine}$$

7.17) FeS_2: $55.9 + 2(32.1) = 120$ g/mol

$$1.43 \times 10^6 \text{ g } \cancel{Fe} \times \frac{120 \text{ g } FeS_2}{55.9 \text{ g } \cancel{Fe}} = 3.07 \times 10^6 \text{ g } FeS_2$$

7.18) PbS: $207 + 32.1 = 239$ g/mol

$$9.08 \times 10^5 \text{ g } \cancel{earth} \times \frac{1.85 \text{ g } \cancel{PbS}}{100 \text{ g } \cancel{earth}} \times \frac{207 \text{ g Pb}}{239 \text{ g } \cancel{PbS}} = 1.45 \times 10^4 \text{ g Pb}$$

7.19) $(6.9/86.8)100 = 7.95\%$ Li: $(79.9/86.8)100 = 92.0\%$ Br

7.20) $(40.1/136)100 = 29.5\%$ Ca; $(32.1/136)100 = 23.6\%$ S; $(4 \times 16.0/136)100 = 47.1\%$ O

7.21) $(6 \times 1.0/78.0)100 = 7.7\%$ H; $(6 \times 12.0/78.0)100 = 92.3\%$ C

7.22) $(180/286)100 = 62.9\%$ H_2O

7.23) $68.4 \text{ g } \cancel{LiBr} \times \dfrac{1 \text{ mol LiBr}}{86.8 \text{ g } \cancel{LiBr}} = 0.788$ mol LiBr

7.24) $17.2 \text{ g } \cancel{Cl_2} \times \dfrac{1 \text{ mol } Cl_2}{71.0 \text{ g } \cancel{Cl_2}} = 0.242$ mol Cl_2

7.25) $34.1 \text{ g } \cancel{CaSO_4} \times \dfrac{1 \text{ mol } CaSO_4}{136 \text{ g } \cancel{CaSO_4}} = 0.251$ mol $CaSO_4$

7.26) $0.345 \text{ mol } \cancel{NH_4NO_3} \times \dfrac{80.0 \text{ g } NH_4NO_3}{1 \text{ mol } \cancel{NH_4NO_3}} = 27.6 \text{ g } NH_4NO_3$

7.27) $1.82 \text{ mol } \cancel{C_6H_6} \times \dfrac{78.0 \text{ g } C_6H_6}{1 \text{ mol } \cancel{C_6H_6}} = 142 \text{ g } C_6H_6$

7.28) $0.791 \text{ mol } \cancel{Na_2CO_3 \cdot 10\ H_2O} \times \dfrac{286 \text{ g } Na_2CO_3 \cdot 10\ H_2O}{1 \text{ mol } \cancel{Na_2CO_3 \cdot 10\ H_2O}} = 226 \text{ g } Na_2CO_3 \cdot 10\ H_2O$

7.29) $1.24 \text{ mol } \cancel{Mg} \times \dfrac{6.02 \times 10^{23} \text{ atoms Mg}}{1 \text{ mol } \cancel{Mg}} = 7.46 \times 10^{23}$ atoms Mg

7.30) $0.713 \text{ mol } \cancel{Br_2} \times \dfrac{2 \times 6.02 \times 10^{23} \text{ atoms Br}}{1 \text{ mol } \cancel{Br_2}} = 8.58 \times 10^{23}$ atoms Br

7.31) $29.6 \text{ g } \cancel{Na} \times \dfrac{1 \text{ mol } \cancel{Na}}{23.0 \text{ g } \cancel{Na}} \times \dfrac{6.02 \times 10^{23} \text{ Na atoms}}{1 \text{ mol } \cancel{Na}} = 7.75 \times 10^{23}$ Na atoms

7.32) $3.40 \text{ g Ca} \times \dfrac{1 \text{ mol Ca}}{40.1 \text{ g Ca}} \times \dfrac{6.02 \times 10^{23} \text{ Ca atoms}}{1 \text{ mol Ca}} = 5.10 \times 10^{22}$ Ca atoms

7.33) $0.004\,61 \text{ g I}_2 \times \dfrac{1 \text{ mol I}_2}{254 \text{ g I}_2} \times \dfrac{2 \text{ mol I}}{1 \text{ mol I}_2} \times \dfrac{6.02 \times 10^{23} \text{ I atoms}}{1 \text{ mol I}}$
$= 2.19 \times 10^{19}$ I atoms

7.34) $0.521 \text{ mol H}_2\text{S} \times \dfrac{6.02 \times 10^{23} \text{ H}_2\text{S molecules}}{1 \text{ mol H}_2\text{S}} = 3.14 \times 10^{23}$ H$_2$S molecules

7.35) $0.0626 \text{ mol Br}_2 \times \dfrac{6.02 \times 10^{23} \text{ Br}_2 \text{ molecules}}{1 \text{ mol Br}_2} = 3.77 \times 10^{22}$ Br$_2$ molecules

7.36) $12.4 \text{ g N}_2 \times \dfrac{1 \text{ mol N}_2}{28.0 \text{ g N}_2} \times \dfrac{6.02 \times 10^{23} \text{ N}_2 \text{ molecules}}{1 \text{ mol N}_2} = 2.67 \times 10^{23}$ N$_2$ molecules

7.37) $6.45 \text{ g CO} \times \dfrac{1 \text{ mol CO}}{28.0 \text{ g CO}} \times \dfrac{6.02 \times 10^{23} \text{ CO molecules}}{1 \text{ mol CO}} = 1.39 \times 10^{23}$ CO molecules

7.38) $13.6 \text{ g LiBr} \times \dfrac{1 \text{ mol LiBr}}{86.8 \text{ g LiBr}} \times \dfrac{6.02 \times 10^{23} \text{ LiBr units}}{1 \text{ mol LiBr}} = 9.43 \times 10^{22}$ LiBr units

7.39) $2.35 \times 10^{21} \text{ Ba atoms} \times \dfrac{1 \text{ mol Ba}}{6.02 \times 10^{23} \text{ Ba atoms}} = 3.90 \times 10^{-3}$ mol Ba

7.40) $1.09 \times 10^{23} \text{ Br}_2 \text{ molecules} \times \dfrac{1 \text{ mol Br}_2}{6.02 \times 10^{23} \text{ Br}_2 \text{ molecules}} = 0.181$ mol Br$_2$

7.41) $7.06 \times 10^{23} \text{ He atoms} \times \dfrac{1 \text{ mol He}}{6.02 \times 10^{23} \text{ He atoms}} \times \dfrac{4.00 \text{ g He}}{1 \text{ mol He}} = 4.69$ g He

7.42) $4.06 \times 10^{22} \text{ O}_2 \text{ molecules} \times \dfrac{1 \text{ mol O}_2}{6.02 \times 10^{23} \text{ O}_2 \text{ molecules}} \times \dfrac{32.0 \text{ g O}_2}{1 \text{ mol O}_2} = 2.16$ g O$_2$

7.43) $1.19 \times 10^{20} \text{ KI units} \times \dfrac{1 \text{ mol KI}}{6.02 \times 10^{23} \text{ KI units}} \times \dfrac{166 \text{ g KI}}{1 \text{ mol KI}} = 0.0328$ g KI

7.44) 6 and 10 are both divisible by 2, so the empirical formula of C$_6$H$_{10}$ is C$_3$H$_5$. C$_7$H$_{10}$ cannot be "reduced," and is therefore an empirical formula. It may be—and, in fact, is—a molecular formula.

7.45–7.49)

	ELEMENT	GRAMS	MOLES	MOLE RATIO	FORMULA RATIO	EMPIRICAL FORMULA
7.45)	C	40.0	3.33	1	1	
	H	6.7	6.7	2	2	CH$_2$O
	O	53.3	3.33	1	1	
7.46)	C	25.3	2.11	1	3	
	H	3.5	3.5	1.66	5	C$_3$H$_5$
7.47)	Na	32.4	1.41	2	2	
	S	22.6	0.704	1	1	Na$_2$SO$_4$
	O	45.0	2.81	3.99	4	
7.48)	Mg	13.2	0.543	1	1	
	Ca	21.8	0.544	1	1	
	C	13.0	1.08	2	2	MgCaC$_2$O$_6$
	O	52.1	3.26	6	6	

	ELEMENT	GRAMS	MOLES	MOLE RATIO	FORMULA RATIO	EMPIRICAL FORMULA
7.49)	H	5.00	5.00	1	1	HF—Molar
	F	95.00	5.00	1	1	weight = 20.0 g/mol formula units

$$\frac{40.0 \text{ g compound}}{1 \text{ mol compound}} \times \frac{1 \text{ mol EF units}}{20.0 \text{ g compound}} = \frac{2 \text{ mol EF units}}{1 \text{ mol compound}}$$

	ELEMENT	GRAMS	MOLES	MOLE RATIO	FORMULA RATIO	EMPIRICAL FORMULA
7.50)	P	10.3	0.332	1	2	P_2O_3
	O	8.0	0.50	1.5	3	

$$\frac{220 \text{ g compound}}{1 \text{ mol compound}} \times \frac{1 \text{ mol EF units}}{110 \text{ g compound}} = \frac{2 \text{ mol EF units}}{1 \text{ mol compound}}$$

Molecular formula is therefore P_4O_6.

CHAPTER 8

ANSWERS TO EQUATIONS BALANCING EXERCISE

1) $4 \text{ Na} + O_2 \rightarrow 2 \text{ Na}_2O$
2) $H_2 + Cl_2 \rightarrow 2 \text{ HCl}$
3) $4 \text{ P} + 3 O_2 \rightarrow 2 P_2O_3$
4) $KClO_4 \rightarrow KCl + 2 O_2$
5) $Sb_2S_3 + 6 \text{ HCl} \rightarrow 2 \text{ SbCl}_3 + 3 H_2S$
6) $2 \text{ NH}_3 + H_2SO_4 \rightarrow (NH_4)_2SO_4$
7) $CuO + 2 \text{ HCl} \rightarrow CuCl_2 + H_2O$
8) $Zn + Pb(NO_3)_2 \rightarrow Zn(NO_3)_2 + Pb$
9) $2 \text{ AgNO}_3 + H_2S \rightarrow Ag_2S + 2 \text{ HNO}_3$
10) $2 \text{ Cu} + S \rightarrow Cu_2S$
11) $2 \text{ Al} + 2 H_3PO_4 \rightarrow 3 H_2 + 2 \text{ AlPO}_4$
12) $2 \text{ NaNO}_3 \rightarrow 2 \text{ NaNO}_2 + O_2$
13) $Mg(ClO_3)_2 \rightarrow MgCl_2 + 3 O_2$
14) $2 H_2O_2 \rightarrow 2 H_2O + O_2$
15) $2 \text{ BaO}_2 \rightarrow 2 \text{ BaO} + O_2$
16) $H_2CO_3 \rightarrow H_2O + CO_2$
17) $Pb(NO_3)_2 + 2 \text{ KCl} \rightarrow PbCl_2 + 2 \text{ KNO}_3$
18) $2 \text{ Al} + 3 Cl_2 \rightarrow 2 \text{ AlCl}_3$
19) $4 \text{ P} + 5 O_2 \rightarrow 2 P_2O_5$
20) $NH_4NO_2 \rightarrow N_2 + 2 H_2O$
21) $3 H_2 + N_2 \rightarrow 2 \text{ NH}_3$
22) $Cl_2 + 2 \text{ KBr} \rightarrow Br_2 + 2 \text{ KCl}$
23) $BaCl_2 + (NH_4)_2CO_3 \rightarrow BaCO_3 + 2 \text{ NH}_4Cl$
24) $MgCO_3 + 2 \text{ HCl} \rightarrow MgCl_2 + CO_2 + H_2O$
25) $2 \text{ P} + 3 I_2 \rightarrow 2 \text{ PI}_3$
26) $2 \text{ PbO}_2 \rightarrow 2 \text{ PbO} + O_2$
27) $2 \text{ Al} + 6 \text{ HCl} \rightarrow 2 \text{ AlCl}_3 + 3 H_2$
28) $Fe_2(SO_4)_3 + 3 \text{ Ba(OH)}_2 \rightarrow 3 \text{ BaSO}_4 + 2 \text{ Fe(OH)}_3$

29) $2 \text{Al} + 3 \text{CuSO}_4 \rightarrow \text{Al}_2(\text{SO}_4)_3 + 3 \text{Cu}$
30) $2 \text{KClO}_3 \rightarrow 2 \text{KCl} + 3 \text{O}_2$
31) $3 \text{Mg} + \text{N}_2 \rightarrow \text{Mg}_3\text{N}_2$
32) $2 \text{C}_6\text{H}_{14} + 19 \text{O}_2 \rightarrow 12 \text{CO}_2 + 14 \text{H}_2\text{O}$
33) $3 \text{FeCl}_2 + 2 \text{Na}_3\text{PO}_4 \rightarrow \text{Fe}_3(\text{PO}_4)_2 + 6 \text{NaCl}$
34) $\text{Li}_2\text{O} + \text{HOH} \rightarrow 2 \text{LiOH}$
35) $2 \text{HgO} \rightarrow 2 \text{Hg} + \text{O}_2$
36) $\text{CaSO}_4 \cdot 2 \text{H}_2\text{O} \rightarrow \text{CaSO}_4 + 2 \text{H}_2\text{O}$
37) $2 \text{C}_3\text{H}_7\text{CHO} + 11 \text{O}_2 \rightarrow 8 \text{CO}_2 + 8 \text{H}_2\text{O}$
38) $\text{NaHCO}_3 + \text{HCl} \rightarrow \text{NaCl} + \text{H}_2\text{O} + \text{CO}_2$
39) $\text{Bi}(\text{NO}_3)_3 + 3 \text{NaOH} \rightarrow \text{Bi}(\text{OH})_3 + 3 \text{NaNO}_3$
40) $\text{FeS} + 2 \text{HBr} \rightarrow \text{FeBr}_2 + \text{H}_2\text{S}$
41) $\text{Zn}(\text{OH})_2 + \text{H}_2\text{SO}_4 \rightarrow \text{ZnSO}_4 + 2 \text{HOH}$
42) $\text{P}_4\text{O}_{10} + 6 \text{H}_2\text{O} \rightarrow 4 \text{H}_3\text{PO}_4$
43) $\text{C}_4\text{H}_9\text{OH} + 6 \text{O}_2 \rightarrow 4 \text{CO}_2 + 5 \text{H}_2\text{O}$
44) $\text{CaC}_2 + 2 \text{H}_2\text{O} \rightarrow \text{C}_2\text{H}_2 + \text{Ca}(\text{OH})_2$
45) $3 \text{CaCO}_3 + 2 \text{H}_3\text{PO}_4 \rightarrow \text{Ca}_3(\text{PO}_4)_2 + 3 \text{CO}_2 + 3 \text{H}_2\text{O}$
46) $\text{PCl}_5 + 4 \text{H}_2\text{O} \rightarrow \text{H}_3\text{PO}_4 + 5 \text{HCl}$
47) $\text{CaI}_2 + \text{H}_2\text{SO}_4 \rightarrow 2 \text{HI} + \text{CaSO}_4$
48) $\text{C}_3\text{H}_7\text{COOH} + 5 \text{O}_2 \rightarrow 4 \text{CO}_2 + 4 \text{H}_2\text{O}$
49) $\text{Mg}(\text{CN})_2 + 2 \text{HCl} \rightarrow 2 \text{HCN} + \text{MgCl}_2$
50) $(\text{NH}_4)_2\text{S} + \text{HgBr}_2 \rightarrow 2 \text{NH}_4\text{Br} + \text{HgS}$

QUESTIONS AND PROBLEMS

8.1)   Two molecules of benzene react with fifteen molecules of oxygen to produce twelve molecules of carbon dioxide and six molecules of water.

8.2) $2 \text{Ca}(s) + \text{O}_2(g) \rightarrow 2 \text{CaO}(s)$

8.3) $4 \text{P}(s) + 5 \text{O}_2(g) \rightarrow \text{P}_4\text{O}_{10}(s)$; also $\text{P}_4(s) + 5 \text{O}_2(g) \rightarrow \text{P}_4\text{O}_{10}(s)$

8.4) $2 \text{K}(s) + \text{F}_2(g) \rightarrow 2 \text{KF}(s)$

8.5) $\text{Si}(s) + 2 \text{Cl}_2(g) \rightarrow \text{SiCl}_4(s)$

8.6) $3 \text{Mg}(s) + \text{N}_2(g) \rightarrow \text{Mg}_3\text{N}_2(s)$

8.7) $2 \text{NaCl}(s) \rightarrow 2 \text{Na}(s) + \text{Cl}_2(g)$

8.8) $2 \text{HgO}(s) \rightarrow 2 \text{Hg}(l) + \text{O}_2(g)$

8.9) $\text{H}_2\text{CO}_3(aq) \rightarrow \text{H}_2\text{O}(l) + \text{CO}_2(g)$

8.10) $2 \text{H}_2\text{O}_2(l) \rightarrow 2 \text{H}_2\text{O}(l) + \text{O}_2(g)$

8.11) $\text{C}_3\text{H}_8(g) + 5 \text{O}_2(g) \rightarrow 3 \text{CO}_2(g) + 4 \text{H}_2\text{O}(l)$

8.12) $2 \text{C}_2\text{H}_2(g) + 5 \text{O}_2(g) \rightarrow 4 \text{CO}_2(g) + 2 \text{H}_2\text{O}(l)$

8.13) $2 \text{CH}_3\text{CHO}(l) + 5 \text{O}_2(g) \rightarrow 4 \text{CO}_2(g) + 4 \text{H}_2\text{O}(l)$

8.14) $\text{C}_{12}\text{H}_{22}\text{O}_{11}(s) + 12 \text{O}_2(g) \rightarrow 12 \text{CO}_2(g) + 11 \text{H}_2\text{O}(l)$

8.15) $\text{Mg}(s) + \text{H}_2\text{SO}_4(aq) \rightarrow \text{MgSO}_4(aq) + \text{H}_2(g)$

8.16) $\text{Ba}(s) + 2 \text{HOH}(l) \rightarrow \text{Ba}(\text{OH})_2(aq) + \text{H}_2(g)$

8.17) $\text{Zn}(s) + 2 \text{AgNO}_3(aq) \rightarrow 2 \text{Ag}(s) + \text{Zn}(\text{NO}_3)_2(aq)$

8.18) $\text{Mg}(s) + \text{NiCl}_2(aq) \rightarrow \text{Ni}(s) + \text{MgCl}_2(aq)$

8.19) $\text{Br}_2(l) + 2 \text{NaI}(aq) \rightarrow \text{I}_2(aq) + 2 \text{NaBr}(aq)$

8.20) $\text{AgNO}_3(aq) + \text{KBr}(aq) \rightarrow \text{AgBr}(s) + \text{KNO}_3(aq)$

8.21) $\text{Pb}(\text{NO}_3)_2(aq) + \text{CuSO}_4(aq) \rightarrow \text{PbSO}_4(s) + \text{Cu}(\text{NO}_3)_2(aq)$

8.22) $MgCl_2(aq) + 2\,NaF(aq) \rightarrow 2\,NaCl(aq) + MgF_2(s)$

8.23) $Na_2S(aq) + 2\,AgNO_3(aq) \rightarrow 2\,NaNO_3(aq) + Ag_2S(s)$

8.24) $2\,NaOH(aq) + MgBr_2(aq) \rightarrow 2\,NaBr(aq) + Mg(OH)_2(s)$

8.25) $KOH(aq) + HNO_3(aq) \rightarrow KNO_3(aq) + HOH(\ell)$

8.26) $Mg(OH)_2(s) + 2\,HCl(aq) \rightarrow MgCl_2(aq) + 2\,HOH(\ell)$

8.27) $Zn(s) + H_2O(g) \rightarrow ZnO(s) + H_2(g)$

8.28) $BaO(s) + H_2O(\ell) \rightarrow Ba(OH)_2(aq)$

8.29) $3\,FeO(s) + 2\,Al(s) \rightarrow 3\,Fe(s) + Al_2O_3(s)$

8.30) $Fe_2O_3(s) + 3\,CO(g) \rightarrow 2\,Fe(s) + 3\,CO_2(g)$

CHAPTER 9

9.1) $3.91 \text{ mol CO}_2 \times \dfrac{1 \text{ mol CaCO}_3}{1 \text{ mol CO}_2} = 3.91 \text{ mol CaCO}_3$

9.2) $0.284 \text{ mol HCl} \times \dfrac{1 \text{ mol CO}_2}{2 \text{ mol HCl}} = 0.142 \text{ mol CO}_2$

9.3) $0.462 \text{ mol CaCO}_3 \times \dfrac{1 \text{ mol CO}_2}{1 \text{ mol CaCO}_3} \times \dfrac{44.0 \text{ g CO}_2}{1 \text{ mol CO}_2} = 20.3 \text{ g CO}_2$

9.4) $54.1 \text{ g CaCO}_3 \times \dfrac{1 \text{ mol CaCO}_3}{100.1 \text{ g CaCO}_3} \times \dfrac{1 \text{ mol CaCl}_2}{1 \text{ mol CaCO}_3} = 0.540 \text{ mol CaCl}_2$

9.5) $95.2 \text{ g NaCl} \times \dfrac{1 \text{ mol NaCl}}{58.5 \text{ g NaCl}} \times \dfrac{1 \text{ mol Na}_2\text{SO}_4}{2 \text{ mol NaCl}} \times \dfrac{142 \text{ g Na}_2\text{SO}_4}{1 \text{ mol Na}_2\text{SO}_4}$
$$= 116 \text{ g Na}_2\text{SO}_4$$

9.6) $1.09 \text{ g AgBr} \times \dfrac{1 \text{ mol AgBr}}{188 \text{ g AgBr}} \times \dfrac{2 \text{ mol Na}_2\text{S}_2\text{O}_3}{1 \text{ mol AgBr}} \times \dfrac{158 \text{ g Na}_2\text{S}_2\text{O}_3}{1 \text{ mol Na}_2\text{S}_2\text{O}_3}$
$$= 1.83 \text{ g Na}_2\text{S}_2\text{O}_3$$

9.7) $12 \text{ g KHC}_4\text{H}_4\text{O}_6 \times \dfrac{1 \text{ mol KHC}_4\text{H}_4\text{O}_6}{188 \text{ g KHC}_4\text{H}_4\text{O}_6} \times \dfrac{1 \text{ mol NaHCO}_3}{1 \text{ mol KHC}_4\text{H}_4\text{O}_6} \times$
$$\dfrac{84.0 \text{ g NaHCO}_3}{1 \text{ mol NaHCO}_3} = 5.4 \text{ g NaHCO}_3$$

9.8) $2.69 \text{ g MnO}_2 \times \dfrac{1 \text{ mol MnO}_2}{86.9 \text{ g MnO}_2} \times \dfrac{1 \text{ mol Cl}_2}{1 \text{ mol MnO}_2} \times \dfrac{71.0 \text{ g Cl}_2}{1 \text{ mol Cl}_2} = 2.20 \text{ g Cl}_2$

9.9) $45.0 \text{ g CHCl}_3 \times \dfrac{1 \text{ mol CHCl}_3}{119 \text{ g CHCl}_3} \times \dfrac{3 \text{ mol Cl}_2}{1 \text{ mol CHCl}_3} \times \dfrac{71.0 \text{ g Cl}_2}{1 \text{ mol Cl}_2} = 80.5 \text{ g Cl}_2$

9.10) $175 \text{ g C}_{17}\text{H}_{35}\text{COONa} \times \dfrac{1 \text{ mol C}_{17}\text{H}_{35}\text{COONa}}{306 \text{ g C}_{17}\text{H}_{35}\text{COONa}} \times$
$$\dfrac{3 \text{ mol NaOH}}{3 \text{ mol C}_{17}\text{H}_{35}\text{COONa}} \times \dfrac{40.0 \text{ g NaOH}}{1 \text{ mol NaOH}} = 22.9 \text{ g NaOH}$$

9.11) $255 \text{ g Cl}_2 \times \dfrac{1 \text{ mol Cl}_2}{71.0 \text{ g Cl}_2} \times \dfrac{1 \text{ mol O}_2}{2 \text{ mol Cl}_2} \times \dfrac{32.0 \text{ g O}_2}{1 \text{ mol O}_2} = 57.5 \text{ g O}_2$

9.12) $6.90 \text{ g CO}_2 \times \dfrac{1 \text{ mol CO}_2}{44.0 \text{ g CO}_2} \times \dfrac{1 \text{ mol C}_6\text{H}_{12}\text{O}_6}{6 \text{ mol CO}_2} \times \dfrac{180 \text{ g C}_6\text{H}_{12}\text{O}_6}{1 \text{ mol C}_6\text{H}_{12}\text{O}_6} = 4.70 \text{ g C}_6\text{H}_{12}\text{O}_6$

9.13) $48.3 \text{ g (CaSO}_4)_2 \cdot \text{H}_2\text{O} \times \dfrac{1 \text{ mol (CaSO}_4)_2 \cdot \text{H}_2\text{O}}{290 \text{ g (CaSO}_4)_2 \cdot \text{H}_2\text{O}} \times$

$$\dfrac{2 \text{ mol CaSO}_4 \cdot 2\text{H}_2\text{O}}{1 \text{ mol (CaSO}_4)_2 \cdot \text{H}_2\text{O}} \times \dfrac{172 \text{ g CaSO}_4 \cdot 2\text{ H}_2\text{O}}{1 \text{ mol CaSO}_4 \cdot 2\text{ H}_2\text{O}} = 57.3 \text{ g CaSO}_4 \cdot 2\text{ H}_2\text{O}$$

9.14) $12\ 800 \text{ g N}_2\text{O} \times \dfrac{1 \text{ mol N}_2\text{O}}{44.0 \text{ g N}_2\text{O}} \times \dfrac{1 \text{ mol NH}_4\text{NO}_3}{1 \text{ mol N}_2\text{O}} \times \dfrac{80.0 \text{ g NH}_4\text{NO}_3}{1 \text{ mol NH}_4\text{NO}_3}$

$$= 23\ 300 \text{ g NH}_4\text{NO}_3$$

9.15) $5.00 \text{ tons SO}_2 \times \dfrac{1 \text{ ton mol SO}_2}{64.1 \text{ tons SO}_2} \times \dfrac{4 \text{ ton mol FeS}_2}{8 \text{ ton mol SO}_2} \times$

$$\dfrac{120 \text{ tons FeS}_2}{1 \text{ ton mol FeS}_2} \times \dfrac{100 \text{ tons ore}}{12.1 \text{ tons FeS}_2} = 38.7 \text{ tons ore}$$

9.16) $3.94 \text{ g Zn} \times \dfrac{1 \text{ mol Zn}}{65.4 \text{ g Zn}} \times \dfrac{1 \text{ mol H}_2}{1 \text{ mol Zn}} \times \dfrac{24.5 \text{ L H}_2}{1 \text{ mol H}_2} = 1.48 \text{ L H}_2$

9.17) $5.19 \times 10^4 \text{ g CH}_4 \times \dfrac{1 \text{ mol CH}_4}{16.0 \text{ g CH}_4} \times \dfrac{1 \text{ mol C}_2\text{H}_2}{2 \text{ mol CH}_4} \times \dfrac{26.0 \text{ g C}_2\text{H}_2}{1 \text{ mol C}_2\text{H}_2}$

$$= 4.22 \times 10^4 \text{ g C}_2\text{H}_2 \text{ theoretical}$$

$$\dfrac{3.81 \times 10^4 \text{ g}}{4.22 \times 10^4 \text{ g}} \times 100 = 90.3\% \text{ yield}$$

9.18) $7.00 \times 10^6 \text{ g Cu}_2\text{S} \times \dfrac{1 \text{ mol Cu}_2\text{S}}{159 \text{ g Cu}_2\text{S}} \times \dfrac{2 \text{ mol Cu}}{1 \text{ mol Cu}_2\text{S}} \times$

$$\dfrac{63.5 \text{ g Cu theoretical}}{1 \text{ mol Cu}} \times \dfrac{61.2 \text{ g Cu actual}}{100 \text{ g Cu theoretical}} = 3.42 \times 10^6 \text{ g Cu actual}$$

9.19) $325 \text{ kg NaOH actual} \times \dfrac{100 \text{ kg NaOH theoretical}}{92.0 \text{ kg NaOH actual}} \times \dfrac{1 \text{ mol NaOH}}{0.0400 \text{ kg NaOH}} \times$

$$\dfrac{1 \text{ mol Na}_2\text{CO}_3}{2 \text{ mol NaOH}} \times \dfrac{0.106 \text{ kg Na}_2\text{CO}_3}{1 \text{ mol Na}_2\text{CO}_3} = 468 \text{ kg Na}_2\text{CO}_3$$

9.20) $3.85 \text{ g CH}_4 \times \dfrac{1 \text{ mol CH}_4}{16.0 \text{ g CH}_4} \times \dfrac{1 \text{ mol CHCl}_3}{1 \text{ mol CH}_4} \times \dfrac{120 \text{ g CHCl}_3}{1 \text{ mol CHCl}_3}$

$$= 28.9 \text{ g CHCl}_3 \text{ theoretical}$$

$$\dfrac{23.0 \text{ g CHCl}_3 \text{ actual}}{28.9 \text{ g CHCl}_3 \text{ theoretical}} \times 100 = 79.6\% \text{ yield}$$

9.21) $15.1 \text{ g C}_7\text{H}_6\text{O}_3 \times \dfrac{1 \text{ mol C}_7\text{H}_6\text{O}_3}{138 \text{ g C}_7\text{H}_6\text{O}_3} \times \dfrac{1 \text{ mol C}_9\text{H}_8\text{O}_4}{1 \text{ mol C}_7\text{H}_6\text{O}_3} \times \dfrac{180 \text{ g C}_9\text{H}_8\text{O}_4 \text{ theoretical}}{1 \text{ mol C}_9\text{H}_8\text{O}_4} \times$

$$\dfrac{76.1 \text{ g C}_9\text{H}_8\text{O}_4 \text{ actual}}{100 \text{ g C}_9\text{H}_8\text{O}_4 \text{ theoretical}} = 15.0 \text{ g C}_9\text{H}_8\text{O}_4$$

9.22) $5.00 \times 10^4 \text{ g CaCN}_2 \text{ actual} \times \dfrac{100 \text{ g CaCN}_2 \text{ theoretical}}{72.6 \text{ g CaCN}_2 \text{ actual}} \times$

$$\dfrac{1 \text{ mol CaCN}_2}{80.1 \text{ g CaCN}_2} \times \dfrac{1 \text{ mol CaC}_2}{1 \text{ mol CaCN}_2} \times \dfrac{64.1 \text{ g CaC}_2}{1 \text{ mol CaC}_2} = 5.51 \times 10^4 \text{ g CaC}_2$$

9.23) $40.0 \text{ g NH}_3 \times \dfrac{1 \text{ mol NH}_3}{17.0 \text{ g NH}_3} \times \dfrac{6 \text{ mol H}_2\text{O}}{4 \text{ mol NH}_3} \times \dfrac{18.0 \text{ g H}_2\text{O}}{1 \text{ mol H}_2\text{O}} = 63.5 \text{ g H}_2\text{O}$

$$40.0 \text{ g O}_2 \times \dfrac{1 \text{ mol O}_2}{32.0 \text{ g O}_2} \times \dfrac{6 \text{ mol H}_2\text{O}}{3 \text{ mol O}_2} \times \dfrac{18.0 \text{ g H}_2\text{O}}{1 \text{ mol H}_2\text{O}} = 45.0 \text{ g H}_2\text{O maximum yield}$$

$$40.0 \text{ g O}_2 \times \dfrac{1 \text{ mol O}_2}{32.0 \text{ g O}_2} \times \dfrac{4 \text{ mol NH}_3}{3 \text{ mol O}_2} \times \dfrac{17.0 \text{ g NH}_3}{1 \text{ mol NH}_3} = 28.3 \text{ g NH}_3 \text{ g used}$$

40.0 g NH_3 at start $-$ 28.3 g NH_3 used $= 11.7$ g NH_3 left

9.24) $19.6 \text{ g C} \times \dfrac{1 \text{ mol C}}{12.0 \text{ g C}} \times \dfrac{1 \text{ mol Zn}}{1 \text{ mol C}} \times \dfrac{65.4 \text{ g Zn}}{1 \text{ mol Zn}} = 107$ g Zn maximum yield

$135 \text{ g ZnO} \times \dfrac{1 \text{ mol ZnO}}{81.4 \text{ g ZnO}} \times \dfrac{1 \text{ mol Zn}}{1 \text{ mol ZnO}} \times \dfrac{65.4 \text{ g Zn}}{1 \text{ mol Zn}} = 108$ g Zn

$19.6 \text{ g C} \times \dfrac{1 \text{ mol C}}{12.0 \text{ g C}} \times \dfrac{1 \text{ mol ZnO}}{1 \text{ mol C}} \times \dfrac{81.4 \text{ g ZnO}}{1 \text{ mol ZnO}} = 133$ g ZnO used

135 g ZnO at start $-$ 133 g ZnO used $= 2$ g ZnO left

9.25) $3.00 \text{ ton } S_2Cl_2 \times \dfrac{1 \text{ ton mol } S_2Cl_2}{135 \text{ ton } S_2Cl_2} \times \dfrac{1 \text{ ton mol } CCl_4}{2 \text{ ton mol } S_2Cl_2} \times \dfrac{154 \text{ ton } CCl_4}{1 \text{ ton mol } CCl_4}$

$= 1.71$ ton CCl_4 maximum yield

$1.00 \text{ ton } CS_2 \times \dfrac{1 \text{ ton mol } CS_2}{76.2 \text{ ton } CS_2} \times \dfrac{1 \text{ ton mol } CCl_4}{1 \text{ ton mol } CS_2} \times \dfrac{154 \text{ ton } CCl_4}{1 \text{ ton mol } CCl_4}$

$= 2.02$ ton CCl_4

$3.00 \text{ ton } S_2Cl_2 \times \dfrac{1 \text{ ton mol } S_2Cl_2}{135 \text{ ton } S_2Cl_2} \times \dfrac{1 \text{ ton mol } CS_2}{2 \text{ ton mol } S_2Cl_2} \times \dfrac{76.2 \text{ ton } CS_2}{1 \text{ ton mol } CS_2}$

$= 0.847$ ton CS_2 used

1.00 ton CS_2 at start $-$ 0.847 ton CS_2 used $= 0.15$ ton CS_2 left

9.26) $MgCl_2 \cdot 6\, H_2O \rightarrow Mg;$ $MgCl_2 \cdot 6\, H_2O \rightarrow 2\, AgCl$

$4.01 \text{ g AgCl} \times \dfrac{1 \text{ mol AgCl}}{144 \text{ g AgCl}} \times \dfrac{1 \text{ mol Mg}}{2 \text{ mol AgCl}} \times \dfrac{24.3 \text{ g Mg}}{1 \text{ mol Mg}} = 0.339$ g Mg

$4.01 \text{ g AgCl} \times \dfrac{1 \text{ mol AgCl}}{144 \text{ g AgCl}} \times \dfrac{1 \text{ mol } MgCl_2 \cdot 6\, H_2O}{2 \text{ mol AgCl}} \times \dfrac{203 \text{ g } MgCl_2 \cdot 6\, H_2O}{1 \text{ mol } MgCl_2 \cdot 6\, H_2O}$

$= 2.83$ g $MgCl_2 \cdot 6\, H_2O$

$\dfrac{2.83 \text{ g } MgCl_2 \cdot 6\, H_2O}{5.25 \text{ g sample}} \times 100 = 53.9\%\ MgCl_2 \cdot 6\, H_2O$

CHAPTER 10

10.1) K ·   · P̈ :   : B̈r :

10.2) Ga :   · Pb :

10.3) F, Cl, Br, I, At.

10.4) Na^+, Mg^{2+}, and Al^{3+} are isoelectronic with Ne; S^{2-} and Cl^- are isoelectronic with Ar; less common, P^{3-}.

10.5) O^{2-} and F^-; less common, N^{3-}.

10.6) S^{2-}, K^+, Ca^{2+}; less common, P^{3-} and Sc^{3+}.

10.7) (a) Krypton, $Z = 36$; (b) selenide ion, Se^{2-}; (c) Rb^+.

10.8) $K \cdot$ ⟶ : S̈ : → K^+ $\left[: \ddot{S} : \right]^{2-}$ K^+ K_2S
       $K \cdot$ +

10.9) S^{2-}.

10.10) Ca^{2+}, K^+, Cl^-, Br^-.

10.11) For noble gas atoms, and the monatomic ions that are isoelectronic with them, size decreases with increasing atomic number. Larger number of protons in nucleus gives it greater attracting force for electrons and pulls them in more closely.

10.12) Ions are formed when neutral atoms lose or gain electrons. The electron(s) that is(are) lost by one atom is(are) gained by another, effectively a transfer of electrons from one atom to another. The attraction between the ions produced makes up the "ionic bond." Covalent bonds are formed when a pair of electrons is shared by the two bonded atoms. Effectively the electrons belong to both atoms, spending some time near each nucleus.

10.13) K—Cl bond is ionic, formed by "transferring" an electron from a potassium atom to a chlorine atom. Cl—Cl bond is covalent, formed by two chlorine atoms sharing a pair of electrons.

10.14) $:\ddot{C}l\cdot + \cdot\ddot{I}: \rightarrow :\ddot{C}l:\ddot{I}:$

10.15) Nonmetal atoms are usually one, two or possibly three or four short of an "octet" of electrons, and achieve that octet most easily by gaining the missing electrons. When two nonmetal atoms combine, the easiest way for both atoms to reach the octet is to gain each other's electrons, or share them, forming a covalent bond. If the second atom is a metal, however, it has one, two or possibly three electrons more than an octet. It reaches the octet by giving its electrons to the nonmetal, becoming a positive ion itself, and making the nonmetal atom a negative ion. The two atoms form an ionic bond.

10.16) The energy of the system decreases.

10.17) A bond that has a symmetrical distribution of electrical charge is nonpolar; if the charge distribution is unsymmetrical the bond is polar. Bonds between identical atoms are completely nonpolar.

10.18) Cl—Cl; Br—Cl; I—Cl; F—Cl. The Cl—Cl bond is nonpolar; the other bonds are polar.

10.19) In Br—Cl and I—Cl, chlorine is the more electronegative; in F—Cl, fluorine is the negative pole.

10.20) "Electronegativity is a measure of the relative ability of two atoms to attract the pair of electrons forming a single covalent bond between them." Because noble gases do not normally form bonds, electronegativity numbers are not assigned to them.

10.21) A multiple bond exists when two atoms are bonded by two or three pairs of electrons, forming two or three covalent bonds. A bond with one pair of electrons is a single bond, two pairs are a double bond, and three pairs are a triple bond.

CHAPTER 11

11.1)   H—$\ddot{Br}$:   H—$\ddot{S}$:   H—$\ddot{P}$—H
                          | |
                          H   H

        HBr   H$_2$S   PH$_3$

11.2) $:\ddot{F}—\ddot{O}:$ $:C≡O:$ $\left[\begin{array}{c} :\ddot{O}: \\ | \\ :\ddot{O}—S—\ddot{O}: \\ | \\ :\ddot{O}: \end{array}\right]^{2-}$
              |
          $:\ddot{F}:$

        OF$_2$ CO   SO$_4{}^{2-}$

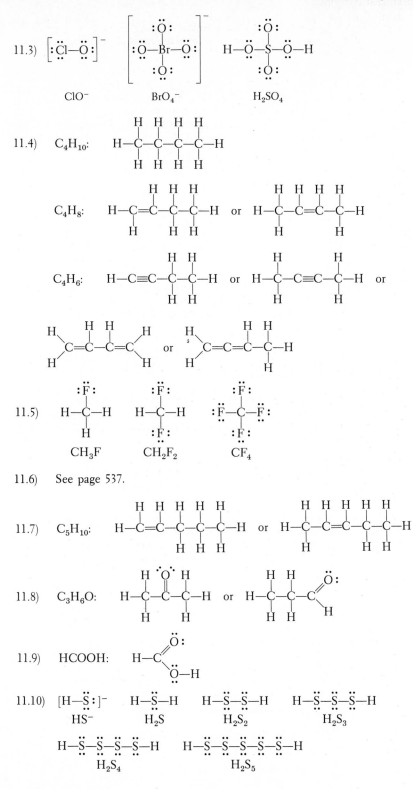

11.3) $[:\ddot{C}l-\ddot{O}:]^-$ $\left[:\ddot{O}-\overset{\overset{\overset{\displaystyle :\ddot{O}:}{|}}{|}}{Br}-\ddot{O}:\right]^-$ $H-\ddot{O}-\overset{\overset{\displaystyle :\ddot{O}:}{\|}}{\underset{\underset{\displaystyle :\ddot{O}:}{\|}}{S}}-\ddot{O}-H$

ClO⁻   BrO₄⁻   H₂SO₄

11.4) C₄H₁₀: H—C—C—C—C—H

C₄H₈: H—C=C—C—C—H or H—C—C=C—C—H

C₄H₆: H—C≡C—C—C—H or H—C—C≡C—C—H or

        C=C—C=C   or   C=C=C—C—H

11.5)   H—C—H   H—C—H   :F—C—F:
        CH₃F   CH₂F₂   CF₄

11.6) See page 537.

11.7) C₅H₁₀: H—C=C—C—C—C—H or H—C—C=C—C—C—H

11.8) C₃H₆O: H—C—C—C—H or H—C—C—C

11.9) HCOOH: H—C

11.10) [H—S̈:]⁻   H—S̈—H   H—S̈—S̈—H   H—S̈—S̈—S̈—H
        HS⁻   H₂S   H₂S₂   H₂S₃

        H—S̈—S̈—S̈—S̈—H   H—S̈—S̈—S̈—S̈—S̈—H
            H₂S₄   H₂S₅

11.11) If the total number of electrons is *odd* they cannot arrange themselves so that each atom is surrounded by eight electrons, an *even* number.

11.12) The Lewis diagram for AsI_3 is at the right. The formula AsI_5 suggests that the central arsenic atom forms five bonds involving five electron pairs, or ten electrons. This is two more electrons than the eight in an octet.

11.13)

SUBSTANCE	ELECTRON PAIR GEOMETRY	MOLECULAR GEOMETRY
BeH_2	Linear	Linear
CF_4	Tetrahedral	Tetrahedral
OF_2	Tetrahedral	Bent

11.14)

IO_4^-	Tetrahedral	Tetrahedral
ClO_2^-	Tetrahedral	Bent
CO_3^{2-}	Trigonal planar	Trigonal planar

11.15) C in C_2H_5OH — Tetrahedral — Tetrahedral

11.16) N in CH_3NH_2 — Tetrahedral — Trigonal pyramid

11.17) C in C_2H_4 — Trigonal planar — Trigonal planar

11.18) C in HCHO — Trigonal planar — Trigonal planar

11.19) Bonds are distributed symmetrically around the carbon atom, so the total charge distribution in the molecule is symmetrical.

11.20) HCl, with an electronegativity difference of 0.9, is more polar than HI, with an electronegativity difference of 0.4. The halogen end is more negative in both molecules.

11.21) H_2O is more polar than H_2S. Both bond and molecular polarity decrease for the hydrides of column 6A from oxygen to tellurium. H_2Te has zero electronegativity difference in its bonds, and the molecule is essentially nonpolar.

11.22)

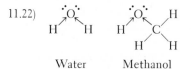

Water   Methanol

Bond angles around oxygen atoms are approximately equal, according to electron repulsion principle. H—O bond is much more polar than C—O bond (electronegativity differences 1.4 vs 0.4). Bonding electrons are therefore displaced more toward oxygen in the water molecule, which is the more polar of the two.

CHAPTER 12

12.1) Copper; krypton; manganese; nitrogen.

12.2) Na; H or H_2; Si; Pb.

12.3) Calcium ion; chromium(III) ion; zinc ion; phosphide ion; bromide ion.

12.4) Li^+; NH_4^+; N^{3-}; F^-; Hg^{2+}.

12.5) Nitric acid; sulfurous acid; $HClO_4$; H_2SeO_4.

12.6) SO_4^{2-}; ClO_2^-; iodate ion; hypobromite ion.

12.7) HCO_3^-; $H_2PO_4^-$; hydrogen sulfate ion.

12.8)   K$_2$S; Cu(NO$_3$)$_2$; NaHCO$_3$.

12.9)   Magnesium sulfite; aluminum fluoride; lead carbonate [or lead(II) carbonate].

12.10) S-l_____ Br$_3$; HI.

_____ate; sulfur trioxide.

_____Cl; SiF$_6$.

_____ydrogen sulfide (or dihydrogen sulfide).

_____ydrogen phosphate ion); copper(II) oxide;

_____carbonate; sodium dichromate.

_____gI$_2$; Al(OH)$_3$.

_____odate; selenic acid.

_____oSO$_4$; Pb(NO$_3$)$_2$.

_____sodium chlorite.

_____CrCl$_2$; HC$_2$H$_3$O$_2$.

_____ilver nitrate.

_____O; H$_2$S(aq).

_____isulfur decafluoride.

_____nate; In$_2$Se$_3$; Hg$_2$(SCN)$_2$.

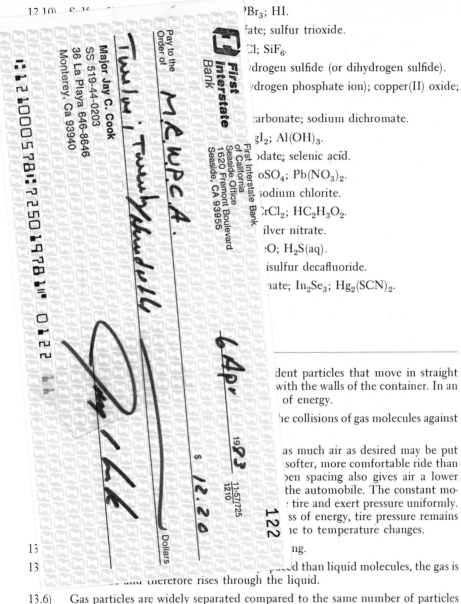

_____dent particles that move in straight _____with the walls of the container. In an _____of energy.

_____he collisions of gas molecules against

_____as much air as desired may be put _____softer, more comfortable ride than _____pen spacing also gives air a lower _____the automobile. The constant mo- _____ tire and exert pressure uniformly. _____ss of energy, tire pressure remains _____e to temperature changes.

13_____   ng.

13_____   _____ed than liquid molecules, the gas is _____ and therefore rises through the liquid.

13.6)   Gas particles are widely separated compared to the same number of particles close together in the liquid state.

13.7)   When gas particles are pushed close to each other, the intermolecular attractions become significant. The molecules are no longer independent. The ideal gas model is violated, so the gas does not behave ideally.

13.8)   Particles move in straight lines until they hit something—eventually the walls of the container, thereby filling it.

13.9)   Pressure, volume, temperature, quantity.

13.10) 1 atmosphere is the name for the pressure that will support a column of mercury 760 millimeters high. This is the pressure exerted at sea level on a "normal" day. The *torr* is a name for one millimeter of mercury as a pressure measurement.

13.11) $P_g = 747$ mm Hg $+ (729 - 263)$ mm Hg $= 1213$ mm Hg

13.12) 64.9 cm Hg $\times \dfrac{1 \text{ atm}}{76.0 \text{ cm Hg}} = 0.854$ atm

13.13) $445 \text{ atm} \times \dfrac{14.7 \text{ pounds/in.}^2}{1 \text{ atm}} = 6.54 \times 10^3$ pounds/in.^{2}

13.14) 3.15 L $\times \dfrac{0.940 \text{ atm}}{6.26 \text{ atm}} = 0.473$ L

13.15) 9.50 atm $\times \dfrac{59.8 \text{ L}}{(59.8 + 3.16) \text{ L}} = 9.02$ atm

13.16) Absolute zero is the temperature at which the average translational kinetic energy of particles is zero. This presumably means no particle movement.

13.17) $328 + 273 = 601$ K

13.18) $90 - 273 = -183°C$

13.19) $85\,600$ L $\times \dfrac{(25 + 273) \text{ K}}{(15 + 273) \text{ K}} = 88\,600$ L

13.20) 912 torr $\times \dfrac{(25 + 273) \text{ K}}{(68 + 273) \text{ K}} = 802$ torr

13.21) 73.4 mL $\times \dfrac{824 \text{ torr}}{749 \text{ torr}} \times \dfrac{(23 + 273) \text{ K}}{(43 + 273) \text{ K}} = 75.6$ mL

13.22) 1.62 L $\times \dfrac{1.02 \text{ atm}}{4.86 \text{ atm}} \times \dfrac{(3 + 273) \text{ K}}{(23 + 273) \text{ K}} = 0.317$ L

13.23) Standard temperature and pressure. These are arbitrary standards at which gases may be compared. Standard temperature is $0°C$; standard pressure is 760 mm Hg, or 1 atmosphere.

13.24) 47.9 mL $\times \dfrac{718 \text{ torr}}{760 \text{ torr}} \times \dfrac{273 \text{ K}}{(26 + 273) \text{ K}} = 41.3$ mL

13.25) 46.9 mL $\times \dfrac{760 \text{ torr}}{738 \text{ torr}} \times \dfrac{(273 + 24) \text{ K}}{273 \text{ K}} = 52.5$ mL

13.26) $V = \dfrac{nRT}{P} = \dfrac{0.16 \text{ mol}}{751 \text{ torr}} \times (22 + 273) \text{ K} \times \dfrac{62.4 \text{ L torr}}{\text{mol K}} = 3.92$ L

13.27) $n = \dfrac{PV}{RT} = \dfrac{9.40 \text{ atm}}{0.0821 \dfrac{\text{L atm}}{\text{mol K}}} \times \dfrac{2.55 \text{ L}}{(273 + 20) \text{ K}}$

$= 9.40 \text{ atm} \times \dfrac{\text{mol K}}{0.0821 \text{ L atm}} \times \dfrac{2.55 \text{ L}}{(273 + 20) \text{ K}} = 0.996$ mol

13.28) $MW = \dfrac{gRT}{PV} = \dfrac{5.89 \text{ g}}{745 \text{ torr}} \times \dfrac{(22 + 273) \text{ K}}{2.68 \text{ L}} \times \dfrac{62.4 \text{ L torr}}{\text{mol K}} = 54.3$ g/mol

13.29) $P = \dfrac{gRT}{V(MW)} = \dfrac{25.0 \text{ g } SO_2}{2.15 \text{ L}} \times \dfrac{0.0821 \text{ L atm}}{\text{mol K}} \times \dfrac{1 \text{ mol } SO_2}{64.1 \text{ g } SO_2} \times 293 \text{ K} = 4.36$ atm

13.30) $T = \dfrac{PV(MW)}{gR} = \dfrac{24.0 \text{ atm}}{624 \text{ g } O_2} \times 21.2 \text{ L} \times \dfrac{32.0 \text{ g } O_2}{1 \text{ mol } O_2} \times \dfrac{\text{mol K}}{0.0821 \text{ L atm}} = 318 \text{ K} = 45°C$

13.31) $g = \dfrac{PV(MW)}{RT} = \dfrac{798 \text{ torr}}{282 \text{ K}} \times 1.75 \text{ m}^3 \times \dfrac{1000 \text{ L}}{1 \text{ m}^3} \times \dfrac{0.004\ 00 \text{ kg He}}{1 \text{ mol He}} \times$

$$\dfrac{\text{mol K}}{62.4 \text{ L torr}} = 0.317 \text{ kg He}$$

13.32)

ELEMENT	GRAMS	MOLES	MOLE RATIO
C	85.6	7.13	1
H	14.4	14.4	2

Empirical formula: CH_2
Molar weight of empirical formula: 14.0 g/mol

$$MW = \dfrac{gRT}{VP} = \dfrac{1.69 \text{ g}}{1 \text{ L}} \times \dfrac{62.4 \text{ L torr}}{\text{mol K}} \times \dfrac{298 \text{ K}}{750 \text{ torr}} = 41.9 \text{ g/mol}$$

$$\dfrac{41.9 \text{ g}}{1 \text{ mol}} \times \dfrac{1 \text{ mol formula units}}{14.0 \text{ g}} = 3 \text{ mol formula units/mol}$$

Molecular formula, C_3H_6

13.33) The volume occupied by one mole.

13.34) $16.9 \text{ L} \times \dfrac{1 \text{ mol}}{22.4 \text{ L}} = 0.754 \text{ mol}$

13.35) $\dfrac{1.83 \text{ g}}{1 \text{ L}} \times \dfrac{22.4 \text{ L}}{1 \text{ mol}} = 41.0 \text{ g/mol}$

13.36) $\dfrac{0.937 \text{ g}}{0.744 \text{ L}} \times \dfrac{22.4 \text{ L}}{1 \text{ mol}} = 28.2 \text{ g/mol}$

13.37) $\dfrac{6.49 \text{ g NH}_3}{9.44 \text{ L}} \times \dfrac{1 \text{ mol NH}_3}{17.0 \text{ g NH}_3} = 0.0404 \text{ mol/L}$

Alternative method: $\dfrac{n}{V} = \dfrac{P}{RT} = \dfrac{745 \text{ torr}}{295 \text{ K}} \times \dfrac{1 \text{ mol K}}{62.4 \text{ L torr}} = 0.0405 \text{ mol/L}$

13.38) $2.65 \text{ g HgO} \times \dfrac{1 \text{ mol HgO}}{217 \text{ g HgO}} \times \dfrac{1 \text{ mol O}_2}{2 \text{ mol HgO}} \times \dfrac{22.4 \text{ L O}_2}{1 \text{ mol O}_2} = 0.137 \text{ L O}_2$

13.39) $85.0 \text{ mL H}_2 \times \dfrac{1 \text{ mol H}_2}{22\ 400 \text{ mL H}_2} \times \dfrac{1 \text{ mol Mn}}{1 \text{ mol H}_2} \times \dfrac{54.9 \text{ g Mn}}{1 \text{ mol Mn}} = 0.209 \text{ g Mn}$

13.40) $3.26 \text{ g Mg} \times \dfrac{1 \text{ mol Mg}}{24.3 \text{ g Mg}} \times \dfrac{1 \text{ mol O}_2}{2 \text{ mol Mg}} = 0.0671 \text{ mol O}_2$

$$V = 0.0671 \text{ mol O}_2 \times \dfrac{(25 + 273) \text{ K}}{752 \text{ torr}} \times \dfrac{62.4 \text{ L torr}}{\text{K mol}} = 1.66 \text{ L O}_2$$

13.41) $n = 4.16 \text{ L H}_2 \times \dfrac{1.65 \text{ atm}}{(243 + 273) \text{ K}} \times \dfrac{1 \text{ K mol}}{0.0821 \text{ L atm}} = 0.162 \text{ mol H}_2$

$$0.162 \text{ mol H}_2 \times \dfrac{1 \text{ mol CuO}}{1 \text{ mol H}_2} \times \dfrac{79.5 \text{ g CuO}}{1 \text{ mol CuO}} = 12.9 \text{ g CuO}$$

13.42) Equal volumes of gases at the same temperature and pressure contain equal numbers of molecules. Such volumes are therefore in the same ratio as the ratio of their numbers of molecules. Equation coefficients, which represent the ratio of numbers of molecules, must also correspond to the equivalent ratio of volumes.

13.43) $1.28 \text{ L O}_2 \times \dfrac{2 \text{ L H}_2}{1 \text{ L O}_2} = 2.56 \text{ L H}_2$

13.44) $28.3 \, \cancel{L \, H_2O} \times \dfrac{1.46 \, \cancel{atm}}{2.19 \, \cancel{atm}} \times \dfrac{(460 + 273) \, \cancel{K}}{(540 + 273) \, \cancel{K}} \times \dfrac{1 \, L \, CO}{1 \, \cancel{L \, H_2O}} = 17.0 \, L \, CO$

13.45) The total pressure exerted by a gaseous mixture is the sum of the partial pressures of the components of the mixture: $P = p_1 + p_2 + p_3 + \ldots$ See Equation 13.29 and related text (p. 295) for explanation.

13.46) $P = 0.364 \, atm + 0.108 \, atm + 0.529 \, atm = 1.001 \, atm$

13.47) $P_{H_2} = 751 \, torr - 40.0 \, torr = 711 \, torr$

13.48) $755 \, torr \times \dfrac{0.93 \, mol \, Ar}{100 \, mol \, total} = 7.0 \, torr = $ partial pressure of argon

CHAPTER 14

14.1) Because of the distances between molecules in the gas phase, a given quantity occupies a much larger volume as a gas than as a liquid. Density is mass/volume. Therefore the larger volume of the gas yields a lower density.

14.2) Air can be compressed by pushing the widely spaced molecules closer together. Liquid molecules are already close, and therefore cannot be compressed.

14.3) Substances with strong intermolecular attractions tend to have low vapor pressures. The strong attractions make the evaporation rate slow at a given temperature. A relatively low vapor concentration—and thus a low equilibrium vapor pressure—is therefore sufficient to make the condensation rate equal to the evaporation rate.

14.4) Molar heat of vaporization is the amount of heat required to vaporize one mole of a liquid.

14.5) Motor oil is more viscous than water. From this it may be predicted that intermolecular attractions are stronger in motor oil, as strong attractions lead to internal resistance to liquid flow, which is the property called viscosity.

14.6) Surface tension is greater in mercury than in water, indicating stronger intermolecular attractions.

14.7) Soap reduces intermolecular attractions so the soapy water is able to penetrate fabrics and cleanse them throughout.

14.8) Gas.

14.9) Of the three compounds, only N_2O is a liquid at $-90°C$, and therefore only N_2O possesses the property of viscosity. If a solid is considered more "viscous" than a liquid, NO_2 is the most viscous.

14.10) Dipole forces are electrostatic attractions between polar molecules, the negatively charged region of one molecule being attracted to the positively charged region of another molecule. Dispersion, or London, forces are weaker electrostatic attractions between molecules that are generally nonpolar, but which become "temporary dipoles" through instantaneously shifting electron clouds. Hydrogen bonds are exceptionally strong intermolecular attractions between the hydrogen that is bonded to a strongly electronegative element in a polar molecule and the negative region, often the same highly electronegative element, of an adjacent molecule. The hydrogen atom becomes a small, highly concentrated region of positive charge that can approach the negatively charged region of the nearby molecule quite closely.

14.11) HBr and NF_3: dipole; C_2H_2: dispersion; C_2H_5OH: hydrogen bonds.

14.12) The high melting points of ionic compounds suggest correctly that ionic bonds are stronger than dipole forces. Both are electrical in character, ionic forces arising from a nearly complete transfer of electrons, and dipole forces from nonsymmetrical distribution of electrical charge within molecules.

14.13) CCl_4, because it is larger, as suggested by its higher molecular weight.

14.14) H_2S, because it is slightly more polar.

14.15) Fluorine, oxygen, and nitrogen. The electronegativity difference between hydrogen and these elements is large enough to shift the bonding electron pair away from the hydrogen atom. If the molecule is polar, a hydrogen bond develops between the hydrogen atom of one molecule and the fluorine, oxygen, or nitrogen atom of another.

14.16) (a) Dispersion forces; (b) dipole forces.

14.17) (a), (c), (d).

14.18) C_6H_{14}, a larger molecule with higher molecular weight than C_3H_8, will have stronger intermolecular attractions and therefore higher melting and boiling points.

14.19) SO_2 molecules are larger than CO_2 molecules, and would therefore be expected to have stronger intermolecular attractions. The stronger the intermolecular attractions, the lower the vapor pressure. The prediction therefore is that CO_2 has the higher vapor pressure.

14.20) Opposing changes are occurring at equal rates.

14.21) Evaporation at constant rate begins immediately when liquid is introduced. At that time condensation rate is zero. Net rate of increase in vapor concentration is a maximum, so rate of vapor pressure increase is a maximum at start. Later condensation rate is more than zero, but less than evaporation rate. Net rate of increase in vapor concentration is less than initially, so rate of vapor pressure increase is less than initially. At equilibrium, evaporation and condensation rates are equal. Vapor concentration and therefore vapor pressure remain constant.

14.22) All of the liquid evaporated before the vapor concentration was high enough to yield a condensation rate equal to the evaporation rate. At lower than equilibrium vapor concentration, the vapor pressure is lower than the equilibrium vapor pressure.

14.23) Concentration is measured in moles/liter, n/V. Solving ideal gas equation for concentration, $\dfrac{n}{V} = \dfrac{p}{RT} = \left(\dfrac{1}{RT}\right) p$. At constant temperature 1/RT is a constant, so pressure and concentration are proportional.

14.24) (a) Second and third boxes have greatest pressure—the equilibrium pressure. (b) First box probably has least, having all evaporated before reaching equilibrium. Only possible exception is if vapor pressure just reached equilibrium pressure as last molecule evaporated in first box.

14.25) 93 torr, the vapor pressure of the water at which equilibrium is reached.

14.26) Boiling point is temperature at which vapor bubbles form spontaneously anywhere in a liquid. This is the temperature at which the vapor pressure equals the pressure above the liquid. See Figure 14.10.

14.27) Gas, because vapor pressure is greater than surrounding pressure.

14.28) High boiling liquids have strong intermolecular attractions, and therefore require high energy to escape from the liquid to form a gas. Evaporation rate is therefore slow, quickly equaled by condensation rate at low vapor concentration, or vapor pressure.

14.29) More energy is required to vaporize X, so it would have the higher boiling point and lower vapor pressure.

14.30) Crystalline solids have rigid and precise geometric structures because of highly ordered and recurring patterns, whereas amorphous solids have some freedom to move because of irregular structure.

14.31) A, molecular; B, metallic.

CHAPTER 15

15.1) When the sugar is first added to the water, the crystals are suspended and eventually settle. The mixture is a suspension. After sufficient time the sugar will all dissolve, with the crystal breaking into individual molecules. At this time the mixture is a solution. In between there is a point at which the particles are too small to be seen, but still consist of many molecules. At this time the mixture is a colloid.

15.2) Solute is customarily dispersed in solvent. When two substances are dissolved in each other in nearly equal quantities, either may be considered the solvent and the other the solute.

15.3) If solute A is very soluble, its 10 gram per 100 gram of solvent concentration may be quite *dilute* compared to the possible concentration. If solute B is only slightly soluble, its 5 grams per 100 grams of solvent may be close to its maximum solubility, and therefore *concentrated*. Dilute and concentrated are relative terms that may be used to compare different concentrations of a given solute-solvent mixture, but not different mixtures.

15.4) Drop a small amount of solute into the solution. If the solution is unsaturated, the solute will dissolve; if saturated, it will simply settle to the bottom; if supersaturated, it will probably cause additional crystallization.

15.5) Any quantity units of solute over any quantity units of solvent may be used to express solubility.

15.6) Miscibility, miscible, and immiscible generally refer to a solution of one liquid in another. Their meanings correspond with solubility, soluble, and insoluble as applied to solutions in general.

15.7) Attractions between solute particles and solvent molecules are the primary forces promoting solution. If they are stronger than interparticle attractions within the solute and within the solvent, solution will occur.

15.8) If more solute than that required to reach solubility is placed into a given amount of solvent, the solute will dissolve until the solution is saturated. Dissolving and precipitation continue to occur between the saturated solution and the yet undissolved solute, the system being in a state of equilibrium. By pouring off the solution, you isolate a saturated solution from the excess solute.

15.9) Stirring increases the net dissolving rate.

15.10) Carbon tetrachloride because both solute and solvent are nonpolar.

15.11) Cyclopentane. Cyclopentane is nonpolar and methanol is polar. If the spot failed to dissolve in a polar solvent (water) there is a better chance it will dissolve in a nonpolar solvent than a second polar solvent.

15.12) The statement is true only to the extent that an increase in air pressure increases the partial pressure of carbon dioxide. Solubility of a gas depends upon the partial pressure of the solute gas, not the total gas pressure above the solution.

15.13) $\dfrac{2.78 \text{ g NaCl}}{32.4 \text{ g solution}} \times 100 = 8.58\% \text{ NaCl}$

15.14) $75.0 \text{ g solution} \times \dfrac{4.00 \text{ g H}_3\text{BO}_3}{100 \text{ g solution}} = 3.00 \text{ g H}_3\text{BO}_3$

15.15) $\dfrac{16.9 \text{ g C}_6\text{H}_{12}\text{O}_6}{0.0875 \text{ kg H}_2\text{O}} \times \dfrac{1 \text{ mol C}_6\text{H}_{12}\text{O}_6}{180 \text{ g C}_6\text{H}_{12}\text{O}_6} = 1.07 \text{ m C}_6\text{H}_{12}\text{O}_6$

15.16) $\dfrac{10.0 \text{ g NaCl}}{0.0900 \text{ kg H}_2\text{O}} \times \dfrac{1 \text{ mol NaCl}}{58.5 \text{ g NaCl}} = 1.90 \text{ m NaCl}$

15.17) $0.0500 \text{ kg H}_2\text{O} \times \dfrac{1.50 \text{ mol HCOOH}}{1 \text{ kg H}_2\text{O}} \times \dfrac{46.0 \text{ g HCOOH}}{1 \text{ mol HCOOH}} = 3.45 \text{ g HCOOH}$

15.18) $0.150 \text{ L} \times \dfrac{0.125 \text{ mol AgNO}_3}{1 \text{ L}} \times \dfrac{170 \text{ g AgNO}_3}{1 \text{ mol AgNO}_3} = 3.19 \text{ g AgNO}_3$

15.19) $0.750 \text{ L} \times \dfrac{0.480 \text{ mol H}_2\text{C}_2\text{O}_4 \cdot 2\,\text{H}_2\text{O}}{1 \text{ L}} \times \dfrac{126 \text{ g H}_2\text{C}_2\text{O}_4 \cdot 2\,\text{H}_2\text{O}}{1 \text{ mol H}_2\text{C}_2\text{O}_4 \cdot 2\,\text{H}_2\text{O}} = 45.4 \text{ g H}_2\text{C}_2\text{O}_4 \cdot 2\,\text{H}_2\text{O}$

15.20) $\dfrac{16.2 \text{ g (NH}_4)_2\text{SO}_4}{0.3000 \text{ L}} \times \dfrac{1 \text{ mol (NH}_4)_2\text{SO}_4}{132 \text{ g (NH}_4)_2\text{SO}_4} = 0.409 \text{ M (NH}_4)_2\text{SO}_4$

15.21) $\dfrac{56.2 \text{ g CuSO}_4 \cdot 5\,\text{H}_2\text{O}}{0.5000 \text{ L}} \times \dfrac{1 \text{ mol CuSO}_4 \cdot 5\,\text{H}_2\text{O}}{249.5 \text{ g CuSO}_4 \cdot 5\,\text{H}_2\text{O}} = 0.451 \text{ M CuSO}_4 \cdot 5\,\text{H}_2\text{O}$

15.22) $\dfrac{1.06 \text{ g soln}}{1 \text{ mL soln}} \times \dfrac{22.0 \text{ g NH}_4\text{Cl}}{100 \text{ g soln}} \times \dfrac{1 \text{ mol NH}_4\text{Cl}}{53.5 \text{ g NH}_4\text{Cl}} \times \dfrac{1000 \text{ mL}}{1 \text{ L}} = 4.36 \text{ M NH}_4\text{Cl}$

15.23) $0.0453 \text{ L} \times \dfrac{0.378 \text{ mol KNO}_3}{1 \text{ L}} = 0.0171 \text{ mol KNO}_3$

15.24) $1.24 \text{ mol H}_2\text{SO}_4 \times \dfrac{1000 \text{ mL}}{18 \text{ mol H}_2\text{SO}_4} = 69 \text{ mL}$

15.25) $0.150 \text{ L} \times \dfrac{0.633 \text{ mol (NH}_4)_2\text{CO}_3}{1 \text{ L}} \times \dfrac{1}{0.450 \text{ L diluted}} = 0.211 \text{ M (NH}_4)_2\text{CO}_3$

15.26) $0.500 \text{ L} \times \dfrac{0.450 \text{ mol H}_2\text{SO}_4}{1 \text{ L}} \times \dfrac{1000 \text{ mL}}{18.0 \text{ mol H}_2\text{SO}_4} = 12.5 \text{ mL}$

15.27) An acid with two ionizable hydrogens may lose only one in one reaction, but both in another reaction. When it loses one, there is one equivalent per mole; when it loses both, there are two equivalents per mole.

15.28) $\dfrac{65.6 \text{ g KOH}}{1.50 \text{ L}} \times \dfrac{1 \text{ mol KOH}}{56.1 \text{ g KOH}} \times \dfrac{1 \text{ eq KOH}}{1 \text{ mol KOH}} = 0.780 \text{ N KOH}$

15.29) 3 equivalents per mole. $\dfrac{192 \text{ g H}_3\text{C}_6\text{H}_5\text{O}_7}{3 \text{ eq H}_3\text{C}_6\text{H}_5\text{O}_7} = 64.0 \text{ g/eq}$

15.30) 40.0 g/eq

15.31) $\dfrac{11.9 \text{ g H}_3\text{PO}_4}{0.100 \text{ L}} \times \dfrac{2 \text{ eq H}_3\text{PO}_4}{98.0 \text{ g H}_3\text{PO}_4} = 2.43 \text{ eq/L} = 2.43 \text{ N H}_3\text{PO}_4$

15.32) 2 equivalents per mole. $\dfrac{98.0 \text{ g H}_2\text{SO}_4}{3 \text{ eq H}_2\text{SO}_4} = 49.0 \text{ g/eq}$

15.33) $\dfrac{164 \text{ g Na}_3\text{PO}_4}{2 \text{ eq Na}_3\text{PO}_4} = 82.0 \text{ g/eq}$

15.34) $\dfrac{0.284 \text{ mol Na}_3\text{PO}_4}{1 \text{ L}} \times \dfrac{2 \text{ eq}}{1 \text{ mol}} = 0.568 \text{ N Na}_3\text{PO}_4$

15.35) $0.0500\,\cancel{L} \times \dfrac{0.114\text{ eq HCl}}{1\,\cancel{L}} = 0.005\,70\text{ eq HCl}$

15.36) $0.0500\,\cancel{L} \times \dfrac{0.114\,\cancel{\text{eq HCl}}}{1\,\cancel{L}} \times \dfrac{36.5\text{ g HCl}}{1\,\cancel{\text{eq HCl}}} = 0.208\text{ g HCl}$

15.37) $0.750\,\cancel{L} \times \dfrac{0.250\,\cancel{\text{eq}}\text{ H}_2\text{C}_2\text{O}_4}{1\,\cancel{L}} \times \dfrac{90.0\text{ g}}{2\,\cancel{\text{eq}}} = 8.44\text{ g H}_2\text{C}_2\text{O}_4$

Dissolve 8.44 g $H_2C_2O_4$ and dilute to 750 mL.

15.38) $\dfrac{90.0\text{ g H}_2\text{C}_2\text{O}_4}{2\text{ eq}} = 45.0\text{ g H}_2\text{C}_2\text{O}_4/\text{eq}$

15.39) $0.005\,00\,\cancel{\text{eq H}_3\text{PO}_4} \times \dfrac{1\text{ L}}{0.200\,\cancel{\text{eq H}_3\text{PO}_4}} = 0.0250\text{ L} = 25.0\text{ mL}$

15.40) $K_2CrO_4 + BaCl_2 \rightarrow BaCrO_4 + 2\,KCl$

$0.0500\,\cancel{L} \times \dfrac{0.424\,\cancel{\text{mol BaCl}_2}}{1\,\cancel{L}} \times \dfrac{1\,\cancel{\text{mol BaCrO}_4}}{1\,\cancel{\text{mol BaCl}_2}} \times \dfrac{253\text{ g BaCrO}_4}{1\,\cancel{\text{mol BaCrO}_4}} = 5.36\text{ g BaCrO}_4$

15.41) $Na_3PO_4 + 3\,AgNO_3 \rightarrow Ag_3PO_4 + 3\,NaNO_3$

$2.10\text{ g }\cancel{\text{Na}_3\text{PO}_4} \times \dfrac{1\,\cancel{\text{mol Na}_3\text{PO}_4}}{164\,\cancel{\text{g Na}_3\text{PO}_4}} \times \dfrac{3\,\cancel{\text{mol AgNO}_3}}{1\,\cancel{\text{mol Na}_3\text{PO}_4}} \times \dfrac{1000\text{ mL}}{0.246\,\cancel{\text{mol AgNO}_3}} = 156\text{ mL}$

15.42) $2.4\,\cancel{\text{L H}_2} \times \dfrac{1\,\cancel{\text{mol H}_2}}{22.4\,\cancel{\text{L H}_2}} \times \dfrac{6\,\cancel{\text{mol NaOH}}}{3\,\cancel{\text{mol H}_2}} \times \dfrac{1000\text{ mL}}{6.2\,\cancel{\text{mol NaOH}}} = 35\text{ mL}$

15.43) $MgCl_2 + 2\,NaOH \rightarrow Mg(OH)_2 + 2\,NaCl$

$0.0500\,\cancel{L} \times \dfrac{0.240\,\cancel{\text{mol MgCl}_2}}{1\,\cancel{L}} \times \dfrac{1\,\cancel{\text{mol Mg(OH)}_2}}{1\,\cancel{\text{mol MgCl}_2}} \times \dfrac{58.3\text{ g Mg(OH)}_2}{1\,\cancel{\text{mol Mg(OH)}_2}}$

$= 0.700\text{ g Mg(OH)}_2$

$0.0500\,\cancel{L} \times \dfrac{0.420\,\cancel{\text{mol NaOH}}}{1\,\cancel{L}} \times \dfrac{1\,\cancel{\text{mol Mg(OH)}_2}}{2\,\cancel{\text{mol NaOH}}} \times \dfrac{58.3\text{ g Mg(OH)}_2}{1\,\cancel{\text{mol Mg(OH)}_2}}$

$= 0.612\text{ g Mg(OH)}_2\text{ maximum yield}$

15.44) $61.2\,\cancel{\text{kg NaHCO}_3} \times \dfrac{1\,\cancel{\text{kmol NaHCO}_3}}{84.0\,\cancel{\text{kg NaHCO}_3}} \times \dfrac{1\,\cancel{\text{kmol CO}_2}}{1\,\cancel{\text{kmol NaHCO}_3}} \times \dfrac{22.4\text{ kL CO}_2}{1\,\cancel{\text{kmol CO}_2}}$

$= 16.3\text{ kL CO}_2 = 1.63 \times 10^4\text{ L CO}_2.$

$140\,\cancel{\text{gal}} \times \dfrac{3.785\,\cancel{L}}{1\,\cancel{\text{gal}}} \times \dfrac{1.13\,\cancel{\text{mol HNO}_3}}{1\,\cancel{L}} \times \dfrac{1\,\cancel{\text{mol CO}_2}}{1\,\cancel{\text{mol HNO}_3}} \times \dfrac{22.4\text{ L CO}_2}{1\,\cancel{\text{mol CO}_2}}$

$= 1.34 \times 10^4\text{ L CO}_2\text{ maximum yield}$

15.45) $5.34\,\cancel{\text{g KHC}_8\text{H}_4\text{O}_4} \times \dfrac{1\,\cancel{\text{mol KHC}_8\text{H}_4\text{O}_4}}{204\,\cancel{\text{g KHC}_8\text{H}_4\text{O}_4}} \times \dfrac{1\text{ mol Na}_2\text{CO}_3}{2\,\cancel{\text{mol KHC}_8\text{H}_4\text{O}_4}} \times \dfrac{1}{0.0325\text{ L}}$

$= 0.403\text{ M Na}_2\text{CO}_3$

15.46) $HCl + NaOH \rightarrow HOH + NaCl$

$0.0100\,\cancel{L} \times \dfrac{0.862\,\cancel{\text{mol NaOH}}}{1\,\cancel{L}} \times \dfrac{1\text{ mol HCl}}{1\,\cancel{\text{mol NaOH}}} \times \dfrac{1}{0.0168\text{ L}} = 0.513\text{ M HCl}$

15.47) $0.0149\,\cancel{L} \times \dfrac{0.518\,\cancel{\text{mol AgNO}_3}}{1\,\cancel{L}} \times \dfrac{1\text{ mol NiCl}_2}{2\,\cancel{\text{mol AgNO}_3}} \times \dfrac{1}{0.0100\text{ L}} = 0.386\text{ M NiCl}_2$

15.48) $0.0342\,\cancel{L} \times \dfrac{0.0271\,\cancel{\text{mol KMnO}_4}}{1\,\cancel{L}} \times \dfrac{5\,\cancel{\text{mol FeCl}_2}}{1\,\cancel{\text{mol KMnO}_4}} \times \dfrac{55.9\text{ g Fe}}{1\,\cancel{\text{mol FeCl}_2}} \times \dfrac{100}{0.504\text{ g ore}}$

$= 51.4\%\text{ Fe}$

15.49) $0.512 \text{ g } \cancel{H_2C_2O_4} \times \dfrac{2 \text{ eq}}{90.0 \text{ g } \cancel{H_2C_2O_4}} \times \dfrac{1}{0.0162 \text{ L}} = 0.702 \text{ N NaOH}$

15.50) $\dfrac{25.0 \times 0.324}{21.9} = 0.370 \text{ N NaOH}$

15.51) $\dfrac{0.452 \text{ g base}}{0.0318 \cancel{\text{ L HCl}}} \times \dfrac{1 \cancel{\text{ L HCl}}}{0.169 \text{ eq}} = 84.1 \text{ g/eq}$

15.52) A colligative property of a solution is one that is independent of the identity of the solute. Specific gravity, with opposite effects from different solutes, is not a colligative property.

15.53) $\dfrac{96.1 \text{ g } \cancel{C_2H_6O_2}}{0.100 \text{ kg } H_2O} \times \dfrac{1 \text{ mol } C_2H_6O_2}{62.0 \text{ g } \cancel{C_2H_6O_2}} = 15.5 \text{ m } C_2H_6O_2$

$\Delta T_b = 0.52 \times 15.5 = 8.1°C; \ T_b = 100.0 + 8.1 = 108.1°C$

15.54) $\dfrac{4.34 \text{ g } \cancel{C_6H_4Cl_2}}{0.0700 \text{ kg naphthalene}} \times \dfrac{1 \text{ mol } C_6H_4Cl_2}{147 \text{ g } \cancel{C_6H_4Cl_2}} = 0.422 \text{ m } C_6H_4Cl_2$

$\Delta T_f = (-6.9)(0.423) = -2.9°C; \ T_f = 80.2 - 2.9 = 77.3°C$

15.55) $m = \dfrac{\Delta T}{K_b} = \dfrac{(100.89 - 100.00)}{0.52} = 1.71 \text{ m}$

CHAPTER 16

16.1) A solution of A will be a conductor of electricity, whereas a solution of B will be a nonconductor. A yields ions in solution; B yields molecules.

16.2) A conducting solution must contain ions. "Electricity" is carried through the solution by positive ions moving toward the negative electrode and negative ions toward the positive electrode.

16.3) LiF: $Li^+(aq) + F^-(aq)$ $Mg(NO_3)_2$: $Mg^{2+}(aq) + 2 NO_3^-(aq)$
$FeCl_3$: $Fe^{3+}(aq) + 3 Cl^-(aq)$

16.4) NH_4NO_3: $NH_4^+(aq) + NO_3^-(aq)$ $Ca(OH)_2$: $Ca^{2+}(aq) + 2 OH^-(aq)$
Li_2SO_3: $2 Li^+(aq) + SO_3^{2-}(aq)$

16.5) $HCHO_2$: $HCHO_2(aq)$ HCl: $H^+(aq) + Cl^-(aq)$
$HC_4H_4O_6$: $HC_4H_4O_6(aq)$

16.6) HI: $H^+(aq) + I^-(aq)$ H_2SO_4: $2 H^+(aq) + SO_4^{2-}(aq)$
$H_3C_2H_5O_7$: $H_3C_2H_5O_7(aq)$

16.7) $Ba(s) + Zn^{2+}(aq) \rightarrow Ba^{2+}(aq) + Zn(s)$

16.8) No reaction.

16.9) $3 Mg(s) + 2 Al^{3+}(aq) \rightarrow 3 Mg^{2+}(aq) + 2 Al(s)$

16.10) $Ba^{2+}(aq) + CO_3^{2-}(aq) \rightarrow BaCO_3(s)$

16.11) $Co^{2+}(aq) + 2 OH^-(aq) \rightarrow Co(OH)_2(s)$

16.12) $Fe^{2+}(aq) + S^{2-}(aq) \rightarrow FeS(s)$

16.13) No reaction.

16.14) $Mg^{2+}(aq) + 2 F^-(aq) \rightarrow MgF_2(s)$
$3 Zn^{2+}(aq) + 2 PO_4^{3-}(aq) \rightarrow Zn_3(PO_4)_2(s)$

16.15) $H^+(aq) + C_6H_5O_7^-(aq) \rightarrow HC_6H_5O_7(aq)$

16.16) $H^+(aq) + OH^-(aq) \rightarrow HOH(\ell)$

16.17) $H^+(aq) + C_4H_4O_6^-(aq) \rightarrow HC_4H_4O_6(aq)$

16.18) $2\,H^+(aq) + CO_3^{2-}(aq) \rightarrow H_2O(\ell) + CO_2(g)$

16.19) $NH_4^+(aq) + OH^-(aq) \rightarrow NH_3(aq) + HOH(\ell)$

16.20) $2\,Ag^+(aq) + CO_3^{2-}(aq) \rightarrow Ag_2CO_3(s)$

16.21) $Al^{3+}(aq) + PO_4^{3-}(aq) \rightarrow AlPO_4(s)$

16.22) $Ca(s) + 2\,Ag^+(aq) \rightarrow Ca^{2+}(aq) + 2\,Ag(s)$

16.23) $2\,H^+(aq) + SO_3^{2-}(aq) \rightarrow H_2O(\ell) + SO_2(aq)$

16.24) $H^+(aq) + OH^-(aq) \rightarrow HOH(\ell)$

16.25) $AgCl(s) + I^-(aq) \rightarrow AgI(s) + Cl^-(aq)$

16.26) $2\,Li(s) + Cu^{2+}(aq) \rightarrow 2\,Li^+(aq) + Cu(s)$

16.27) $H^+(aq) + C_7H_5O_2^-(aq) \rightarrow HC_7H_5O_2(aq)$

16.28) $NH_4^+(aq) + OH^-(aq) \rightarrow NH_3(aq) + HOH(\ell)$

16.29) $Ca^{2+}(aq) + 2\,F^-(aq) \rightarrow CaF_2(s)$

16.30) $2\,K(s) + Zn^{2+}(aq) \rightarrow 2\,K^+(aq) + Zn(s)$

16.31) $2\,H^+(aq) + Ba(OH)_2(s) \rightarrow 2\,HOH(\ell) + Ba^{2+}(aq)$

16.32) $2\,H^+(aq) + Cu(OH)_2(s) \rightarrow 2\,HOH(\ell) + Cu^{2+}(aq)$

16.33) $2\,Al(s) + 6\,OH^-(aq) \rightarrow 3\,H_2(g) + 2\,AlO_3^{3-}(aq)$

16.34) $H_3PO_4(aq) + OH^-(aq) \rightarrow HOH(\ell) + H_2PO_4^-(aq)$
        $H_3PO_4(aq) + 2\,OH^-(aq) \rightarrow 2\,HOH(\ell) + HPO_4^{2-}(aq)$

CHAPTER 17

17.1) Acids, H^+; bases, OH^-.

17.2) Arrhenius acid is source of H^+, or a proton; Brönsted-Lowry acid is a proton donor. Concepts are in agreement. Arrhenius base is a source of OH^- ion; Brönsted-Lowry base is a proton acceptor. OH^- is one of many examples of Brönsted-Lowry bases, so Arrhenius is more limited concept.

17.3) HOH and H_2CO_3; H_2O and CO_3^{2-}.

17.4) Forward: acid, HNO_2; base, CN^-. Reverse: acid, HCN; base, NO_2^-

17.5) HNO_2 and NO_2^-; CN^- and HCN.

17.6) HSO_4^- and SO_4^{2-}; $C_2O_4^{2-}$ and $HC_2O_4^-$.

17.7) $H_2PO_4^-$ and HPO_4^{2-}; HCO_3^- and H_2CO_3

17.8) A Lewis acid must have an empty valence orbital, capable of forming a covalent bond by accepting an electron pair from a Lewis base. This unshared electron pair characterizes a Lewis base.

17.9) The hydrogen ion is capable of accepting an electron pair, and the hydroxide ion is capable of donating an electron pair to form a chemical bond. These Arrhenius acid and base ions therefore satisfy the Lewis concept of acids and bases.

17.10)

$$AlCl_3 \quad + \quad Cl^- \quad \rightarrow \quad AlCl_4^-$$

Aluminum in $AlCl_3$ is surrounded by three electron pairs. It is therefore capable of accepting a fourth pair, which qualifies it as a Lewis acid. The chloride ion has four unshared pairs of valence electrons. When one of these (see color electron pair) becomes a bonding pair with an aluminum chloride molecule (see color bond in $AlCl_4^-$), the chloride ion has served as a Lewis base.

17.11) The sulfur in SO_3 acts as a Lewis acid by accepting an electron pair from the oxide ion, a Lewis base.

17.12) A strong acid donates protons readily, a weak acid reluctantly. Strong acids are at the top of Table 17.1, weak acids at the bottom.

17.13) CO_3^{2-}; $H_2PO_4^-$; SO_4^{2-}; Br^-.

17.14) HI; $H_2C_2O_4$; HSO_3^-; NH_4^+; H_2O.

17.15) $HC_7H_5O_2(aq) + SO_4^{2-} \rightleftharpoons C_7H_5O_2^-(aq) + HSO_4^-(aq)$; reverse.

17.16) $H_2C_2O_4(aq) + NH_3(aq) \rightleftharpoons HC_2O_4^-(aq) + NH_4^+(aq)$; forward.

17.17) $H_3PO_4(aq) + CN^-(aq) \rightleftharpoons H_2PO_4^-(aq) + HCN(aq)$; forward.

17.18) $H_2BO_3^-(aq) + NH_4^+(aq) \rightarrow H_3BO_3(aq) + NH_3(aq)$; reverse.

17.19) $HPO_4^{2-}(aq) + HC_2H_3O_2(aq) \rightarrow H_2PO_4^-(aq) + C_2H_3O_2^-(aq)$; forward.

17.20) Water ionizes very slightly and does not produce enough ions to light an ordinary conductivity device. With a sufficiently sensitive detector, water displays a very weak conductivity.

17.21) Strongly acidic, pH less than 4; weakly acidic, 4-6; neutral, or close to neutral, 6-8; weakly basic, 8-10; strongly basic, above 10. Ranges are arbitrary. If pH of a solution is x, then hydrogen ion concentration is 10^{-x} mol/L.

17.22) $[OH^-] = \dfrac{10^{-14}}{10^{-6}} = 10^{-8}$; pOH = 8;

pH $= 14 - 8 = 6$; $[H^+] = 10^{-6}$; close to neutral.

17.23) pH $= 14 - 3 = 11$; $[H^+] = 10^{-11}$;

$[OH^-] = \dfrac{10^{-14}}{10^{-11}} = 10^{-3}$; pOH = 3; strongly basic.

17.24) $[H^+] = 10^{-1}$; $[OH^-] = \dfrac{10^{-14}}{10^{-1}} = 10^{-13}$;

pOH = 13; pH $= 14 - 13 = 1$; strongly acidic.

17.25) pOH = 9; pH $= 14 - 9 = 5$;

$[H^+] = 10^{-5}$; $[OH^-] = \dfrac{10^{-14}}{10^{-5}} = 10^{-9}$; weakly acidic.

17.26) $[OH^-] = \dfrac{1.0 \times 10^{-14}}{3.4 \times 10^{-9}} = 2.9 \times 10^{-6}$

pOH $= -\log (2.9 \times 10^{-6}) = 5.54$
pH $= 14.00 - 5.54 = 8.46$
$[H^+] = 10^{-8.46} = 3.5 \times 10^{-9}$

17.27) pH $= 14.00 - 11.82 = 2.18$
$[H^+] = 10^{-2.18} = 6.6 \times 10^{-3}$

$[OH^-] = \dfrac{1.0 \times 10^{-14}}{6.6 \times 10^{-3}} = 1.5 \times 10^{-12}$

pOH $= -\log (1.5 \times 10^{-12}) = 11.82$

17.28) pOH $= -\log 0.26 = 0.59$
pH $= 14.00 - 0.59 = 13.41$
$[H^+] = 10^{-13.41} = 3.9 \times 10^{-14}$

$[OH^-] = \dfrac{1.0 \times 10^{-14}}{3.9 \times 10^{-14}} = 0.26$

17.29) $[H^+] = 10^{-12.05} = 8.9 \times 10^{-13}$

$[OH^-] = \dfrac{1.0 \times 10^{-14}}{8.9 \times 10^{-13}} = 1.1 \times 10^{-2}$

pOH $= -\log (1.1 \times 10^{-2}) = 1.96$
pH $= 14.00 - 1.96 = 12.04$

CHAPTER 18

18.1) Oxidation is loss of electrons, or increase in oxidation number; reduction is gain of electrons or decrease in oxidation number. From the standpoint of the first definition, electrons lost by one species must be gained by the other. Oxidation and reduction therefore must be simultaneous processes.

18.2) (a), (b), and (c), oxidation; (d), reduction.

18.3) Oxidation.

18.4) Reduction.

18.5) $Ni^2 + 2e^- \rightarrow Ni$
$\underline{\qquad\qquad Mg \rightarrow Mg^{2+} + 2e^-}$
$Ni^{2+} + Mg \rightarrow Ni + Mg^{2+}$

18.6) $PbO_2 + SO_4^{2-} + 4\,H^+ + 2e^- \rightarrow PbSO_4 + 2\,H_2O$
$\underline{\qquad\qquad\quad Pb + SO_4^{2-} \rightarrow PbSO_4 + 2e^-}$
$PbO_2 + 2\,SO_4^{2-} + 4\,H^+ + Pb \rightarrow 2\,PbSO_4 + 2\,H_2O$

18.7) $+2, -1, +1, +5$.

18.8) $+5, -3, +7, +3$.

18.9) (a) Copper reduced from $+2$ to 0; (b) cobalt reduced from $+3$ to $+2$.

18.10) (a) Sulfur oxidized from $+4$ to $+6$; (b) P oxidized from -3 to 0.

18.11) (a) F oxidized from -1 to 0; (b) Mn reduced from $+6$ to $+4$.

18.12) Hydrogen is the reducing agent, and copper oxide is the oxidizing agent.

18.13) BrO_3 is the oxidizing agent, HNO_2 the reducing agent.

18.14) Reactions are similar in that each transfers a subatomic particle, a proton transfer for acid-base and an electron transfer for redox. Further similarities are in use of terms "strong" and "weak," and the fact that some species can function in either role—i.e., acid or base, or oxidizer or reducer. A significant difference is that all acid-base reactions are single proton transfer reactions, although there may be successive proton transfer reactions, whereas many redox reactions involve transfer of two or more electrons.

18.15) Ag^+ is a stronger oxidizer than H^+, based on their relative positions in Table 18.1. This means that Ag^+ has a stronger attraction for electrons than does H^+.

18.16) Al, H_2, Fe^{2+}, Cl^-.

18.17) $Ni + Zn^{2+} \rightleftharpoons Ni^{2+} + Zn$; reverse.

18.18) $2\,Fe^{3+} + Co \rightleftharpoons 2\,Fe^{2+} + Co^{2+}$; forward.

18.19) $\frac{1}{2}\,O_2 + 2\,H^+ + Ca \rightleftharpoons H_2O + Ca^{2+}$; forward.

18.20) $SO_4^{2-} + 4\,H^+ + 2e^- \rightarrow SO_2 + 2\,H_2O$
$$\underline{\quad\quad (Ag \rightarrow Ag^+ + e^-)2 \quad\quad}$$
$SO_4^{2-} + 4\,H^+ + 2\,Ag \rightarrow SO_2 + 2\,H_2O + 2\,Ag^+$

18.21) $NO_3^- + 10\,H^+ + 8e^- \rightarrow NH_4^+ + 3\,H_2O$
$$\underline{\quad\quad (Zn \rightarrow Zn^{2+} + 2e^-)4 \quad\quad}$$
$NO_3^- + 10\,H^+ + 4\,Zn \rightarrow NH_4^+ + 3\,H_2O + 4\,Zn^{2+}$

18.22) $Cr_2O_7^{2-} + 14\,H^+ + 6e^- \rightarrow 2\,Cr^{3+} + 7\,H_2O$
$$\underline{\quad\quad (Fe^{2+} \rightarrow Fe^{3+} + e^-)6 \quad\quad}$$
$Cr_2O_7^{2-} + 14\,H^+ + 6\,Fe^{2+} \rightarrow 2\,Cr^{3+} + 7\,H_2O + 6\,Fe^{3+}$

18.23) $(MnO_4^- + 4\,H^+ + 3e^- \rightarrow MnO_2 + 2H_2O)2$
$$\underline{\quad\quad (2\,I^- \rightarrow I_2 + 2e^-)3 \quad\quad}$$
$2\,MnO_4^- + 8\,H^+ + 6\,I^- \rightarrow 2\,MnO_2 + 4\,H_2O + 3\,I_2$

18.24) $2\,BrO_3^- + 12\,H^+ + 10e^- \rightarrow Br_2 + 6\,H_2O$
$$\underline{\quad\quad (2\,Br^- \rightarrow Br_2 + 2e^-)5 \quad\quad}$$
$2\,BrO_3^- + 12\,H^+ + 10\,Br^- \rightarrow 6\,Br_2 + 6\,H_2O$
$BrO_3^- + 6\,H^+ + 5\,Br^- \rightarrow 3\,Br_2 + 3\,H_2O$

CHAPTER 19

19.1) $2400 \text{ kcal} \times \dfrac{1000 \text{ cal}}{1 \text{ kcal}} \times \dfrac{4.184 \text{ J}}{1 \text{ cal}} = 1.00 \times 10^7 \text{ J}$

19.2) $40 \text{ ft}^2 \times \dfrac{30.5^2 \text{ cm}^2}{1^2 \text{ ft}^2} \times \dfrac{8.1 \text{ kJ}}{\text{cm}^2 \text{ min}} \times \dfrac{60 \text{ min}}{1 \text{ hr}} \times \dfrac{1 \text{ kcal}}{4.18 \text{ kJ}} = 4.33 \times 10^6 \text{ kcal}$

19.3) $325 \text{ g} \times \dfrac{3.5 \text{ J}}{\text{g} \,^\circ\text{C}} \times (-19 - 22)\,^\circ\text{C} \times \dfrac{1 \text{ kJ}}{1000 \text{ J}} = -46.6 \text{ kJ};$
47 kJ, must be removed.   [11 kcal]

19.4) $\dfrac{94.7 \text{ J}}{5.624 \text{ g}} \times \dfrac{1}{(32.4 - 18.6)\,^\circ\text{C}} = 1.22 \text{ J/g} \,^\circ\text{C}$ [0.29 cal/g $^\circ$C]

19.5) $m = \dfrac{Q}{c\,\Delta T} = \dfrac{245 \text{ J}}{14.9^\circ\text{C}} \times \dfrac{\text{g} \,^\circ\text{C}}{0.38 \text{ J}} = 43 \text{ g}$

19.6) The total heat flow in the calorimeter, made up of the energy absorbed by the water and the energy lost by the metal, equals zero.

$100 \text{ g} \times \dfrac{4.18 \text{ J}}{\text{g} \,^\circ\text{C}} \times (29.6 - 23.1)\,^\circ\text{C} + 39.0 \text{ g} \times \dfrac{X \text{ J}}{\text{g} \,^\circ\text{C}} \times (29.6 - 98.2)\,^\circ\text{C} = 0$

$2.7 \times 10^3 \text{ J} - 2.68 \times 10^3 X \text{ J} = 0$

$X = \dfrac{2.7 \times 10^3}{2.68 \times 10^3} = 1.0 \text{ J/g} \,^\circ\text{C}$ [0.24 cal/g $^\circ$C]

19.7) $666 \, \cancel{g} \times \dfrac{0.498 \, \cancel{J}}{\cancel{g} \, ^\circ\cancel{C}} \times (27.0 - 70.0) \, ^\circ C \times \dfrac{1 \, kJ}{1000 \, \cancel{J}} = -14.3 \, kJ;$

14.3 kJ, were lost. [3.41 kcal]

19.8) $3785 \, \cancel{g} \times \dfrac{2.26 \, kJ}{\cancel{g}} = 8.55 \times 10^3 \, kJ$ $[2.04 \times 10^3 \, kcal]$

19.9) $25.0 \, \cancel{g} \times \dfrac{310 \, \cancel{J}}{\cancel{g}} \times \dfrac{1 \, kJ}{1000 \, \cancel{J}} = 7.75 \, kJ$ [1.85 kcal]

19.10) $\dfrac{9.71 \, \cancel{kJ}}{50.0 \, g} \times \dfrac{1000 \, J}{1 \, \cancel{kJ}} = 194 \, J/g$ [46.4 cal/g]

19.11) $52.3 \, \cancel{kJ} \times \dfrac{1 \, \cancel{mol \, NH_3}}{21.8 \, \cancel{kJ}} \times \dfrac{17.0 \, g \, NH_3}{1 \, \cancel{mol \, NH_3}} = 40.8 \, g \, NH_3$

19.12) $75.0 \, \cancel{g} \times \dfrac{137 \, \cancel{J}}{1 \, \cancel{g}} \times \dfrac{1 \, kJ}{1000 \, \cancel{J}} = 10.3 \, kJ$ [2.46 kcal]

19.13) $6.28 \, \cancel{kJ} \times \dfrac{1000 \, \cancel{J}}{1 \, \cancel{kJ}} \times \dfrac{1 \, g}{87.9 \, \cancel{J}} = 71.4 \, g$

19.14) $\dfrac{23.9 \, kJ}{65.8 \, \cancel{g \, C_7H_8}} \times \dfrac{92.0 \, \cancel{g \, C_7H_8}}{1 \, mol \, C_7H_8} = 33.4 \, kJ/mol$ [7.99 kcal/mol]

19.15) Horizontally, energy, or heat; vertically, temperature.

19.16) D E F.

19.17) G.

19.18) The solid substance melts completely at constant temperature K.

19.19) Energy = O − N.

19.20) $225 \, \cancel{g} \times \dfrac{0.004 \, 18 \, kJ}{\cancel{g} \, ^\circ\cancel{C}} \times (0 - 16) \, ^\circ\cancel{C} + 225 \, \cancel{g} \times \dfrac{-0.335 \, kJ}{\cancel{g}} +$

$225 \, \cancel{g} \times \dfrac{0.0021 \, kJ}{\cancel{g} \, ^\circ\cancel{C}} \times (-12 - 0) \, ^\circ C = -15 \, kJ - 75.4 \, kJ - 5.67 \, kJ = -96 \, kJ$

Freezer must remove 96 kJ [23 kcal].

19.21) $500 \, \cancel{g} \times \dfrac{1.99 \, J}{\cancel{g} \, ^\circ\cancel{C}} \times (-196 + 209) \, ^\circ\cancel{C} + 500 \, \cancel{g} \times \dfrac{47.7 \, J}{1 \, \cancel{g}} +$

$500 \, \cancel{g} \times \dfrac{1.00 \, J}{\cancel{g} \, ^\circ C} \times (23 + 196) \, ^\circ C = 13 \times 10^3 \, J + 23.9 \times 10^3 \, J + 110 \times 10^3 \, J$

$= 147 \times 10^3 \, J = 147 \, kJ$ [35.1 kcal]

19.22) $C_3H_8 + 5 \, O_2 \rightarrow 3 \, CO_2 + 4 \, H_2O + 2220 \, kJ$
$C_3H_8 + 5 \, O_2 \rightarrow 3 \, CO_2 + 4 \, H_2O;$ $\Delta H = -2220 \, kJ$

19.23) $CaO + H_2O \rightarrow Ca(OH)_2 + 65.3 \, kJ$
$CaO + H_2O \rightarrow Ca(OH)_2;$ $\Delta H = -65.3 \, kJ$

19.24) $2 \, Al_2O_3 + 3 \, C + 2160 \, kJ \rightarrow 3 \, CO_2 + 4 \, Al$
$2 \, Al_2O_3 + 3 \, C \rightarrow 3 \, CO_2 + 4 \, Al;$ $\Delta H = +2160 \, kJ$

19.25) $4.00 \times 10^3 \, \cancel{g \, C_4H_{10}} \times \dfrac{1 \, \cancel{mol \, C_4H_{10}}}{58.0 \, \cancel{g \, C_4H_{10}}} \times \dfrac{5770 \, kJ}{2 \, \cancel{mol \, C_4H_{10}}} = 1.99 \times 10^5 \, kJ$

$[4.78 \times 10^4 \, kcal]$

19.26) $5.5 \times 10^4 \, \cancel{kJ} \times \dfrac{1 \, \cancel{mol \, C_6H_{14}}}{4140 \, \cancel{kJ}} \times \dfrac{86.0 \, g \, C_6H_{14}}{1 \, \cancel{mol \, C_6H_{14}}} = 1.14 \times 10^3 \, g \, C_6H_{14}$

19.27) $1.82 \times 10^7 \; kJ \times \dfrac{4 \; mol \; Al}{2160 \; kJ} \times \dfrac{0.027 \; kg \; Al}{1 \; mol \; Al} = 910 \; kg \; Al$

CHAPTER 20

20.1) Rates of opposing changes are equal in an equilibrium.

20.2) An equilibrium must be "closed" in the sense that neither matter nor energy may enter or leave the system. If these conditions are met, there may be an equilibrium in an open beaker.

20.3) Only a small portion of the fluoride ion present will be converted to HF, or most of the HF is converted to F^-. HF and/or OH^- concentrations must be very small compared to the concentration of F^-.

20.4) Sufficient energy and proper orientation.

20.5) Negative; see Figure 19.5.

20.6) See Figure 20.3. The activated complex position is at the highest point of the curve.

20.7) Activation energy is the minimum energy particles must have, usually in the form of kinetic energy, to participate in a reaction-producing collision. It also represents the increase in potential energy required to reach the activated complex point and pass the potential energy barrier. At a given temperature, the reaction with the lower activation energy proceeds faster because a larger portion of the reactant molecules have sufficient energy to meet the activation energy requirement.

20.8) At any instant the particles making up a sample of matter have different kinetic energies, and these energies change from one instant to the next. Over a period of time, however, the *average* kinetic energy, expressed by temperature, remains constant. Only those particles with the highest kinetic energies are capable of reaction-producing collisions.

20.9) A catalyst is a substance that increases the rate of a chemical reaction by lowering the activation energy. It does this by furnishing a different reaction path, or reaction mechanism. This increases the fraction of the total sample that is able to participate in reaction-producing collisions.

20.10) The reaction rate increases if the concentration of A increases, and decreases if the concentration of B decreases. Both reasons; reaction rates vary directly with reactant concentrations.

20.11) At the beginning, when both reactants are at their highest concentrations.

20.12) Nitrogen and hydrogen concentrations decrease, while ammonia concentration increases.

20.13) Le Chatelier's Principle says that the shift will be in the direction that will partially counteract the initial change. The shift must therefore be in the direction that would reduce [NO], which is in the reverse direction.

20.14) If Cl_2 is removed, the shift is in the direction in which Cl_2 will be produced. That direction is forward.

20.15) An increase in $[C_2H_3O_2^-]$ would be counteracted by a shift in the direction in which the ion is consumed—the reverse direction.

20.16) If volume of a gaseous equilibrium is reduced, the equilibrium will shift in the direction of fewer molecules. In the reverse direction two molecules combine to form one.

20.17) An increase in volume causes the equilibrium to shift in the direction of more molecules. A shift in the forward direction has one gaseous molecule producing two.

20.18) If heat is added, the equilibrium will shift in the direction in which heat is absorbed—in the reverse direction.

20.19) Heating the system would cause a shift in the direction in which heat is absorbed—in the forward direction, thereby favoring the formation of HI.

20.20) (a) and (d), forward; (b) and (c), reverse.

20.21) $\dfrac{[SO_3]^2}{[SO_2]^2[O_2]}$

20.22) $\dfrac{[CH_4][H_2S]^2}{[H_2]^4[CS_2]}$

20.23) $[Cd^{2+}][OH^-]^2$

20.24) $\dfrac{[H^+][NO_2^-]}{[HNO_2]}$

20.25) $\dfrac{[Ag^+][CN^-]^2}{[Ag(CN)_2^-]}$

20.26) An equilibrium equation may be written in two directions that yield equilibrium constants that are reciprocals of each other.

20.27) Because the equilibrium constant is very small, the reaction must be favored in the reverse direction.

20.28) $HC_2H_3O_2 \rightleftharpoons H^+ + C_2H_3O_2^-$. The equilibrium constant, $\dfrac{[H^+][C_2H_3O_2^-]}{[HC_2H_3O_2]}$, will be small. A weak acid ionizes only slightly, producing very small concentrations of the species in the numerator of the equilibrium constant expression. The unionized acid concentration in the denominator will remain almost unchanged, very large compared to the numerator.

20.29) (a) Forward. HCl is a strong acid and ionizes almost completely. The equilibrium constant is large.
(b) Reverse. $BaSO_4$ is very slightly soluble, yielding extremely small ion concentrations. The equilibrium constant is small.

20.30) $K = \dfrac{[SO_3]^2}{[SO_2]^2[O_2]} = \dfrac{0.60^2}{(0.90)^2(0.45)} = 0.99$

20.31) $HF(aq) \rightleftharpoons H^+(aq) + F^-(aq)$

I	1.0	0	0
R	−0.0265	+0.0265	+0.0265
E	1.0	0.0265	0.0265

$K = \dfrac{[H^+][F^-]}{[HF]} = \dfrac{(0.0265)^2}{1.0}$
$= 7.0 \times 10^{-4}$

20.32) $H_2(g) + I_2(g) \rightleftharpoons 2\,HI(g)$

I	0	0	3.00
R	+y	+y	−2y
E	y	y	3.00 − 2y

$[H_2] = [I2] = 0.327$
$[HI] = 3.00 - 2(0.327) = 2.35$

$K = \dfrac{[HI]^2}{[H_2][I_2]} = 51.5$

$\dfrac{(3.00 - 2y)^2}{y^2} = 51.5$

$\dfrac{3.00 - 2y}{y} = 7.18$

$y = 0.327$

20.33) $HC_3H_5O_2 \rightleftharpoons H^+ + C_3H_5O_2^- \qquad K_a = \dfrac{[H^+][C_3H_5O_2^-]}{[HC_3H_5O_2]} = 1.4 \times 10^{-5}$

$$[HC_3H_5O_2] = 11.8 \text{ g } HC_3H_5O_2 \times \frac{1 \text{ mol } HC_3H_5O_2}{74.0 \text{ g } HC_3H_5O_2} \times \frac{1}{V} = 0.159 \text{ mol/V}$$

$$[C_3H_5O_2^-] = 32.5 \text{ g } KC_3H_5O_2 \times \frac{1 \text{ mol } KC_3H_5O_2}{112 \text{ g } KC_3H_5O_2} \times \frac{1}{V} = 0.290 \text{ mol/V}$$

$$[H^+] = K_a \times \frac{[HC_3H_5O_2]}{[C_3H_5O_2^-]} = 1.4 \times 10^{-5} \times \frac{0.159/V}{0.290/V} = 7.7 \times 10^{-6}$$

$$pH = -\log(7.7 \times 10^{-6}) = 5.11$$

20.34) $CaCO_3(s) \rightleftharpoons Ca^{2+}(aq) + CO_3^{2-}(aq)$ $\quad K_{sp} = [Ca^{2+}][CO_3^{2-}] = 8.7 \times 10^{-9}$

Let s = solubility of $CaCO_3 = [Ca^{2+}] = [CO_3^{2-}]$: $\quad s^2 = 8.7 \times 10^{-9}$
$$s = 9.3 \times 10^{-5}$$

20.35) $MgF_2(s) \rightleftharpoons Mg^{2+}(aq) + 2 F^-(aq)$ $\quad K_{sp} = [Mg^{2+}][F^-]^2 = 6.5 \times 10^{-9}$

Let s = solubility of $MgF_2 = [Mg^{2+}]; [F^-] = 2s$: $\quad (s)(2s)^2 = 6.5 \times 10^{-9}$
$$s = 1.2 \times 10^{-3}$$

CHAPTER 21

21.1) Radioactivity is the spontaneous emission of rays, or particles, from the nucleus of an atom.

21.2) Alpha particle, or α particle, a helium nucleus, $_2^4He$; beta particle, or β particle, an electron, $_{-1}^0e$; and gamma ray, or γ-ray, a high energy electromagnetic ray, similar to an X-ray.

21.3) The gas is ionized.

21.4) A cloud chamber is a vessel containing air that is supersaturated in a vapor. As radioactive emissions pass through the chamber, they ionize the air. The vapor then condenses on the ions, leaving a visible "track" behind the emitted particle.

21.5) Transmutation of an element is the change of an atom from one element to another, resulting from a change in the number of protons in the nucleus of the atom.

21.6) Emission of an alpha particle reduces the atomic number by 2 and reduces the mass number by 4.

21.7) $_{82}^{212}Pb \rightarrow _{83}^{212}Bi + _{-1}^0e$; $_{90}^{231}Th \rightarrow _{91}^{231}Pa + _{-1}^0e$

21.8) $_{90}^{228}Th \rightarrow _{88}^{224}Ra + _2^4He$; $_{86}^{222}Rn \rightarrow _{84}^{218}Po + _2^4He$

21.9) The half-life of a radioactive substance is the time required for half of the sample to decay. The fraction of a sample remaining after the passage of six half-lives is $(1/2)^6$, or $1/64$.

21.10) $n = 66 \text{ years} \times \dfrac{1 \text{ half-life}}{22 \text{ years}} = 3 \text{ half-lives}$

$R = 100 \times (1/2)^3 = 12.5 \text{ g}$

21.11) $2014 - 1964 = 50 \text{ years}$; $n = {}^{50}/_{28} = 1.8 \text{ half-lives (two significant figures)}$
$R = 600 \times (1/2)^{1.8} = 1.7 \times 10^2 \text{ g}$

21.12) $n = \dfrac{\log\ (0.960/0.120)}{0.301} = 3.00$ half-lives

3.00 half-lives $\times \dfrac{10.0\ \text{min}}{1\ \text{half-life}} = 30.0$ min

21.13) $n = \dfrac{\log\ (1/0.135)}{0.301} = 2.89$ half-lives

2.89 half-lives $\times \dfrac{5.72 \times 10^3\ \text{yr}}{1\ \text{half-life}} = 1.65 \times 10^4$ years

21.14) $n = \dfrac{\log\ (8.64/2.16)}{0.301} = 2.00$ half-lives; $\dfrac{12.25\ \text{hr}}{2.00\ \text{half-lives}} = 6.13$ hr/half-life

21.15) $n = \dfrac{\log\ (100/97.72)}{0.301} = 0.0333$ half-life

$\dfrac{1.00\ \text{year}}{0.0333\ \text{half-life}} = 30.0$ years/half-life

21.16) Chemical properties depend upon the electrons outside the nucleus, which are the same for all isotopes of an element. Nuclear change depends upon a particular isotope of an element. In radioactivity, for example, one isotope of an element may emit alpha particles, another isotope may emit beta particles, and a third isotope may be stable.

21.17) UCl_4. Because of the lower molar weight of UCl_4, there are more moles of uranium in 100 grams of UCl_4 than in 100 grams of UBr_4. Only the radioactive element, uranium, contributes to radioactivity. The radioactivity of 0.10 mole of UCl_4 will be the same as that of 0.10 mole of UBr_4, inasmuch as both samples contain the same number of moles of uranium.

21.18) Lead is the stable end product of natural radioactive decay series. It is constantly being produced in natural radioactivity.

21.19) Nuclear bombardment involves directing a nuclear particle to strike another nucleus, producing a nuclear reaction.

21.20) Cyclotron and linear accelerator, both mentioned in text. Others are the betatron and synchrocyclotron.

21.21) The transuranium elements are the elements having atomic numbers greater than 92. They appear as the actinium series of elements at the bottom of the periodic tables in this text.

21.22) $^{99}_{43}\text{Tc}$

21.23) $^{239}_{94}\text{Pu}$

21.24) $^{59}_{27}\text{Co}$

21.25) A fission reaction is one in which a large nucleus splits into two nuclei of intermediate mass.

21.26) The conversion of mass to energy in a nuclear reaction results in the enormous amount of energy resulting from a nuclear change. This energy is potential in character, stored as "binding energy" that holds the positively charged protons in the nucleus together.

21.27) The major advantage of nuclear energy as a source of electricity is the abundant supply of fuel. Furthermore, the fuel cost is, at the time of this writing, competitive with fossil fuels and more dependable than fossil fuels.

21.28) A fusion reaction is one in which two light nuclei combine to form a nucleus of greater mass, in contrast with the fission reaction in which a heavy nucleus splits into two lighter nuclei.

CHAPTER 22

22.1) Excluding carbonates, cyanides, and the oxides of carbon, plus a few other substances, organic chemistry is the chemistry of carbon. Inorganic chemistry is the chemistry of the other elements plus the above exclusions.

22.2) A compound consisting of hydrogen and carbon.

22.3) A hydrocarbon containing only single bonds, with each carbon bonded to four other atoms, is an alkane. A homologous series is one in which each member differs from those next to it by the same structural unit. Alkanes differ by a $—CH_2—$ unit, having the general formula C_nH_{2n+2}.

22.4) A condensed, or line, formula lists on a single line the structural components of an organic compound. It is better than molecular formulas in conveying some idea of structure, class of organic compound, and distinction between possible isomers of compounds with identical molecular formulas.

22.5)
$$H—\overset{\overset{\displaystyle H}{|}}{\underset{\underset{\displaystyle H}{|}}{C}}—\overset{\overset{\displaystyle H}{|}}{\underset{\underset{\displaystyle H}{|}}{C}}—\overset{\overset{\displaystyle H}{|}}{\underset{\underset{\displaystyle H}{|}}{C}}—\overset{\overset{\displaystyle H}{|}}{\underset{\underset{\displaystyle H}{|}}{C}}—\overset{\overset{\displaystyle H}{|}}{\underset{\underset{\displaystyle H}{|}}{C}}—\overset{\overset{\displaystyle H}{|}}{\underset{\underset{\displaystyle H}{|}}{C}}—\overset{\overset{\displaystyle H}{|}}{\underset{\underset{\displaystyle H}{|}}{C}}—\overset{\overset{\displaystyle H}{|}}{\underset{\underset{\displaystyle H}{|}}{C}}—\overset{\overset{\displaystyle H}{|}}{\underset{\underset{\displaystyle H}{|}}{C}}—H$$

$CH_3CH_2CH_2CH_2CH_2CH_2CH_2CH_2CH_3$ or $CH_3(CH_2)_7CH_3$

22.6) Isomerism is the existence of two or more compounds of the same molecular formula, but different structural formulas and different physical and chemical properties. Of the compounds listed (a) and (e) are isomers, both having the molecular formula C_9H_{20}.

22.7) Alkane hydrocarbon minus one hydrogen. Named by alkane prefix designating number of carbon atoms, plus -yl suffix.

22.8)

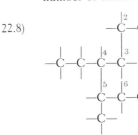

Octane. It is possible to count out an eight-carbon chain.

22.9)

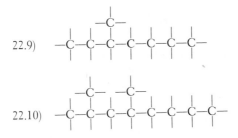

22.10)

22.11) 3-ethylpentane.

22.12) 3-chloroheptane.

22.13) 1-bromo-4,5-dichlorohexane.

22.14)

```
      Br
      |
   —C—C—
      |
      Br
```

22.15)

```
   Br       Br Br
   |        |  |
  —C—C—C—C—C—
   |  |  |  |  |
   Br
```

22.16)

```
   Br Br
   |  |
  —C—C—C—C—
   |  |  |  |
   Cl Br
```

22.17) A double bond between two carbons; C_nH_{2n}.

22.18) $H—C\equiv C—H$; acetylene, or ethyne.

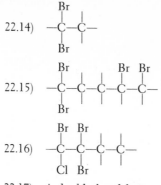

```
      H
      |
H—C≡C—C—H
      |
      H
```

```
      H  H
      |  |
H—C≡C—C—C—H
      |  |
      H  H
```

```
   H        H
   |        |
H—C—C≡C—C—H
   |        |
   H        H
```

Propyne   1-butyne   2-butyne

22.19) The double bond is between the first and second carbons in 1-hexene, the second and third in 2-hexene, and the third and fourth in 3-hexene.

22.20) Different structural arrangements around a double bond. See page 544.

22.21) *cis*-2-pentene.

22.22)

```
   H  H  H  H  H  H
   |  |  |  |  |  |
H—C—C—C—C—C=C    1-hexene
   |  |  |  |     |
   H  H  H  H     H
```

22.23)

```
   H  H  H  H  H
   |  |  |  |  |
H—C—C—C—C=C    1-pentene
   |  |  |     |
   H  H  H     H
```

22.24) In an addition reaction one bond of a double or triple bond opens and new atoms are added by bonding to adjacent carbons without displacing other atoms. In a substitution reaction in a saturated compound, atoms already bonded must be displaced to make room for new atoms.

22.25) $C_3H_6 + H_2 \rightarrow C_3H_8$

22.26) The chemical combination of two or more monomers. The number so combined may be many thousands.

22.27)

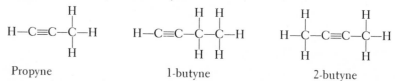

```
   H  Cl   H  Cl   H  Cl        H  Cl H  Cl H  Cl
   |  |    |  |    |  |         |  |  |  |  |  |
   C=C  +  C=C  +  C=C    →   —C—C—C—C—C—C—
   |  |    |  |    |  |         |  |  |  |  |  |
   H  H    H  H    H  H        H  H  H  H  H  H
```

22.28) Aromatic compounds have ring structures; aliphatic compounds have open chain structures.

22.29) (a) and (c), *m*-dichlorobenzene or 1,3-dichlorobenzene; (b) *p*-dichlorobenzene, or 1,4-dichlorobenzene

22.30) Hydroxyl group, —OH; R—OH, or ROH.

22.31) $H-\overset{\underset{H}{|}}{\underset{H}{\overset{H}{|}}}C-\overset{\underset{H}{|}}{\overset{H}{|}}C-\overset{\underset{H}{|}}{\overset{H}{|}}C-\overset{\underset{H}{|}}{\overset{H}{|}}C-\overset{\underset{H}{|}}{\overset{H}{|}}C-OH$

22.32) In a secondary alcohol, the carbon to which the hydroxyl group is attached is bonded to two other carbon atoms, giving a minimum of three carbon atoms in the molecule.

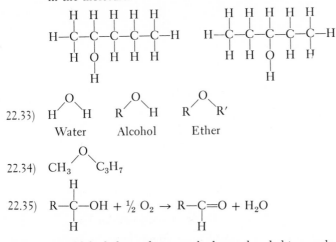

22.33)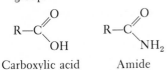

       Water   Alcohol   Ether

22.34) $CH_3\overset{O}{\diagup}\!\!\diagdown C_3H_7$

22.35) $R-\overset{\underset{H}{|}}{\overset{H}{|}}C-OH + \tfrac{1}{2} O_2 \rightarrow R-\overset{|}{C}=O + H_2O$
                                      H

22.36) An aldehyde has at least one hydrogen bonded to a carbonyl group, whereas a ketone has two alkyl groups bonded to the carbonyl group. Aldehydes and ketones both have a carbonyl group.

22.37) $RCOOH;\ -C\overset{\diagup\!\!O}{\diagdown\!\!OH}$

22.38) Carboxyl groups are polar, and hydrogen bonding is present. This leads to relatively strong intermolecular attractions and therefore high boiling points.

22.39) $RCOOH \rightarrow RCOO^- + H^+$

22.40) An amine is a substituted ammonia in which one or more hydrogens of ammonia are replaced by alkyl groups.

22.41) Methylpropylamine.

22.42) In an amide the —OH part of the carboxyl group is replaced by a —NH$_2$ group:

$R-C\overset{\diagup\!\!O}{\diagdown\!\!OH}$ $R-C\overset{\diagup\!\!O}{\diagdown\!\!NH_2}$

Carboxylic acid   Amide

index

649